Coherence and Control in Chemistry

University of Leeds, United Kingdom
25–27 July 2011

FARADAY DISCUSSIONS
Volume 153, 2011

RSC Publishing

The Faraday Division of the Royal Society of Chemistry, previously the Faraday Society, founded in 1903 to promote the study of sciences lying between Chemistry, Physics and Biology.

EDITORIAL STAFF

Editor
Philip Earis

Deputy editor
Jane Hordern

Senior publishing editor
Nicola Nugent

Publishing editors
Carrie Mowatt, Amaya Camara-Campos

Publishing assistants
Hannah Porter, Claire Sissen

Publisher
Niamh O'Connor

Faraday Discussions (Print ISSN 1359-6640, Electronic ISSN 1364-5498) is published 4 times a year by the Royal Society of Chemistry, Thomas Graham House, Science Park, Milton Road, Cambridge, UK CB4 0WF. Volume 153 ISBN-13: 978 1 84973 2383

2011 annual subscription price: print+electronic £669, US $1,247; electronic only £602, US $1,122. Customers in Canada will be subject to a surcharge to cover GST. Customers in the EU subscribing to the electronic version only will be charged VAT. All orders, with cheques made payable to the Royal Society of Chemistry, should be sent to RSC Distribution Services, c/o Portland Customer Services, Commerce Way, Colchester, Essex, UK CO2 8HP.
Tel +44 (0) 1206 226050;
E-mail sales@rscdistribution.org

If you take an institutional subscription to any RSC journal you are entitled to free, site-wide web access to that journal. You can arrange access *via* Internet Protocol (IP) address at www.rsc.org/ip. Customers should make payments by cheque in sterling payable on a UK clearing bank or in US dollars payable on a US clearing bank. Periodicals postage is paid at Rahway, NJ and at additional mailing offices. Airfreight and mailing in the USA by Mercury Airfreight International Ltd., 365 Blair Road, Avenel, NJ 07001, USA.

US Postmaster: send address changes to *Faraday Discussions*, c/o Mercury Airfreight International Ltd., 365 Blair Road, Avenel, NJ 07001. All despatches outside the UK by Consolidated Airfreight.

PRINTED IN THE UK

Faraday Discussions documents a long-established series of *Faraday Discussion* meetings which provide a unique international forum for the exchange of views and newly acquired results in developing areas of physical chemistry, biophysical chemistry and chemical physics.

SCIENTIFIC COMMITTEE, Volume 153

Chair
Professor Ben Whitaker (University of Leeds, UK)

Professor Regina de Vivie-Riedle (Ludwig-Maximilians-Universitat Munchen, Germany)
Professor Thomas Weinacht (Stony Brook University, USA)
Professor Helen Fielding (University College London, UK)
Dr Mike Bearpark (Imperial College London, UK)

FARADAY STANDING COMMITTEE ON CONFERENCES

Chair
D E Heard (Leeds, UK)

W A Brown (UCL, UK)
I Hamley (Reading, UK)
J Hirst (Nottingham, UK)
A Mount (Edinburgh, UK)

© The Royal Society of Chemistry 2011. Apart from fair dealing for the purposes of research or private study, or criticism or review, as permitted under the Copyright, Designs and Patents Act 1988 and Related Rights Regulations 2003, this publication may only be reproduced, stored or transmitted, in any form or by any means, with the prior permission in writing of the Publishers or in the case of reprographic reproduction in accordance with the terms of licences issued by the Copyright Licensing Agency in the UK. US copyright law applicable to users in the USA. The Royal Society of Chemistry takes reasonable care in the preparation of this publication but does not accept liability for the consequences of any errors or omissions.

Royal Society of Chemistry:
Registered Charity No. 207890.

⊚The paper used in this publication meets the requirements of ANSI/NISO Z39.48-1992 (Permanence of Paper).

Coherence and Control in Chemistry

Faraday Discussions

www.rsc.org/faraday_d

A General Discussion on Coherence and Control in Chemistry was held at the University of Leeds, Leeds, United Kingdom on 25th, 26th and 27th July 2011.

RSC Publishing is a not-for-profit publisher and a division of the Royal Society of Chemistry. Any surplus made is used to support charitable activities aimed at advancing the chemical sciences. Full details are available from www.rsc.org

CONTENTS

ISSN 1359-6640; ISBN 978-1-84973-238-3

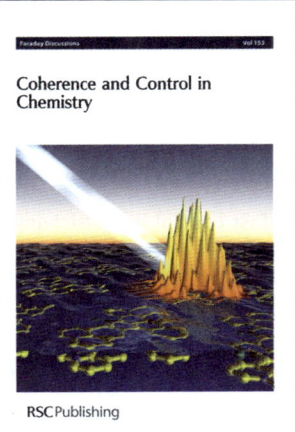

Cover
See Miller *et al.*, *Faraday Discuss.*, 2011, **153**, 27–39.

This figure depicts the excitation of the dye molecule Rhodamine 110 with phase modulated femtosecond laser pulses, appearing nearly white, that leads to beautiful interferences in the induced coherences and enhanced population transfer—as depicted by the 2D spectra arising as peaks from the excited molecule.

Image reproduced by permission of Joerg M. Harms from *Faraday Discuss.*, 2011, **153**, 27.

INTRODUCTORY LECTURE

9 **Ultrafast laser control of electron dynamics in atoms, molecules and solids**
   Matthias Wollenhaupt and Thomas Baumert

PAPERS AND DISCUSSIONS

27 **Coherently-controlled two-dimensional spectroscopy: Evidence for phase induced long-lived memory effects**
    Valentyn I. Prokhorenko, Alexei Halpin and R. J. Dwayne Miller

41 **Electronic energy transfer in model photosynthetic systems: Markovian *vs.* non-Markovian dynamics**
    Navinder Singh and Paul Brumer

51	**Coherent control of single molecules at room temperature** Daan Brinks, Richard Hildner, Fernando D. Stefani and Niek F. van Hulst
61	**Exploring the role of phase modulation on photoluminescence yield** D. G. Kuroda, C. P. Singh, Z. Peng and V. D. Kleiman
73	**General discussion**
93	**Extracting dynamics of excitonic coherences in congested spectra of photosynthetic light harvesting antenna complexes** Justin R. Caram and Gregory S. Engel
105	**Multiconfigurational Ehrenfest approach to quantum coherent dynamics in large molecular systems** Dmitrii V. Shalashilin
117	**The influence of the optical pulse shape on excited state dynamics in provitamin D_3** Kuo-Chun Tang and Roseanne J. Sension
131	**Wavepacket and potential reconstruction by four-wave mixing spectroscopy: preliminary application to polyatomic molecules** David Avisar and David J. Tannor
149	**Entanglement in interference-based quantum control: the wave function is not enough** Moshe Shapiro and Paul Brumer
159	**Searching for pathways involving dressed states in optimal control theory** Philipp von den Hoff, Markus Kowalewski and Regina de Vivie-Riedle
173	**Photoelectron photoion coincidence imaging of ultrafast control in multichannel molecular dynamics** C. Stefan Lehmann, N. Bhargava Ram, Daniel Irimia and Maurice H. M. Janssen
189	**General discussion**
213	**A General control mechanism of energy flow in the excited state of polyenic biochromophores** Tiago Buckup, Jürgen Hauer, Judith Voll, Regina Vivie-Riedle and Marcus Motzkus
227	**Coherent control of vibrational transitions: Discriminating molecules in mixtures** A. C. W. van Rhijn, A. Jafarpour, M. Jurna, H. L. Offerhaus and J. L. Herek
237	**Coherent control of the motion of complex molecules and the coupling to internal state dynamics** Paul Venn and Hendrik Ulbricht
247	**Combining dissociative ionization pump–probe spectroscopy and *ab initio* calculations to interpret dynamics and control through conical intersections** Spiridoula Matsika, Congyi Zhou, Marija Kotur and Thomas C. Weinacht
261	**Nonadiabatic *ab initio* molecular dynamics including spin–orbit coupling and laser fields** Philipp Marquetand, Martin Richter, Jesús González-Vázquez, Ignacio Sola and Leticia González
275	**Dynamic stark control: model studies based on the photodissociation of IBr** Cristina Sanz-Sanz, Gareth W. Richings and Graham A. Worth
293	**General discussion**

321	**From molecular control to quantum technology with the dynamic Stark effect** Philip J. Bustard, Guorong Wu, Rune Lausten, Dave Townsend, Ian A. Walmsley, Albert Stolow and Benjamin J. Sussman
343	**Controlled redistribution of vibrational population by few-cycle strong-field laser pulses** William A. Bryan, C. R. Calvert, R. B. King, J. B. Greenwood, W. R. Newell and I. D. Williams
361	**Control of coherent excitation of neon in the extreme ultraviolet regime** Jürgen Plenge, Andreas Wirsing, Christopher Raschpichler, Bernhard Wassermann and Eckart Rühl
375	**Optical manipulation of coherent phonons in superconducting $YBa_2Cu_3O_{7-\delta}$ thin films** Yasuaki Okano, Hiroyuki Katsuki, Yoshihiro Nakagawa, Hiroshi Takahashi, Kazutaka G. Nakamura and Kenji Ohmori
383	**Femtosecond coherent control of thermal photoassociation of magnesium atoms** Leonid Rybak, Zohar Amitay, Saieswari Amaran, Ronnie Kosloff, Michał Tomza, Robert Moszynski and Christiane P. Koch
395	**General discussion**

CONCLUDING REMARKS

415	**A perspective on controlling quantum phenomena** Herschel Rabitz

ADDITIONAL INFORMATION

419	**Poster titles**
421	**List of participants**
423	**Index of contributors**

PAPER

Ultrafast laser control of electron dynamics in atoms, molecules and solids

Matthias Wollenhaupt and Thomas Baumert*

Received 5th September 2011, Accepted 19th September 2011
DOI: 10.1039/c1fd00109d

Exploiting coherence properties of laser light together with quantum mechanical matter interferences in order to steer a chemical reaction into a pre-defined target channel is the basis of coherent control. The increasing availability of laser sources operating on the time scale of molecular dynamics, *i.e.* the femtosecond regime, and the increasing capabilities of shaping light in terms of amplitude, phase and polarization also on the time scale of molecular dynamics brought the temporal aspect of this field to the fore. Since the last *Faraday Discussion* (*Faraday Discussion 113, Stereochemistry and control in molecular reaction dynamics*) devoted to this topic more than a decade ago a tremendous cross-fertilization to neighbouring "quantum technology disciplines" in terms of experimental techniques and theoretical developments has occurred. Examples are NMR, quantum information, ultracold molecules, nonlinear spectroscopy and microscopy and extreme nonlinear optics including attosecond-science. As pointed out by the organizers, this meeting brings us back to chemistry and aims to assess recent progress in our general understanding of coherence and control in chemistry and to define new avenues for the future. To that end we will in the Introductory lecture first shortly review some aspects of coherent control. This will not be fully comprehensive and is mainly meant to give some background to current experimental efforts of our research group in controlling (coherent) electronic excitations with tailored light fields. Examples and perspectives for the latter will be given.

1 Introduction

Coherent control is a fascinating facet of femtochemistry.[1,2] Traditionally femtochemistry deals with laser-based real-time observations of molecular dynamics by making use of light pulses that are short in comparison to the molecular time scale. Coherent control goes beyond this *ansatz*. Here one seeks to actively exert microscopic control over molecular dynamics at the quantum level on intrinsic time scales. The goal is to steer any type of light-induced molecular processes from an initial state to a pre-defined target state with high selectivity and with high efficiency. Progress in this fast expanding research field is documented in recent textbooks,[3–5] review articles[6–22] and special issues.[23–27]

Suitable tools to achieve this goal are shaped femtosecond optical laser pulses in amplitude, phase and polarization (see Fig. 1 (a)), where different shaping techniques are reviewed in the literature.[11,28–31] Precision pulse shaping down to the zeptosecond regime has been reported recently, opening the perspective of controlling electron dynamics with unprecedented precision (see Fig. 1 (b)).[32]

Universität Kassel, Institut für Physik und CINSaT, Heinrich-Plett-Str. 40, 34132 Kassel, Germany. E-mail: baumert@physik.uni-kassel.de; Fax: +49 561 804 4453; Tel: +49 561 804 4452

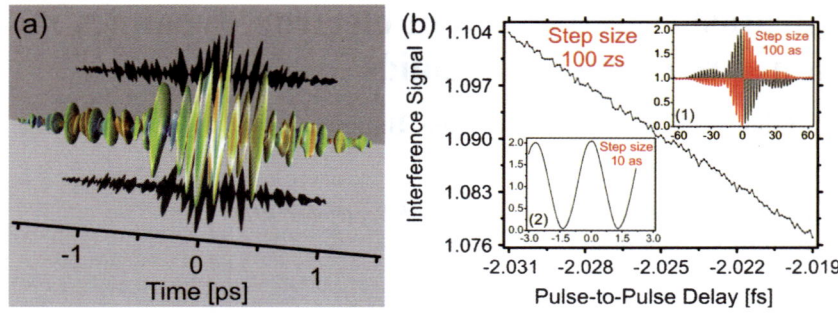

Fig. 1 (a) Example of an optimal laser pulse—shaped in amplitude, phase and polarization—employed to maximize the K_2^+ ion yield from Resonance Enhanced Multi Photon Ionization (REMPI) of potassium dimers K_2 exploiting the vectorial properties of light–matter interaction in a feedback learning loop (adapted from ref. 48). In the three-dimensional representation the electric field amplitudes are indicated by the sizes of the corresponding ellipses and the instantaneous frequencies are indicated by colors. The black shadows represent the amplitude envelopes of the two orthogonal polarization components. (b) Example of zeptosecond precision pulse shaping. Measured interference signal of two time delayed pulses created by a high precision pulse shaper. Inset (1) shows the interferogram with a step size of 100 as, whereas inset (2) shows a zoom into the interferogram with a step size of 10 as. The main figure shows a measurement within the time delay interval from −2.031 fs to −2.0199 fs measured with a step size of 100 zs. The obtained time resolution (2 σ) is 300 zeptoseconds (1 zs = 1 zeptosecond = 10^{-3} attosecond = 10^{-6} femtosecond, figure adapted from ref. 32).

Optimized light fields can be found for example by employing adaptive feedback learning loops[33–39] or by fine tuning the parameters of physically motivated pulse shapes,[40–44] where experimentally determined quantum control landscapes[45,46] can help identifying underlying physical mechanisms especially in the strong-field regime. Shifting the focus to spectroscopy, the understanding of different dynamical mechanisms can lead to sets of physically motivated pulse shapes, where the parameters of these pulse shapes can be adapted to the molecules under study either systematically or evolutionary. This spectroscopy *ansatz* would go beyond the typical pulse sequence spectroscopy with Fourier limited pulses and it was suggested to name this approach quantum control spectroscopy.[18,47]

In contrast to weak-field (perturbative) quantum control schemes where the population of the initial state is approximately constant during the interaction with the external light field, the strong-field (non perturbative) regime is characterized by efficient population transfer. Adiabatic strong-field techniques such as rapid adiabatic passage (RAP) or stimulated Raman adiabatic passage[49,50] are employed for instance with laser pulses in the picosecond[11,51–53] to nanosecond domain allowing for population transfer with unit efficiency in quantum state systems. Only recently were these techniques transferred to the femtosecond regime. For example, selectivity based on (dynamic Stark shifted) RAP combined with high efficiency was demonstrated in an atomic ladder system with the help of chirped laser pulses,[54] and piecewise adiabatic passage was demonstrated in an atomic two level system with chirped pulse sequences.[55] Furthermore, it was shown that the effects of the dynamic Stark shift reducing the excitation efficiency can be compensated with temporally structured pulses.[56] Switching the electronic population to different final states with high efficiency *via* selective population of dressed states (SPODS) is a further fundamental resonant strong-field effect as the only requirement is the use of intense ultrashort pulses exhibiting time varying phases such as phase jumps[57–60] or chirps.[43,61]

The modification of the electronic potentials due to the interaction with the electric field of the laser pulse has another important aspect pertaining to molecules, as the nuclear motion can be significantly altered in light induced potentials. An experimental review devoted to the topic of small molecules in intense laser fields focusing mainly on H_2 excitation and fragmentation dynamics is given in ref. 62. Experimental examples for modifying the course of reactions of neutral molecules after an initial excitation *via* altering the potential surfaces can be found in ref. 63 and 64 where the amount of initial excitation on the molecular potential can be set *via* Rabi type oscillations.[65] Nonresont interaction with an excited vibrational wavepacket can in addition change the population of the vibrational states.[66]

Although a high degree of excitation can be achieved *via* Rabi oscillations, this approach is not attractive for efficient coherent control schemes as the resonant Rabi oscillation period is proportional to the scalar product of the electric dipole moment times the electric field of the laser pulse. As a consequence, different excitation levels are achieved due to the intensity distribution within the focal area of a typical Gaussian laser beam and due to different orientations of the molecules in a typical isotropic sample. This is why the above mentioned adiabatic strong-field approaches are especially important as they are robust against these effects.

Conceptually most of the mechanisms underlying coherent control have been demonstrated in the gas phase, where the latest highlight is quantum control of bond formation in a catalytic surface reaction.[67] However, as relevant chemistry, biology and medicine is typically taking place in the liquid phase, laser control of dissolved molecules is most promising for applications. The field is reviewed in ref. 16 where direct control of ground state vibrational excitation,[68] control of energy flow in large light harvesting molecules,[69] control of isomerization processes[70,71] and optical discrimination of molecules with nearly identical absorption profiles[72-74] are prominent examples. Examples of robust and efficient electronic excitation of molecules especially in the liquid phase are rare (see ref. 75 and 76 for theoretical discussions related to earlier experiments[36,77]).

In this Introductory lecture we focus on (coherent) electronic excitation with shaped laser pulses. We first discuss our experiments devoted to create designer electron wave packets in the continuum by making use of the electronic structure of atoms together with polarization shaped laser pulses. We present our tomography method to reconstruct the three-dimensional electron distributions and hint at possible applications. We then focus on coherent strong-field excitation. This is the regime beyond perturbative descriptions of light matter interactions and below the regime where the ionization probability reaches unity. It is the regime where Rabi cycling is important and we highlight the importance of controlling resonant processes in this regime. Two examples are given. We start discussing experiments in the gas phase, where a first laser pulse creates a charge oscillation with maximum coherence. In the region of valence electron excitations, these charge oscillations are on the order of one femtosecond and making use of these coherences for reaction control requires a second pulse which interacts with this coherence with attosecond precision. We then turn to experiments in the liquid phase where we discuss strong-field scenarios for adiabatic population transfer. These are derived from simulations based on our strong-field experiments making use of physically motivated pulse parameterizations for adiabatic strong-field interactions. Finally we briefly present our experiments demonstrating control of ionization processes in dielectrics for material processing on the nanometre scale. Typical excitation densities are in the range of 10^{21} electrons cm^{-3} and collision times on the order of one femtosecond prevent coherence introduced from the light field from being exploited. Nevertheless, we show that shaping temporal asymmetric profiles address multi photon ionization and electron-electron impact ionization in a different fashion.

2. Tomography of designer electron wave packets

In this chapter the main focus is to make use of the electronic structure of matter together with polarization shaped laser pulses in order to create designed electron wave packets in the continuum. The techniques developed, open new routes to chemical analytics as well as to the measurement of photoelectron angular distributions in the molecular frame.

The basis of this approach was the demonstration of interferences of free electron wave packets generated by a pair of identical, time-delayed, femtosecond laser pulses which ionized excited atomic potassium.[78] In that experiment two different schemes were investigated: threshold electrons produced by one-photon ionization with parallel laser polarization and above threshold ionization electrons produced by a two-photon transition with crossed laser polarization. As the measurement does not provide knowledge on the "which way information" *i.e.* whether the system is ionized by the first laser pulse or the second, double pulse photo ionization is a Young's double slit experiment in the time domain and naturally interferences have to occur. The time evolution of these interferences can be understood in a Wigner description[79] and the interferences can be used for pulse characterization[79] also in the attosecond regime.[80]

Making use of resonance enhanced multi photon ionization (REMPI) with amplitude, phase and polarization shaped laser pulses, the continuum can be structured by addressing different quantum mechanical states during the REMPI process as for example: the central wavelength and the spectral width of the laser will in a perturbative regime address different final states due to energy conservation and selection rules, where the number of possibilities is rapidly increasing with the number of photons; in addition non-perturbative fields induce Stark shifts that are reflected in the resonant case in the Autler Towns Splitting[81] and can be controlled with amplitude[32,57] and phase shaped[59,82] laser pulses; in the non-resonant case dynamic Stark shifting gives access to contributions from off-resonant states that can be selectively controlled for example with chirped laser pulses[54] as well.

Therefore, by tailoring the state of polarization, a high degree of control on the angular and energy distributions of ultrashort free-electron wave packets is obtained by interference of multiple excitation and ionization pathways.[83] The generated photoelectron angular distributions (PADs) were recorded with the velocity map imaging (VMI) technique.[84,85] These PADs are so called Abel projections of (in general) complex-shaped wave packets. If the detector plane contains an axis of symmetry—as for example linearly polarized light with the polarization vector parallel to the detector plane or circularly polarized light with the direction of the propagation of the light parallel to the detector plane—the three-dimensional distribution can be derived *via* different inversion methods from the two-dimensional projections.[86–89] For polarization shaped laser pulses these approaches do not work. However, measurements of PADs by rotating the complex polarization-shaped laser pulse delivers all information required for tomographic techniques to reconstruct the three-dimensional electron wave packets.[90] With the help of this technique designed complex-shaped three-dimensional electron wave packets are accessible to direct measurements. Two examples of such designer electron wave packets are given. In Fig. 2 an adaptively optimized designer electron wave packet is displayed together with its tomographic reconstruction. Fig. 3 shows a reconstruction from excitation with a "v" and polarization shaped pulse (see Fig. caption).[91]

In the recent past PADs have proven to be essential to analyze ionization dynamics and (neutral) molecular dynamics[85,92] because they contain highly differential information. The tomography approach presented could be used for example to determine radial phase shifts of partial waves in ionization processes or to give better understanding in strong-field ionization.[93] Combining the tomography approach with molecular alignment techniques[94,95] would give direct access to molecular frame PADs complementing coincidence techniques[96,97] that can give this information in the case of direct fragmentation processes. Another route to obtain molecular frame PADs could make use of alignment *via* excitation by exploiting the vectorial properties of light matter interaction. For such an approach polarization shaped laser pulses are again attractive as demonstrated for optimizing REMPI processes in the potassium dimer.[48] In that experiment—due to the

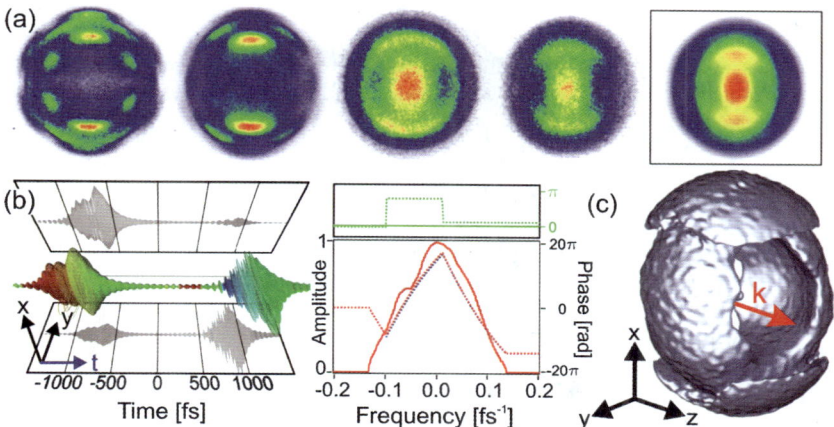

Fig. 2 Adaptive optimization of photoelectron angular distribution (PAD) measured by velocity map imaging (VMI) from resonance enhanced multi photon ionization (REMPI) of potassium atoms with polarization shaped femtosecond laser pulses employing an evolutionary algorithm. (a) The target PAD (boxed) is iteratively approximated by optimization of the spectral phase of the laser pulse. Results during the optimization process are shown from left to right for the individuals no. 1, 130, 260 and the optimized result at no. 597. (b, left): Optimized polarization shaped femtosecond laser pulse which yields the target PAD in a three-dimensional representation and (b, right) optimized pulse in the frequency domain. The pulse parameters are: central wavelength of 790 nm, FWHM pulse duration of 30 fs and peak intensity of about 3×10^{11} W cm^{-2}. For the optimization, the spectral phase modulation function was parameterized by piecewise linear functions in order to delay different spectral bands with respect to each other and, in addition, a relative phase between both polarization components was applied in order to control the ellipticity of individual spectral bands. (c) Three-dimensional tomographic reconstruction of the designed free electron wave packet which optimally reproduces the two-dimensional target PAD. Analysis of the pulse shape reveals that the optimized wave packet is created by two time-delayed slightly elliptically polarized sub pulses containing different spectral components (adapted from M. Krug PhD thesis[91]).

symmetry of the electronic states involved in the REMPI process—the light had to adapt its polarization state to parallel and perpendicular transitions during the excitation resulting in complex polarization shaped pulses (see Fig. 1 (a)).

Now lets turn to analytics. Given the argument that angular and energy distributions of ultrashort free-electron wave packets are obtained by interference of multiple excitation and ionization pathways being sensitive to the electronic structure, the combination of polarization shaping and velocity map imaging could be a sensitive analytic tool for molecular recognition in the gas phase where especially chiral recognition is attractive.[98] To that end we have started experiments on randomly oriented enantiomers of camphor and fenchone, and preliminary data were presented in the Introductory lecture. We ionized the corresponding enantiomers *via* a 2 + 1 REMPI process with circular polarized light and observed asymmetries in the forward-backward direction of ejected electrons. These experiments build on the pioneering synchrotron work on photoelectron circular dichroism by Laurent Nahon and Ivan Powis with coworkers[99] where a striking forward-backward electron ejection asymmetry was found in one-photon ionization of camphor enantiomers with circular polarized light. Note that sensitive chiral recognition is a prerequisite for chiral purification schemes based on coherent control techniques. For the latter see for example the work by Moshe Shapiro and coworkers and references therein.[100]

3. Control out of states of maximum electronic coherence

In this section the main focus is on a specific strong-field effect *i.e.* photon locking which is one realization of control *via* selective population of dressed states

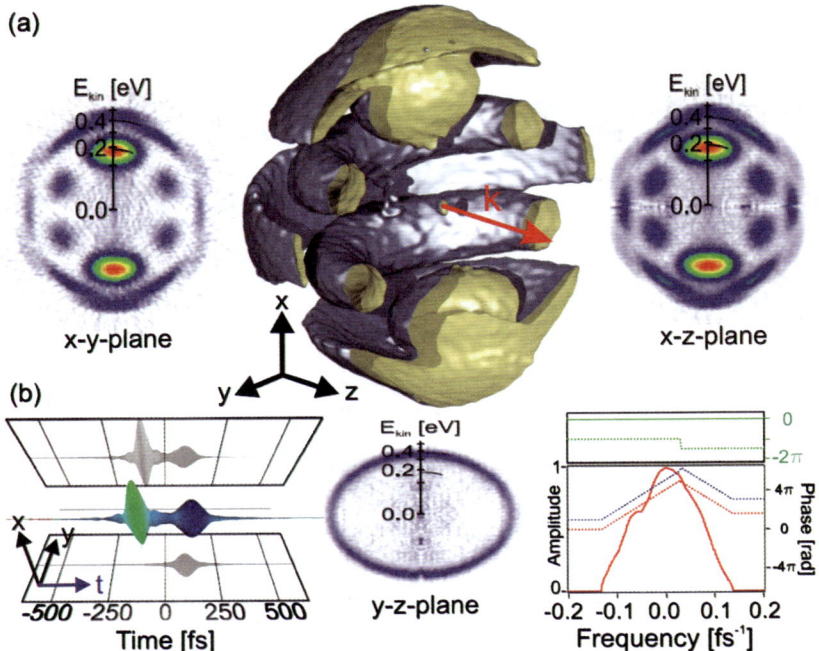

Fig. 3 Example of a designed electron wave packet in the continuum resulting from REMPI of potassium atoms employing combined "v"-shaped spectral phase modulation and polarization shaping.[83] (a) Three-dimensional tomographic reconstruction of a designed free electron wave packet obtained by multiple measurements of PADs with the complex polarization-shaped laser pulse rotated about the propagation vector using a $\lambda/2$ plate (see ref. 90 for further details on the tomography technique). The cuts through the origin in the x-y-plane, the x-z-plane and the y-z-plane illustrate the rich structure of the wave packet. The f-orbital type structure of the wave packet is attributed to the linearly polarized pulse in the x direction whereas the "hat" results from the circularly polarized delayed blue detuned pulse. (b) Pulse shape in time and frequency domain (the central wavelength is 790 nm, the FWHM pulse duration 30 fs and the peak intensity about 4×10^{12} W cm^{-2}). The "v"-shaped spectral phase advances the intense red spectral band and retards the weaker blue spectral band. Due to the additional phase jump, the first pulse is linearly polarized along the x-axis whereas the second pulse is circularly polarized (adapted from M. Krug PhD thesis[91]).

(SPODS). In analogy to vibrational wave packet control an oscillating charge distribution with maximum amplitude *i.e.* a state of maximum electronic coherence is created with a first laser pulse and by timing a second laser pulse with sub cycle precision (down to the zeptosecond regime, see Fig. 1 (a)) switching of population to different final states in atoms and molecules with high efficiency is demonstrated (see Fig. 4, Fig. 5 and Fig. 6). SPODS can also be used to measure decoherence phenomena as suggested in ref. 101 and beautifully exploited on single molecule spectroscopy.[102]

Motivation for these investigations stems from the enormous success of closed loop adaptive control experiments (see introduction) often resulting in complicated and not uniquely defined pulses. In many cases the first excitation band in such scenarios is reached after non-resonant absorption of a few photons, requiring already strong-fields. Taking into account that during a multi-photon process regions with an increasing density of electronic states are reached, suggests that strong-field coupling effects are present under such excitation conditions. Since the resonant control pathways will always dominate the controlled dynamics, resonant control scenarios will be important when looking for mechanisms. Increasing

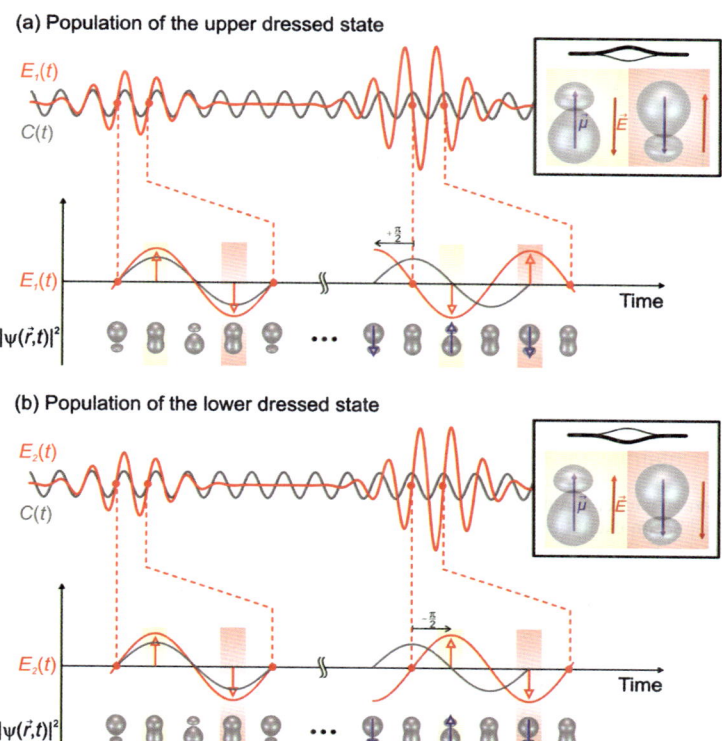

Fig. 4 Coherent control by photon locking employing a two pulse sequence displayed in a spatio-temporal picture illustrating population of the upper dressed state (a) and the lower dressed state (b). A comparison of a reference carrier oscillation $C(t)$ (grey sinusoidal curve with constant amplitude) in phase with the first weaker pulse of the driving pulse sequence $E(t)$ (red) highlights the phase shift of the second pulse in the sequence. The first pulse creates a state of maximum coherence exemplified by the electronic wave packet $|\psi(r, t)|^2$, *i.e.* a charge oscillation with maximum amplitude. The asymmetry in the charge distribution gives rise to an oscillating electric dipole moment μ (blue arrow), initially following the driving force with a phase shift of $\pi/2$. By timing the second pulse with sub cycle precision its phase relative to the induced atomic charge oscillation is controlled in order to realize photon locking. In scenario (a) the phase of the second pulse in the sequence $E_1(t)$ is shifted by $+\pi/2$. This carrier phase jump (introduced by suitable *spectral* phase modulation) shifts the electric field and the induced charge oscillation *out of phase*, *i.e.* during the second pulse both vectors E and μ oscillate anti parallel throughout. In this configuration the populations are *locked* and, hence, the dipole oscillation remains unaltered. This applies also to scenario (b), in which the carrier phase jump of $-\pi/2$ shifts the electric field *in phase* with the dipole moment such that subsequently both vectors oscillate parallel during the second pulse. The anti parallel (parallel) configuration maximizes (minimizes) the energy $W = -\mu E$ of the interacting system which is equivalent to selective population of the upper (lower) dressed state. SPODS *via* photon locking is an example of coherent control by tailoring of the phase of the laser field with respect to the ultrafast electron dynamics. A detailed discussion of SPODS in the spatio-temporal picture is presented in ref. 43.

significance for resonant control schemes is also given when ultra broad spectra for coherent control are employed. These come along with the ability to produce and shape shorter and shorter pulses.

In general, strong laser fields give rise to an energy splitting of the resonant state into two (so called dressed) states in the order of $\hbar\Omega$, where Ω describes the Rabi frequency. The decisive step in switching among different final electronic states is realized by the manipulation of dressed state energies and dressed state populations.

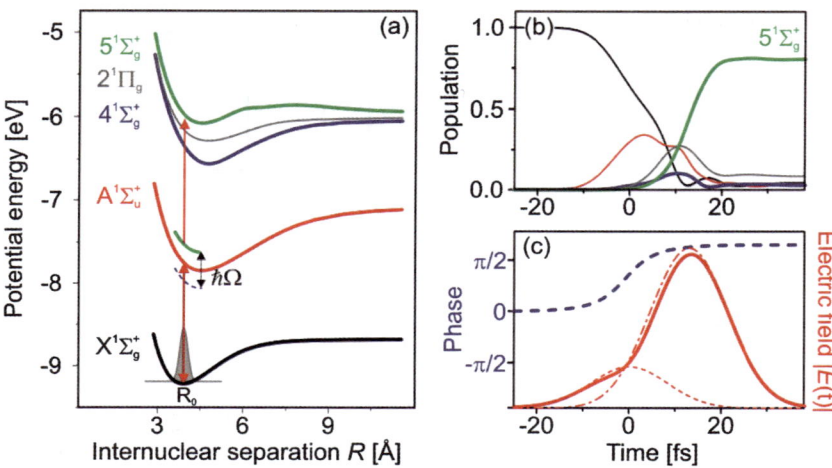

Fig. 5 Wave packet simulation on control of population in potassium dimers K$_2$ by SPODS. (a) Multi-photon excitation scheme for potassium dimers and (b) time evolution of the population in the X $^1\Sigma_g^+$-state (black), the A $^1\Sigma_u^+$-state (red), the 4 $^1\Sigma_g^+$-state (blue), the 2 $^1\Pi_g$-state (grey) and the 5 $^1\Sigma_g^+$-state (green). The first part of the pulse creates a superposition state of the X $^1\Sigma_g^+$ and the A $^1\Sigma_u^+$ states. During the second part of the pulse the X $^1\Sigma_g^+$ and the A $^1\Sigma_u^+$ states are locked in a state of maximum coherence. The optical phase controls which of the dressed states (indicated at R_0) energetically separated by $\hbar\Omega$ is selectively populated. Absorption of another photon leads to population transfer to one of the (non-resonant) states 4 $^1\Sigma_g^+$ and 5 $^1\Sigma_g^+$. Selective population of the upper dressed state with subsequent transition to the 5 $^1\Sigma_g^+$-state is illustrated. (c) Absolute value of the envelope of the electric field of the pulse sequence $|E(t)|$ (red) and its temporal phase function (blue dashed). The sequence consists of two pulses (red dashed) with a FWHM of 14.1 fs and a central wavelength of 830 nm separated by of 12.1 fs. The second stronger pulse has a peak intensity of approximately 7×10^{11} W cm^{-2} and exhibits a relative phase jump of approximately $\pi/2$. At this delay and phase the upper dressed state is selectively populated leading to resonant excitation of the 5 $^1\Sigma_g^+$-state with about 80% efficiency (adapted from ref. 104).

By suitable shaping of the driving laser field, it is possible to populate one of these two (dressed) states, *i.e.* to realize selective population of dressed states (SPODS). Effectively, the population of a single dressed state amounts to a controlled energy shift of the resonant state into a desired direction. By the variation of the laser intensity the energy splitting can be controlled, and thus a particular target state among the manifold of final states is addressed. Experimentally strong-field coherences in the potassium 4p–4s transition were excited with shaped laser pulses and the selective population of dressed states was monitored during the interaction by a perturbative two-photon ionization process into the continuum with the help of photoelectron spectroscopy. Dressed state population is reflected in the amplitude of the corresponding Autler–Towns component, where energy shifts of the order of several hundred meV have been observed[57] highlighting again the importance for strong-field control of chemical reactions. In a multitude of experiments on the potassium atom it was found that continuous temporal phase variations[43,61] as well as temporal phase discontinuities[43,45,57,58,82,101] lead to SPODS and only so called "real" pulses (pulses with a constant temporal phase except for π-jumps which merely represent a change of the sign of the envelope) do not exhibit this effect[61] because the pulse area theorem applies. In such cases, which are challenging to realize experimentally, weak-field control schemes can be extended to the strong-field regime.[103]

Robustness of SPODS, *i.e.* insensitivity to reasonable changes in the pulse energy, was found for chirped pulses[43,61] as well as for pulses with step phase modulation[59]

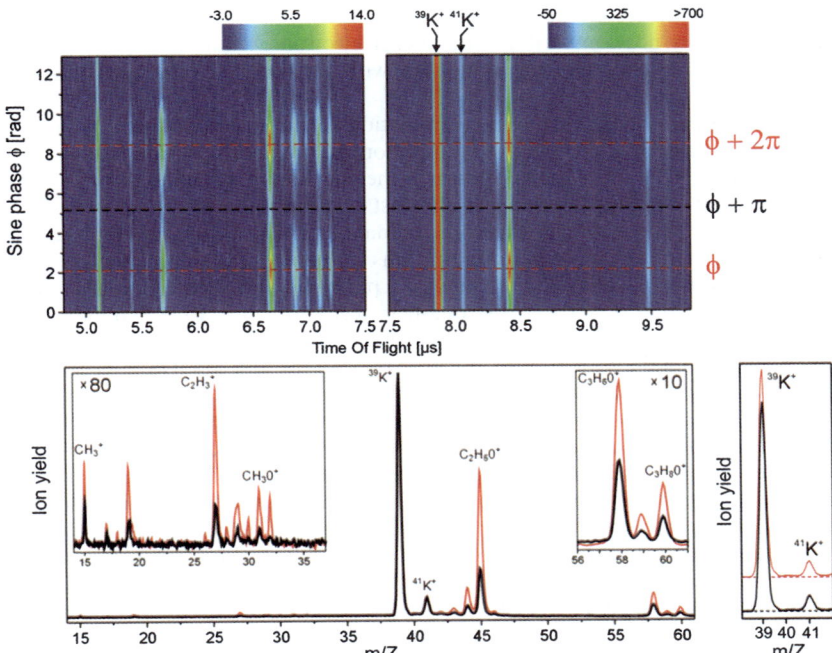

Fig. 6 Control of the fragmentation of isopropyl alcohol (C_3H_8O) using a pulse sequence. Upper panel: False color representation of TOF—mass spectra from the dissociation of isopropyl alcohol. Pulse sequences are created by a sinusoidal spectral phase modulation function $\varphi(\omega) = 0.5 \sin[50\ \text{fs}\ (\omega - 2.4\ \text{fs}^{-1}) + \phi]$ of a 10 μJ, 30 fs, 785 nm femtosecond laser pulse. The phase ϕ is varied within 4π. Lower panels: sections through the mass spectra at $\phi \approx 2$ rad (red) and $\phi \approx 5.5$ rad (black) exhibit a pronounced variation of the molecular ion yield. At the mass of $m = 45$ u ($C_2H_5O^+$) a variation in the molecular ion yield by a factor of 3 is observed. The insets show a magnification of the spectra. The atomic ion yield of $^{39}K^+$ and $^{41}K^+$ measured simultaneously as a reference shows no variation with the phase (right). The observed phase dependence hints to ultrafast switching of coherent electronic excitation in a molecule by SPODS (adapted from ref. 109).

and an extension of SPODS to a three-state system was studied recently.[60] The conclusion is that coherent electronic excitation of resonant states with shaped laser pulses leads in general to SPODS in this atomic model system. Many different pulse shapes lead to comparable dressed state energy shifts and dressed state population. If operative in larger systems as well this could explain why in closed loop experiments optimzed pulses often do not exhibit a unique structure.

In order to investigate this aspect, photon locking—being one specific realization of SPODS—was studied also with respect to molecules. This strong-field quantum control scheme—based on concepts originally developed in NMR (spin-locking)[105]—makes explicit use of temporal phase changes within the pulse and was demonstrated experimentally using nanosecond laser pulses.[106,107] We extend these techniques to the femtosecond time scale with relevant applications to coherent control. The main physical picture behind photon locking is as follows: a first resonant interaction creates an oscillating charge distribution with maximum amplitude and by timing a second interaction with sub cycle precision switching of population to different final states is achieved (see Fig. 4 and Fig. 5). As the charge oscillations are in the order of one femtosecond for typical valence excitations control has to be performed with attosecond precision. Sub 10 as precision in such an electron control scheme has recently been demonstrated.[32] Direct control of valence bond electron dynamics with attosecond pulses seems to be difficult, as the corresponding photon energy is usually not compatible with valence electron excitation. However, direct monitoring of valence electron dynamics with attosecond techniques has been achieved.[108]

Regarding molecules, wave packet simulations on a generic diatomic molecule[58] and on the potassium dimer[104] have been performed, implementing a photon locking pulse sequence. The simulations confirmed the selectivity and tunability of SPODS in the presence of nuclear motion and ultrafast efficient population transfer to target states (see Fig. 5). Preliminary experimental data on the potassium dimer were presented in the Introductory lecture. Indications for control out of coherently excited electronic states with subcycle precision were presented and data on intensity variations confirmed that the expected strong-field scenario is operative. No indications of weak-field control *via* higher order spectral interferences were observed in accordance with the findings on atoms.[57] Further indications that SPODS in general and photon locking as a specific example are important strong-field mechanisms come from theoretical considerations: in paper 11 of this meeting (DOI: 10.1039/c1fd00031d) Philipp von den Hoff, Markus Kowalewski and Regina de Vivie-Riedle concluded after investigating selective excitation in the potassium dimer with the help of optical control theory (OCT): "the SPODS mechanism is an optimal solution in the OCT search space. From the properties of the OCT algorithm it is known that high quality control and robust solutions are found even for complex quantum systems including a large number of control variables. In this sense the SPODS can be regarded as a robust way to control the selective population of higher lying electronic states, opening a wide spectrum of applications ranging form reaction control within molecules up to discrimination between different molecules in a mixture." When studying the excitation of ground-surface vibrational motion while minimizing radiation damage Ronnie Kosloff, Audrey Dell Hemmerich and David Tannor found that photon locking is a key ingredient in that control scenario.[110]

For applications to chemistry, a validation of such a coherent control strategy on larger molecules is helpful. As a first step, we investigated the mass spectra—measured with a time of flight spectrometer—from dissociation of isopropyl alcohol using a photon locking sequence.[109] The results shown in Fig. 6 show pronounced variations in the molecular ion yield upon variations of the temporal phase in the pulse sequence. In these experiments the simultaneously measured ion yield from potassium atoms showed no significant variations. This result confirms that no spectral/spatial cross-sensitivities are introduced by our pulse shaper. The observations show that control of the molecular dynamics of isopropyl alcohol is exerted by the optical phase of the shaped pulse. The phase dependence of the signal is hinting to a SPODS mechanism, however, systematic studies on the intensity dependence for a final proof have not been performed so far.

4. Adiabatic population transfer on molecules in solution

In this section we show that physically motivated pulse parameterizations based on frequency sweeps together with temporal pulse envelope asymmetries help to identify possible mechanisms for adiabatic population transfer in molecules. Joint wave packet motion (JOMO) and eigenstate preparation are presented as strong-field processes complementing frequency ordering followed by pump dump scenarios[111] and picosecond ultrabroadband positive chirp approaches.[75]

Rapid adiabatic passage is a well known concept in atomic physics where the complete population from one state is transferred to another state.[50] In the simplest case a detuned laser of sufficient pulse area sweeps through a resonance in order to achieve complete population transfer. Experimental realisations down to the picosecond[51] and femtosecond[54,61] regime have been reported and extensions to artificial atoms, *i.e.* quantum dots were demonstrated as well.[112]

As already mentioned in the introduction, a high degree of excitation can also be achieved *via* Rabi oscillations, however, this approach is not always attractive for efficient coherent control schemes as the resonant Rabi oscillation period is proportional to the scalar product of the electric dipole moment times the electric field of the laser pulse. As a consequence, different excitation levels are achieved due to the intensity distribution within the focal area of a typical Gaussian laser beam and in case of molecules due to different orientations in a typical isotropic sample. This is why adiabatic strong-field approaches are especially important as they are robust against these effects.

Efficient and robust population transfer in molecular systems would have many exciting experimental applications ranging from life science to fundamental research. In life science the simultaneous excitation of all fluorophores within a focused laser pulse could trigger efficient localized photoreactions within a cell, lead to brighter images in laser-based fluorescence microscopy or even to enhanced resolution. In fundamental research all kinds of excited electronic state spectroscopy would benefit.

This is why strong-field excitation of molecules especially in solution has already attracted interest. A prominent experiment was an experiment by Shank and coworkers[77] studying fluorescence from laser dyes after chirped excitation in the weak- and strong-field regime. Whereas in the weak-field no changes in the fluorescence yield were observed, in the strong-field a fluorescence suppression was observed for negative chirp. As an explanation a frequency ordered pump–dump mechanism was offered based on theoretical considerations of Sandy Ruhman and Ronnie Kosloff.[111] Another example is the feedback optimization of fluorescence in a laser dye, where an enhancement of fluorescence was also found for positive chirped pulses by Kent Wilson and coworkers.[36] In the light of these experiments a theoretical publication by Wilson and coworkers[75] was published. In that work it was found that nearly complete electronic population inversion of molecules can be achieved with intense positively chirped broadband laser pulses, as a combined result of vibrational coherence and adiabatic inversion. Strong-field quantum calculations demonstrated inversion probabilities of up to 99%. In addition the results were shown to be robust with respect to changes in light field parameters as well as to thermal and condensed phase conditions. Similar conclusions were drawn by Ronnie Kosloff and coworkers taking a different approach: they found that a field of sufficiently high chirp rate imposes a certain relative phase between a ground and excited state wave function of a two-level system that explains the unidirectionality of the population transfer from the ground to the excited state in atomic and molecular systems.

These findings motivated our experiments on efficient and robust population transfer in sensitizer dyes.[113] Important natural pulse parameters to study strong-field effects in order to induce adiabatic population transfer are frequency sweeps and time varying intensity profiles. This is why we parameterized our excitation pulses in terms of the instantaneous frequency and temporal envelope asymmetries, *i.e.* in technical terms group delay dispersion (GDD) and third order dispersion (TOD) (see Fig. 7). We investigated two photosensitizer dyes in solution under the same experimental conditions being prepared in the triplet ground state. Excitation within the triplet system was followed by intersystem crossing and the corresponding singlet fluorescence was monitored as a measure of population transfer in the triplet system. We recorded control landscapes with respect to the fluorescence intensity on both dyes by a systematic variation of laser pulse shapes combining GDD and TOD. In the strong-field regime we found highly structured topologies with large areas of maximum or minimum population transfer that were insensitive over a certain range of applied laser intensities thus demonstrating robustness. One example is displayed in Fig. 7.

We then compared our experimental results to vibrational wave packet simulations for a molecular two-state system, where for generality two parameterized

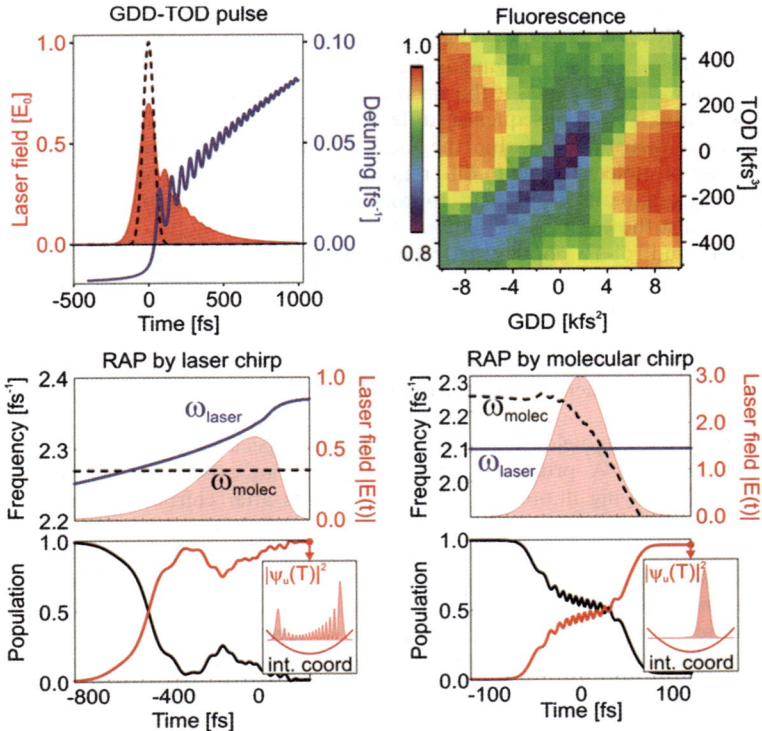

Fig. 7 Strong-field control of population transfer in a sensitizer dye in solution. Upper left: Example of a shaped femtosecond laser pulse obtained by spectral phase modulation using the phase function $\varphi(\omega) = \phi_2/2\,(\omega-\omega_0)^2 + \phi_3/6\,(\omega-\omega_0)^3$ with a GDD of $\phi_2 = 4$ kfs^2 and a TOD of $\phi_3 = 200$ kfs^3. The initial unshaped laser pulse (dashed) has a FWHM of 60 fs. Due to the combined action of GDD and TOD both the shape of the pulse envelope (red shaded) and the instantaneous laser frequency (blue) are controlled (see ref. 113 for a comprehensive discussion of GDD-TOD pulse shapes). Upper right: Measured fluorescence signal from strong-field excitation (peak intensity of the bandwidth limited pulse of about 60 GW cm^{-2}) of porphyrazine molecules as a function of GDD and TOD. A highly structured topology is observed with large areas of maximum and minimum population transfer. Lower panels: Two examples of control of population transfer in molecule. In the "RAP by laser chirp" scenario the laser frequency ω_{laser} (blue) sweeps over the molecular transition ω_{molec} (black) such that at the crossing of both curves population is adiabatically transferred from the ground sate (black) into the excited state (red). Note that after the interaction at time $T \approx 400$ fs population is transferred to the upper state with almost unit efficiency and the system is in a vibrational eigenstate of the upper potential. Alternatively, in the "RAP by molecular chirp" a bandwidth limited pulse with a constant instantaneous frequency ω_{laser} (blue) drives the molecular dynamics such that the molecular transition frequency ω_{molec} (black) sweeps over the laser frequency due to the wave packet motion inducing almost complete population transfer. In this scenario the wave packet dynamics is characterized by Joint Motion (JOMO) of the ground- and the excited-state wave packet (see ref. 113 for a detailed discussion of JOMO). Accordingly, after the interaction at time $T \approx 150$ fs, a vibrational wave packet is formed in the upper potential.

one-dimensional harmonic potentials were considered. Calculated control landscapes based on the same pulse parameters were in good accordance with experimental data for both sensitizer dyes, while different detunings of the laser central frequency to the dye absorption bands are accounted for by appropriate laser detunings in simulations. We identified areas with complete population transfer and nearly complete population return inside the landscapes, both being robust over a wide range of intensity variations. By modelling decoherence in a simple approach, good agreement of the measured and simulated landscapes implied that coherent control of population transfer in sensitizer dyes can take place in the liquid phase, i.e. in the presence of decoherence in accordance with the findings by Kent Wilson and coworkers.[75]

The good agreement motivated us to analyze the physical mechanisms controlling the final state populations further in simulations. We found that atom-like interpretations of adiabatic interactions are possible. Two scenarios are presented in Fig. 7.

In one scenario the wave packet dynamics are characterized by a coupling of the ground state and excited state wave packets, inducing a joint wave packet motion (JOMO), leading to a well-defined joint internuclear distance $R(t)$. The time-dependent molecular transition frequency induced by JOMO allows for efficient population transfer when a crossing with the instantaneous frequency of the laser occurs ("RAP by molecular chirp"). Such a crossing can even be achieved when a bandwidth-limited pulse with constant instantaneous laser frequency is used. The displayed off-resonant excitation was not in the parameter space of our experiment. However, coherent population return based on JOMO was within the parameters of the experiment.[113]

The second scenario is based on the excitation of a single vibrational eigenstate by narrowband interaction in the Franck–Condon region during the weak starting part of the laser pulse. A "RAP by laser chirp" process is now observed when the instantaneous frequency of the laser is tuned over the molecular resonance during the more intense parts of the pulse.

These findings suggest, that GDD-TOD pulses are a suitable parameterization to study strong-field effects. In the light of possible applications we note, that typical peak intensities of the bandwidth limited pulses in our experiment are approximated to $I = 60$ GW cm^{-2} (taking into account that dye molecules typically provide oscillator strengths in the range of unity, the pulse area reaches 4 times 2π indicating strong-field excitation conditions). Note also that the peak field strength is reduced for shaped pulses (for analytical expressions see ref. 113) and that an applied intensity of 200 GW cm^{-2} is a value that is commonly used as the damage threshold for biological samples (see ref. 114 and references therein).

5. Control of ionization processes in dielectrics

In this last section we briefly highlight the extension of experimental control methodologies to ultrafast laser control of incoherent processes with an emphasis on processing of dielectrics on the nanometre scale. Here, primary processes induced by ultrafast laser radiation involve nonlinear electronic excitation where electron electron collisions at high excitation densities (in the range 10^{21} e cm^{-3} for ablation of dielectrics) destroy any coherence imprinted by the light field. In general, the electronic excitation is followed by energy transfer to the lattice and phase transitions that occur on fast (femtosecond, picosecond) but material-dependent time scales.[115] Optimal energy coupling with the help of suitably shaped temporal pulse envelopes thus gives the possibility to guide the material response towards user-designed directions, offering extended flexibility for quality material processing.[22] We perform prototype studies in the above mentioned spirit on dielectrics, water and metals. In dielectrics we have observed different thresholds for material processing with temporally asymmetric pulse shapes that we attributed to control of different ionization processes i.e. multi-photon ionization and avalanche ionization (see Fig. 8).[116–118] The resulting nanometre scale structures were one order of magnitude below the diffraction limit. Recently we extended our studies to investigate the dynamics of free electron plasma created by femtosecond pulses in a thin water jet[119] to a direct observation of the free electron density after excitation with temporally shaped laser pulses by using spectral interference techniques.[120,121] Exploiting polarization-dependent near field effects[122] is an alternative route to nanoscale

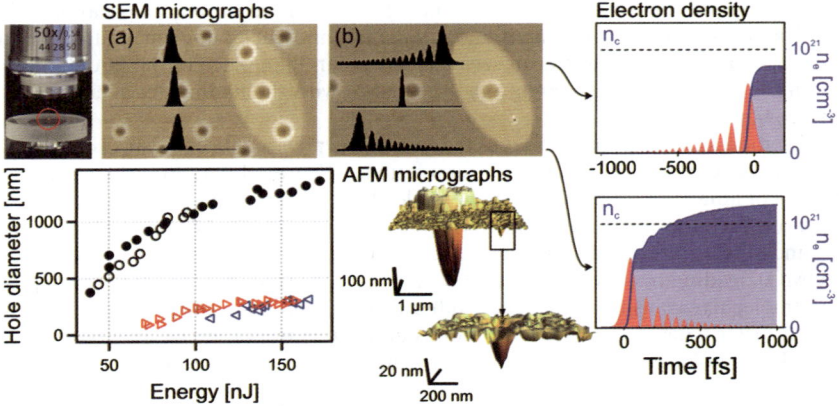

Fig. 8 Control of ionization processes in fused silica *via* asymmetrically shaped femtosecond pulses. Upper left: In our materials processing platform, a femtosecond laser pulse is focused using a microscope objective. Due to the nonlinear interaction the laser induced plasma is highly localized. Scanning electron microscope (SEM) micrographs of a measurement pattern: for an applied energy and focal position, a triplet of laser pulses is highlighted by the ellipse. Negative, zero, and positive TOD were used. Normalized temporal intensity profiles are sketched for comparison between different TODs. (a) Low TOD ($\phi_3 = \pm 2.5 \times 10^4$ fs^3, $E = 77$ nJ) results in negligible differences between created structures. (b) High positive TOD ($\phi_3 = +6 \times 10^5$ fs^3, $E = 71$ nJ) results in a change of structure size and threshold energy (adapted from ref. 117). Diameters of ablation structures for fused silica with ($\phi_3 = 0$ fs^3, ○ and ●) for two completely independent measurements confirm the reproducibility of our setup), for ($\phi_3 = +6 \times 10^5$ fs^3 (red) and ($\phi_3 = -6 \times 10^5$ fs^3 (blue)). Exemplified topology of the large structure (top) due to the unshaped pulse and the small structure (bottom) obtained by the shaped pulse measured *via* AFM. Note that this small structure is an order of magnitude below the diffraction limit. Right: Transient free electron density n_e (solid blue lines) as modeled by a simple rate equation describing multi photon ionization with k photons (σ_k) and avalanche ionization (α) *via* dn_e/d$t = \sigma_k I^k + \alpha n_e I$, together with the contributions provided by photo ionization alone $\sigma_k I^k$ (light blue shaded area) and avalanche photo ionization alone $\alpha n_e I$ (dark blue shaded area) and the corresponding electrical field (red shaded area) of the modulated pulses proportional to $I^{1/2}$. Note that the free electron density exceeds the critical value for material ablation n_c only for a positive value of ϕ_3. This observation has been discussed in a *seed and heat* mechanism based on a refined ionization model employing a multiple rate equation approach.[116]

material processing of dielectrics[123,124] Minimizing the spatial structure and at the same time maximizing the spectrochemical sensitivity for fs-LIBS[125,126] (LIBS = laser induced breakdown spectroscopy) completes our experiments exploiting temporal pulse tailoring for material processing.

6. Conclusions

In this Introductory lecture we first summarized recent developments in coherent control before we focused on our experiments on (coherent) electronic excitation with shaped femtosecond laser pulses with respect to recent developments.

We started by discussing how to make use of the electronic structure of matter together with polarization shaped laser pulses in order to create designed electron wave packets in the continuum and showed how to reconstruct the created three-dimensional electron distributions *via* tomography. We pointed out that the techniques developed open new routes to molecular recognition with an emphasis on chiral recognition in the gas phase as well as to the measurement of photoelectron angular distributions in the molecular frame.

We then turned to control of coherent electronic excitation in the Rabi cycling regime with shaped laser pulses. We highlighted the importance of controlling resonant processes under such strong-field conditions and gave two examples. In gas phase experiments we hinted at a general strong-field scheme based on selective population of dressed states (SPODS) and especially to the consequences of controlling electronic coherences (being in the order of one femtosecond for typical valence excitations) with attosecond precision. By this, population to different final states in atoms and molecules can be switched with high selectivity and efficiency. In the liquid phase we discussed strong-field scenarios for adiabatic population transfer. In our experiments on porphyrazines we used physically motivated pulse parameterizations for adiabatic strong-field interactions. These were based on frequency sweeps together with temporal pulse envelope asymmetries. Comparison to simulations helped us to identify possible mechanisms for adiabatic population transfer in molecules. Joint wave packet motion (JOMO) and eigenstate preparation are presented as adiabatic strong-field processes complementing the well known frequency ordering followed by pump–dump scenarios and picosecond ultrabroadband positive chirp approaches.

Finally we briefly presented our experiments demonstrating control of incoherent ionization processes in dielectrics for material processing on the nanometre scale. We showed that shaping temporal asymmetric profiles address multi photon ionization and electron-electron impact ionization in a different fashion leading to different observed ablation thresholds for pulses with the same fluence, same statistical pulse duration and same focusing conditions. By this we point to a further extension of ultrafast coherent control methodologies in material processing: here optimal energy coupling with the help of suitably shaped temporal pulse envelopes gives the possibility to guide the material response towards user-designed directions, offering extended flexibility for quality material processing.

7. Acknowledgements

T.B. would like to thank Ben Whitaker, Mike Bearpark, Regina de Vivie-Riedle, Helen Fielding and Thomas Weinacht from the Scientific Committee for the invitation to give this introductory lecture and for putting such a stimulating program together. T.B. is also grateful for the lovely conference dinner including the traditional Faraday Loving Cup ceremony. Besides the authors mentioned on our publications the work presented here is mainly based on the PhD work of Tim Bayer, Lars Englert, Jens Köhler and Marc Krug with substantial experimental support by Cristian Sarpe. Financial support by DFG and the EU-ITN-FASTQUAST is acknowledged as well.

References

1 A. H. Zewail, *J. Phys. Chem.*, 2000, **104**, 5660–5694.
2 J. Manz and L. Woeste, *Femtosecond Chemistry*, VCH, Weinheim, 1995.
3 S. A. Rice and M. Zhao, *Optical control of molecular dynamics*, Wiley, New York, 2000.
4 M. Shapiro and P. Brumer, *Principles of the Quantum Control of Molecular Processes*, John Wiley & Sons, Hoboken, New Jersey, 2003.
5 D. Tannor, *Introduction to Quantum Mechanics: A Time-Dependent Perspective*, Palgrave Macmillan Publishers Limited, Houndmills, Basingstoke, Hampshire, England, 2007.
6 D. J. Tannor and S. A. Rice, *Adv. Chem. Phys.*, 1988, **70**, 441–523.
7 M. Shapiro and P. Brumer, *Int. Rev. Phys. Chem.*, 1994, **13**, 187–229.
8 T. Baumert, J. Helbing, and G. Gerber, in *Advances in Chemical Physics - Photochemistry: Chemical Reactions and their control on the Femtosecond Time Scale*, ed. I. Prigogine and S. A. Rice, John Wiley & Sons, Inc., New York, 1997, pp. 47–77.
9 H. Rabitz, R. de Vivie-Riedle, M. Motzkus and K. Kompa, *Science*, 2000, **288**, 824–828.
10 M. Shapiro and P. Brumer, *Rep. Prog. Phys.*, 2003, **66**, 859–942.
11 D. Goswami, *Phys. Rep.*, 2003, **374**, 385–481.
12 M. Dantus and V. V. Lozovoy, *Chem. Rev.*, 2004, **104**, 1813–1859.
13 V. Bonacic-Koutecky and R. Mitric, *Chem. Rev.*, 2005, **105**, 11–65.
14 T. Brixner, T. Pfeifer, G. Gerber, M. Wollenhaupt, and T. Baumert, in *Femtosecond Laser Spectroscopy*, (P. Hannaford, Ed), pp. 225–266. Springer Verlag, (2005).

15 M. Wollenhaupt, V. Engel and T. Baumert, *Annu. Rev. Phys. Chem.*, 2005, **56**, 25–56.
16 P. Nuernberger, G. Vogt, T. Brixner and G. Gerber, *Phys. Chem. Chem. Phys.*, 2007, **9**, 2470–2497.
17 J. Werschnik and E. K. U. Gross, *J. Phys. B: At., Mol. Opt. Phys.*, 2007, **40**, R175–R211.
18 W. Wohlleben, T. Buckup, J. L. Herek and M. Motzkus, *ChemPhysChem*, 2005, **6**, 850–857.
19 Y. Silberberg, *Annu. Rev. Phys. Chem.*, 2009, **60**, 277–292.
20 K. Ohmori, *Annu. Rev. Phys. Chem.*, 2009, **60**, 487–511.
21 C. Brif, R. Chakrabarti and H. Rabitz, *New J. Phys.*, 2010, **12**, 075008.
22 R. Stoian, M. Wollenhaupt, T. Baumert, and I. V. Hertel, in *Laser Precision Microfabrication*, ed. K. Sugioka, M. Meunier, and A. Piqué, Springer-Verlag, Berlin Heidelberg, 2010, pp. 121–144.
23 *Advances in Chemical Physics: Chemical Reactions and Their Control on the Femtosecond Time Scale, XXth Solvay Conference on Chemistry*, ed. P. Gaspard and I. Burghardt, John Wiley & Sons Inc., New York, 1997, vol. 101.
24 J. L. Herek, *J. Photochem. Photobiol., A*, 2006, **180**, 225.
25 H. Fielding, M. Shapiro and T. Baumert, *J. Phys. B: At., Mol. Opt. Phys.*, 2008, **41**, 070201–070201-1.
26 H. Rabitz, *New J. Phys.*, 2009, **11**, 105030.
27 H. Fielding and M. A. Robb, *Phys. Chem. Chem. Phys.*, 2010, **12**, 15569.
28 A. M. Weiner, *Rev. Sci. Instrum.*, 2000, **71**, 1929–1960.
29 M. Wollenhaupt, A. Assion, and T. Baumert, in *Springer Handbook of Lasers and Optics*, ed. F. Träger, Springer Science + Business Media, New York, 2007, pp. 937–983.
30 D. B. Strasfeld, S.-H. Shim and M. T. Zanni, *Adv. Chem. Phys.*, 2009, **141**, 1–28.
31 A. Monmayrant, S. Weber and B. Chatel, *J. Phys. B: At., Mol. Opt. Phys.*, 2010, **43**, 103001-103001-34.
32 J. Köhler, M. Wollenhaupt, T. Bayer, C. Sarpe and T. Baumert, *Opt. Express*, 2011, **19**, 11638–11653.
33 R. S. Judson and H. Rabitz, *Phys. Rev. Lett.*, 1992, **68**, 1500–1503.
34 T. Baumert, T. Brixner, V. Seyfried, M. Strehle and G. Gerber, *Appl. Phys. B: Lasers Opt.*, 1997, **65**, 779–782.
35 D. Meshulach, D. Yelin and Y. Silberberg, *Opt. Commun.*, 1997, **138**, 345–348.
36 C. J. Bardeen, V. V. Yakovlev, K. R. Wilson, S. D. Carpenter, P. M. Weber and W. S. Warren, *Chem. Phys. Lett.*, 1997, **280**, 151–158.
37 A. Assion, T. Baumert, M. Bergt, T. Brixner, B. Kiefer, V. Seyfried, M. Strehle and G. Gerber, *Science*, 1998, **282**, 919–922.
38 R. J. Levis and H. A. Rabitz, *J. Phys. Chem. A*, 2002, **106**, 6427–6444.
39 C. Daniel, J. Full, L. Gonzáles, C. Lupulescu, J. Manz, A. Merli, S. Vajda and L. Wöste, *Science*, 2003, **299**, 536–539.
40 T. Hornung, R. Meier and M. Motzkus, *Chem. Phys. Lett.*, 2000, **326**, 445–453.
41 A. Bartelt, A. Lindinger, C. Lupulescu, S. Vajda and L. Wöste, *Phys. Chem. Chem. Phys.*, 2003, **5**, 3610–3615.
42 S. Fechner, F. Dimler, T. Brixner, G. Gerber and D. J. Tannor, *Opt. Express*, 2007, **15**, 15387–15401.
43 T. Bayer, M. Wollenhaupt and T. Baumert, *J. Phys. B: At., Mol. Opt. Phys.*, 2008, **41**, 074007–074007-13.
44 S. Ruetzel, C. Stolzenberger, F. Dimler, D. J. Tannore and T. Brixner, *Phys. Chem. Chem. Phys.*, 2011, **13**, 8627–8636.
45 M. Wollenhaupt, A. Präkelt, C. Sarpe-Tudoran, D. Liese and T. Baumert, *J. Mod. Opt.*, 2005, **52**, 2187–2195.
46 H. A. Rabitz, M. M. Hsieh and C. M. Rosenthal, *Science*, 2004, **303**, 1998–2001.
47 M. Motzkus and T. Baumert, *Symposium Quantum Control Spectroscopy (at spring meeting of Deutsche Physikalische Gesellschaft)*, 2010.
48 T. Brixner, G. Krampert, T. Pfeifer, R. Selle, G. Gerber, M. Wollenhaupt, O. Graefe, C. Horn, D. Liese and T. Baumert, *Phys. Rev. Lett.*, 2004, **92**, 208301–208301-4.
49 N. V. Vitanov, T. Halfmann, B. W. Shore and K. Bergmann, *Annu. Rev. Phys. Chem.*, 2001, **52**, 763–809.
50 B. W. Shore, *acta physica slovaca*, 2008, **58**, 243–486.
51 J. S. Melinger, S. R. Gandhi, A. Hariharan, J. X. Tull and W. S. Warren, *Phys. Rev. Lett.*, 1992, **68**, 2000–2003.
52 I. R. Sola, J. Santamaria and V. S. Malinovsky, *Phys. Rev. A: At., Mol., Opt. Phys.*, 2000, **61**, 043413–043413-7.
53 V. S. Malinovsky and J. L. Krause, *Eur. Phys. J. D*, 2001, **14**, 147–155.
54 M. Krug, T. Bayer, M. Wollenhaupt, C. Sarpe-Tudoran, T. Baumert, S. S. Ivanov and N. V. Vitanov, *New J. Phys.*, 2009, **11**, 105051.

55 S. Zhdanovich, E. A. Shapiro, M. Shapiro, J. W. Hepburn and V. Milner, *Phys. Rev. Lett.*, 2008, **100**, 103004–103004-4.
56 C. Trallero-Herrero, J. L. Cohen and T. Weinacht, *Phys. Rev. Lett.*, 2006, **96**, 063603–063603-4.
57 M. Wollenhaupt, A. Assion, O. Bazhan, C. Horn, D. Liese, C. Sarpe-Tudoran, M. Winter and T. Baumert, *Phys. Rev. A: At., Mol., Opt. Phys.*, 2003, **68**, 015401–015401-4.
58 M. Wollenhaupt, D. Liese, A. Präkelt, C. Sarpe-Tudoran and T. Baumert, *Chem. Phys. Lett.*, 2006, **419**, 184–190.
59 T. Bayer, M. Wollenhaupt, C. Sarpe-Tudoran and T. Baumert, *Phys. Rev. Lett.*, 2009, **102**, 023004-1-023004-4.
60 M. Wollenhaupt, T. Bayer, N. V. Vitanov and T. Baumert, *Phys. Rev. A: At., Mol., Opt. Phys.*, 2010, **81**, 053422–053422-9.
61 M. Wollenhaupt, A. Präkelt, C. Sarpe-Tudoran, D. Liese and T. Baumert, *Appl. Phys. B: Lasers Opt.*, 2006, **82**, 183–188.
62 J. H. Posthumus, *Rep. Prog. Phys.*, 2004, **67**, 623–665.
63 T. Frohnmeyer, M. Hofmann, M. Strehle and T. Baumert, *Chem. Phys. Lett.*, 1999, **312**, 447–454.
64 B. J. Sussman, D. Townsend, M. Y. Ivanov and A. Stolow, *Science*, 2006, **314**, 278–281.
65 T. Baumert, V. Engel, C. Meier and G. Gerber, *Chem. Phys. Lett.*, 1992, **200**, 488–494.
66 H. Goto, H. Katsuki, H. Ibrahim, H. Chiba and K. Ohmori, *Nat. Phys.*, 2011, **7**, 383–385.
67 P. Nuernberger, D. Wolpert, H. Weiss and G. Gerber, *Proc. Natl. Acad. Sci. U. S. A.*, 2010, **107**, 10366–10370.
68 D. B. Strasfeld, S.-H. Shim and M. T. Zanni, *Phys. Rev. Lett.*, 2007, **99**, 038102–038102-4.
69 J. L. Herek, W. Wohlleben, R. Cogdell, D. Zeidler and M. Motzkus, *Nature*, 2002, **417**, 533–535.
70 G. Vogt, G. Krampert, P. Niklaus, P. Nuernberger and G. Gerber, *Phys. Rev. Lett.*, 2005, **94**, 068305–068305-4.
71 V. I. Prokhorenko, A. M. Nagy, S. A. Waschuk, L. S. Brown, R. R. Birge and R. J. D. Miller, *Science*, 2006, **313**, 1257–1261.
72 T. Brixner, N. H. Damrauer, P. Niklaus and G. Gerber, *Nature*, 2001, **414**, 57–60.
73 M. Roth, L. Guyon, J. Roslund, V. Boutou, F. Courvoisier, J.-P. Wolf and H. Rabitz, *Phys. Rev. Lett.*, 2009, **102**, 253001–253001-4.
74 J. Petersen, R. Mitric, V. Bonacic-Koutecky, J.-P. Wolf, J. Roslund and H. Rabitz, *Phys. Rev. Lett.*, 2010, **105**, 073003.
75 J. Cao, C. J. Bardeen and K. R. Wilson, *Phys. Rev. Lett.*, 1998, **80**, 1406–1409.
76 J. Vala and R. Kosloff, *Opt. Express*, 2001, **8**, 238–245.
77 G. Cerullo, C. J. Bardeen, Q. Wang and C. V. Shank, *Chem. Phys. Lett.*, 1996, **262**, 362–368.
78 M. Wollenhaupt, A. Assion, D. Liese, C. Sarpe-Tudoran, T. Baumert, S. Zamith, M. A. Bouchene, B. Girard, A. Flettner, U. Weichmann and G. Gerber, *Phys. Rev. Lett.*, 2002, **89**, 173001–173001-4.
79 M. Winter, M. Wollenhaupt and T. Baumert, *Opt. Commun.*, 2006, **264**, 285–292.
80 F. Lindner, M. G. Schätzel, H. Walther, A. Baltuska, E. Goulielmakis, F. Krausz, D. B. Milosevic, D. Bauer, W. Becker and G. G. Paulus, *Phys. Rev. Lett.*, 2005, **95**, 040401–040401-4.
81 S. H. Autler and C. H. Townes, *Phys. Rev.*, 1955, **100**, 703–722.
82 M. Wollenhaupt, A. Präkelt, C. Sarpe-Tudoran, D. Liese, T. Bayer and T. Baumert, *Phys. Rev. A: At., Mol., Opt. Phys.*, 2006, **73**, 063409–063409-15.
83 M. Wollenhaupt, M. Krug, J. Köhler, T. Bayer, C. Sarpe-Tudoran and T. Baumert, *Appl. Phys. B: Lasers Opt.*, 2009, **95**, 245–259.
84 A. T. J. B. Eppink and D. H. Parker, *Rev. Sci. Instrum.*, 1997, **68**, 3477–3484.
85 *Imaging in Molecular Dyanmics - Technology and Applications*, ed. B. Whitaker, Cambridge University Press, Cambridge, 2003.
86 C. Bordas, F. Pauling, H. Helm and D. L. Huestis, *Rev. Sci. Instrum.*, 1996, **67**, 2257–2268.
87 M. J. J. Vrakking, *Rev. Sci. Instrum.*, 2001, **72**, 4084–4089.
88 V. Dribinski, A. Ossadtchi, V. A. Mandelshtam and H. Reisler, *Rev. Sci. Instrum.*, 2002, **73**, 2634–2642.
89 G. A. Garcia, L. Nahon and I. Powis, *Rev. Sci. Instrum.*, 2004, **75**, 4989–4996.
90 M. Wollenhaupt, M. Krug, J. Köhler, T. Bayer, C. Sarpe-Tudoran and T. Baumert, *Appl. Phys. B: Lasers Opt.*, 2009, **95**, 647–651.
91 M. Krug, PhD thesis, Kohärente Kontrolle winkelaufgelöster Photoelektronenspektren, Universität Kassel, 2010.
92 A. Stolow, A. E. Bragg and D. M. Neumark, *Chem. Rev.*, 2004, **104**, 1719–1757.
93 C. Smeenk, L. Arissian, A. Staude, D. M. Villeneuve and P. B. Corkum, *J. Phys. B: At., Mol. Opt. Phys.*, 2009, **42**, 165402.

94 H. Stapelfeldt and T. Seideman, *Rev. Mod. Phys.*, 2003, **75**, 543–557.
95 C. Horn, M. Wollenhaupt, M. Krug, T. Baumert, R. de Nalda and L. Banares, *Phys. Rev. A: At., Mol., Opt. Phys.*, 2006, **73**, 031401–031401-4.
96 J. Ullrich, R. Moshammer, A. Dorn, R. Dörner, L. Ph, H. Schmidt and H. Schmidt-Böcking, *Rep. Prog. Phys.*, 2003, **66**, 1463–1545.
97 A. Vredenborg, W. G. Roeterdink and M. H. M. Janssen, *Rev. Sci. Instrum.*, 2008, **79**, 063108.
98 *Chiral Recognition in the Gas Phase*, ed. A. Zehnacker, CRC Press, Boca Raton FL USA, 2010.
99 L. Nahon, G. A. Garcia, C. J. Harding, E. Mikajlo and I. Powis, *J. Chem. Phys.*, 2006, **125**, 114309.
100 X. Li and M. Shapiro, *J. Chem. Phys.*, 2010, **132**, 194315.
101 M. Wollenhaupt, A. Präkelt, C. Sarpe-Tudoran, D. Liese and T. Baumert, *J. Opt. B: Quantum Semiclassical Opt.*, 2005, **7**, S270–S276.
102 R. Hildner, D. Brinks and N. F. van Hulst, *Nature Physics*, 2010, 1–6.
103 N. Dudovich, T. Polack, A. Péer and Y. Silberberg, *Phys. Rev. Lett.*, 2005, **94**, 083002–083002-4.
104 M. Wollenhaupt and T. Baumert, *J. Photochem. Photobiol., A*, 2006, **180**, 248–255.
105 S. R. Hartmann and E. L. Hahn, *Phys. Rev.*, 1962, **128**, 2053.
106 E. T. Sleva, I. M. Xavier Jr. and A. H. Zewail, *J. Opt. Soc. Am. B*, 1985, **3**, 483–487.
107 Y. S. Bai, A. G. Yodh and T. W. Mossberg, *Phys. Rev. Lett.*, 1985, **55**, 1277–1280.
108 E. Goulielmakis, Z.-H. Loh, A. Wirth, R. Santra, N. Rohringer, V. S. Yakovlev, S. Zherebtsov, T. Pfeifer, A. M. Azzeer, M. F. Kling, S. R. Leone and F. Krausz, *Nature*, 2010, **466**, 739–743.
109 M. Wollenhaupt, T. Bayer, A. Klumpp, C. Sarpe-Tudoran, and T. Baumert, in *Springer Proceedings in Physics 127 (Physics and Engineering of New Materials)*, ed. D. T. Cat, A. Pucci, and K. Wandelt, Springer, 2008, pp. 327–335.
110 R. Kosloff, A. D. Hammerich and D. Tannor, *Phys. Rev. Lett.*, 1992, **69**, 2172–2175.
111 S. Ruhman and R. Kosloff, *J. Opt. Soc. Am. B*, 1990, **7**, 1748–1752.
112 C.-M. Simon, T. Belhadj, B. Chatel, T. Amand, A. Lemaitre, O. Krebs, P. A. Dalgarno, R. J. Warburton, X. Marie and B. Urbaszek, *Phys. Rev. Lett.*, 2011, **106**, 166801.
113 J. Schneider, M. Wollenhaupt, A. Winzenburg, T. Bayer, J. Köhler, R. Faust and T. Baumert, *Phys. Chem. Chem. Phys.*, 2011, **13**, 8733–8746.
114 K. E. Sheetz and J. Squier, *J. Appl. Phys.*, 2009, **105**, 051101.
115 B. Rethfeld, K. Sokolowski-Tinten, D. von der Linde and S. I. Anisimov, *Appl. Phys. A: Mater. Sci. Process.*, 2004, **79**, 767–769.
116 L. Englert, B. Rethfeld, L. Haag, M. Wollenhaupt, C. Sarpe-Tudoran and T. Baumert, *Opt. Express*, 2007, **15**, 17855–17862.
117 L. Englert, M. Wollenhaupt, L. Haag, C. Sarpe-Tudoran, B. Rethfeld and T. Baumert, *Appl. Phys. A: Mater. Sci. Process.*, 2008, **92**, 749–753.
118 M. Wollenhaupt, L. Englert, A. Horn and T. Baumert, *J. Laser Micro/Nanoeng.*, 2009, **4**, 144–151.
119 C. Sarpe-Tudoran, A. Assion, M. Wollenhaupt, M. Winter and T. Baumert, *Appl. Phys. Lett.*, 2006, **88**, 261109–261109-3.
120 E. Tokunaga, A. Terasaki and T. Kobayashi, *Opt. Lett.*, 1992, **17**, 1131–1133.
121 V. V. Temnov, K. Sokolowski-Tinten, P. Zhou, A. El-Khamhawy and D. von der Linde, *Phys. Rev. Lett.*, 2006, **97**, 237403–237403-4.
122 M. Aeschlimann, M. Bauer, D. Bayer, T. Brixner, F. J. García de Abajo, W. Pfeiffer, M. Rohmer, C. Spindler and F. Steeb, *Nature*, 2007, **446**, 301–304.
123 F. Hubenthal, R. Morarescu, L. Englert, L. Haag, T. Baumert and F. Träger, *Appl. Phys. Lett.*, 2009, **95**, 063101–063101-3.
124 R. Morarescu, L. Englert, B. Kolaric, P. Damman, R. A. L. Vallée, T. Baumert, F. Hubenthal and F. Träger, *J. Mater. Chem.*, 2011, **21**, 4076–4081.
125 A. Assion, M. Wollenhaupt, L. Haag, F. Mayorov, C. Sarpe-Tudoran, M. Winter, U. Kutschera and T. Baumert, *Appl. Phys. B: Lasers Opt.*, 2003, **77**, 391–397.
126 W. Wessel, A. Brückner-Foit, J. Mildner, L. Englert, L. Haag, A. Horn, M. Wollenhaupt and T. Baumert, *Eng. Fract. Mech.*, 2010, **77**, 1874–1883.

PAPER

Coherently-controlled two-dimensional spectroscopy: Evidence for phase induced long-lived memory effects

Valentyn I. Prokhorenko,[*a] Alexei Halpin[b] and R. J. Dwayne Miller[*ab]

Received 23rd May 2011, Accepted 25th May 2011
DOI: 10.1039/c1fd00095k

Using low-intensity phase-shaped excitation pulses we used two-dimensional (2D) electronic spectroscopy to follow the time dependence of the coherent correlations imposed on a solvated organic dye (Rhodamine 101 in methanol) at room temperature. Shaping of the excitation pulses strongly affects both the real and imaginary parts of the 2D-spectra, especially at small waiting times. In particular, the periodic phase modulation of the excitation pulses appears as a two-dimensional grid-like modulation in the correlation spectrum corresponding to the waiting time $T = 0$. By increasing the waiting time, this modulation quickly disappears in ω_t space. However, it is still present in ω_τ space even at very long waiting times (≥ 80 ps) where the inhomogeneous broadening is significantly reduced, and reaches its stationary value of $\sim 16\%$. The resonant nature of this induced modulation at long waiting time allows us to conclude that phase shaping of the excitation induces a long-lived memory in solvated organic dyes that is associated with coherent population transfer.

1 Introduction

Multidimensional spectroscopy has become a powerful tool for the investigation of complex quantum systems, especially molecules and molecular assemblies. Since its first realization using rf (NMR), the excitation range has been extended to the IR-domain and later to the visible and UV, which allows direct study of molecular electronic transitions or resonances (MER) in molecular aggregates. The most common technique in the visible range is two-dimensional electronic photon-echo spectroscopy (2D-PE), where the 2D spectrum corresponds to the 2D Fourier transform of a third-order induced polarization generated at a fixed waiting time T (the delay between the second and third excitation pulses) and irradiated in the "photon echo" direction $\vec{k}_e = \vec{k}_2 + \vec{k}_3 - \vec{k}_1$, where **1** and **2** are the excitation pulses that label the excitation axis by virtue of the τ-dependent induced polarization, and **3** is the "reading" pulse. This kind of 2D spectroscopy in the visible was realized for the first time in 1998 by David Jonas et al.[1] The full impact of this relatively new spectroscopy is just beginning to become apparent. With the advent of coherent control methodology into multidimensional spectroscopies, it is now possible to manipulate coherences in a similar fashion to spin states in NMR. Only in this case, rather than

[a] Max Planck Research Department for Structural Dynamics, Department of Physics, University of Hamburg, Center of Free Electron Laser Science, DESY, Notkestr. 85, 22607 Hamburg, Germany. E-mail: valentyn.prokhorenko@mpsd.cfel.de; dmiller@lphys.chem.utoronto.ca; Fax: +49 408998 5364; Tel: +49 408998 6217
[b] University of Toronto, The Departments of Chemistry and Physics, 80 St. George St., Toronto, ON, M5S3H6, Canada

manipulating coherences to gain information on structure as in NMR, it is now possible manipulate coherences to provide information on couplings that affect dynamics and even steer the chemical dynamics. In this respect, the influence of multidimensional spectroscopies in the optical domain is still in its infancy. The full potential of this approach depends on the development of experimental methodologies to facilitate the manipulation of coherences and pulse sequences to the level achieved in NMR. Great progress has been made on this front such that it is now possible to determine the cross-correlations between different transitions in complex quantum systems and their interaction with the environment, tracing energy transfer pathways, as well as the investigation of structural changes in molecular aggregates, supramolecular assemblies, DNA and proteins. However, understanding optical 2D spectra is not as straightforward as in the NMR-case, and often requires intense theoretical work with numerical modeling. In this regard, it is interesting to note that the first attempt to model the 2D spectrum of an excitonically-coupled molecular assembly was performed by Mukamel *et al.* more than ten years ago,[2,3] in advance of the first theoretical investigations of 2D-PE electronic spectroscopy for the single molecule case by Jonas and co-workers.[4,5] The nature of the field interactions and connection to the 2D-spectra needed to be fully worked out to make the connection to excitonically coupled systems. Similar issues will arise in the treatment of coherence transfer where complex pulses are used to manipulate the state preparation in multilevel systems as will be discussed below.

Since that time an enormous number of theoretical and experimental papers devoted to 2D-PE electronic spectroscopy have been published (some of them can be found in the recent monograph ref. 6). The outline of these investigations can be summarized as follows:

• Unlike in NMR, the molecules cannot be described as pure two-level systems.

• Unlike in NMR or even the IR-domain, molecular electronic transitions interact strongly with the environment (the bath), which leads to significant complication of the 2D-spectra. Besides inhomogeneous broadening, there are strong dephasing processes of the electronic transitions due to their coupling to bath modes. The spectra become broadened and unstructured, such that the cross-correlation features cannot be fully resolved even at low temperatures,[7] and sometimes the positions are inferred in order to follow the associated dynamics. (see, *e.g.*, refs 8 and 9).

• Coupling of electronic transitions to vibrational molecular modes leads to further complication of the 2D-spectra. These modes appear in spectra as additional cross-peaks,[10] and can mask cross-peaks that originate due to excitonic interactions and/or interfere with them.

In principle, the combination of coherent control protocols with 2D-spectroscopy can help overcome these difficulties and increase the resolution of 2D-spectra. As it was stated already in 2001 by Jonas and co-authors, "combining current pulse shaping technologies with noncollinear geometries should ultimately allow optical 2D experiments in which the sample can be modified or perturbed to highlight specific properties, similar to the complex NMR sequences used today".[5] On the theoretical front, the possibility to control relative amplitudes of different peaks in 2D-spectra was numerically demonstrated by Abramavicius and Mukamel[11] using tailored excitation pulses. These authors investigated the simplest excitonically coupled system (*H*-type dimer) for the specific case of 2D-spectroscopy corresponding to the heterodyne-detected two-pulse photon echo method. In the *H*-dimer the lowest exciton state is not observable in the absorption spectrum since its transition dipole moment is zero, yet this "hidden" state can be visualized in the 2D-spectrum created with specifically shaped excitation pulses (found by using a genetic search algorithm). On the experimental front, before performing 2D coherently-controlled spectroscopy of complicated excitonically-coupled systems, it is necessary to first experimentally investigate the effect of coherent control on 2D-spectra for the simplest system, *e.g.*, a monomer system in solution.

The first experimental realization of coherently-controlled electronic 2D spectroscopy was recently made by Prokhorenko et al.[12] in the visible using acousto-optic (AO) and deformable mirror (DM) based pulse shapers. In this scheme, the AO-shaper controls the phase and spectral amplitude of the excitation pulses **1** and **2** with high spectral resolution (less than 1 nm), while the DM-shaper[13] controls the phase of the reading pulse **3** with relatively low spectral resolution. To test the approach, we performed coherent control studies of the 2D-spectra of the organic dye Rhodamine 101 dissolved in methanol (Rh101/MeOH) using the same shaped pulses obtained earlier in the coherent control experiments of population transfer of this molecule in the weak excitation limit.[14] We found that applying these optimal pulses leads to the appearance of a distinct structure in the 2D-spectra[12,15] and concluded that the system–bath interaction (electronic transition–solvent environment) can be controlled and suppressed for a couple of hundred fs. As a result, a fine structure appears in the correlation 2D-spectra (measured at waiting time $T = 0$). Note that in this experiment both the amplitude and the phase of the excitation pulses were tailored.

Here, we present the results of an extended investigation of coherent control effects on the 2D-spectra of Rh101/MeOH using excitation pulses with different shapes, and focus attention on the phase shaping that does not alter the intensity spectrum of the pulses. Surprisingly, the phase shaping of the excitation pulses affects not only the correlation spectra, but also the 2D-spectra measured at very long waiting times (tens of ps) where we cannot expect the presence of any phase memory for the state preparation in the investigated medium based on the known time scales for the system response†. The phase-only shaping introduces a feature reminiscent of amplitude-only shaping although the spectral amplitude of the excitation fields are constant across the spectra. This unexpected observation is found to be related to resonance conditions of the generated pulse profile with the dominant low frequency Franck–Condon mode modulating the transition probability that points to some association with coherent population transfer effects found earlier.

2 Experimental details

All the experiments reported here were performed at room temperature using the 2D coherently-controlled setup described earlier in ref. 12 with minor modification (for increasing the spectra acquisition speed we replaced the previous camera with a Tec5 camera, Germany) using P-polarized beams. The sample, Rh101/MeOH, with an optical density of 0.25 in the maximum of the absorbance, was circulated through a flow cell of 400 μm path length with 150 μm fused silica optical windows. The spectra of the sample (absorption and fluorescence) together with excitation (unshaped) and reading pulses are shown in Fig. 1. Note that the width of the excitation pulse spectrum (≈ 40 nm) was limited by the acceptance bandwidth of the AO-shaper. The spectra of the excitation and reading pulses were adjusted for maximal overlap with the absorption spectrum of the sample. All pulses were generated from the same non-collinear optical parametric amplifier (NOPA) and then were split into excitation pulses, traveling through the AO-shaper, and a reading pulse, controlled by the DM-shaper. In the experiments reported here the DM-shaper was adjusted to give a nearly transform-limited (TL) reading pulse with a duration of 11 fs FWHM, and only shaping of the excitation pulses (**1**, **2**) was performed using the aforementioned AO shaper. Without shaping, the duration of the excitation pulses was 22 fs FWHM, which is close to the TL limit. For characterization of the pulse profiles (with and without shaping), the third-order frequency-resolved optical gating (FROG) method,[16] described in detail in ref. 14, was used

---

† The typical electronic dephasing time for solvated organic dyes at room temperature is 20–80 fs, and dephasing of vibrational transitions occurs within $\sim$1 ps.

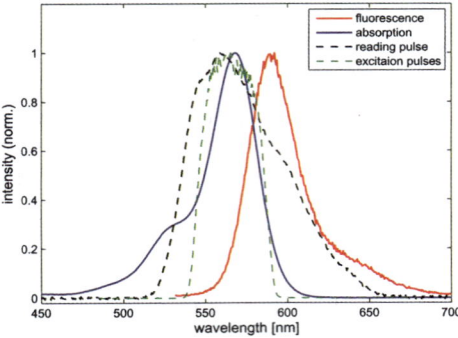

Fig. 1 Absorption and fluorescence spectra of Rh101 in MeOH together with the excitation (unshaped) and reading pulses.

together with the retrieval algorithm from the commercial program FROG3 (Femtosoft Technologies).

The experiments were carried out in the linear excitation regime where the magnitude of the induced PE-polarization P_{PE} scales linearly with respect to the excitation energy (same for the pump–probe signal) by keeping the energy of the reading pulse at a constant level: $|P_{PE}| \propto |E_1||E_2||E_3|$, $|E_3|$ = const. Here $E_{1,\,2}$ are the amplitudes of the electrical fields for the excitation pulses; their product is proportional to the excitation energy (at fixed pulse duration). Fig. 2 shows the measured magnitude of the PE-signal induced with unshaped pulses vs. excitation energy for waiting times $T = 0$ and $T = 80$ ps (the energy of the reading pulse was fixed at 5 nJ). As can be seen, for both waiting times, the PE-response is still linear up to 5 nJ of excitation energy. The deviation from a linear dependence (shown by dotted lines) at higher excitation is due to saturation of the absorption in Rh101/MeOH. Therefore, in the experiments we used excitation pulses with energies 2–4 nJ to avoid saturation effects. For control of the 2D-spectra, we used the previously found optimally shaped amplitude and phase pulses from ref. 14, and additionally 3 different shapes of excitation pulses: linearly-chirped (both positive and negative) with a rate of 400 fs², with periodic modulation of the spectral amplitude, and with periodic modulation of the spectral phase $\phi(\omega) = 1.23 \sin(2\pi\omega/\omega_m)$, where ω_m is the modulation frequency. Such phase modulation leads to an appearance in the time domain of several sub-pulses equally spaced by $\Delta t = 33357/\omega_m$, where ω_m is given in cm^{-1} and Δt in fs. As an example, Fig. 3 shows the temporal profiles of pulses for a phase modulation frequency of 220 cm^{-1} ($\Delta t = 150$ fs) retrieved from the measured FROG-traces. This shaping was previously used in coherent control experiments performed by Motzkus

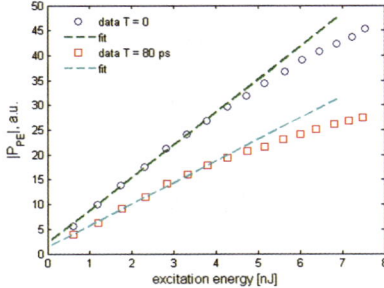

Fig. 2 Power dependence of the photon echo signal (at fixed energy of the reading pulse). Offset around zero excitation is due to influence of the scattered reading pulse.

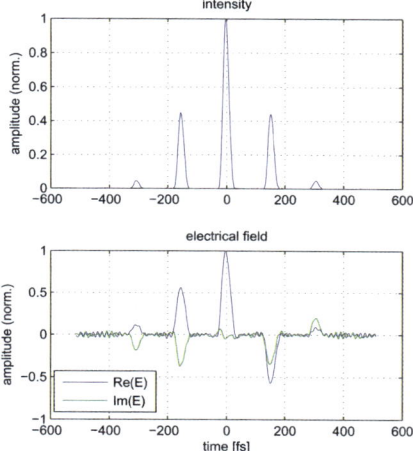

Fig. 3 Temporal profile of excitation pulse with periodically modulated phase at a frequency of 220 cm^{-1}.

and co-authors (see, *e.g.*, ref. 17). The spectra of the shaped pulses are shown in Fig. 4. It is important to note that the AO-shaper employed has significant cross-talk between amplitude and phase-modulation of spectral phase that leads to appearance of some residual modulation in the spectrum of the pulse. Therefore, a special correction program was developed to remove this parasitic modulation in the intensity spectra of the pulses, and used low RF-power to drive the AO (20–30% from the maximum). Fig. 4 compares the spectra of pulses measured with and without phase modulation with $\omega_m = 220$ cm^{-1}; the difference in spectral profiles is negligibly small ($\leq 1.5\%$) and no residual modulation is observed. Note that the spectra of the chirped pulses were also identical to the spectrum of the TL-pulse (not shown). In comparing the different 2D-experiments (with and without phase shaping), the energies of the excitation pulses were kept equal to within 1–1.5%.

Depending on the experiment, the data were collected with delay steps $\Delta \tau$ of 1 fs (non-shaped pulses) and 2 or 4 fs (shaped pulses), respectively; at each delay 40–100 spectra were averaged. For the transient grating measurements, the delay between the excitation pulses **1, 2** was fixed at $\tau = 0$, and the scan was performed for the

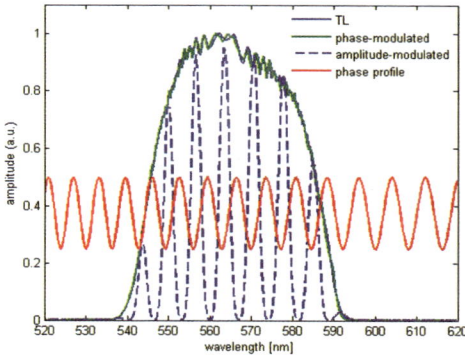

Fig. 4 Intensity spectra of excitation pulses and the phase profile for a modulation frequency of 220 cm^{-1} (upshifted for better viewing).

waiting time T with steps of 4 fs. Creation of the 2D-spectra from the measured data and their "phasing" were performed according to procedures described in ref. 12. However, in order to ensure the independence of the observed features by applying pulse shaping, we also compare the 2D-PE power spectra $|S_{PE}|^2 = S_{PE}S_{PE}^*$ whose profiles are fully independent on possible uncertainties in determining the absolute phases for reconstructing the 2D spectra.

3 Results and discussion

3.1 Optimal pulses from the population transfer experiments

In our previous paper[12] we showed only correlation 2D-spectra (at $T = 0$) created with the optimal excitation pulses found in coherent control experiments of the population transfer in Rh101/MeOH, reported in ref. 14. These spectra displayed sharp features in both ω_τ and ω_t domains (corresponding to Fourier transformations along the time delay between the excitation pulses τ and time t, respectively). By increasing the waiting time T, these features become less sharp in the ω_t domain; however, they are still present in the ω_τ domain of the 2D-spectra. Fig. 5 shows the 2D-spectra measured at $T = 10$ ps with (left) and without (center) phase modulation. On first glance, there is no difference in the shapes of the 2D-spectra at this waiting time; however, comparison of their profiles along ω_τ shows quite significant differences in their amplitudes, especially for the spectral peak located at ≈ 18000 cm^{-1} (555 nm), whose amplitude is reduced by $\approx 40\%$ in the case of both amplitude and phase shaping.

There is no closed analytical expression for the correlation 2D-spectra, but at long waiting times the 2D-spectrum of a molecule, in the approximation of a thin optical layer can be described by the following equation (obtained from eqn. 17–19 in ref. 5):

$$S_{PE}(\omega_\tau, \omega_t) = \tilde{E}_1^*(\omega_\tau)\tilde{E}_2(\omega_\tau)\tilde{E}_3(\omega_t)\sigma_a(\omega_\tau) \\ \times \{\sigma_a(\omega_t) + \sigma_f(\omega_t) - iKK[\sigma_a(\omega_t) + \sigma_f(\omega_t)]\} \quad (1)$$

where $\tilde{E}_i(\omega)$ are spectral profiles of the electrical pulses, $\sigma_{a,f}(\omega)$ are the absorption and fluorescence profiles of a molecule, and the symbol KK stands for the Kramers–Kronig transformation. If the excitation pulses **1** and **2** have equal spectral shapes $\tilde{E}_1(\omega) = \tilde{E}_2(\omega) \equiv \tilde{E}_e(\omega)$, then the 2D-spectrum in the ω_τ domain will be proportional to the *intensity* profile of the excitation pulse $I_e(\omega) = \tilde{E}_e(\omega)\tilde{E}_e^*(\omega)$, and any amplitude modulation of its spectrum will modulate the 2D-spectrum along ω_τ, in full agreement with observations (the spectral profiles of the optimal pulses used can be found in ref. 14). In contrast, phase modulation should have no effect at long waiting times since the intensity spectrum of the excitation pulse $I_e(\omega)$ is phase-independent. However, the clear presence of a difference in the 2D-spectra created with- and without phase shaping allows us to conclude that at long waiting times, long enough to ensure full dephasing of the excited electronic state (≈ 50 fs)

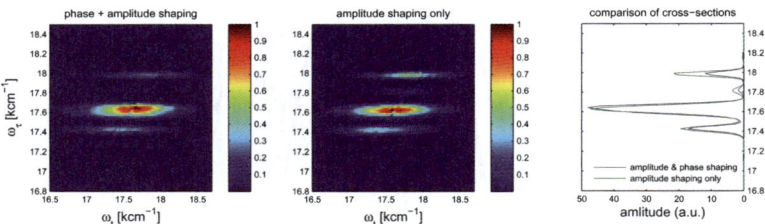

Fig. 5 2D power spectra measured at waiting time $T = 10$ ps using optimal excitation pulses from ref. 14 with and without phase modulation (left and center panels), and comparison of their vertical cross-sections (right panel). Note that 1 kcm^{-1} = 1000 cm^{-1}.

and vibrational modes (~1 ps), the 2D-spectra can be controlled using phase shaping. In order to clarify the influence of phase shaping (separate the effects of phase shaping and the combination of both phase and amplitude shaping effects), we performed additional experiments using phase-shaping-only-chirped pulses as the simplest coherent control tool, and pulses with periodically-modulated phases.

3.2 Chirped pulses

Applying a positive or negative chirp to TL excitation pulses in the frequency domain $\tilde{E}_e(\omega) = |A_e(\omega)|e^{i\phi(\omega - \omega_0)^2}$ with the chirp rate ϕ of 400 fs² stretches the excitation pulse duration from 22 fs to 80–90 fs FWHM. The excitation of Rh101/MeOH with chirped pulses seriously affects the shape of the correlation spectra and, as can be seen from Fig. 6, they look like "optically" distorted 2D-spectra compared to those created with TL-pulses (the lowest row in Fig. 6). Such distortion of 2D-spectra due to chirp in the excitation pulses was already pointed out in ref. 18. However, the 2D-spectra measured at long waiting time ($T = 40$ ps) do not show any difference in their shapes as compared to the excitation with TL-pulses. It is noticeable that at this waiting time the inhomogeneous broadening is already relaxed due to spectral diffusion; the rotation of the imaginary part of the 2D-spectrum, tilted initially by 45° to an almost vertical position, is direct evidence that Rh101/MeOH has only homogeneous broadening at this timescale (as it follows from eqn (1) and the known time scale for spectral diffusion). Comparison of the 2D-spectra profiles along the coordinates ω_τ, ω_t at $T = 40$ ps shows that the chirped pulses nevertheless affect their amplitudes (Fig. 7)—by excitation with negatively-chirped pulses the magnitude of the power 2D-spectrum decreases by $\approx 16\%$, but a positively-chirped pulse does not significantly change the 2D-spectral amplitude. Thus, by comparison with our previous experiment where both phase and amplitude were shaped, we can conclude that phase shaping alone affects the 2D-spectrum at long waiting times (we scanned T up to 80 ps delay and observed the same difference in 2D-spectra). The theoretical expression, eqn (1), cannot explain this effect. It was derived according to the Loring-Mukamel theory for the third-order nonlinear response function for a two-level molecule. In this theory the electronic states have only diagonal coupling to the bath (T_2-process or dephasing), the off-diagonal coupling, which is responsible for the population relaxation (T_1-process), is not taken into account. This simplified model for the coupling to the bath does not

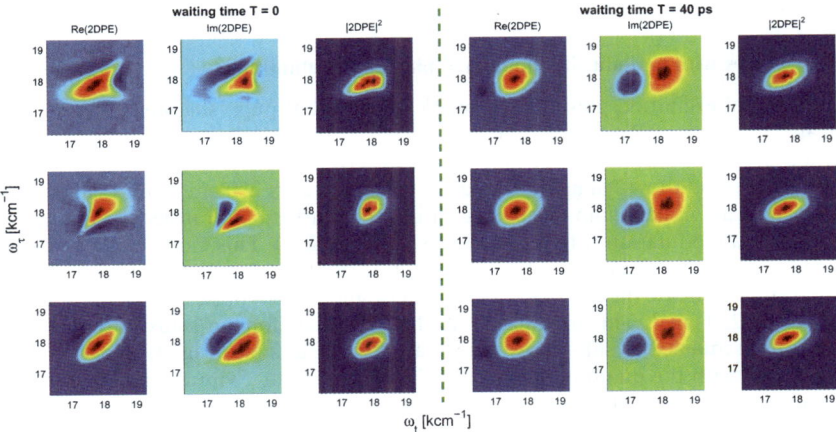

Fig. 6 2D spectra created with chirped excitation pulses at waiting time $T = 0$ and 40 ps (as indicated). Top row-positive chirp +400 fs², middle row-negative chirp −400 fs². For comparison, the 2D spectra created with TL pulses are shown in the bottom row.

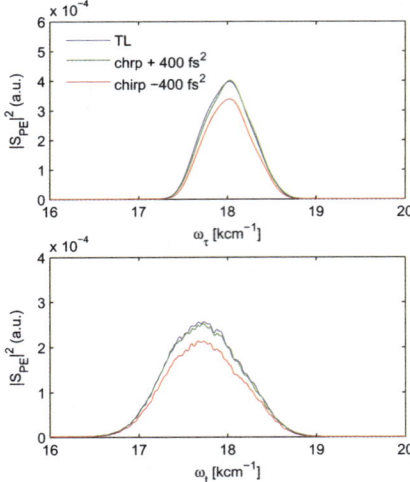

Fig. 7 Comparison of the cross-sections of 2D spectra created with TL- and chirped pulses and measured at waiting time $T = 40$ ps.

include the relative contributions to the relaxation dynamics of the different nuclear fluctuations (intramolecular and solvent) on the induced polarization during the field interaction time. In this respect, this theory cannot capture properly the phase dependent aspects of the induced polarization and affect on the long time dynamics using shaped pulses.

The experimental observations, together with our previous experiment, are in line with the results of coherent control of the population transfer in the weak field limit, reported in ref. 14 and more recently, in ref. 19. This work found that manipulation of the phase of the excitation pulses can increase or suppress the amount of population in the excited state of solvated organic molecules. Thus, a possible explanation for the change in the amplitudes of the 2D-spectra at long waiting times by excitation with chirped pulses is the induced change in the efficiency of the $S_0 \rightarrow S_1$ population transfer in solvated molecules. Given the nonlinear nature of the generated PE signal, even within the linear limit of excitation or weak field limit, this effect will be magnified in a 2D spectrum.

3.3 Pulses with periodically modulated phases and amplitudes

Chirping of excitation pulses is the simplest method of pulse shaping; more sophisticated (and very useful in coherent control experiments) is the periodic phase modulation. In particular, from the pump–probe experiments[14] it is known that in Rh101/MeOH there is a low-lying vibrational state located at ≈ 220 cm^{-1}. This vibrational mode appears in the pump–probe[14] (see Fig. 4 therein) and in transient-grating (TG) signals as a modulation with a period of $\sim$150 fs (Fig. 8). This mode has been implicated in the enhanced coherent population transfer found earlier. Therefore, it was interesting to create the 2D-spectra using phase-modulated pulses with a modulation period of 220 cm^{-1}. This modulation in the frequency domain leads to the appearance of a sequence of sub-pulses spaced apart by 150 fs (see Fig. 3) that will be in resonance with the quantum beats caused by the 220 cm^{-1} vibrational state. Fig. 9 compares the 2D-spectra measured at waiting times $T = 0$ and 40 ps using periodically shaped pulses. Periodic phase modulation of the excitation pulses leads to appearance of a fine grid-like structure in the correlation spectrum ($T = 0$, upper row); the depth of the modulation in the power spectrum reaches almost 100% along the ω_τ coordinate and 70% along ω_t. The modulations along ω_t and ω_τ in the 2D

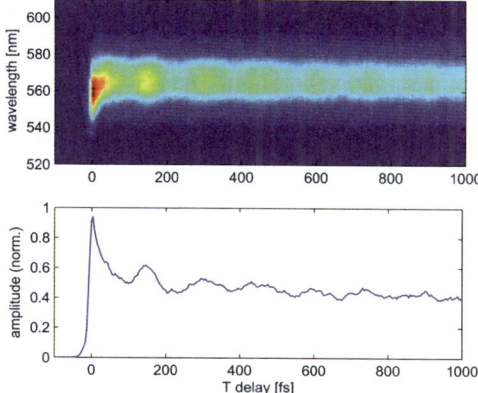

Fig. 8 Spectrally-resolved transient grating signal (top) from Rh101/MeOH and its integrated amplitude (bottom).

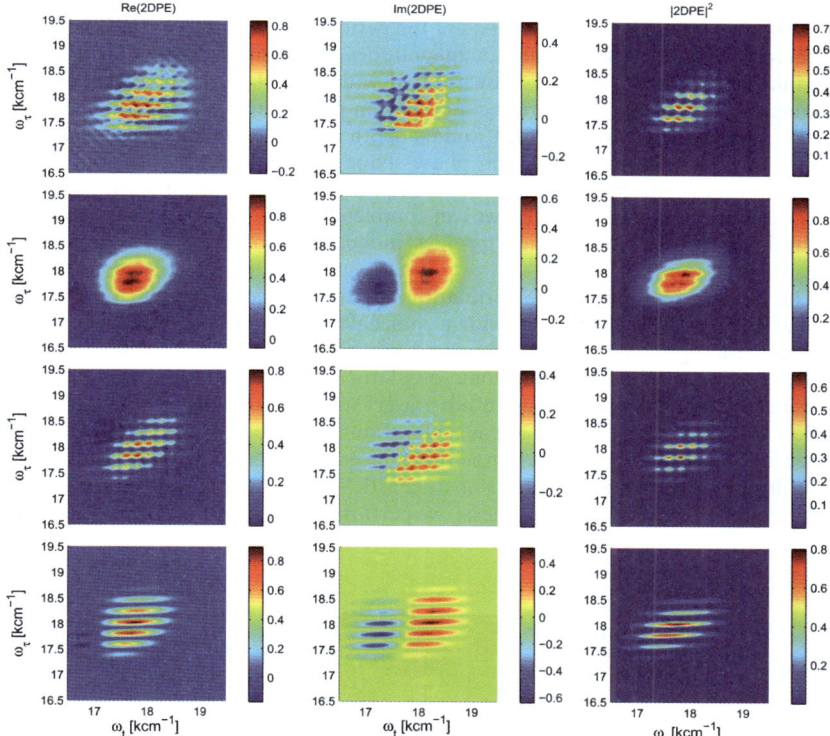

Fig. 9 2D spectra created using periodically *phase-modulated* pulses measured at $T = 0$ (top row) and $T = 40$ ps (2nd row), and with periodically *amplitude-modulated* pulses measured at $T = 0$ (3rd row) and $T = 40$ ps (last row). In all cases the modulation frequency was 220 cm^{-1}.

power spectrum have identical periods (close to 220 cm^{-1}) but shifted in phase by π. In control experiments with a thin glass plate instead of the Rh101/MeOH sample, we did not observe any modulation in the 2D-spectrum along the ω_t coordinate using shaped pulses. Therefore, we can conclude that the modulation in the

2D-spectrum in the ω_t domain arises due to correlation between time variables t and τ coupled *via* phase memory (both electronic and vibrational) in the photoinduced coherences for Rh101/MeOH. Using periodic amplitude shaping (see spectrum in Fig. 4) also causes similar modulations in the 2D correlation spectra in both dimensions (third row in Fig. 9). At long waiting times the modulation is still present but only along the ω_τ coordinate with a modulation depth of 100%, in accord with the theoretical predictions given by eqn (1). In this latter case, the spectral amplitude of the excitation pulses is modulated and this should appear along the excitation axis.

By increasing the waiting time T, the structure of the 2D-spectrum created with the phase-shaped excitation pulses becomes more complicated (not shown), and after $T \sim 5$ ps the modulation along ω_t fully disappears. However, the 2D-spectra still display this periodic modulation in the ω_τ domain even at very long times. Fig. 9 (second row) shows the 2D-spectrum measured at $T = 40$ ps; as can be seen, both real and imaginary parts of the 2D-spectrum are modulated. The modulation pattern can be clearly seen in the differential 2D-spectrum (Fig. 10(B)) which corresponds to the difference between the power spectra measured without and with phase modulation. Fig. 10(A) compares the profiles of the 2D-spectra along ω_τ; besides periodic modulation, phase shaping leads (as in the case of chirped pulses) to an overall decrease in the absolute magnitude of the 2D-spectrum by approximately 20%. From the differential spectrum integrated over ω_t and plotted in Fig. 10(C), we see that the magnitude of the periodic is about 50–70% of the overall difference in the spectral modulation magnitudes. Surprisingly, the frequency of this modulation (190 ± 20 cm^{-1}) is downshifted with respect to the phase modulation frequency (220 cm^{-1}). This periodic modulation persists up to 80 ps (the maximum available delay for T in our setup), and its depth remains unchanged within 5–80 ps (Fig. 11). According to eqn (1), no phase shape effect should be observed at long waiting times except for the possible (residual) modulation in the intensity spectrum of the excitation pulses; however, we can completely exclude this possibility since the excitation spectra have identical profiles independent of the phase shaping (see Fig. 4).

The observed pattern in the correlation spectra plotted in Fig. 9 (top) are very similar the calculated 2D-spectra for a molecule weakly coupled to the bath with one underdamped vibrational frequency[4] (see also more recent calculations[20] for a weakly damped displaced harmonic oscillator). Our first thought was that these phase-shaped pulses act as a "quantum eraser" and suppress the system–bath interaction so that the low-lying 220 cm^{-1} vibrational state becomes visible in the 2D-spectrum. Basically, the interaction of the induced polarization with the applied field is sufficient to overcome the effects of stochastic solvent fluctuations on the induced polarization. To check this hypothesis, we performed a series of measurements using

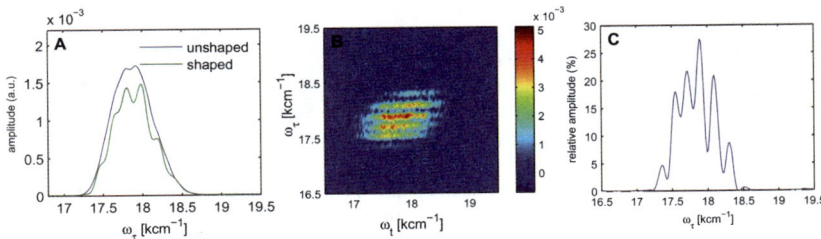

Fig. 10 (A) Comparison of vertical cross-sections of 2D spectra measured at $T = 40$ ps and created with and without phase modulation (see Fig. 9). (B) Differential 2D-spectrum at waiting time $T = 40$ ps obtained by substraction of the 2D-spectrum excited with periodically phase modulated pulses from the 2D-spectrum measured with TL-pulses (from Fig. 9). (C) Integrated over the ω_t differential 2D-spectrum shown in panel (B); the modulation frequency along ω_τ corresponds to ~190 cm^{-1} while the phase shaping frequency was 220 cm^{-1}.

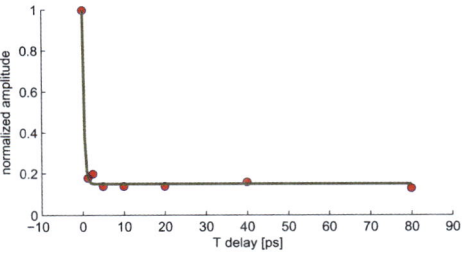

Fig. 11 Waiting time dependence of the modulation depth in 2D spectra excited with phase-shaped pulses having 220 cm^{-1} modulation frequency.

excitation pulses with different modulation frequencies ranging from 160 to 280 cm^{-1} and found that they also induce effects similar to the Fig. 9 grid-like structures in the 2D correlation spectra with similar modulation depths (not shown). However, at long waiting times the modulation in the 2D-spectra along ω_τ clearly exhibits a resonance character, *i.e.*, the magnitude of the pattern in the 2D-spectrum is maximal if the phase of the excitation pulses is modulated with the frequency of 220 cm^{-1}, corresponding to the frequency of the low-lying vibrational mode in Rh101/MeOH. Fig. 12 shows the changes in the modulation depth of 2D-spectra measured at $T = 40$ ps along ω_τ *vs.* modulation frequency. The modulation in the 2D-spectrum is 2–3 times higher at 220 cm^{-1}, and the FWHM of the observed resonance can be estimated to be ≈ 50 cm^{-1} which roughly corresponds to the lifetime of the 220 cm^{-1} Rh101 vibrational state (900 fs according to ref. 14). The resonance nature of the induced modulation in the 2D-spectra at long waiting times allows us to conclude that the phase shaping of the excitation pulses induces a long-lived phase memory in solvated dyes, which appears to originate from the quantum interference of the shaped excitation field and induced polarization with the associated vibrational coherences. The presence of this resonance in manifesting the observed effect is also a direct indication of the irrelevance of possible residual inhomogeneous broadening that might survive for solvated organic dyes after tens of ps. The effect is related to the resonantly driven vibronic coherences in the molecular system.

In our numerical simulations‡ of the 2D-spectra performed for both static and dynamic inhomogeneous broadening (the Brownian oscillator model), we can nicely reproduce all the observed features in the coherently controlled 2D-spectra but only for waiting times up to ~1 ps. At waiting times longer than the dephasing times of electronic and vibrational states, we do not find any influence of phase shaping on the 2D-spectra profiles (these results will be published separately). Our simulations are based on the third-order response function developed by Mukamel and co-workers (see details in ref. 21) which, as was already mentioned above, includes only the diagonal system–bath interaction responsible for the dephasing processes only. It is necessary to take into account the off-diagonal terms into the system–bath interaction Hamiltonian that leads to the appearance of irreversible relaxation

---

‡ The simulation program is based on the conventional third-order response theory developed by Mukamel and co-workers within the assumption of a two-level electronic state model (excited-stated absorption in Rh101 was not taken into account); the interaction with the bath is given by the so-called line-shape function $g(t)$. For calculation of $g(t)$, we used a model system–bath correlation function that includes a low-frequency overdamped mode (*i.e.*, phonon wing) and underdamped 220 cm^{-1} mode. The spectral width and amplitude was adjusted to mimic the measured TG-signal (Fig. 8). The overall Huang-Rhys factor was set to ~4, which is typical for solvated organic dyes at room temperature. Detailed expressions for $g(t)$, the third-order response function and the third-order induced polarization (photon echo signal) were taken from ref. 21.

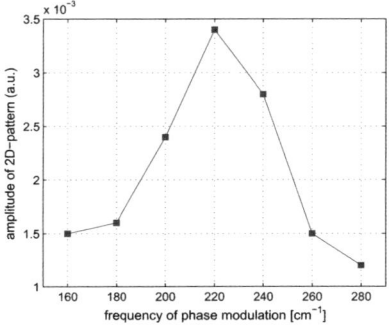

Fig. 12 Modulation depth in 2D spectra measured at a waiting time $T = 40$ *vs.* the phase modulation frequency of the excitation pulses.

of populations (T_1-processes). These terms and specific inclusion of the shaped excitation fields can probably explain the control of 2D-spectra at long waiting times, as in case of the population control in the weak field regime.[22] In any case, we think that both these effects—the population control at long delay times and control of 2D-spectra at long waiting times—are closely related.

4 Conclusions

We experimentally demonstrated coherent control of the 2D electronic spectra of a solvated dye using various tailored excitation pulses and found that the shapes of 2D-spectra are very sensitive to the manipulation of the phase and amplitude of the excitation pulses, especially at zero waiting time (correlation 2D-spectra). In particular, periodically phase-modulated excitation pulses cause modulation of 2D-spectra in the ω_t domain only at waiting times smaller than (or comparable to) the characteristic dephasing times since the correlation between the time variables t and τ is limited by the molecular phase memory. This effect can be used to increase the sensitivity in the study of dephasing relaxation processes in molecules, molecular aggregates (excitonically-coupled systems), and semiconductors.

Most interestingly, we found an effect of coherent control in 2D-spectra along the excitation axis, measured at very long waiting times, which cannot be explained using existing theoretical models for photon echo formation in open quantum systems. Phase shaping affects the overall magnitude of the 2D-spectrum and in the specific case of periodic phase modulation also modulates its shape. The most pronounced effect is the induced memory along the ω_τ coordinate for long waiting times. This effect is reminiscent of amplitude modulation of the excitation pulses although the spectral amplitudes are constant. This effect is manifest through a resonant coupling to a low frequency vibrational mode that most strongly modulates the optical transition through a time dependent Franck–Condon effect. The fact that this induced change in 2D-spectra along ω_τ survives at long T illustrates the conserved coherence or phase memory along this time coordinate. The electric field interaction is evidently sufficient to suppress the normal destruction of coherence during the course of the τ period. This feature is highly relevant to our previous work on the weak field control of photoisomerization in bacteriorhodopsin[23] where the optimal pulses exerting phase dependent control over the isomerization coordinate were longer than the coherence determined with impulsive excitation conditions. This explanation needs to be followed up with more detailed theoretical treatments. The clear resonance character of the observed effect allows us to assume that both the ability to coherently control population transfer and to observe long lived phase memory effects are related and should be sensitive to quantum interference of the shaped electric fields with the molecular vibronic states.

To conclude, we would like to stress the importance of these results for coherent control of 2D-spectra for more complicated quantum systems, e.g., controlling off diagonal features in the 2D spectra of excitonically-coupled molecular aggregates as tests of coherent transport, or to enhance such off-diagonal features. The coherent or phase-dependent effects giving rise to modulation in the correlation spectra or long lived, field dependent, phase memory effects will need to be taken into account.

Acknowledgements

The research was supported by the Natural Sciences and Engineering Research of Canada. VIP and RJDM acknowledge support through the Max Planck Society, University of Hamburg and CFEL DESY.

References

1 J. D. Hybl, A. W. Albrecht, S. M. G. Faeder and D. M. Jonas, *Chem. Phys. Lett.*, 1998, **297**, 307–313.
2 W. M. Zhang, V. Chernyak and S. Mukamel, in *Ultrafast Phenomena XI*, ed. T. Elsaesser, J. G. Fujimoto, D. A. Wiersma and W. Zinth, Springer-Verlag, Berlin, 1998, pp. 663–665.
3 W. M. Zhang, V. Chernyak and S. Mukamel, *J. Chem. Phys.*, 1999, **110**, 5011–5029.
4 S. M. G. Faeder and D. M. Jonas, *J. Phys. Chem. A*, 1999, **103**, 10489–10505.
5 J. D. Hybl, A. A. Ferro and D. M. Jonas, *J. Chem. Phys.*, 2001, **115**, 6606–6622.
6 M. Cho, *Two-dimensional optical spectroscopy*, CRC Press, 2009.
7 T. Brixner, J. Stenger, H. M. Vaswani, M. Cho, R. E. Blankenship and G. R. Fleming, *Nature*, 2005, **434**, 625–628.
8 E. Collini1, C. Y. Wong1, K. E. Wilk, P. M. G. Curmi, P. Brumer and G. D. Scholes, *Nature*, 2010, **463**, 644–647.
9 G. Panitchayangkoon, D. Hayes, K. A. Fransted, J. R. Caram, E. Harel, J. Wen, R. E. Blankenship and G. S. Engel, *Proc. Natl. Acad. Sci. U. S. A.*, 2010, **107**, 12766–12770.
10 N. Christensson, F. Milota, J. Hauer, J. Sperling, O. Bixner, A. Nemeth and H. F. Kauffmann, *J. Phys. Chem. B*, 2011, **115**, 5383–5391.
11 D. Abramavicius and S. Mukamel, *J. Chem. Phys.*, 2004, **120**, 8373–8378.
12 V. I. Prokhorenko, A. Halpin and R. J. D. Miller, *Opt. Express*, 2009, **17**, 9764–9779.
13 R. B. E. Zeek, M. M. Murnane, H. C. Kapteyn, S. Backus and G. Vdovin, *Opt. Lett.*, 2000, **25**, 587–589.
14 V. I. Prokhorenko, A. M. Nagy and R. J. D. Miller, *J. Chem. Phys.*, 2005, **122**, 184502–184513.
15 R. J. D. Miller, A. Paarmann and V. I. Prokhorenko, *Acc. Chem. Res.*, 2009, **42**, 1442–1451.
16 R. Trebino, K. W. DeLong, D. N. Fittinghoff, J. N. Sweetser, M. A. Krumbügel, B. A. Richman and D. J. Kane, *Rev. Sci. Instrum.*, 1997, **68**, 3277–3296.
17 J. Hauer, H. Skenderovic, K. L. Kompa and M. Motzkus, *Chem. Phys. Lett.*, 2006, **421**, 523–528.
18 N. Christensson, Y. Avlasevich, A. Yartsev, K. Müllen, T. Pascher and T. Pullerits, *J. Chem. Phys.*, 2010, **132**, 174508–174518.
19 P. van der Walle, M. T. W. Milder, L. Kuipers and J. L. Herek, *Proc. Natl. Acad. Sci. U. S. A.*, 2009, **106**, 7714–7717.
20 D. Egorova, M. F. Gelin and W. Domcke, *J. Chem. Phys.*, 2007, **126**, 074314–074325.
21 S. Mukamel, *Principles of nonlinear optical spectroscopy*, Oxford University Press, New York, 1st edn, 1995.
22 V. I. Prokhorenko, A. M. Nagy, L. S. Brown and R. J. D. Miller, *Chem. Phys.*, 2007, **341**, 296–309.
23 V. I. Prokhorenko, A. M. Nagy, S. A. Waschuk, L. S. Brown, R. R. Birge and R. J. D. Miller, *Science*, 2006, **313**, 1257–1261.

PAPER

Electronic energy transfer in model photosynthetic systems: Markovian vs. non-Markovian dynamics

Navinder Singh[ab] and Paul Brumer[*b]

Received 9th March 2011, Accepted 15th April 2011
DOI: 10.1039/c1fd00038a

A simple numerical algorithm for solving the non-Markovian master equation in the second Born approximation is developed and used to propagate the traditional dimer system that models electronic energy transfer in photosynthetic systems. Specifically, the coupled integro-differential equations for the reduced density matrix are solved by an efficient auxiliary function method in both the energy and site representations. In addition to giving exact results to this order, the approach allows us to access the range of the reorganization energy and decay rates of the phonon auto-correlation function for which the Markovian Redfield theory and the second-order approximation is useful. For example, the use of Redfield theory for $\lambda > 10$ cm^{-1} in Fenna–Mathews–Olson (FMO) type systems is shown to be fundamentally inaccurate.

1 Introduction

The investigation of physical and chemical mechanisms of photosynthesis has a long history.[1-6] Motivation stems from both the desire to understand natural light-induced processes and the technological goal of fabricating artificial photosynthetic devices for solar energy use. One current goal is to understand the mechanisms of photosynthesis, with international focus recently directed towards understanding the mechanism and efficiency of the process by which an exciton, created by light absorption, "migrates" to the reaction center. Particularly relevant in this regard are recent observations of long-lived quantum coherence effects (often as long as 600 fs) in the transport of electronic energy in such photosynthetic systems,[7-10] for which detailed theoretical understanding is currently being sought.[11-18]

In several systems of interest, a central role is played by the energy transport wire, i.e., the FMO protein which is a trimer made of identical subunits containing seven BChl (bacteriochlorophyll) molecules each. The electronic energy transfers from one BChl molecule to the other, thus moving it from the antenna to the reaction center. The system can be divided into two coupled subsystems: (1) excitation-carrying BChl molecules, and (2) the surrounding protein bath. For such model systems, quantum dynamics can be readily analyzed in two limiting cases defined by the relative contributions of the inter BChl-molecule (system) coupling V, responsible for electronic energy transfer, and system-bath coupling constant λ, responsible for decoherence. These parameters define two important time scales: the excitation transfer time scale $\tau_{transfer} \equiv \hbar/V$, and the decoherence time scale $\tau_{deco} \equiv \hbar/\lambda$. If the system-bath coupling is very weak and $\tau_{deco} \gg \tau_{transfer}$, the system is almost closed and the

[a]Physical Research Laboratory, Navrangpura, Ahmedabad, 380009, India
[b]Chemical Physics Theory Group, Department of Chemistry, University of Toronto, Toronto, Ontario, M5S 3H6, Canada. E-mail: pbrumer@chem.utoronto.ca; Fax: +01 4169785325; Tel: +01 4169783569

Schrödinger equation can be used to study the dynamics. In the opposite case $\tau_{deco} \ll \tau_{transfer}$ (strong system-bath coupling), the system is open, the decoherence rate is very fast, the dynamics is almost incoherent and a simple Pauli type master equation description suffices. These limiting regimes are well understood. Real light harvesting systems, however, fall between these extremes.

A standard approach used to treat this intermediate regime is to use the second Born quantum master equation,[19] a perturbative master equation up to second order in system-bath interaction with weak system-bath coupling, plus its Markovian approximation (*e.g.*, as in the Redfield master equation). Recently, two new approaches have been studied for arbitrary coupling regimes. One is based on weakening the system-bath coupling removal of system-bath interaction and repartitioning the Hamiltonian term using a polaron transformation, followed by the standard second Born master equation.[20] The second approach is based on a reduced hierarchy equation of Kubo and Tanimura, starting from the path integral approach for quantum dissipative systems.[21,22]

It has recently been pointed out in the literature[23] that Markovian Redfield theory cannot explain the long coherence times observed in the recent 2-D photon echo experiments.[7–10] Specifically, the approach taken by Ishizaki–Fleming[23] is to use the Markovian Redfield theory *a priori* and then to examine the consequences. In the results, one observes unphysical behavior when compared to standard Förster theory. This unphysical behavior arises, as discussed in this reference, from the Markovian and secular approximations.

In this paper we introduce a simple method to solve the second-order approximation in the second Born quantum master equation without doing the Markov approximation on the slowly decaying envelope of the density matrix. This approach contrasts with, for example, ref. 23 in which Markovian Redfield theory is used *a priori*, and its consequences analyzed. Our approach allows, by comparison with the results using the Markov approximation, a reliable determination of the range of reorganization energy λ and decay rate of the phonon auto-correlation function, over which one can use the Markovian Redfield theory and the second-order approximation. In addition, this approach permits, by examining the size of the fourth-order term, an estimate of the range of validity of the second-order approximation. Although the method developed here is applicable to general systems, for computational simplicity we study, as do others, the dimer system.

For comparison purposes we note, for example, a sample method for numerical propagation of the Nakajima–Zwanzig projection operator formalism developed in ref. 24. A key feature of that method is a special parametrization of the bath spectral density, leading to a set of coupled equations for primary and N auxiliary density matrices. These coupled master equations are solved numerically by representing the density operator in an eigen representation or on a coordinate space grid, using the Fourier method to calculate the action of the kinetic and potential energy operators, and a combination of split operator and Cayley implicit method to compute the time evolution. The method is general but far more complex than that developed herein, which takes advantage of the exponential bath correlation function often used in photosynthetic light harvesting models. The result is a simple set of ordinary differential equations, which is easy to solve.

In the following section (Section 2) we outline the basic model for a dimer. In Section 3 we introduce the second Born quantum master equation and phonon correlation function, diagonalize the Hamiltonian and cast the master equation into both the site and energy representations. Section 4 gives a new auxiliary function method of solving these equations, and an analysis of results in the Markovian approximation is provided. Discussion of the results and the underlying physical picture is given at the end of Section 5. In Section 5.1 we comment on the regime of validity of the second-order approximation in the master equation by estimating the order of magnitude of the fourth-order term. The last section provides a brief summary.

2 The model: dimer system

Consider a model dimer system given by the following standard Frenkel exciton Hamiltonian:[23]

$$H_{tot} = H^{el} + H^{reorg} + H^{ph} + H^{el-ph} \tag{1}$$

$$H^{el} = \sum_{n=1}^{2} \varepsilon_n^0 |n\rangle\langle n| + J(|1\rangle\langle 2| + |2\rangle\langle 1|) \tag{2}$$

$$H^{reorg} = \sum_{n=1}^{2} \lambda_n |n\rangle\langle n|, \quad \lambda_n = \sum_i \hbar\omega_i d_{ni}^2 / 2 \tag{3}$$

$$H^{ph} = \sum_{n=1}^{2} h_n^{ph}, \quad h_n^{ph} = \sum_i \hbar\omega_i (p_i^2 + q_i^2)/2 \tag{4}$$

$$H^{el-ph} = \sum_{n=1}^{2} V_n u_n, \quad V_n = |n\rangle\langle n|, \quad u_n = -\sum_i \hbar\omega_i d_{ni} q_i \tag{5}$$

Here $|n\rangle$ represents the state in which only the nth site is excited and all others are in the ground state. The quantity ε_n^0 is the excited electronic energy of the nth site in the absence of phonons, and J is the electronic coupling between the sites which is responsible for EET. The ground state energies of the donor and acceptor are set equal to zero and λ_j is the reorganization energy of the jth site that is dissipated in the bath after the electronic transition occurs. The quantity d_{ji} is the dimensionless displacement of the equilibrium configuration of the ith phonon mode between the ground and the excited electronic state of the jth site, and q_i, p_i are the dimensionless coordinates and momenta of the ith phonon mode of frequency ω_i.

3 The second-Born quantum master equation

The method of projection operators used to obtain open system master equations is well known.[24] With the help of projection operators one can obtain the following quantum master equation for the reduced density matrix of the system in the second Born approximation, which is valid when system-bath coupling is weak as compared to the characteristic energy scale of the system [see, e.g., May and Kuhn[19]].

$$\frac{\partial \rho^I(t)}{\partial t} = -\frac{i}{\hbar} \sum_{j=1}^{2} \langle u_j \rangle \left[V_j^I, \rho^I \right] - \frac{1}{\hbar^2} \sum_{i,j=1}^{2} \int_0^t d\tau \Big(C_{ij}(t-\tau) \left[V_i^I(t), V_j^I(\tau)\rho^I(\tau) \right]$$

$$- C_{ij}^*(t-\tau) \left[V_i^I(t), \rho^I(\tau) V_j^I(\tau) \right] \Big) \tag{6}$$

Here, the interaction representation has been used, which is defined for system operators as,

$$\hat{O}^I(t) = U_S^{\dagger}(t) \hat{O} U_S(t) \tag{7}$$

where $U_S(t) = \exp\left(-\frac{i}{\hbar}\hat{H}_s t\right)$ is the time evolution operator, and $\hat{H}_s = \sum_{n=1}^{2} (\varepsilon_n^0 + \lambda_n)|n\rangle\langle n| + J(|1\rangle\langle 2| + |2\rangle\langle 1|)$ is the system Hamiltonian. Here, the bath is assumed to be a continuum of harmonic oscillators, and the bath correlation functions are defined as

$$C_{ij}(t) \equiv \langle u_i(t)u_j(0)\rangle - \langle u_i\rangle\langle u_j\rangle. \tag{8}$$

Below, the canonical average of the bath operators, $\langle u_j \rangle$, which involve the averaging over the product of displacement and bath position co-ordinates is taken to be zero. The above master equation [eqn (6)] is also termed the time *convolution equation* and can be obtained from the Nakajima-Zwanzig equation with a zeroth-order approximation to the time evolution operator in the kernel.[24]

Converting this master equation [eqn (6)] back to the Schrödinger representation gives

$$\frac{\partial \rho(t)}{\partial t} = -\frac{i}{\hbar}[H_s, \rho(t)]$$
$$-\frac{1}{\hbar^2}\sum_{i,j=1}^{2}\int_0^t d\tau (C_{ij}(t-\tau)[V_i, U_s(t-\tau)V_j\rho(\tau)U_s^\dagger(t-\tau)]$$
$$-C_{ij}^*(t-\tau)[V_i, U_s(t-\tau)\rho(\tau)V_j U_s^\dagger(t-\tau)]). \tag{9}$$

We consider the case where the characteristics of the bath as seen by both the sites are the same, and there is no bath correlation between the sites. The bath correlation function is then of the form $C_{ij}(t) = C(t)\delta_{ij}$, where

$$C(t) = \int_{-\infty}^{+\infty} \frac{d\omega}{2\pi} C(\omega) e^{-i\omega t}. \tag{10}$$

$$C(\omega) = 2\hbar(1 + n(\omega))J(\omega). \tag{11}$$

Assuming the Drude–Lorentz model for the spectral density $J(\omega) = 2\lambda \frac{\omega\gamma}{\omega^2 + \gamma^2}$ where λ is the reorganization energy, and assuming the high temperature approximation $\left(\frac{\hbar\omega}{k_B T} \ll 1\right)$, as is appropriate for the systems like the FMO model [see ref. 21, 22 and 23], we obtain the correlation function as,

$$C(t) = \frac{2\lambda}{\beta} e^{-\gamma t}, \beta = \frac{1}{k_B T} \tag{12}$$

3.1 Explicit site representation of the non-Markovian master equation

For the explicit site representation we need the eigensystem of the Hamiltonian. Assuming $\lambda_1 = \lambda_2 \equiv \lambda$ the eigenvalues E_i and eigenvectors $|e_i\rangle$ for the system Hamiltonian

$$H_s = \sum_{n=1}^{2}\left(\varepsilon_n^0 + \lambda_n\right)|n\rangle\langle n| + J(|1\rangle\langle 2| + |2\rangle\langle 1|) \tag{13}$$

can easily be obtained as

$$E_{1,2} = \frac{1}{2}\left(\varepsilon_1^0 + \varepsilon_2^0 + 2\lambda \mp \sqrt{\left(\varepsilon_1^0 + \varepsilon_2^0 + 2\lambda\right)^2 - 4\left(\varepsilon_1^0\varepsilon_2^0 - J^2 + \lambda(\varepsilon_1^0 + \varepsilon_2^0) + \lambda^2\right)}\right)$$

$$|e_1\rangle = \frac{1}{\sqrt{\alpha_1^2 + 1}}\begin{pmatrix}\alpha_1 \\ 1\end{pmatrix}, \quad |e_2\rangle = \frac{1}{\sqrt{\alpha_2^2 + 1}}\begin{pmatrix}\alpha_2 \\ 1\end{pmatrix}$$

$$\alpha_{1,2} = \frac{1}{2J}\left(\Delta \mp \sqrt{\Delta^2 + 4J^2}\right), \Delta = \varepsilon_1^0 - \varepsilon_2^0. \tag{14}$$

Here the column vectors denote components in the site basis, and the energy eigenkets are both orthogonal (since $\alpha_1\alpha_2 = -1$) and normalized. With lengthy but straightforward calculations, eqn (9) for the reduced density operator can be written explicitly in the site representation, using eqn (10) and (11), and as a set of coupled integro-differential delay equations

$$\frac{dx(t)}{dt} = -2\frac{J}{\hbar}y_2(t)$$

$$\frac{dy_1(t)}{dt} = \frac{\Delta}{\hbar}y_2(t) - \frac{4\lambda}{\beta\hbar^2}e^{-\gamma t}\int_0^t d\tau e^{\gamma\tau}[\eta_1\cos(E_{12}(t-\tau))y_1(\tau) + \eta_2\sin(E_{12}(t-\tau))y_2(\tau)]$$

$$\frac{dy_2(t)}{dt} = -\frac{\Delta}{\hbar}y_1(t) - \frac{J}{\hbar}(1-2x(t)) - \frac{4\lambda}{\beta\hbar^2}e^{-\gamma t}\int_0^t d\tau e^{\gamma\tau}[-\eta_2\sin(E_{12}(t-\tau))y_1(\tau)$$

$$+\eta_3\cos(E_{12}(t-\tau))y_2(\tau) + 2\Omega y_2(\tau)] \quad (15)$$

with $\eta_1 = 1$, $\eta_2 = -\dfrac{\Delta}{\sqrt{\Delta^2 + 4J^2}}$,

$\eta_3 = \dfrac{\Delta^2}{\Delta^2 + 4J^2}$, $E_{12} = E_1 - E_2 = -\dfrac{\sqrt{\Delta^2 + 4J^2}}{\hbar}$, $\Omega = \dfrac{2J^2}{\Delta^2 + 4J^2}$.

Here $x(t) \equiv \rho^{11}(t) \equiv \langle 1|\hat{\rho}(t)|1\rangle$ (site), $y_1(t) \equiv \text{Re}[\rho_{12}(t)]$, and $y_2(t) \equiv \text{Im}[\rho_{12}(t)]$, with subscripts denoting the sites.

3.2 Energy representation of the non-Markovian master equation

The kets $|e_{1,2}\rangle$ in eqn (14) are the eigenstates of the Hamiltonian H_s. The equation for a general element of the reduced density matrix in energy representation

$$\rho^e_{ab}(t) \equiv \langle e_a|\hat{\rho}(t)|e_b\rangle, \quad (16)$$

is obtained from eqn (9) and (10) as

$$\frac{d\rho^e_{ab}(t)}{dt} = -i\omega_{ab}\rho^e_{ab} - \frac{1}{\hbar^2}\sum_{i,c,d=1}^{2}\int_0^t d\tau C(t-\tau)$$

$$\left[V_i^{ac}V_i^{cd}e^{-i\omega_{cb}(t-\tau)}\rho^e_{db}(\tau) - V_i^{ac}V_i^{db}e^{-i\omega_{ad}(t-\tau)}\rho^e_{cd}(\tau)\right]$$

$$-C^*(t-\tau)\left[V_i^{ac}V_i^{db}e^{-i\omega_{cb}(t-\tau)}\rho^e_{cd}(\tau) - V_i^{cd}V_i^{db}e^{-i\omega_{ad}(t-\tau)}\rho^e_{ac}(\tau)\right], \quad (17)$$

with $\omega_{ab} = (E_a - E_b)/\hbar$ and

$$V_1^{ac} = \frac{\alpha_a\alpha_c}{\sqrt{\alpha_a^2+1}\sqrt{\alpha_c^2+1}}, \quad V_2^{ac} = \frac{1}{\sqrt{\alpha_a^2+1}\sqrt{\alpha_c^2+1}}. \quad (18)$$

Results in the energy representation, using eqn (12), are discussed below.

4 Method of solution: non-Markovian

To obtain a solution for the non-Markovian case, we first convert the coupled integro-differential equations in the site representation [eqn (15)] to a larger number of coupled ordinary differential equations, a transformation made possible by the exponential form of the correlation function. The resultant coupled ordinary differential equations can be easily solved numerically. This transformation is performed as follows. First, for computational simplicity we put $\tau' = \gamma\tau$ in eqn (15) and then $\gamma t = t'$ in the resulting equations, and define three auxiliary functions $f_i(t')$:

$$f_1(t') \equiv \int_0^{t'} d\tau' e^{\tau'} \left[\cos\left[\frac{E_{12}}{\gamma}(t'-\tau')\right] \tilde{y}_1(\tau') + \eta_2 \sin\left[\frac{E_{12}}{\gamma}(t'-\tau')\right] \tilde{y}_2(\tau') \right],$$

$$f_2(t') \equiv \int_0^{t'} e^{\tau'} \tilde{y}_2(\tau') d\tau',$$

$$f_3(t') \equiv \int_0^{t'} d\tau' e^{\tau'} \left[-\eta_2 \sin\left[\frac{E_{12}}{\gamma}(t'-\tau')\right] \tilde{y}_1(\tau') + \eta_3 \cos\left[\frac{E_{12}}{\gamma}(t'-\tau')\right] \tilde{y}_2(\tau') \right]. \quad (19)$$

Here, $\tilde{y}_1(t') \equiv y_1(t'/\gamma)$, $\tilde{y}_2(t') \equiv y_2(t'/\gamma)$ and we also define $\tilde{x}(t') \equiv x(t'/\gamma)$. We then obtain six coupled ordinary differential equations, three from eqn (15) and three from differentiating the three auxiliary functions, giving:

$$\dot{\tilde{x}}(t') = -\frac{2J}{\gamma\hbar} \tilde{y}_2(t'),$$

$$\dot{\tilde{y}}_1(t') = \frac{\Delta}{\gamma\hbar} \tilde{y}_2(t') - \frac{4\lambda}{\beta\gamma^2\hbar^2} e^{-t'} f_1(t'),$$

$$\dot{\tilde{y}}_2(t') = -\frac{\Delta}{\gamma\hbar 2} \tilde{y}_1(t') - \frac{J}{\gamma\hbar 2} + 2\frac{J}{\gamma\hbar 2} \tilde{x}(t') - \frac{8\lambda}{\beta\gamma^2\hbar^2} \Omega e^{-t'} f_2(t') - \frac{4\lambda}{\beta\gamma^2\hbar^2} e^{-t'} f_3(t'),$$

$$\dot{f}_1(t') - e^{t'} \tilde{y}_1(t') = e^{t'} \tilde{y}_1(t') + \frac{E_{12}}{\gamma} e^{t'} \eta_2 \tilde{y}_2(t') - \left(\frac{E_{12}}{\gamma}\right)^2 f_1(t'),$$

$$\dot{f}_2(t') = e^{t'} y_2(t),$$

$$\dot{f}_3(t') - e^{t'} \eta_3 \dot{\tilde{y}}_2(t') = e^{t'} \eta_3 \tilde{y}_2(t') - \frac{E_{12}}{\gamma} \eta_2 \gamma e^{t'} \tilde{y}_1(t') - \left(\frac{E_{12}}{\gamma}\right)^2 f_3(t'), \quad (20)$$

where overdots denote derivatives with respect to t'. These equations can be efficiently solved numerically.

For comparison with other studies, results are given below for the particular initial conditions: $\rho_{11}(0) = \tilde{x}(0) = 1$, $\tilde{y}_1(0) = \tilde{y}_2(0) = 0$, $f_1(0) = f_2(0) = f_3(0) = \dot{f}_1(0)\dot{f}_3(0) = 0$. These initial conditions (corresponding to all the population being on site 1, and no coherences), are those which have been used extensively in previous investigations [see, e.g., ref. 23] but are somewhat unphysical, because they lack initial coherences which become important in photo-excitation. This problem of initial conditions and state preparation will be treated with a more plausible model in a future publication.[25]

The numerical solution with full non-Markovian form is plotted later below.

5 Energy representation and Markovian limit

To consider the Markov approximation, we note that it is particularly simple to invoke this approximation in the energy representation. Hence, below we first utilize

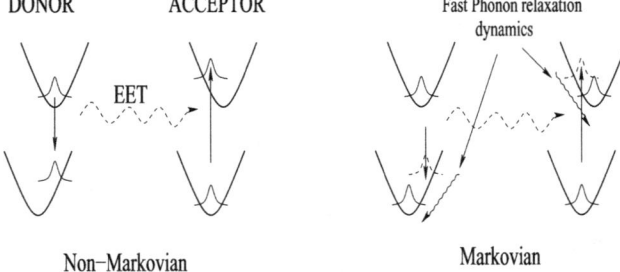

Fig. 1 Qualitative physical picture. Non-Markovian regime is marked by slow dissipation of the reorganization energy. (Errata: The non-Markovian picture should be redrawn so as to conserve energy.)

the energy basis and then convert the result back to the site representation for comparison with the non-Markovian solution.

As discussed in ref. 23, the approximate physical picture is given in Fig. 1. The non-Markovian regime is marked by slow dissipation of the reorganization energy, *i.e.*, the slow decay of the phonon correlation function as compared to relaxation dynamic time scale, – the decay of the envelope part of the density matrix. Transitions occur in accord with the vertical Franck–Condon principle. In the Markovian regime phonon relaxation is very fast (large γ) as compared to the decay of the envelope of the density matrix. The wiggly lines in Fig. 1 show the process of phonon relaxation. Thus, phonons remain effectively in equilibrium during the process of EET in the Markovian regime.

The Markov approximation can be performed when the time scale on which the envelope of the density matrix decays is much longer than the decay time of the phonon correlation function.[19] Then one can introduce the following approximation:

$$\rho_{ab}^e(t-\tau) \equiv e^{-i\omega_{ab}(t-\tau)}\tilde{\rho}_{ab}^e(t-\tau) \simeq e^{-i\omega_{ab}(t-\tau)}\tilde{\rho}_{ab}^e(t) = e^{i\omega_{ab}\tau}\rho_{ab}^e(t). \qquad (21)$$

To obtain the equations in the Markov approximation, eqn (17) is first converted to dimensionless form with $\tau' = \gamma\tau$ and $t' = \gamma t$. Putting $t - \tau = \tau'$ in the resulting equation of energy representation and then implementing the above approximation on the density matrix elements allows the time integration to be performed easily for the case of exponential phonon correlation function [eqn (11)]. The result is the set of Markovian equations:

$$\dot{\tilde{\rho}}_{ab}^e(t') = -i\bar{\omega}_{ab}\tilde{\rho}_{ab}^e(t')$$

$$-\frac{2\lambda}{\beta\hbar^2\gamma^2}\sum_{i,c,d}\left(\frac{V_i^{ac}V_i^{cd}}{1-i\bar{\omega}_{dc}}\tilde{\rho}_{db}^e(t') - \frac{V_i^{ac}V_i^{db}}{1+i\bar{\omega}_{db}}\tilde{\rho}_{cd}^e(t')\right)$$

$$+\frac{2\lambda}{\beta\hbar^2\gamma^2}\sum_{i,c,d}\left(\frac{V_i^{ac}V_i^{db}}{1-i\bar{\omega}_{ca}}\tilde{\rho}_{cd}^e(t') - \frac{V_i^{cd}V_i^{db}}{1+i\bar{\omega}_{cd}}\tilde{\rho}_{ac}^e(t')\right). \qquad (22)$$

Here $\rho_{ab}^e(t'/\gamma) \equiv \tilde{\rho}_{cd}^e(t'), \bar{\omega}_{ab} \equiv \frac{\omega_{ab}}{\gamma}$. Eqn (22) constitutes a system of coupled ordinary differential equations that can be solved with given initial conditions.

The results can be transformed back to the site representation using the transformation

$$\rho_{ij}(t) = \langle i|\rho(t)|j\rangle = \sum_{a,b}\langle i|e_a\rangle\rho_{ab}^e\langle e_b|j\rangle, \qquad (23)$$

where ρ_{ij} is in site representation and ρ_{ab}^e is in energy representation, and $i, j, a, b \in \{1, 2\}$. Eqn (23) constitutes four linear equations that provides the exact relationship between the representations.

Fig. 2 compares the solution for the Markovian master equation to the non-Markovian results for the standard weak electronic coupling parameter values in photosynthetic EET: $\gamma^{-1} = 100$ fs, $J = 20$ cm^{-1}, $\Delta = 100$ cm^{-1}, $T = 300$ K. The initial excitation is assumed to be on site one. The Markovian approximation is seen to be very good for $\lambda < 1$ cm^{-1}, fair for $\lambda = 2$ cm^{-1} and invalid for reorganization energies $\lambda > 10$ cm^{-1}.

To explore regimes of validity of the Markovian approximation for other values of the physical constants, we present sample results in Fig. 3 and 4, obtained by varying (J, λ) and (γ^{-1}, λ), keeping $\Delta = 100$ cm^{-1}. The results show that the Markovian approximation is poor for large λ and for small J and for large γ^{-1} and small λ. Other parameter values can be readily examined using this approach.

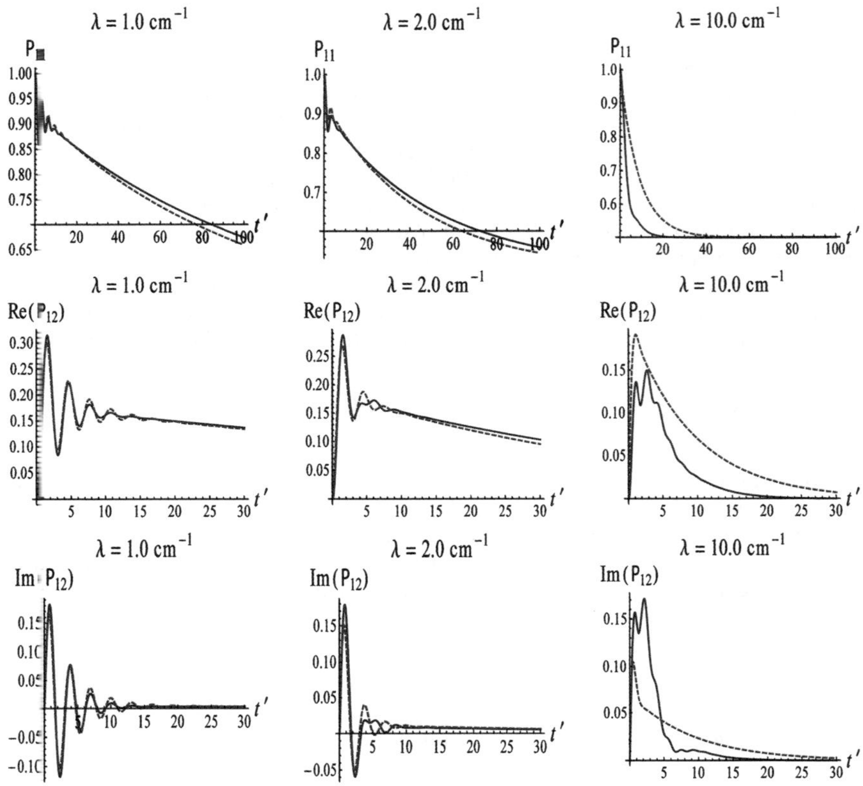

Fig. 2 Time evolution of population on site 1 [$\rho_{11}(t)$] and the coherences [$\rho_{12}(t)$] [solid curve is the non-Markovian solution and dotted curve is the Markovian approximation], for various values of λ (in cm^{-1}). The breakdown of the Markov approximation at $\lambda = 10$ cm^{-1} is clearly visible from temporal behavior of both real and imaginary parts of ρ_{12} at $\lambda = 10$. The time t' is dimensionless, with 10 units on this scale being equivalent to one ps. The other parameters are the standard values given in the text.

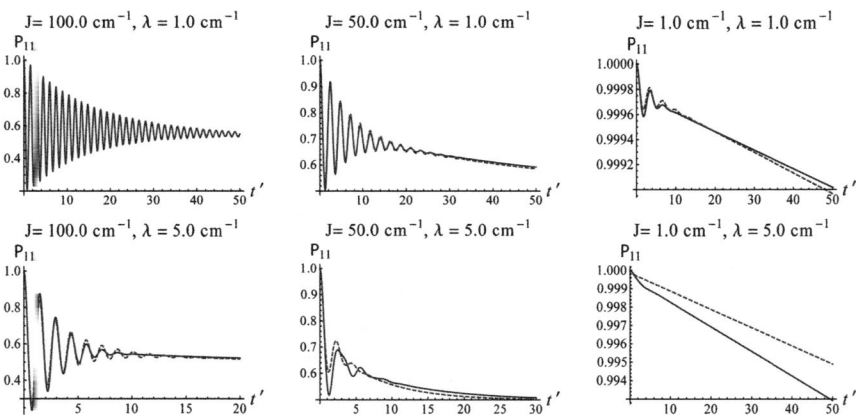

Fig. 3 Time evolution of population on site 1 for various values of the reorganization energy λ and inter-site coupling J (solid curve is the non-Markovian solution and dotted curve is the Markovian approximation). The level separation $\Delta = 100$ cm^{-1} and $\gamma^{-1} = 100$ fs. It is clear that Markovian approximation is poor for large λ and small J. Time t' is the dimensionless time, 10 units on this scale are equivalent to one ps.

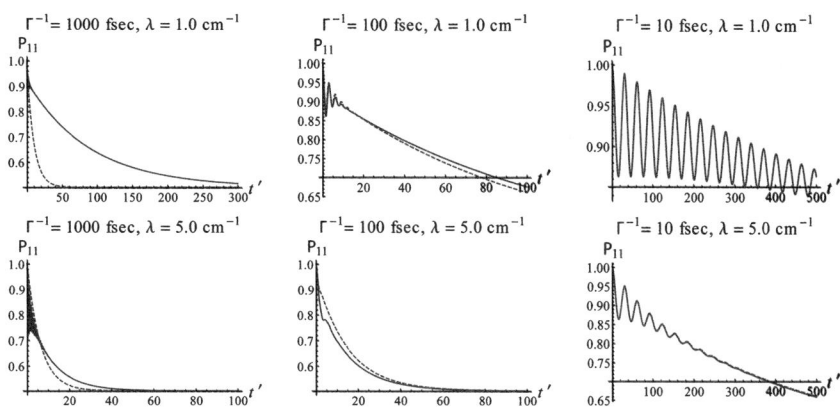

Fig. 4 Time evolution of population on site 1 for various values of the reorganization energy λ and phonon relaxation time γ^{-1} (solid curve is the non-Markovian solution and dotted curve is the Markovian approximation). The level separation $\Delta = 100$ cm^{-1} and $J = 20$ cm^{-1}. It is clear that Markovian approximation is poor for large relaxation times γ^{-1} and small λ. t' is the scaled time, as in the figure above (here γ^{-1} is denoted Γ^{-1}).

5.1 The fourth-order term

A generalization of this approach[25] allows us to estimate the relative magnitude of the fourth-order term in the expansion of the master equation and to gain insight into the range over which the second-order Born approximation is valid. Such an analysis[25] shows that the ratio of the fourth-order term to the second-order term is parametrized by $\frac{\lambda}{\gamma^2 \beta \hbar^2}$, which is of the order of 0.074 for $\lambda = 1$ cm^{-1}, and ~0.74 for $\lambda = 10$ cm^{-1}. This suggests that the second-order approximation for the master equation is good for $\lambda \sim 1$ for the standard set of parameters ($\gamma^{-1} = 100$ fs, $J = 20$ cm^{-1}, $\Delta = 100$ cm^{-1}, $T = 300$ K). However, for $\lambda \geq 10$, the fourth-order term cannot be neglected.

6 Summary

We have introduced a new direct method for solving the second-order Born master equation for the exponential bath correlation functions often adopted for electronic energy transfer in models of photosynthetic light harvesting systems. The method has allowed a direct assessment of the range of reorganization energies over which the Markovian approximation is valid. A generalization of the method allows determination of the relative sizes of the fourth-order and second-order terms in the Born expansion, allowing an assessment of the range of validity of the second-order expansion. Dynamics with reorganization energies $\lambda \geq 10$ cm^{-1}, with other parameters characteristic of model light harvesting systems, clearly fall well outside the range of validity of these approximations.

Acknowledgements

Financial support from the U.S. Air Force Office of Scientific Research under grant number FA9550-10-1-0260 and from the Natural Sciences and Engineering Research Council of Canada are gratefully acknowledged. We thank Prof. G. Scholes for numerous discussions on electronic energy transfer.

References

1 T. Forster, *Ann. Phys.*, 1948, **437**, 55.
2 R. Silbey, *Annu. Rev. Phys. Chem.*, 1976, **27**, 203.

3 H. V. Amerongen, L. Valkunas and R. V. Grondelle, *Photosynthetic Excitons*, World Scientific, Singapore, 2002.
4 R. E. Blankenship, *Molecular Mechanisms of Photosynthesis*, Blackwell Science, Oxford, 2002.
5 V. M. Kenkre and R. S. Knox, *Phys. Rev. B: Solid State*, 1974, **9**, 5279.
6 V. M. Kenkre, *Phys. Rev. B: Solid State*, 1975, **12**, 2150.
7 G. S. Engel, T. R. Calhoun, E. L. Read, T. K. Ahn, T. Mancal, Y.-C. Cheng, R. E. Blankenship and G. R. Fleming, *Nature*, 2007, **446**, 782.
8 H. Lee, Y.-C. Cheng and G. R. Fleming, *Science*, 2007, **316**, 1462.
9 G. Panitchayangkoon, D. Hayes, K. A. Fransted, J. R. Caram, E. Harel, J. Wen, R. E. Blankenship and G. S. Engel, *Proc. Natl. Acad. Sci. U. S. A.*, 2010, **107**, 12766.
10 E. Collini, C. Y. Wong, K. E. Wilk, P. M. G. Curmi, P. Brumer and G. D. Scholes, *Nature*, 2010, **463**, 644.
11 M. B. Plenio and S. F. Huelga, *New J. Phys.*, 2008, **10**, 113019.
12 M. Sarovar, Y.-C. Cheng and K. B. Whaley, *Phys. Rev. E: Stat., Nonlinear, Soft Matter Phys.*, 2009, **83**, 011906.
13 M. Sarovar, A. Ishizaki, G. R. Fleming and K. B. Whaley, *Nat. Phys.*, 2010, **6**, 462.
14 A. Ishizaki and G. R. Fleming, *New J. Phys.*, 2010, **12**, 055004.
15 F. Caruso, A. W. Chin, A. Datta, S. F. Huelga and M. B. Plenio, *Phys. Rev. A: At., Mol., Opt. Phys.*, 2010, **81**, 062346.
16 F. Fassioli and A. Olaya-Castro, *New J. Phys.*, 2010, **12**, 085006.
17 M. Mohseni, P. Rebentrost, S. Lloyd and A. Aspuru-Guzik, *J. Chem. Phys.*, 2008, **129**, 174106.
18 P. Rebentrost, M. Mohseni and A. Aspuru-Guzik, *J. Phys. Chem. B*, 2009, **113**, 9942.
19 V. May and O. Kuhn, *Charge and Energy Transfer Dynamics in Molecular Systems*, Wiley-VCH, New York, 2004.
20 S. Jang, Y.-C. Cheng, D. R. Reichman and J. D. Eaves, *J. Chem. Phys.*, 2008, **129**, 101104.
21 A. Ishizaki and G. R. Fleming, *J. Chem. Phys.*, 2009, **130**, 234111.
22 Y. Tanimura and R. Kubo, *J. Phys. Soc. Jpn.*, 1989, **58**, 101.
23 A. Ishizaki and G. R. Fleming, *J. Chem. Phys.*, 2009, **130**, 234110.
24 H.-P. Breuer and F. Petruccione, *The Theory of Open Quantum Systems*, Oxford University Press, Oxford, 2002.
25 N. Singh and P. Brumer, J. Chem. Phys., *submitted*.

PAPER

Coherent control of single molecules at room temperature

Daan Brinks,[a] Richard Hildner,[a] Fernando D. Stefani[b] and Niek F. van Hulst*[ac]

Received 3rd May 2011, Accepted 6th June 2011
DOI: 10.1039/c1fd00087j

The detection of individual molecules allows to unwrap the inhomogeneously broadened ensemble and reveal the spatial disorder and temporal dynamics of single entities. During 20 years of increasing sophistication this approach has provided valuable insights into biomolecular interactions, cellular processes, polymer dynamics, *etc*. Unfortunately the detection of fluorescence, *i.e.* incoherent spontaneous emission, has essentially kept the time resolution of the single molecule approach out of the range of ultrafast coherent processes. In parallel coherent control of quantum interferences has developed as a powerful method to study and actively steer ultrafast molecular interactions and energy conversion processes. However the degree of coherent control that can be reached in ensembles is restricted, due to the intrinsic inhomogeneity of the synchronized subset. Clearly the only way to overcome spatio-temporal disorder and achieve key control is by addressing individual units: *coherent control of single molecules*. Here we report the observation and manipulation of vibrational wave-packet interference in individual molecules at ambient conditions. We show that adapting the time and phase distribution of the optical excitation field to the dynamics of each molecule results in a superior degree of control compared to the ensemble approach. Phase reversal does invert the molecular response, confirming the control of quantum coherence. Time-phase maps show a rich diversity in excited state dynamics between different, yet chemically identical, molecules. The presented approach is promising for single-unit coherent control in multichromophoric systems. Especially the role of coherence in the energy transfer of single antenna complexes under physiological conditions is subject of great attention. Now the role of energy disorder and variation in coupling strength can be explored, beyond the inhomogeneously broadened ensemble.

1. Introduction

The detection of individual molecules is a widely applied method to discern and follow molecular photodynamics beyond the intrinsic inhomogeneities of the ensemble.[1] The applications of "single-molecule detection" stretch across all fields, from physics, to chemistry and biology: *e.g.* the nature of dark states in molecules[2] or quantum dots;[3] efficient single photon sources;[4] detection of nanoscale optical fields;[5] the dynamics of supramolecular complexes[6] and polymers;[7] DNA and protein interactions;[8] natural light harvesting systems;[9] cell membrane

[a]*ICFO—Institut de Ciencies Fotoniques, Mediterranean Technology Park, 08860 Castelldefels (Barcelona), Spain. E-mail: Niek.vanHulst@ICFO.es*
[b]*Dept. de Física, Fac. de Ciencias Exactas y Naturales, Universidad de Buenos Aires, Buenos Aires, Argentina*
[c]*ICREA—Institució Catalana de Recerca i Estudis Avançats, 08015 Barcelona, Spain*

organisation[10] and intracellular processes.[11] Generally, in all these experiments the single molecular fluorescence is detected by high quantum efficiency photon counters, typically with signals of 10^3 (up to 10^6) counts per second and detector response times of 40–200 ps. As a result real-time dynamics can be recorded with sub-millisecond time resolution. Using time-correlated-single-photon-counting (TCSPC) and pulsed lasers also faster processes, such as fluorescence rates and charge transfer dynamics, can be recorded with $\sim$100 ps time resolution, yet with integration times of milliseconds or even longer.

The dynamic range of single molecule detection is fundamentally limited by detection of the "slow" nanosecond spontaneous emission. Particularly the ultrafast fs-ps regime is a true challenge. In fact, ultrafast spectroscopy of ensembles, with methods such as transient pump–probe spectroscopy, relies on absorption detection. Detection of a single molecule in absorption, with its minute cross-section at room temperature, is an equally challenging task. Yet progress is being made. Recently single molecules were detected with photothermal contrast, *i.e.* the heating of their environments by single-molecule absorption could be detected as change in local refractive index contrast.[12] In parallel, first individual molecules were detected at room temperature in direct absorption, mainly by ultimate reduction of the background noise.[13,14] Though this technique is promising, the absorption contrast is yet very limited for ultrafast experiments. The inverse process, stimulated emission detection, is also being explored and might reach single molecule sensitivity in the near future.[15]

Several years ago we started to explore routes towards ultrafast single-molecule detection.[16] Individual molecules were excited by a delayed pulse pair, while recording the fluorescence. A femtosecond pulse exciting a molecule can at most result in *one* fluorescence photon, while the molecule will have decayed to the ground state before the next pulse arrives. Thus to make two delayed pulses interact with one and the same molecule it is imperative to operate close to saturation conditions, whereas in the linear regime only the pulse repetition rate is effectively doubled. Unfortunately a single fluorophore driven at saturation conditions will bleach very fast. By optimizing the laser repetition rate to the saturation conditions of the molecule we managed to establish workable settings and record ultrafast traces of single molecules.[17] The average photon count rate corresponded to half the laser pulse repetition rate, confirming a balance between stimulated absorption and emission: saturation indeed. At short delay times (<300 fs) we observed transients indicative of intramolecular vibrational relaxation and dephasing at room temperature. We extended the method to multichromophoric complexes, revealing the effect of conformational disorder on fs photodynamics, for the first time at the level of single complexes.[18]

From these experiments we deduced two important lessons for future research. First: exciting single molecules at saturation conditions at room temperature should be avoided, as it limits severely the time window of observation. Second: the carrier envelope offset phase of the delayed pulses should be controlled to get insight into coherent phenomena. The logical solution is: *coherent control of a single molecule*, which is the subject of this communication. The combined use of a broad band pulse and a pulse shaper allows versatile generation of pulse pair sequences by spectral modulation of one-and-the-same pulse. Such coherent experiments can be carried out in the linear excitation regime, thus relaxing substantially the conditions for single molecule detection. Moreover a pulse shaper allows for rapid electronic control of the spectral phase and amplitude, allowing phase scans before photo bleaching of the molecule.

In recent years the observation of coherences at room temperature is receiving great attention.[19–23] Particularly the discovery of long-lived electronic coherences, up to 200–300 fs, in various photosynthetic complexes (Fenna-Matthews-Olson, FMO; marine cryptophyte algae) has generated strong efforts, both in experiment and theory, to understand their origin and explore their potential role in biological function. The occurrence of coherences in pigment–protein complexes at physiological conditions challenges the common notion that interactions with the local environment universally lead to decoherence; possibly the protein scaffolds protect electronic coherences.[21–23] Even the possibility of entanglement in light-harvesting complexes was proposed.[24–26] So far experimental approaches have concentrated on 2D-electronic-spectroscopy on ensembles of pigment-protein complexes at various temperatures. Unfortunately any observed coherence is a spatial and temporal average of an inhomogeneous distribution and therefore hard to relate to particular functionality in complex natural systems. Consequently coherent ultrafast detection of individual complexes will be important to unravel the role of coherence in the efficiency of natural photosynthetic complexes.

Here we will describe our experimental approach, illustrated with step-by-step results. The observation of vibrational wave-packets of individual organic molecules at room temperature is shown.[27] Most importantly superior coherent control is achieved by addressing an individual molecule as a well-defined single quantum system.[27,28]

2. Molecule and method

As molecule of study we used a higher rylene homologue: DiNaphtho-Quaterrylene-bis(Dicarbox-Imide)†, in short DN-QDI (Fig. 1). Rylenes are π-conjugated systems with outstanding chemical, thermal and photochemical stability, forming excellent building blocks for (excitonic) energy transfer systems. Terrylene-imides are particularly suited for single molecule studies because they exhibit high fluorescence quantum yields of 40–99% combined with good photo-stability.[29] DN-QDI has a fluorescence lifetime of about 3 ns and a quantum efficiency of 40%. Its core expansion with additional naphthalene units shifts the absorption spectrum

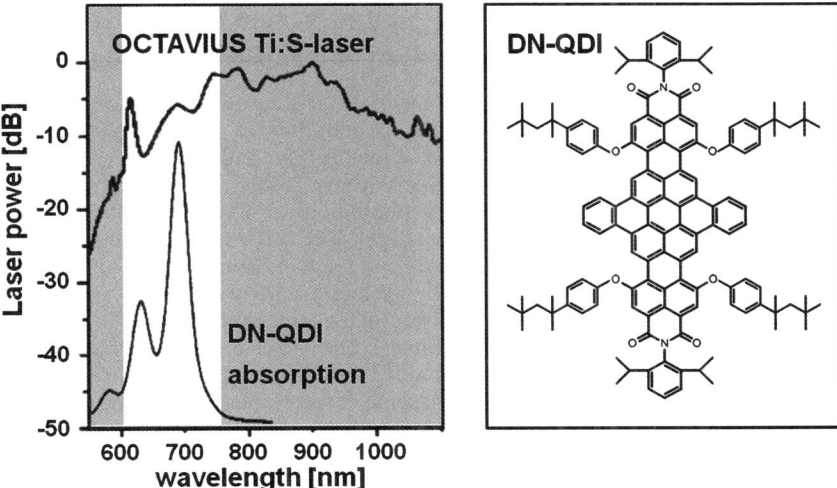

Fig. 1 (left) Broad band spectrum (log-scale) of the 7 fs Octavius Ti:Sa laser used for excitation, and absorption spectrum (linear scale) of the molecule DN-QDI, with maximum at 700 nm and clear vibrational progression; (right) molecular structure of the red fluorophore DN-QDI, DiNaphtho-Quaterrylenebis(Dicarbox-Imide)†.

† N,N9- bis-(N-2,6-diisopropylphenyl)-1,6,11,16-tetrakis[4-(1,1,3,3-tetramethylbutyl)-phenoxy]-8,9:18,19-dinaphthoquaterrylene-3,4:13,14-bis(dicarboximide)

to the near-infrared up to 750 nm with a maximum at 700 nm (in toluene solution).[30] It exhibits a prominent vibrational progression (C–C stretch, $\sim$1380 cm^{-1}); see Fig. 1.

Single-molecule samples were prepared by dissolving DN-QDI together with poly (methyl-metacrylate), PMMA, in toluene and spin-coating this solution onto standard microscopy glass cover-slips. With a concentration of about 10^{-9} Molar of DN-QDI thin layers ($\sim$40 nm) were prepared with less than one molecule per μm^2.

To achieve phase control of a single molecule it is important to coherently excite a large part of the absorption band. Here we drive the molecule with the output of a 85 MHz repetition rate mode-locked Octavius Titanium:Sapphire-laser (Octavius-85M, Menlo Systems, IdestaQE[31]). The Octavius spectrum stretches from $\sim$550 nm wavelength to deep into the infrared, spanning an "octave" in frequency, providing a 7 fs pulse when Fourier limited (Fig. 1). Here we selected a spectral bandwidth of 120 nm (15 fs) around the central wavelength of 676 nm (white band in Fig. 1), thus covering nearly the entire DN-QDI absorption spectrum, and interacting with a manifold of vibrational levels in the electronically excited state.

For coherent control we employed a 4f-pulse shaper based on a Spatial Light Modulator (SLM) for dispersion control and pulse shaping. The 4f-pulse shaper was adapted from the Multiphoton Intrapulse Interference Phase Scan (MIIPS-Box of Biophotonics Solutions Inc.).[32] The MIIPS system was modified to operate at wavelengths below 700 nm, with the second harmonic spectrum being detected in the sample plane. The shaper is designed in a double-pass configuration,[33] with a mirror at the end of the beam path reflecting the light back for a second pass through the shaper. This double-pass configuration minimises spatio-temporal coupling[34] and allows larger phase distortions to be compensated. For the experiments using 15 fs pulses and the SLM for pulse shaping, the lowest central laser wavelength with sufficient spectral intensity across the entire 120 nm band to allow for accurate phase compensation and pulse shaping was 676 nm. In all experiments pulses were first compressed to their transform limit of 15 fs in the sample plane.

3. Single molecules excited by a delayed femtosecond pulse pair

Single molecules were detected in a home-built *epi*-fluorescence confocal microscope. The shaped fs-laser light was first spatially filtered in a lens-pinhole-lens combination, directed towards the confocal microscope, and finally focussed onto the single-molecule sample with a 1.3 NA objective (Fluar, Zeiss). The excitation power was permanently recorded with a photodiode at the sample position and kept constant for all settings of the pulse shaper. The single molecular fluorescence was separated from the exciting laser light by a suitable combination of dichroic beam splitters and long-pass filters. Finally the fluorescence was split by a polarising beam splitter and tightly projected onto two, high quantum efficiency, 180 μm area, Avalanche Photo-Diodes (APD, Perkin Elmer). The two channel polarization detection gives direct insight in the orientation of the molecular emission dipole in the focal plane; moreover it provides two independent recordings on the effects of pulse shaping.

In a typical experiment the single molecule sample is scanned using a piezo-electric stage with nanometric position feedback (100 $\times$ 100 μm^2, Mad City Labs) to image the fluorescence and acquire an overview of the position of the molecules. Fig. 2 presents a 100 $\times$ 100 μm^2 image revealing many thousands of individual fluorescent molecules. Next one zooms in on a selected area of the high resolution image to identify the molecular position with nanometric accuracy. Selected single molecules are consecutively brought into the focus of the excitation beam, to record the fluorescence signal as a function of the pulse shape until photo-bleaching. Fig. 2 shows a typical time trace, with the single molecule emitting fluorescence photons during 57 s, until the discrete and irreversible photo-bleaching. In time the pulse shaper is set to generate a fs-pulse-pair with time delay increasing from 0 to 150 fs while

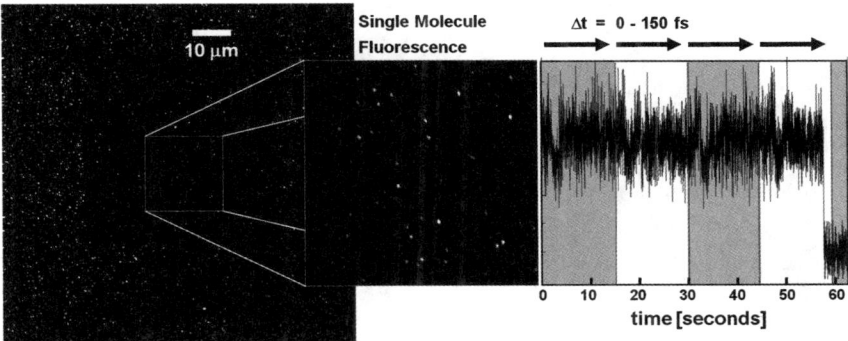

Fig. 2 (left) 100 × 100 μm² overview of an area with thousands of individual fluorescent molecules; (middle) 20 × 20 μm² detail, showing diffraction limited spots of individual molecules, with different fluorescence intensity mainly due to distinct dipole orientation; (right) time trace of a selected molecule, showing fluorescence during 57 s, until discrete photo-bleaching. The molecule was repetitively excited with a delayed double fs-pulse, with delay from 0 to 150 fs (grey-white segments). The delay scan shows as a recurring and controlled variation in the fluorescence intensity.

keeping their relative phase at zero. The pulse sequence is repeated every 15 s (grey-white segments in Fig. 2). Indeed a repetitive and reproducible response can be recognized in the molecular fluorescence signal.

This molecular response basically reflects the variation in excitation probability upon pulse shaping. Here two effects should be distinguished. On the one hand we intent to probe the temporal molecular dynamics by pulse shaping. On the other hand any change in the local excitation conditions will affect the recorded fluorescence of a molecule. To disentangle these effects we monitor the excitation intensity at the focal point. However, overall intensity control is not enough, because 4f-shaper manipulates spectral components in space and thus generally the output beam of any shaper contains spatially distributed spectral (or temporal) components.[34] To avoid spatio-temporal coupling the 4f-shaper is operated in double-pass configuration.[34] Moreover, using the MIIPS system, we make sure to start each experiment with a Fourier-limited pulse, *i.e.* flat spectral phase, in the focus of the confocal microscope.

4. Phase controlled excitation of single molecules

Complete information on both spatial and temporal response is best obtained by scanning full images at various settings of the pulse shaper. Fig. 3 shows a typical example of a set of molecules excited by a delayed pulse pair with increasing delay of $\Delta t = 0$, 21 and 42 fs. The excitation power of the pulse pair is kept constant while the carrier phase difference between each pulse pair is set zero. Diffraction limited spots (~300 nm FWHM) are observed with different fluorescence intensity, sometimes noisy due to the limited signal/noise ratio (about 10) when detecting single molecules. Upon close inspection of the different panels (0-21-42 fs) one observes molecules #1, 2 and 3 to change from dim to bright to dim. In contrast molecules #4, 5 and 6 rather change from bright to dim to bright. Similar other cases can be discerned. Finally certain molecules, such as #7, 8 and 9 show up in the first panel, but have bleached and disappeared in subsequent panels.

To obtain the underlying molecular dynamics we determine the integrated fluorescence intensity of each spot as function of the pulse pair delay. By taking the spatially integrated intensity we are sure to remove any residual spatio-temporal coupling caused by the shaper. Fig. 4a shows the fs time delay (Δt) response for

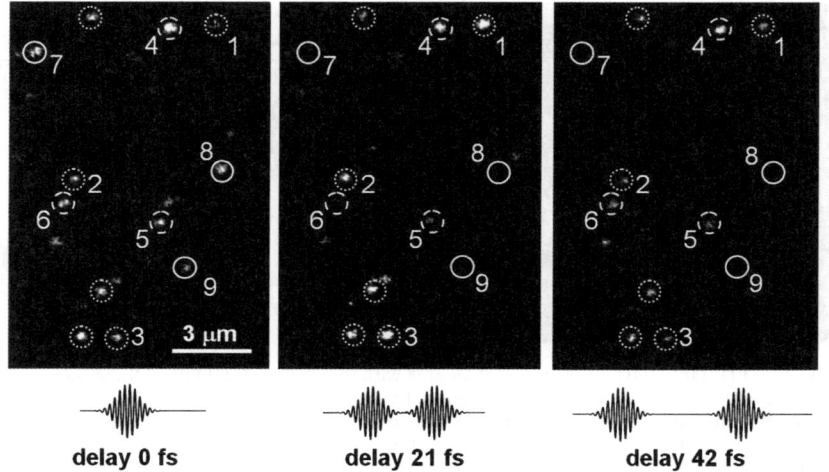

Fig. 3 Fluorescence images of a set of individual DN-QDI molecules excited by a delayed fs pulse pair with 0, 21 and 42 fs delay, respectively. Note the different response amongst molecules: molecules 1, 2 and 3 show dim-bright-dim, while molecules 4, 5 and 6 show bright-dim-bright. Molecules 7, 8, 9 bleached after the first image.

two different molecules. The relative fluorescence is normalized to the fluorescence at long time delay. The single molecule response shows strong oscillatory variations up to 50% of the average signal. The oscillations persist up to ~100 fs with a period of typically 30–40 fs. Interestingly the two presented molecules differ in phase; in fact their response is almost out of phase, for identical excitation conditions.

The oscillations are caused by wave-packet interference: constructive or destructive interference of the excited state wave packets generated by the delayed pulse pair. Fourier analysis shows a vibrational frequency typical for the carbon-carbon stretch at ~1070 cm^{-1}. The difference between the molecules reflects the structural disorder in the molecule-polymer-host system. The inhomogeneously broadened absorption spectrum (Fig. 1) is composed of many individual spectra. The specific spectral width and centre frequency of each molecule compared to the laser spectrum determines the ultrafast response. Thus the two molecules in Fig. 4a have slightly distinct absorption spectra and we are starting to unravel the inhomogeneous broadening. Clearly the average over many individual molecules, reconstituting the

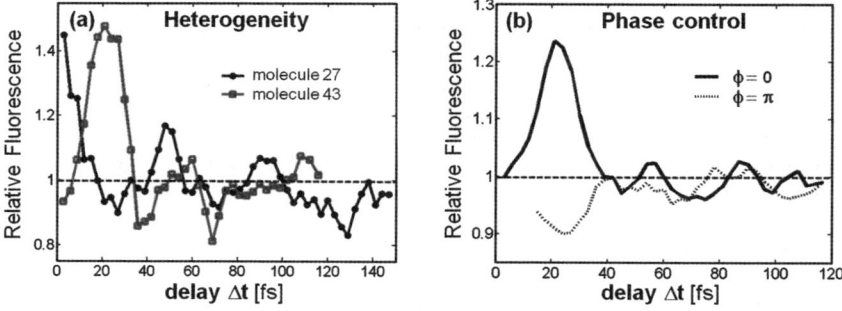

Fig. 4 (a) Integrated fluorescence as function of pulse delay Δt for two different molecules; (b) phase control of single-molecule wave packets: Single molecule fluorescence as function of pulse delay Δt for two in-phase ($\phi = 0$) and two anti-phase ($\phi = \pi$) excitation pulses.

ensemble, will show reduced residual oscillations. In fact in such averaging we find only 10% contrast in wave packet interference, consistent with the expectation from the bulk absorption spectrum. The individual molecules show typically 20 to 60% contrast. Moreover the wave-packet oscillations persist over ~100 fs, longer than the 20–50 fs dephasing time expected from the bulk spectrum. Evidently the selection of a single molecule, with a given conformation of its environment, does lift the ensemble phase averaging effects, allowing larger interference contrasts and therefore superior coherent control.

A crucial element in coherent control is of course the sensitivity to the phase of the optical field. So far we have kept the carrier phase difference between the delayed excitation pulses fixed to zero ($\phi = 0$), while purely varying the delay Δt. Now Fig. 4b also shows the effect of phase: for one and the same molecule the pulse delay response was recorded for both carrier phase difference in-phase ($\phi = 0$) and anti-phase ($\phi = \pi$). Clearly the inverse phase response is observed, confirming the actual phase control. Here it should be noted again that the excitation power has been kept constant in all cases. Thus the decreased response at short time delay (Δt) for the anti-phase ($\phi = \pi$) pulses is a molecular response and not the result of self-interference between the pulse pair.

5. Coherent control of single molecules

Clearly for maximum coherent control each molecule will have its optimal condition in (ϕ, t) phase space of the excitation field. Scanning of time delay or flipping phase by π allows to probe coherent excited state dynamics and gives encouraging clues about the controlled preparation of superposition states, however for complete insight one needs to explore the full (ϕ, t) phase space. Implementation of some feedback algorithm might be the obvious suggestion. However, a single molecule only emits 10^5–10^8 photons before the inevitable photo-bleaching. Thus even a "good molecule" providing 10^6 photocounts and up to a minute observation time, provides no room for feedback. To explore phase space we have designed a series of multiple pulses (four) and systematically varied both their mutual delay time and phase difference. The resulting time-phase response for a chosen molecule is plotted in Fig. 5. Indeed the presented single molecule fluorescence in time-phase space does show clear maxima and minima at certain time-phase combinations. A ~ 50% maximum is observed at $\Delta t = 25$ fs with $\phi = 0.5\ \pi$ and a ~ −50% minimum at $\Delta t = 25$ fs with $\phi = 1.5\ \pi$, again confirming the phase control, when switching to

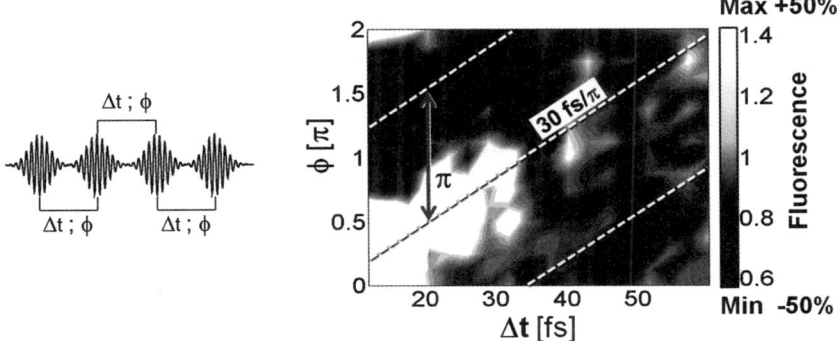

Fig. 5 Coherent control of a single molecule: (left) Excitation by a set of four fs pulses, with controlled time delay (Δt) and carrier envelop phase difference (ϕ) between each consecutive pulse in the sequence; (right) single-molecule time-phase coherent excitation map. The fluorescence intensity is normalized to the average. Dashed lines, separated by π in phase, indicate the progression of maximum and minimum response in phase space.

anti-phase. The ratio between the maximal and minimal response is max/min ≈ 3: a fairly high ratio for coherent control experiments, especially at room temperature. The π-shifted maxima and minima follow time-phase lines with a slope of about 30 fs/π for the chosen molecule. The wave-packet phase evolution can be traced by the optical field, providing an indication of the wave-packet group velocity. Moreover tracing the time-phase line one can deduce a decoherence time of about 40 fs.

6. Conclusions

The presented data clearly establishes the feasibility of coherent control of single molecules at room temperature, thus introducing ultrafast spectroscopy into the realm of individual quantum systems. By tailoring the driving fs field to the excitation spectrum of a chosen molecule superior control is achieved and the coherent dynamics of each individual molecule can be traced.

Here it should be noted that the presented results on single molecule coherent control rely on single photon excitation, performed in the weak field limit. Therefore the measured excitation probability as a function of temporal femtosecond pulse delay has a direct equivalence in the spectral domain, linked by the Fourier principle. Concretely, in the limit of infinite measurements, pure amplitude shaping in the linear regime gives the same information as a frequency scan when performed on a closed quantum system. However, even absorption spectroscopy of a single molecule at room temperature has not been shown yet, as it is very challenging due to the photo-dissociation problem.

More importantly, the DN-QDI molecules are embedded in a PMMA polymer matrix, and the interaction with the polymer host is the main source of the observed heterogeneity. As such the molecule-polymer system represents an open (not-closed) quantum system, for which the linear relation between temporal and spectral linear spectroscopy breaks down.[35] Thus the presented use of shaped femtosecond pulses allows preparing superposition states that cannot be achieved by pure CW excitation, and direct monitoring of the ultrafast dynamics of these states, providing insight beyond linear spectroscopy. As an example we recently presented the observation of Rabi oscillations and optical free-induction decays (OFID) of individual molecules.[36] Ultimately the use of fs pulses opens the way to coherent control of multiphoton excitations, which have no linear spectral equivalent in either open or closed systems.

The presented single molecule sensitivity extends the potential of coherent control into the direction of even more complex systems and environments. Particularly the issue of long-lived coherences in pigment-protein complexes is an obvious case that will benefit on the study of individual complexes. Similarly quantum coherences in single organic molecules could be exploited, for basic quantum optics operations.[36] Furthermore coherences in the excited state dynamics of large conjugated polymeric complexes will be of interest. Also, multi-parameter correlations, such as the relation between molecular conformation and function, can be obtained through single-molecule detection. On another note, the fs dynamics in individual plasmonic nano-particles, nano-antennas or emitter-antenna systems could be controlled.

Despite these interesting prospects one has to be realistic as to practical implications. Firstly, the presented work is based on terrylene derivatives, which are extremely efficient and photostable chromophores, with close-to-unity quantum efficiency and photo-bleaching rate of only 10^{-8}–10^{-7} per optical cycle, *i.e.* ideal for single molecule detection. More common dyes or protein complexes are typically ~100 times less photostable, making ultrafast studies a real challenge. Secondly, it is crucial that the broad band fs pulse is Fourier limited with flat spectral phase in the focus of the confocal microscope. High NA oil immersion objectives introduce extensive dispersion which requires major compensation. Finally, one relies on the detection fluorescence intensity of a molecule: any change in the excitation conditions, spatial or temporal, will affect the recorded signal and can lead to artefacts.

Particularly the spatial shaping, which is concomitant to any phase shaper (spectral or temporal), can cause spatio-temporal coupling, which should be avoided.[34]

Currently we are exploring the ultrafast intra-complex dynamics of electronic excitations in individual light-harvesting (LH2) complexes of purple bacteria.

Acknowledgements

We thank Florian Kulzer and Tim Taminiau for discussions and assistance with the experimental setup, and Klaus Müllen for providing the molecules. We also appreciate technical assistance of Peter Fendel (Menlo Systems, IdestaQE) with the Octavius laser system, and the collaboration with Biophotonics Solutions Inc. in developing a dedicated pulse shaper. Funding by the Spanish ministry of science and innovation MICINN (CSD2007-046-NanoLight.es, MAT2006-08184 and FIS2009-08203), Fundació CELLEX Barcelona and the European Union (FP6 Bio-Light-Touch and ERC Advanced Grant 247330) is gratefully acknowledged.

References

1 W. E. Moerner and L. Kador, *Phys. Rev. Lett.*, 1989, **62**, 2535; M. Orrit and J. Bernard, *Phys. Rev. Lett.*, 1990, **65**, 2716; T. Basché, W. E. Moerner, M. Orrit and U. P. Wild, *Single-Molecule Optical Detection, Imaging and Spectroscopy* (VCH, 1996); W. E. Moerner, *J. Phys. Chem. B*, 2002, **106**, 910; F. Kulzer and M. Orrit, *Annu. Rev. Phys. Chem.*, 2004, **55**, 585.
2 J. A. Veerman, M. F. García-Parajó, L. Kuipers and N. F. van Hulst, *Phys. Rev. Lett.*, 1999, **83**, 2155.
3 M. Kuno, D. P. Fromm, H. F. Hamann, A. Gallagher and D. J. Nesbitt, *J. Chem. Phys.*, 2001, **115**, 1028.
4 B. Lounis and M. Orrit, *Rep. Prog. Phys.*, 2005, **68**, 1129.
5 T. H. Taminiau, F. D. Stefani, F. B. Segerink and N. F. van Hulst, *Nat. Photonics*, 2008, **2**, 234.
6 J. Hernando, J. P. Hoogenboom, E. M. H. P. van Dijk, J. García-López, D. N. Reinhoudt, M. Crego-Calama, N. F. van Hulst and M. F. García-Parajó, *Phys. Rev. Lett.*, 2004, **93**, 236404.
7 D. A. Vanden Bout, W. T. Yip, D. Hu, D. K. Fu, T. M. Swager and P. F. Barbara, *Science*, 1997, **277**, 1074.
8 C. Hofmann, T. J. Aartsma, H. Michel and J. Köhler, *Proc. Natl. Acad. Sci. U. S. A.*, 2003, **100**, 15534; X. Michalet, S. Weiss and M. Jäger, *Chem. Rev.*, 2006, **106**, 1785.
9 A. M. van Oijen, M. Ketelaars, J. Köhler, T. J. Aartsma and J. Schmidt, *Science*, 1999, **285**, 400; M. A. Bopp, A. Sytnik, T. D. Howard, R. J. Cogdell and R. M. Hochstrasser, *Proc. Natl. Acad. Sci. U. S. A.*, 1999, **96**, 11271.
10 T. S. van Zanten, J. Gómez, C. Manzo, A. Cambi, J. Buceta, R. Reigada and M. F. Garcia-Parajo, *Proc. Natl. Acad. Sci. U. S. A.*, 2010, **107**, 15437; T. S. van Zanten, J. Gómez, C. Manzo, A. Cambi, J. Buceta, R. Reigada and M. F. Garcia-Parajo, *Proc. Natl. Acad. Sci. U. S. A.*, 2009, **106**, 18557.
11 J. Elf, G. Li and X. S. Xie, *Science*, 2007, **316**, 1191.
12 A. Gaiduk, M. Yorulmaz, P. V. Ruijgrok and M. Orrit, *Science*, 2010, **330**, 353.
13 S. Chong, W. Min and X. S. Xie, *J. Phys. Chem. Lett.*, 2010, **1**, 3316.
14 P. Kukura, M. Celebrano, A. Renn and V. Sandoghdar, *J. Phys. Chem. Lett.*, 2010, **1**, 3323; M. Celebrano, P. Kukura, A. Renn and V. Sandoghdar, *Nat. Photonics*, 2011, **5**, 95.
15 M. Wei, S. Lu, S. Chong, R. Roy, G. R. Holtom and X. S. Xie, *Nature*, 2009, **461**, 1105.
16 E. M. H. P. van Dijk, J. Hernando, J. García-López, M. Crego-Calama, D. N. Reinhoudt, L. Kuipers, M. F. García-Parajó and N. F. van Hulst, *Phys. Rev. Lett.*, 2005, **94**, 078302.
17 E. M. H. P. van Dijk, J. Hernando, M. F. García-Parajó and N. F. van Hulst, *J. Chem. Phys.*, 2005, **123**, 064703.
18 J. Hernando, E. M. H. P. van Dijk, J. P. Hoogenboom, J. García-López, D. N. Reinhoudt, M. Crego-Calama, M. F. García-Parajó and N. F. van Hulst, *Phys. Rev. Lett.*, 2006, **97**, 216403.
19 H. Lee, Y. C. Cheng and G. R. Fleming, *Science*, 2007, **316**, 1462.
20 G. S. Engel, T. R. Calhoun, E. L. Read, T. K. Ahn, T. Mancal, Y. C. Cheng, R. E. Blankenship and G. R. Fleming, *Nature*, 2007, **446**, 782.
21 G. D. Scholes, *J. Phys. Chem. Lett.*, 2010, **1**, 2.

22 E. Collini, C. Y. Wong, K. E. Wilk, P. M. G. Curmi, P. Brumer and G. D. Scholes, *Nature*, 2010, **463**, 644.
23 G. Panitchayangkoon, D. Hayes, K. A. Fransted, J. R. Caram, E. Harel, J. Wen, R. E. Blankenship and G. S. Engel, *Proc. Natl. Acad. Sci. U. S. A.*, 2010, **107**, 12766.
24 M. Sarovar, A. Ishizaki, G. R. Fleming and K. B. Whaley, *Nat. Phys.*, 2010, **6**, 462.
25 F. Caruso, A. W. Chin, A. Datta, S. F. Huelga and M. B. Plenio, *J. Chem. Phys.*, 2009, **131**, 105106.
26 D. Abramavicius and S. Mukamel, *J. Chem. Phys.*, 2010, **133**, 064510; S. Mukamel, *J. Chem. Phys.*, 2010, **132**, 241105.
27 D. Brinks, F. D. Stefani, F. Kulzer, R. Hildner, T. H. Taminiau, Y. Avlasevich, K. Müllen and N. F. van Hulst, *Nature*, 2010, **465**, 905.
28 R. Hildner, D. Brinks, F. D. Stefani and N. F. van Hulst, *Phys. Chem. Chem. Phys.*, 2011, **13**, 1888.
29 Y. Geerts, H. Quante, H. Platz, R. Mahrt, M. Hopmeier, A. Bohm and K. Müllen, *J. Mater. Chem.*, 1998, **8**, 2357.
30 Y. Avlasevich, S. Müller, P. Erk and K. Müllen, *Chem.–Eur. J.*, 2007, **13**, 6555.
31 See http://www.idestaqe.com/.
32 V. V. Lozovoy, I. Pastirk and M. Dantus, *Opt. Lett.*, 2004, **29**, 775.
33 O. E. Martínez, *IEEE J. Quantum Electron.*, 1987, **QE-23**, 1385.
34 D. Brinks, F. D. Stefani and N. F. van Hulst, in *Ultrafast Phenomena XVI*, ed. P. Corkum, S. de Silvestri, K. A. Nelson, E. Riedle and R. W. Schoenlein, Springer, Berlin, 2009, p. 890.
35 V. I. Prokhorenko, A. M. Nagy, S. A. Waschuk, L. S. Brown, R. R. Birge and R. J. D. Miller, Coherent Control of Retinal Isomerization in Bacteriorhodopsin, *Science*, 2006, **313**, 1257.
36 R. Hildner, D. Brinks and N. F. van Hulst, *Nat. Phys.*, 2011, **7**, 172.

Exploring the role of phase modulation on photoluminescence yield

D. G. Kuroda,[†a] C. P. Singh,[‡a] Z. Peng[ab] and V. D. Kleiman[*a]

Received 17th April 2011, Accepted 7th June 2011
DOI: 10.1039/c1fd00068c

We report an investigation to elucidate the mechanisms of control in phase-sensitive experiments in two molecular systems. A first inspection of optimization procedures yields the same experimental result: increase in the emission efficiency upon excitation by a phase modulated pulse in a two-photon transition. More detailed studies, which include power dependence, spectral response, one and two color pump–probe and pump–pump experiments show that while for one chromophore phase modulation leads to spectral matching between the two-photon cross section and the second order power spectrum for the other it provides a tool to manipulate the wavepacket dynamics in the excited state.

Introduction

In the quest to understand coherent control in complex structures many systems have been studied using closed-loop and open-loop optimization techniques.[1] Most of these experiments provided an early proof-of-principle although more recently there has been successful work using these phenomena to gain insight about molecular properties.[2] In this paper we explore how experiments based on phase modulation of excitation pulses for different systems can yield similar results (*i.e.* photoluminescence efficiency), although the responses arise from very different mechanisms. This manuscript highlights the care needed for interpretation of seemingly similar data. We show how molecular properties influence the control mechanism for increased photoluminescence and by doing so, allow us to learn molecular information about excited state processes.

The results presented correspond to two mechanistically different experiments. In the first instance, we investigate the phase control mechanism for the quantum yield of Rhodamine 6G.[2a,3] Rh6G (Fig. 1, left) is a simple molecule with well-studied electronic structure,[4] making it a good candidate to interpret the components of the optimal laser field and directly relate them to molecular properties. The second system under investigation is a phenylene-ethynylene dendrimer,[5] 2G$_2$-m-OH (Fig. 1, right). Phenylene-ethynylene dendrimers have a high quantum yield for emission and energy transfer to an added trap.[5,6]

In both experiments, photoluminescence follows a two-photon excitation process. By optimally tailoring the excitation pulses, the emission yield is increased; in one experiment as a consequence of phase control modulating spectral components

---

[a] Department of Chemistry, Chemical Physics Center, University of Florida, Gainesville, Florida, USA. E-mail: kleiman@ufl.edu
[b] Department of Chemistry, University of Missouri–Kansas City, Kansas City, MO, 64110-2499, USA

† currently at Department of Chemistry, University of Pennsylvania, Philadelphia, PA 19104–6323, USA.
‡ currently at Raja Ramanna. Centre for Advanced Technology, Indore, India.

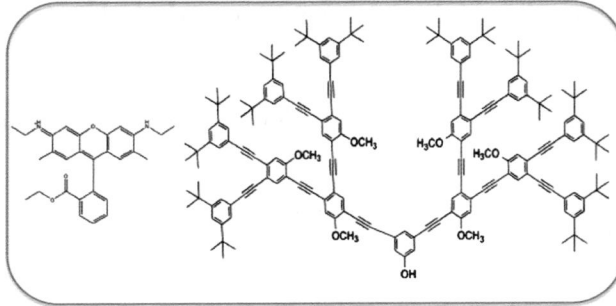

Fig. 1 Chemical structures of Rh6G (left) and 2G$_2$-m-OH (right).

while on the other one because of quantum control of wavepacket dynamics on the excited state.

Experimental methods

The experiments consist of two-photon excitation with tailored near-IR fs pulses. Emission from the lowest lying excited state is measured as a function of phase modulation. The ultrafast laser system is a commercially available Ti:Sapphire oscillator and chirped amplifier system (Spectra Physics, Tsunami® and Spitfire®). This amplified source delivers nearly transform-limited pulses (~45 fs duration, 25 nm FWHM) centered at 800 nm. A fraction of the output energy is used for excitation after passing through the phase modulator apparatus (*vide infra*). At the sample position, the beam is focused to an intensity of ~7 × 10^{11} W cm^{-2}. The emission signal is collected front face at 45 degrees with respect to the excitation beam with a photomultiplier tube to avoid self absorption. Given that we are interested in a phase dependent response we must first remove the intensity dependence of the excitation. This is accomplished by experimentally working with the ratio of two signals that have the same dependence[2n] with intensity. In the Rh6G experiments, the intensity dependence was removed using the simultaneous measurement of 2nd harmonic generation, while for the dendrimer experiments it was accomplished by focusing the 800 nm laser beam on a GaAsP photodiode.[7]

Rhodamine 6G sample was purchased from Exciton and prepared in methanol (spectroscopy grade) with concentrations below 10^{-5} M (OD of 0.25 mm^{-1} at 520 nm) to avoid any aggregation.[8] The synthesis of the dendrimer samples has been described elsewhere.[9] Dendrimer samples were prepared in dichloromethane (spectroscopy grade) solutions at concentrations below 10^{-6} M to avoid any aggregation or excimer formation.[10] Samples are held in a 2 mm optical path quartz cell with a magnetic stirrer to ensure fresh sample for each excitation pulse. The sample's photo-stability is checked by steady state absorption before and after each set of experiments.

Tailoring of the fs pulses is accomplished in a home-made apparatus based on a 4-*f* geometry for a Spatial Light Modulator built in a reflective mode configuration.[11,12] Due to the difficulties in its implementation, there have been only a few examples of reflective-mode configurations with none of them having a truly collinear beam pathway.[11,13] The limitation for the implementation of the reflective mode geometry is that the input and output beams must be separated. In our homemade pulse shaper, a fully collinear and reflective mode configuration is achieved utilizing polarization to separate the input and output beams. One important characteristic of this layout is that the angle of the first order diffraction matches the optical axis of the focusing optics, preventing astigmatism. Additionally, the incident and diffracted beams are constrained to travel in the same horizontal plane to prevent wave-front distortions. The filter mask selected for this apparatus was a liquid crystal based spatial light modulator (CRI Inc. SLM-640-D-VN®). Our dual mask system has a high average transmission (>94%). This setup is configured to produce amplitude and phase modulation using an entrance polarizer (transmission ~80% from 750 to 850nm) and a returning mirror (transmission 95%). Due to the double pass, these components reduce the overall transmission to less than 55%. The overall pulse shaper transmission is 30%, which is mainly determined by all the polarization optics and the efficiency of the grating. Characterization of the (un)modulated pulses is accomplished using the SHG-FROG (Second Harmonic Generation-FROG) technique. A splitter takes a fraction of the beam and propagates it through an optical path equal to that of the excitation beam. Using SHG-FROG we evaluate the pulse phase and compare it to the phase applied on the SLM. To test this characterization a cost function optimizing only fluorescence is used to generate the transform-limited pulse. The pulse obtained from this optimization is independent of which molecule is probed (dendrimer, perylene, Rh6G), and corresponds to a 42–45 fs pulse which is the shortest pulse attainable with the available bandwidth.

In the closed-loop experiments a genetic algorithm was used to find optimal pulses.[14] In the optimization algorithm, the pulse shaper phase is codified by 128 independent parameters, which are transformed into the corresponding 1280 mask pixel phases by linear interpolation between adjacent points. The algorithm starts the iterative search with a population of 80 random individuals, each consisting of 128 genes, and finishes the search optimization after 200 iterative loops.

Due to the large size of the search space, self-consistency and reproducibility is used to infer that the system evolved to an extreme of the searched space. Optimization experiments are reproduced at least three times, each time starting from a different (random) set of 80 pulses. For each experiment the laser stability is followed by simultaneously measuring the raw feedback signals for the unmodulated pulses (phase coefficients in zero phase).

The autocorrelation of the best excitation pulse for all the experiments is examined as an independent check of convergence reproducibility. As an example, for Rh6G, three different experiments present similar complex temporal structures (pulses with similar sub-pulse structure) indicating a convergence towards a non-trivial solution. This is observed in their corresponding SHG-FROG traces, confirming the reproducibility of the found solution.

Results and discussion

Rhodamine 6G

Rh6G is a yellow emitting dye with a xanthane core structure (Fig. 1, left). It has high photo-stability, high extinction coefficient, and high fluorescence quantum yield. Due to these properties Rh6G is commonly used as a laser dye[15] and as a biomolecular probe in microscopy studies.[16] The absorption spectrum of Rh6G shows two distinct absorption maxima at 350 and 530 nm, and a third absorption band at 390 nm with low extinction. The emission spectrum manifests the typical mirror image observed in molecules with similar potential energy surfaces in the ground and first electronic excited states. A fluorescence excitation anisotropy study performed by Eggeling et al.[17] reveals the presence of three distinctive electronic transitions, where the first two are one-photon allowed and the third is one-photon forbidden and two-photon allowed. Rh6G two-photon cross section in the 700 to 850 nm spectral region has been measured by different authors. It presents a sharp change in cross section between 770 and 830 nm.[16,18] Time-resolved experiments show a simple deactivation mechanism in which upon excitation to $S_{n>1}$, the system relaxes to the vibrational ground state of S_1 in a sub picosecond time scale. Coupling among the excited states leads to ultrafast internal conversion,[19] with a quantum yield of emission from S_1 of 95%,[20] and emission lifetime of 3.9 ns.[17]

The $2G_2$-m-OH dendrimer is an efficient light-harvester.[21] Its synthesis has been previously described.[9] $2G_2$-m-OH is a di-dendron consisting of two second generation dendrons coupled through the *meta* positions of a phenyl ring (Fig. 1, right). The two identical and indistinguishable monodendrons are electronically decoupled in the ground state because of their *meta* connectivity. It has been shown previously that the choice of site-substitution on the focal point governs the nature of the optical excitations for the entire molecule.[22] However, within the monodendron, the phenylene-ethynylene units of different length are connected by *ortho*- and *para*- substitution creating extended π conjugated electronic states with broad absorption bands.[6]

The absorption spectra of $2G_2$-m-OH shows a 15 nm red shift compared to the single G_2-OH dendron.[23] This shift is due to an additional phenyl ethynylene unit, which increases the conjugation length in each individual dendron. A better comparison can be made with the steady state spectra of a generation 3 monodendron. In the G_3OH monodendron,[21] the longest linear PE chain has the same number of PE units as the longest linear chain in the $2G_2$-m-OH, and their absorption spectra show similar features (bands, bandwidth, and red-shift). The dendrimer presents a folded spatial arrangement with all the dendritic branches packed in a bouquet-like 3D spatial arrangement.[6] The absorption spectrum of $2G_2$-m-OH has three distinguishable one-photon absorption bands peaked at 320 nm, 363 nm and 411 nm, and its emission peaks at 450 nm with a quantum yield of emission of about 80%.[21]

After excitation near each of these transitions, emission detected at 435 nm shows a lack of rise time dependence on excitation wavelengths ($\tau \sim 500$ fs for all excitation wavelengths),[6] leading us to conclude that energy transfer between the initially excited state and the lowest lying excited state (the emissive state) is the limiting rate, and not vibrational cooling. The fluorescence decay is characterized by a single exponential decay with $\tau = 2.36$ ns.

We evaluate the two-photon excitation response using a technique based on two-photon fluorescence which confirms the featureless cross-section within the spectral FWHM of the excitation source.[2,12] The flat two-photon excitation response obtained agrees with early experiments performed by Melinger and coworkers on a family of similar phenylene-ethynylene dendrimers.[5] That study showed that the two-photon cross section for different phenylene-ethynylene dendrimers varies with the structure and connectivity of the dendrons. Interestingly, up to the third generation they all present a flat two-photon cross section in the region of our excitation source (760 nm to 840 nm) regardless of their internal coupling and conjugation length.[5]

The aim of this work is to control the two-photon induced emission and to evaluate the mechanism responsible for the optimization. Experiments presented here focus on the optimization of fluorescence efficiency, as defined by the ratio of fluorescence intensity to the intensity used to induce it, which is obtained by the second harmonic generation (for Rh6G) or the GaAsP signal (for 2G2-m-OH).

Using a feedback loop with a genetic algorithm the optimal pulse is searched for with a fitness function that includes these intensity ratios. In heuristic searches, the convergence of the optimization is given by the stability of the solution. This stability is represented by the evolution of the average of the fitness for all individuals since, ideally, the average fitness values should asymptotically reach the solution. In these experiments, stability was reached near the 125th generation indicating the presence of a possible maximum in the fluorescence efficiency. The maxima values obtained for the fitness are roughly 11 and 35% (for $2G_2$-m-OH and Rh6G, respectively) higher than the values obtained when exciting with a transform-limited pulse.

A complementary experiment tries to minimize the emission efficiency. From the initial random phase to the optimized pulse, the efficiency ratio for the best individual of each generation decreases. Experiments on Rh6G showed that optimization to a minimum is possible, finding a maximal reduction of the I_{FL}/I_{SHG} ratio of 20%. The autocorrelation of the optimal pulse shows that the minimization of the fluorescence efficiency is not achieved by modulating the pulse to a very long featureless excitation pulse, instead convergence is reached by a relatively short complex pulse. Similar to the fluorescence maximization experiments, the optimal pulses present sub-pulse structure. Comparison between experiments starting with very different pulses shows that different realizations produce the same final result. Interestingly, an optimization to minimize fluorescence emission on 2G$_2$-m-OH does not converge which is a first indication that the control mechanism in both experiments differs and it becomes useful when interpreting the process controlled by phase modulation.

A comparison between the optimal pulse for the maximization or minimization of fluorescence efficiency on Rh6G reveals a counter intuitive result: the pulse that maximizes efficiency is significantly longer than the one that minimizes the efficiency, suggesting solutions specifically designed to match molecular properties of Rh6G. A similar effect has been observed in an analogous study performed by Brixner *et al.*[24]

The present study is based on populating an excited state *via* multi-photon absorption, so a power dependence is used to reveal which singlet state is initially populated. The intensity of the excitation source is enough to produce high order interactions, considering that excited electronic state populations usually have a large and reachable manifold of higher electronic excited states where they can be further transferred. In our experiments the excitation energy is not in resonance with the $S_1 \leftarrow S_0$ transition, so a third interaction will only promote the already excited state to a higher excited state, but it will not increase or decrease the population in the S_1 state. The states that can be reached with this source are only those whose energies are resonant with two or three photon transitions since the near infrared source can only induce the emission of Rh6G through a multi-photon excitation. For every family of excitation pulses (optimal pulse for maximization of emission efficiency, transform-limited pulse and optimal pulse for minimization of emission efficiency), the signal is quadratic on laser intensity suggesting that the state reached with the 800 nm excitation corresponds to a state located at 25000 cm^{-1} above S_0, corresponding to S_2. The possibility of an additional delayed interaction cannot be fully discarded. If a small portion of the excited molecules subsequently interact with a laser pulse, this might provide a change in the dynamics, but the power dependence (within its experimental uncertainty) might not change.

To understand the intensity dependence (and to discard it as possible source of control) it is common to represent the evolution of the optimization process by plotting fluorescence *versus* second harmonic signal space.[24] Fig. 2 (left panel) shows the second harmonic signal and fluorescence intensities for all the tailored pulses tested in the optimization of the fluorescence efficiency, with the red stars indicating the final pulse in each optimization. For reference, we also include an experiment in which only the fluorescence is optimized (equivalent to pulse compression). Data

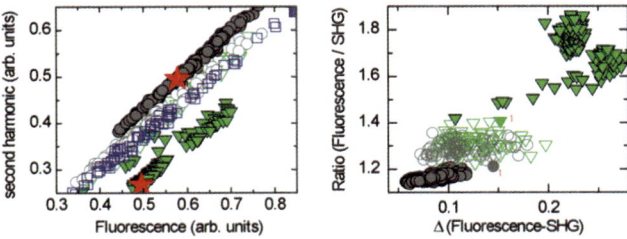

Fig. 2 Fitness values obtained for the best individuals on each generation (filled) and random phase pulses (empty) for maximization of emission efficiency (green triangles), minimization of emission efficiency (grey circles) and fluorescence (blue).

presented in this variable space can be separated in three distinctive regions. The center region corresponds to fitness arising from pulses with random phases and pulses leading to the optimization of fluorescence alone. Below that region are the best individuals corresponding to the optimization in which a maximum of fluorescence over second harmonic signal ratio was sought and above all are the individuals corresponding to an evolution in which the goal was the minimization of the ratio. These three regions denote the physical limits of the search space because each fitness goal is specifically designed to reach the extremes of this space. In addition, it confirms that each optimization reaches a distinct solution (red stars).

Although this representation shows the correlation between the variables and the achieved objective it neither reflects the attained convergence nor the generational evolution of the optimization variables. The region explored by each optimization appears as a delocalized area where more than one solution for each problem can be found. A different representation, plotting the fitness value *versus* the difference in emission and SHG signals provides a better understanding of the evolution of the fitness. The right panel in Fig. 2 shows the new space for the coordinates of the best individual in each generation (solid shapes) and for the coordinates of random initial pulses (open shapes).

The evolution of these variables for both experiments displays different optimization approaches. The maximization exhibits a complex optimization path: starting with a random pulse (see label 1 on Fig. 2, right) there is a region where changes occur mainly on the "difference" axis (horizontally). Consider the situation in which $I_{FL} = a \times I_{SHG}$. During an optimization of the two-photon excitation, in which the pulse must get shorter, I_{SHG} increases and thus the ratio remains constant ($I_{FL}/I_{SHG} = a$) while the difference ($I_{FL} - I_{SHG} = I_{SHG}(a - 1)$) increases linearly with I_{SHG}. This is followed by a simultaneous change in both variables reaching an area in which the difference in signals is not pronounced, but in which the ratio has been improved. The dependence between emission and excitation becomes more complex. In contrast, the minimization shows variations only on the difference axis while keeping the ratio relatively constant. This implies that both variables are changing their magnitude approximately by the same amount in the minimization. Although the experiments start in the same area of the explored space (random phases, open circles, Fig. 2), the convergence zones for each objective are clearly delimited in opposite regions and the convergence zone for the maximization of the fluorescence efficiency is more localized than the zone reached by the minimization. This suggests that the variable space landscape has local extrema zones with different shapes: the maximization falls into a well while the minimization remains in a more shallow area.

To gain some mechanistic understanding of the control process we try an optimization of the final Franck–Condon state localization in the excited state of the chromophore. The fluorescence spectrum is given by the overlap of the wavefunctions of the first electronic excited state with the ground state. A frequency-selected optimization (*e.g.* optimization of the longer-wavelength or shorter-wavelength region) of the emission spectrum could be produced by a perturbation of the potential energy surfaces (strong field control) or by affecting the wave packet evolution. In the second situation, interferences in the vibrational wavepacket evolution promotes the excited state to a *new* region of the multidimensional potential energy surface (PES), inevitably changing the Franck–Condon coefficients and consequently the fluorescence spectrum. Experiments would attempt to optimize selectively the low energy region of the emission spectrum with a new fitness function. A simple ratio of signals F_{red}/F_{total} (where F_{red} and F_{total} correspond to the integrated fluorescence collected with a long pass filter or without it, respectively) cannot be used because as the partial emission is maximized (minimized) it can lead to an undesirable optimization of the error function of the fitness, and not a true fitness optimization. We choose instead to use a fitness function with an invariant fitness error function:

$$\text{fitness} = \left((F_i - F_j)\frac{F_i}{F_j}\right); \Delta\text{fitness} = \left|\frac{2F_i}{F_j} - 1\right|\Delta F_i + \left|-\frac{F_i^2}{F_j^2}\Delta F_j\right|, \text{ where } F_i, \text{ and } F_j$$

correspond to the blue and the red side of the spectrum normalized to the second harmonic signal, selected by the use of band pass filters. The optimization yields a negative result in as much as the FROG traces of the optimal pulses have a strong resemblance to their counterpart on the experiments with the total fluorescence, so optimization of a frequency-selected spectrum is equivalent to optimization of the whole spectrum.

The use of this fitness function yields a "negative" but instructive result. Wavepacket interferometry in the emissive state of Rh6G, whereas the final wavefunction overlap leads to frequency modulated emission, cannot be accomplished.[25]

It is useful to describe the ratio of fluorescence/second harmonic fitness in terms of molecular and laser field properties to understand the molecular manipulation produced in the optimization. Using perturbation analysis, a simple model was first developed by Silberberg et al.[26] and later extended by Gerber et al.[24]

Time dependent perturbation theory can describe the two-photon absorption process. A quantum system originally located in the stationary state $|g\rangle$ can be promoted to the final state $|f\rangle$ with an electric field interaction at time t_0. In the limit of a small perturbation and assuming no resonance with any real intermediate, the amplitude of the final state can be expressed as[26]

$$a_{|f\rangle}(t) = -\frac{1}{\hbar^2}\sum_n \langle f|\mu|n\rangle\langle n|\mu|g\rangle \int_{-\infty}^{t}\int_{-\infty}^{t_1} E(t_1)E(t_2)e^{i(\omega_{fn}t_1 + \omega_{ng}t_2)}dt_1 dt_2 \quad (1)$$

Here, $\langle i|\mu|j\rangle$ are the dipole matrix elements between the $|i\rangle$ and $|j\rangle$ states, and ω_{ij} is the frequency corresponding to the energy difference between those states.

For short time interactions, the frequency domain representation of the two-photon transition probability (away from any intermediate resonance)[26,27] is then given by

$$p_{\text{TPA}} = \frac{1}{\hbar^4}\left|\sum \frac{\langle f|\mu|n\rangle\langle n|\mu|g\rangle}{\omega_{ng} - \omega_{fn}/2}\right|^2 \left|\int_{-\infty}^{\infty} E^2(t)e^{i\omega_{fg}t}dt\right|^2 \quad (2)$$

where the non-molecular contribution is given by the second order power spectrum

$$S^{(2)}(\omega) = \left|\int_{-\infty}^{\infty} E^2(t)e^{i\omega_{fg}t}dt\right|^2$$

$$= \left|\int_{-\infty}^{\infty} A\left(\frac{\omega_{fg}}{2} + \Omega\right)A\left(\frac{\omega_{fg}}{2} - \Omega\right)\exp\left\{i\left[\phi\left(\frac{\omega_{fg}}{2} + \Omega\right) + \phi\left(\frac{\omega_{fg}}{2} - \Omega\right)\right]\right\}d\Omega\right|^2 \quad (3)$$

where $A(w)\exp[i\varphi(w)]$ is the Fourier transform of $E(t)$. It is then clear that changes in the 2nd order power spectrum, $S^{(2)}(w)$, can modulate the transition P_{TPA}.[26]

In the limit of very broad inhomogeneous lines, the transition probability is the sum over all the possible states[2n]

$$P_{\text{TPA}} \approx \int g^{(2)}_{\text{TPA}}(\omega)S^{(2)}(\omega)d\omega \quad (4)$$

Where $g^{(2)}_{\text{TPA}}(\omega)$ encompasses all molecular properties.

The intensity of the second harmonic generation (used to normalized the emission signal) also depends linearly on the 2nd order power spectrum $S^{(2)}(\omega)$.[28]

$$I_{\text{SHG}}(\omega) = g_{\text{SHG}}(\omega)\left|\int_{-\infty}^{\infty} E(\Omega)E(\omega - \Omega)d\Omega\right|^2 \quad (5)$$

where $g_{\text{SHG}}(\omega)$ is a function of the second order susceptibility, crystal length, and group velocity of the fundamental and second harmonic frequencies. The fitness function evaluated in the closed-loop optimization experiments is the ratio

$$\text{fitness} = \frac{I_{FL}}{I_{SHG}} = \frac{N_m \Phi_{FL} \int g^{(2)}_{TPA}(\omega) S^{(2)}(\omega) d\omega}{\int g_{SHG}(\omega) S^{(2)}(\omega) d\omega} \quad (6)$$

where N_m is the number of molecules irradiated and Φ_{FL} is the quantum yield of two-photon induced emission.

Assuming a constant second harmonic generation efficiency across the laser excitation spectrum ($g_{SHG}(\omega) = g_{SHG}$) we obtain a fitness that can be modulated by changes in either the second order power spectrum $S^{(2)}(\omega)$ or the molecular properties encompassed in $g^{(2)}_{TPA}(\omega)$.

The validity of this simple model was demonstrated by the groups of Damrauer[2a] and Joffre.[2a] In both experiments, the two-photon excitation spectrum is the main feature phase-controlled within the optimization process.

The two photon excitation of Rh6G in the spectral region of the excitation presents sharp structure around 800 nm as shown by the groups of Webb[16] and Stelzer,[18a] and more recently by Nag and Goswami.[18b] These experiments an the description above point towards an optimization mechanism for Rh6G controlled by the molecular properties of the excitation step and not by the dynamics on the excited state. In contrast to these results, a very different mechanism will be shown next for the dendrimer system.

Conjugated dendrimer

We have previously used a closed-loop optimization to find a modulated pulse that maximizes emission efficiency in a donor–acceptor system.[2s] In that work, the donor moiety is the 2G$_2$-m- chromophore and the acceptor is a ethynylene-perylene unit. Here, we follow those experiments with the application of similarly modulated pulses to the 2G$_2$-m-OH molecule. The finding that both molecules follow analogous optimization strategies provides hints on excited state dynamics that can be altered by these pulses. The major contribution to coherent control of emission efficiency is the modulation of the wavepacket dynamics created on the excited states of the donor moiety of the light-harvester. Experiments to optimize emission efficiency in the all-conjugated dendritic molecule, 2G$_2$-m-OH, yield a similar result to those experiments in Rh6G only in the sense that an optimal solution is found and that this solution consists of a complex electric field, with a temporal profile longer than a transform-limited pulse.

Using statistical analysis we can mine information from the vast population of pulses tried and discarded throughout the optimizations. In addition, certain features of the optimal pulse can be recognized and used to simplify the solution and propose a physical mechanism on this single-knob control system. This methodology not only reduces the dimensionality of the space of variables but also removes data redundancy by selecting a minimum number of variables that explain most of the phase parameterization and fitness evolution.

In the analysis leading to the phase control of emission efficiency of the complex donor–acceptor molecule and in turn the 2G$_2$-m-OH, the statistical tests[2b,29] provide two variables that help understand the molecular response to phase modulation. One variable points towards a transform-limited contribution to optimize the two-photon excitation process. This initial excitation it essential to the overall process and cannot be neglected when we try to understand phase control of excited states created by multi-photon excitations. The second component is a step function on the spectral phase leading to an additional light–matter interaction whose molecular response is described in Fig. 3. As the size of the step function varies from 0 to 2π, the response from the system goes through a maximum emission efficiency higher than the efficiency obtained with a transform limited pulse (when $\phi = 1.4$ π, the response is ~11% higher than for $\phi = 0$). This particular change of the spectral phase modulates the temporal components yielding a new interaction at a particular time and that interaction is phase sensitive.

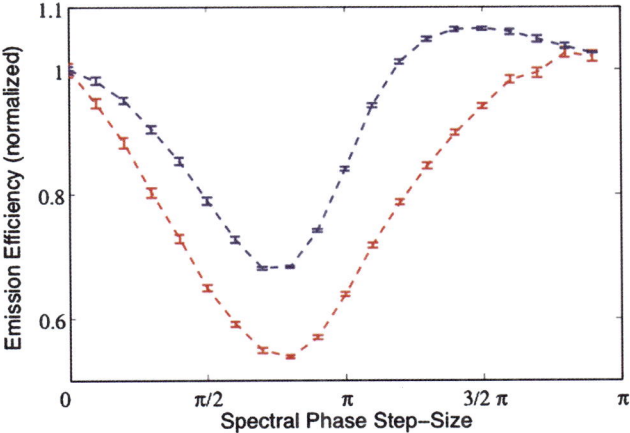

Fig. 3 Emission efficiency of 2G$_2$-m-OH (blue) and or perylene, a molecular system which does not show excited state dynamics control (red).

To understand the physical mechanism responsible for the coherent control of the emission efficiency we perform a series of complementary experiments to show first that an additional interaction is possible and second that that interaction is phase dependent. A two-color (400 + 800 nm) pump–probe transient absorption experiment exposes the presences of an accessible state beyond the initial two-photon absorption (Fig. 4, left panel). That additional interaction and coupling to a higher-energy state is sensitive to the delay between the initial resonant two-photon absorption and the third interaction with the 800 nm pulse. This is shown in Fig. 4, right, where double pulses (800 nm) with varying delays are coded on the excitation pulse and the emission signal is measured. Excitation with a modulated 800 nm double pulses with zero phase relation between pulses shows that for a delay of 70 fs between interactions, the emission efficiency is maximized, beyond the value corresponding to $\Delta t = 0$. These results indicate that a third interaction occurs, with a particular delay, coupling the initially excited state to a higher level. We call this a 2 + 1 mechanism. A power dependence measurement for the modulated and transform-limited pulses shows a near quadratic dependence. This is not unexpected. A quadratic dependence is expected after the initial two-photon excitation. Since only a fraction of the initial excited state is further transferred to a higher lying state, the change in the overall intensity can be within the experimental uncertainty (a 10% increased on the overall emission would correspond to only a power of 2.09 intensity dependence from the integrated fluorescence).

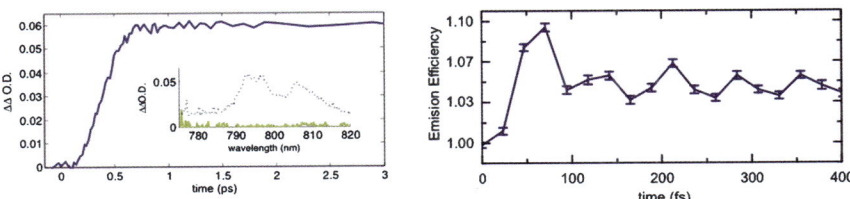

Fig. 4 (Left) Photoinduced absorption at 800 nm after 400 nm excitation, inset shows the spectral response after 0.5 ps. (Right) Emission efficiency after excitation with double pulses (800 nm) with varying delays.

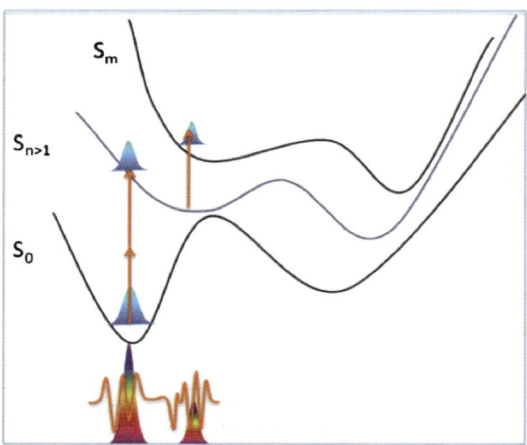

Fig. 5 Cartoon of PES for a 2 + 1 mechanism. The delayed excitation, with proper phase, removes the wavepacket fom a region of coupling to S_0 reducing the non-radiative decay.

The presence of this higher state moves the wavepacket away from the region where there is coupling to a non-radiative decay channel. Thus maintaining the excitation on $S_{>1}$ and allowing for an increased emission efficiency. A cartoon describing a possible mechanism is shown in Fig. 5. The initial excitation reaches S_2 and 70 fs later, while the wavepacket is near the curve crossing, a new interaction occurs with the right relative phase. A partial contribution from this wavepacket is further transfered to $S_{>2}$ and after some time it will reached the right side of the PES, where there is relaxation to S_1, from where emission occurs.

Conclusions

We present a series of experiments to elucidate the mechanism behind phase sensitive control of emission induced by two-photon excitation. Although at first glance the optimization of emission efficiency of both molecular chromophores would suggest a similar mechanism deeper investigations, including second order spectral considerations, frequency resolved optimizations, open loop experiments and pump–probe measurements, show otherwise.

In one case (Rh6G) the control is exerted on the excitation contribution. We conclude that phase modulation leads to spectral matching between the two-photon cross section and the second order power spectrum. By the same token, the lack of frequency structure in the two photon excitation spectrum of 2G$_2$-m-OH indicates that spectral matching cannot be the culprit of the optimization of emission efficiency, instead, the excited state dynamics are controlled by a 2 + 1 mechanism leading to intrinsic coherent control probed by the additional pump–probe experiments.

Acknowledgements

This material is based upon work supported by the National Science Foundation under CHE - 1058638.

References

1 (*a*) R. J. Gordon and S. A. Rice, *Annu. Rev. Phys. Chem.*, 1997, **48**, 601; (*b*) K. Ohmori, *Annu. Rev. Phys. Chem.*, 2009, **60**, 487; (*c*) L. G. C. Rego, L. F. Santos and V. S. Batista,

Annu. Rev. Phys. Chem., 2009, **60**, 293; (*d*) T. Buckup, J. Hauer and M. Motzkus, *New J. Phys.*, 2009, **11**; (*e*) H. Rabitz, *New J. Phys.*, 2009, **11**, 105030; (*f*) P. Nuernberger, G. Vogt, T. Brixner and G. Gerber, *Phys. Chem. Chem. Phys.*, 2007, **9**, 2470.
2 (*a*) M. A. Montgomery and N. H. Damrauer, *J. Phys. Chem. A*, 2007, **111**, 1426; (*b*) M. A. Montgomery, R. R. Meglen and N. H. Damrauer, *J. Phys. Chem. A*, 2006, **110**, 6391; (*c*) T. Okada, I. Otake, R. Mizoguchi, K. Onda, S. S. Kano and A. Wada, *J. Chem. Phys.*, 2004, **121**, 6386; (*d*) M. Bergt, B. Kiefer and G. Gerber, *J. Mol. Struct.*, 1999, **481**, 207; (*e*) C. Daniel, J. Full, L. Gonzalez, C. Lupulescu, J. Manz, A. Merli, S. Vajda and L. Woste, *Science*, 2003, **299**, 536; (*f*) D. Cardoza, M. Baertschy and T. Weinacht, *J. Chem. Phys.*, 2005, **123**, 074315; (*g*) D. Cardoza, B. J. Pearson, M. Baertschy and T. Weinacht, *J. Photochem. Photobiol., A*, 2006, **180**, 277; (*h*) F. Langhojer, D. Cardoza, M. Baertschy and T. Weinacht, *J. Chem. Phys.*, 2005, **122**; (*i*) A. F. Bartelt, T. Feurer and L. Woste, *Chem. Phys.*, 2005, **318**, 207; (*j*) R. de Nalda, C. Horn, M. Wollenhaupt, M. Krug, L. Banares and T. Baumert, *J. Raman Spectrosc.*, 2007, **38**, 543; (*k*) J. Hauer, T. Buckup and M. Motzkus, *Chem. Phys.*, 2008, **350**, 220; (*l*) J. Savolainen, R. Fanciulli, N. Dijkhuizen, A. L. Moore, J. Hauer, T. Buckup, M. Motzkus and J. L. Herek, *Proc. Natl. Acad. Sci. U. S. A.*, 2008, **105**, 7641; (*m*) D. B. Strasfeld, S. H. Shim and M. T. Zanni, *Phys. Rev. Lett.*, 2007, **99**, 038102; (*n*) T. Brixner, N. H. Damrauer, P. Niklaus and G. Gerber, *Nature*, 2001, **414**, 57; (*o*) B. Dietzek, B. Brueggemann, T. Pascher and A. Yartsev, *J. Am. Chem. Soc.*, 2007, **129**, 13014; (*p*) V. I. Prokhorenko, A. M. Nagy, L. S. Brown and R. J. Dwayne Miller, *Chem. Phys.*, 2007, **341**, 296; (*q*) E. C. Carroll, J. L. White, A. C. Florean, P. H. Bucksbaum and R. J. Sension, *J. Phys. Chem. A*, 2008, **112**, 6811; (*r*) M. Greenfield, S. D. McGrane and D. S. Moore, *J. Phys. Chem. A*, 2009, **113**, 2333; (*s*) D. G. Kuroda, C. P. Singh, Z. H. Peng and V. D. Kleiman, *Science*, 2009, **326**, 263; (*t*) J. Roslund, M. Roth, L. Guyon, V. Boutou, F. Courvoisier, J. P. Wolf and H. Rabitz, *Journal of Chemical Physics*, 2011, **134**; (*u*) M. Roth, L. Guyon, J. Roslund, V. Boutou, F. Courvoisier, J. P. Wolf and H. Rabitz, *Phys. Rev. Lett.*, 2009, **102**.
3 J. P. Ogilvie, K. J. Kubarych, A. Alexandrou and M. Joffre, *Opt. Lett.*, 2005, **30**, 911.
4 V. I. Gavrilenko and M. A. Noginov, *J. Chem. Phys.*, 2006, **124**, 044301.
5 Z. H. Peng, J. S. Melinger and V. Kleiman, *Photosynth. Res.*, 2006, **87**, 115.
6 E. Atas, Z. H. Peng and V. D. Kleiman, *J. Phys. Chem. B*, 2005, **109**, 13553.
7 U. Siegner, M. Haiml, J. Kunde and U. Keller, *Opt. Lett.*, 2002, **27**, 315.
8 P. R. Ojeda, I. A. K. Amashta, J. R. Ochoa and I. L. Arbeloa, *Journal of the Chemical Society, Faraday Transactions*, 1988, **84**, 1.
9 Z. H. Peng, Y. C. Pan, B. B. Xu and J. H. Zhang, *J. Am. Chem. Soc.*, 2000, **122**, 6619.
10 (*a*) B. L. Davis, J. S. Melinger, D. McMorrow, Z. H. Peng and Y. C. Pan, *J. Lumin.*, 2004, **106**, 301; (*b*) S. F. Swallen, R. Kopelman, J. S. Moore and C. Devadoss, *J. Mol. Struct.*, 1999, **486**, 585.
11 A. Monmayrant and B. Chatel, *Rev. Sci. Instrum.*, 2004, **75**, 2668.
12 D. G. Kuroda, PhD Thesis, University of Florida, 2008.
13 R. D. Nelson, D. E. Leaird and A. M. Weiner, *Opt. Express*, 2003, **11**, 1763.
14 M. Mitchell, *NetLibrary Inc., An introduction to genetic algorithms*, MIT Press, Cambridge, MA., 1996.
15 W. T. Silfvast, *Laser fundamentals*, Cambridge University Press, Cambridge, Eng.; NY, 1996.
16 M. A. Albota, C. Xu and W. W. Webb, *Appl. Opt.*, 1998, **37**, 7352.
17 C. Eggeling, A. Volkmer and C. A. M. Seidel, *ChemPhysChem*, 2005, **6**, 791.
18 (*a*) A. Fischer, C. Cremer and E. H. K. Stelzer, *Applied Optics*, 1995, **34**, 1989; (*b*) A. Nag and D. Goswami, *Journal of Photochemistry and Photobiology A: Chemistry*, 2009, **206**, 188.
19 G. Angel, R. Gagel and A. Laubereau, *Chemical Physics Letters*, 1989, **156**, 169.
20 M. C. Gazeau, V. Wintgens, P. Valat, J. Kossanyi, D. Doizi, G. Salvetat and J. Jaraudias, *Canadian Journal of Physics*, 1993, **71**, 59.
21 Y. C. Pan, M. Lu, Z. H. Peng and J. S. Melinger, *Journal of Organic Chemistry*, 2003, **68**, 6952.
22 (*a*) O. P. Varnavski, J. C. Ostrowski, L. Sukhomlinova, R. J. Twieg, G. C. Bazan and T. Goodson, *Journal of the American Chemical Society*, 2002, **124**, 1736; (*b*) O. Varnavski, G. Menkir, T. Goodson and P. L. Burn, *Applied Physics Letters*, 2000, **77**, 1120; (*c*) S. Tretiak, V. Chernyak and S. Mukamel, *J. Phys. Chem. B*, 1998, **102**, 3310.
23 J. S. Melinger, B. L. Davis, D. McMorrow, C. Y. Pan and Z. H. Peng, *Journal of Fluorescence*, 2004, **14**, 105.
24 T. Brixner, N. H. Damrauer, B. Kiefer and G. Gerber, *Journal of Chemical Physics*, 2003, **118**, 3692.
25 M. Spanner, C. A. Arango and P. Brumer, *Journal of Chemical Physics*, 2010, **133**.
26 D. Meshulach and Y. Silberberg, *Physical Review A*, 1999, **60**, 1287.

27 B. Broers, L. D. Noordam and H. B. V. Vandenheuvell, *Physical Review A*, 1992, **46**, 2749.
28 W. Demtröder, *Laser spectroscopy: basic concepts and instrumentation*, Springer-Verlag, Berlin; New York, 1981.
29 M. A. Montgomery, R. R. Meglen and N. H. Damrauer, *J. Phys. Chem. A*, 2006, **110**, 6391.

General discussion

Professor Brumer opened the discussion of the paper by Professor Dwayne Miller: Your paper concludes that "the electric field interaction is evidently sufficient to suppress the normal destruction of the coherence during the τ period". This is just a comment that the role of external fields in suppressing decoherence has been the subject of a number of studies by those in the quantum information community (*e.g.* G. Kurizki).[1]

Further, one of our original coherent control scenarios showed that decoherence could be countered by application of external fields.[2]

1 G. Gordon, N. Erez and G. Kurizki, *J. Phys. B*, 2007, **40**, S75.
2 M. Shapiro and P. Brumer, *J. Chem. Phys.*, 1989, **90**, 6179–6186.

Professor Miller replied: I agree that the effect of applied fields on suppressing decoherence is well known at high fields. Especially if one is in the limit of saturation and Rabbi cycling of ground and excited state coherences. The question is what are the relative magnitudes of bath induced collisions on the rms motion of the electrons and instantaneous polarization of the system/molecule in comparison to resonant applied fields in the weak field limit. Even at extremely weak fields, the very act of coupling to the field and resonant transfer to form an excited state leads to a localized change in the electron distribution relative to the ground state. This effect, locally on the electronic polarization of the system, is significant. The question is how does this compare to stochastic fluctuations of the bath. The fluctuations of the cavity and associated field fluctuations at the site in which the system resides are relatively small. We have previously assumed that these bath fluctuations lead to the observed decoherence. Given that the timescales of decoherence, prior to this study, place decoherence on 10–100 fs time scales, the only specific motions of the bath fast enough to lead to decoherence on this timescale are the inertial solvent librations. I could easily imagine that the induced changes in electron distribution from these relatively small amplitude, collective, intermolecular motions is far less than the intramolecular vibrations and Intramolecular Vibrational Energy Redistribution on the electron distribution. Only in the case of dissipative coupling does the process lead to true decoherence. So the question I have is related to the relative magnitudes of the different intramolecular and intermolecular bath fluctuations on the field induced polarization and associated memory. The papers you cite are interesting in their own right but have not addressed this point which I believe captures the essential physics governing decoherence of molecular systems. I do agree that approaches to potentially resolve this point are in place as you state. I hope this problem will be addressed in the near future to provide some theoretical guidance to future experiments to dissect intramolecular dephasing contributions from true decoherence or causal relations in the phase of the excited state wavefunction.

Dr Buckup remarked: The effect of non-transform limited pulses in 2D spectroscopy has been investigated by several works in the last years. In particular, it is well known the side-effects of chirped excitation on the interpretation of 2D electronic spectra[1] or how vibrational coherence can generate "new" features, which resemble electronic coherence signatures.[2] In the case of multipulse excitation, several additional contributions from, for example, non-rephasing signals will overlap with the rephasing pathways of the 2D signal and possibly generate deformations on the 2D spectra not observed with TL pulses. Therefore, the small deformations of 2D signal observed in the paper for long waiting times (T=40ps) when tailored pulses were applied, does not mean *a priori* as suggested in the paper that long coherences are present. How do you distinguish between simple optical artifacts due to use of non TL excitation from real molecular effects due to tailored shaping?

Moreover, the optical signal at T=0 is within the so called shaping window[3] and a simple comparison with the signal coming from a glass substrate is not enough to distinguish all contributions. Nonlinearities of solvents and glass, for example, contains very different contributions and should generate very different 2D spectra at T=0 for tailored pulses. Therefore, it is unclear how the signal at T=0 can be interpreted as performed in the paper when the issues raised above are taken in account.

1 N. Christensson, Y. Avlasevich, A. Yartsev, K. Müllen, T. Pascher and T. Pullerits, *J. Chem. Phys.*, 2010, **132**, 174508.
2 N. Christensson, F. Milota, J. Hauer, J. Sperling, O. Bixner, A. Nemeth, and H. F. Kauffmann, *J. Phys. Chem. B*, 2010, **115**, 5383.
3 T. Buckup, J. Hauer and M. Motzkus, *New J. Phys.*, 2009, **11**, 105049.

Professor Miller replied: Using chirped excitation pulses for measuring 2D-spectra was also investigated in our paper; the observed features in the 2D-spectra cannot be characterized as "side effects" – they originate from the specific chirped field – molecular response/interaction which, in fact, is not the same as in the case of TL-pulses. These features are fully supported by our numerical calculations mentioned in the paper (and shown during the oral presentation) for both "correlation spectra" (T = 0) and for long waiting times (T >= 40 ps) where the effect of chirped excitation pulses vanishes. Note that in our experiments the spectra of the excitation pulses (E1, E2) were restricted in their bandwidths (using an acousto-optic shaper) to not excite higher vibrational states in Rh101 (see corresponding spectra in Fig. 1) so that the vibrational coherences from this state do not contribute to the Photon Echo(PE)-signal as in experiments of Kauffmann *et al.* to which you refer.[1] With respect to your comment on "simple optical artifacts" using shaped excitation pulses, I assume this question relates to the role of self-modulation in the solvent and the optical windows of the cell and possible generation of "parasitic" signal in the same direction as the photon echo that originates from the molecular response. In this regard, we would stress again (see the Experimental Section of the paper) that our measurements were conducted at very low excitation level (2–4 nJ focused into ~50 μm spot) where the signal, measured from a blank cell filled with solvent was approximately two orders of magnitude smaller than the PE-signal from Rh101. Moreover, shaping of the excitation pulses under constant pulse energy leads to a decrease in the instantaneous intensity and thus significantly further decreases the self-modulation effects. Also, it is well known that the non-linear response of dielectric materials (glass and methanol in our case) is almost instantaneous with respect to excitation and relaxes within a few fs (worst case). Therefore, it is hard to expect any contribution of such an "optical artifact" to the PE-signal measured at very long waiting times (>40 ps). However, in order to check for possible

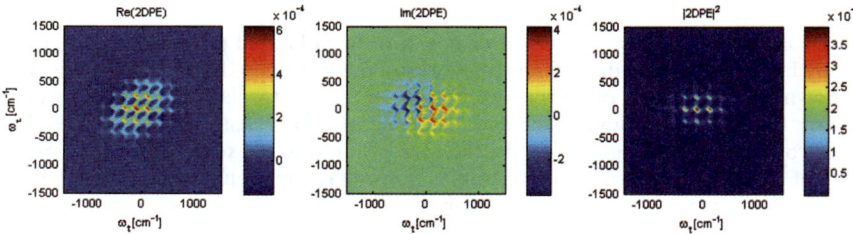

Fig. 1 Calculated 2D-spectum (T=0) for the Rh101 model system by excitation with a periodic phase-shaped pulse with a frequency of 220 cm^{-1} for comparison to the experimental measurements.

contributions of parasitic signals coming from the non-linearity of cell windows and solvent, besides the direct control experiments with a glass plate at a waiting time T = 0 (shown during presentation), we also changed the modulation period to show that the effect is related to the low frequency mode that most strongly modulates the transition probability in the molecular response (Figure 12 in the paper). This control clearly shows the effect is related to the molecular response. In addition, we performed numerical simulations of the 2D-spectra for periodic phase-shaped pulses (as discussed in the paper) using a conventional 3rd order molecular response theory, without including any solvent effects or self-modulation in glass windows. We found excellent agreement between the calculated and measured 2D-spectra. The calculated 2D-spectrum at T = 0 for the Rh101 model system with the parameters listed in the footnote on page 11 of the paper is shown in Fig. 1. As can be seen, these calculations reproduce quite well all the features observed in measured 2D-spectra at T = 0 (compare the experimental results shown in Figure 9 in the paper). Thus, we can safely conclude that in our measurements, performed in the low excitation limit, the contribution of non-linearity of solvent and optical windows is insignificantly small. Further, a contribution can only be construed to be an artifact if the effect is omitted and affects the interpretation. From the extremely good agreement with our theoretical calculations and short time dynamics in the 2D spectra, it is clear we have included the phase modulation effects for non-TL pulses properly. The excitation process is accurately treated such that we can connect long lived frequency correlations to the initial excitation. It is interesting that the use of pulse shaping enables a much more rigorous test of theoretical modeling of 2D spectra than the use of TL pulse as there is more structure in the response function for comparison. The main point of the paper is the clear observation of coherent population transfer during the τ scan (labeling of the excitation axis) that is clearly observable at long time. It is the long time dynamics that are of interest, not the short time, T=0 dynamics, where the modulations are related to electronic coherences in the molecular response, and cannot be related to population transfer *per se* until after decoherence collapses the system response to a population transfer. The effect of shaped pulses on the population becomes a direct observable at long time. In this time interval (T>0), it is possible to observe both the frequency correlation between the excitation field and the population transfer and ensuing spectral diffusion leading to homogeneous broadening. In this context, we are glad for the opportunity for further expand on the distinction between information in 2D spectra within the coherence time and longer waiting times relevant to frequency correlations related to population transfer and spectral diffusion.

1 *J. Phys. Chem. B*, 2010, **115**, 5383.

Professor Weinacht asked: I was wondering about the interpretation of Figure 11 in the paper. It is interesting in that it decays very rapidly to a small but non-zero value at which it stays constant for a long time. That kind of rapid decay followed by a constant level makes me worry a little about background issues contributing to the signal you measure.

Professor Miller replied: Figure 11 in the paper is intended to show the relative magnitude of the effect of using phase modulated pulses on the coherent population transfer. The long term component is not a background contribution but real population transfer that should be constant. This figure is plotting the induced modulation along the τ axis as a function of waiting time T. The large modulations observed at early T are due to resonant and nonresonant coherence effects during the excitation pulse. The pulse durations approach several hundred femtoseconds for the largest chirps and phase modulation. The main point is to illustrate that well outside the coherence time of the system response there are long lived (population) components along the excitation axis that illustrate coherent population transfer has occurred during the τ scan (field on condition) and coherent state preparation. We were only able to show relatively small effects of 5 to 7% with respect to coherent population transfer in our previous studies.[1–3] By using 2D spectroscopy, we can project the coherence onto a second frequency axis to see specifically the frequency correlation of the initial state preparation leading to coherent population transfer. In 2D, the effect of the pulse on population transfer is clearly observable. I should add that this is background-free experiment since we are measuring the difference in 2D-PE spectra obtained with shaped and unshaped pulses having equal energies. If we will take differences between two measurements performed with both unshaped pulses, the resulting signal will be zero. As a control, we tuned the frequency modulation to be resonant and off resonant conditions to be sure there is not a background modulation unrelated to the molecular response, as well as studied the response without the dye. Again the purpose of Figure 11 in the paper is to clearly distinguish the coherent population transfer from the coherence effects during the excitation pulse sequence. The strong modulation observed along the ω_τ axis at long time, T, clearly illustrates the coherent population transfer and shows the effect to be quite significant (16%) compared to the pure electronic coherence effects observed at early T.

1 V. I. Prokhorenko, A. M. Nagy and R. J. D. Miller, *J. Chem. Phys.*, 2005, **122**(18), 184502.
2 V. I. Prokhorenko, A. M. Nagy, S. A. Waschuk, L. S. Brown, R. R. Birge and R. J. D. Miller, *Science*, 2006, **313**(5791), 1257–1261.
3 V. I. Prokhorenko, A. M. Nagy, L. S. Brown and R. J. D. Miller, *Chem. Phys.*, 2007, **341**(1–3), 296–309.

Professor Tannor remarked: You indicated that you started this work because you were puzzled how, based on the 2D vibronic spectroscopy, there could be an apparent 80 fs electronic decoherence time — yet in the rhodopsin work you found that pulse trains of 200 fs produced coherent control. In the present work you show evidence of long-lived memory in vibrational coherence. But I am unclear how vibrational coherence, no matter how strong or how long-lived, can increase the electronic decoherence time. Thus I don't fully understand how this long-lived memory can survive if there is an intrinsic 80 fs electronic decoherence time as you initially supposed.

Professor Miller responded: The apparent decoherence time of bacteriorhodopsin (bR) based on 3 pulse photon echo measurements was reported to be in the 10 fs range (< 50 fs),[1] yet our control pulses were on the order of 200–400 fs. The question we were pondering was what is really being measured in photon echo measurements? Up until present, decoherence based on photon echo measurements were reported as an intrinsic property of the system-bath response much like a lifetime. The reason for this was the underlying assumption that the decoherence was dominated by collisional dephasing with the bath. Decoherence times of a few tens of femtosecond to 100 fs have been reported as a general trend for molecules in solution. What intermolecular bath modes can lead to such fast decoherence? The collective librational motions are the only potential candidates and these relatively small underdamped motions would have to have large coupling coefficients. In the case of bR, the bath is the protein and the motions are much slower and certainly not in the 10 fs range. Clearly, it is motion along the intramolecular coordinate that is dominating the decay in the photon echo signal. This process is related to the fast motion out of the Franck–Condon region and motion of the vibrational wavepacket along the reaction coordinate. The relative phases of the time evolution of this excited state are still related to the initial state preparation. The coherence in this sense will be conserved until a real collision or dissipative process with the bath. The question then is how long lived is this coherence or how long does the vibrational wavepacket stay focused along the reaction coordinate. This information is not captured in a conventional photon echo experiment using a broad bandwidth pulse TL pulse. The 2D experiment allowed us to use the same shaped pulses (chirp, amplitude-phase, and phase modulated pulses) to see explicitly the frequency correlation between the excitation and the degree of population transfer. The enhanced population transfer has been associated with the low frequency mode that dominates the time dependent Franck–Condon factors and shows that as the causal relationship between the excitation pulse. The coherence between the ground and excited state surfaces must persist for time scales much longer than would be inferred from a conventional photon echo and in the case of bR clearly longer than a few tens of femtoseconds. The coherence we observe is generated during the τ scan (field on conditions) and this field interaction also needs to be taken into consideration (see earlier discussion with Professor Brumer). The long lived electronic coherence is not conserved by the vibration coherence *per se*, however if this mode was strongly damped it would contribute to the decay in the coherence. Basically, the point is that dynamics involving the intramolecular bath can dominate the photon echo measurements, as apposed to the intermolecular bath, and the loss in the phase memory along this coordinate is only accessible via coherent control measurements. The coherent control 2D approach that we have introduced gives a nice means to see this effect.

1 J. T. M. Kennis, D. S. Larsen, K. Ohta, M. T. Facciotti, R. M. Glaeser, and G. R. Fleming, *J. Phys. Chem. B*, 2002, **106**, 6067. See Figures 4A and 4B.

Professor Weinacht asked: It seems to me that there is perhaps some confusion arising from the distinction between dephasing and decoherence. If there is rapid dephasing which is followed by decoherence before a revival/rephasing, then it can appear as if the decoherence is much more rapid than it really is. So, for instance, if I consider an anharmonic oscillator which has a period of 10 fs, a dephasing time of 100 fs, a decoherence time of 1000 fs and a revival time of say 10 ps, then a coherent oscillation signal will be washed out at 100 fs, but not yet due to decoherence, rather just dephasing. However, before the wavepacket rephases at 10 ps, coupling to a bath can result in a loss of coherence (decoherence) on a timescale of 1000 fs. Thus there may be coherence out to 1000 fs, but it looks like it is lost by 100 fs.

Professor Miller answered: Yes, this is exactly the point. The conventional view is that the decoherence is determined by stochastic fluctuations of the bath that are uncorrelated to the induced polarization. Decoherence is reported as if it is an intrinsic property of the system-bath interaction. What we have shown is that the decoherence depends on the state preparation. Further, the molecules in question are large enough to have sufficient degrees of freedom to form an intramolecular bath that will lead to the appearance of dephasing within the intramolecular manifold but the state evolution is still causally connected to the initial state preparation until real decoherence occurs through random collisions with the intermolecular bath. In conventional photon echo experiments, short excitation pulses are used that, in addition to the above consideration, invariably lead to the excitation of vibrational wavepackets that move out of the Franck–Condon window and will also lead to a decay in the echo amplitude even though the initial coherence or causal relationship of the phase relationships are still conserved within the vibrational wavepacket. It is only through the effect of shaped optical pulses leading to coherent control of population transfer that the longer lived nature of the quantum coherence can be observed.

Professor Rabitz commented: What information can be read directly from 2D spectra *versus* that requiring the performance of modeling?

Professor Miller answered: The frequency correlation between the excitation and the degree of induced population transfer can be read directly from the 2D spectra. The open questions are the specific intramolecular modes and bath couplings that give rise to the observed frequency correlation. Experimentally, we see a clear connection to the low frequency mode most strongly modulating motion out of the Franck–Condon region for the optical transitions. Modeling of the coupling of this mode to the bath and its influence, as Professor Kosloff and his collaborators have done,[1] would help to understand the mechanism for weak field coherent population transfer. It will be very informative to increase the bandwidth of the pulse shaping to access higher frequency modes as well and modeling would be very helpful in guiding experimental efforts along these lines.

1 G. Katz, M. Ratner and R. Kosloff, *New J. Phys.*, 2010, **12**, 015003.

Professor Rabitz asked: Is it correct that in the solution phase, inherent limitations will arise from the intensity (due to molecular or material damage) thereby restricting the effective order of the Dyson expansion?

Professor Miller replied: I would say yes. There is a peak power limitation. We have found that excitation levels beyond 30 GW/cm^2 lead to multiphoton ionization of the solvent with up to 10% of the excitation absorbed through this process.[1] This observation was for water/buffer. It will be more of an issue for solvents with absorptions that are red shifted and have larger preresonance enhancement. Further, one is generally studying relatively large molecules in solution such that there are large number of electronic levels that will lead to resonant conditions coupling all states and further lead to multiphoton ionization of the molecule itself. It is very difficult if not impossible to selectively excite a particular upper level excited state, for example an S2 level, or coherence between S1 and S2, as all states become mixed. The real problem is the multiphoton ionization under resonant conditions for the molecular system of interest. The threshold will be much lower than the solvent itself. The contribution to the signal from multiphoton ionization of the solvent and target molecule are generally unknown and would be difficult to include in the analysis. For solution phase studies, much lower peak power must be used than gas phase and this will greatly limit the prospects for strong field control.

1 V. I. Prokhorenko, A. Halpin, P. J. M. Johnson, L. S. Brown and R. J. D. Miller, *J. Chem. Phys*, 2011, **134**, 085105.

Professor Engel opened the discussion of the paper by Professor Brumer: You described a very clean, intuitive, and simple model that accurately predicts long-lived electronic zero-quantum coherences (about 250 fs at 277 K, about 1000fs at 77k). Does your model also yield the proper dephasing of single quantum coherences (70 fs at 77 K, 15 fs at 277 K)? If this model does not capture these timescales, could you speculate on how the model could be modified to capture these dynamics?

Professor Brumer responded: The model to which Professor Engel is referring, mentioned at this meeting, has been submitted for publication.[1] The model has been applied to one exciton states in a dimer pair in FMO and in PC645, with great success. The results give analytic expressions for dephasing and relaxation rates in these systems which are in excellent agreement with experiment. They also provide deep insights into the system parameters that result in long-lived coherences.

We have not yet applied this approach to the single exciton coherences or to the coherences reported for other exciton pairs in FMO that Engel reports in this Faraday meeting. However, we plan to do so. In the interim we note that the approach that we advocate utilizes different expressions depending on the system parameters, such as the reorganization energies, the maximum frequency cut-off of the bath, *etc.*

I expect that we will need to invoke these different expressions as we consider different coherences. This work is now underway in our laboratory.

1 L. Pachon and P. Brumer, *quant-ph ArXiv*, 1107.0322.

Dr Shalashilin said: Can your master equation, which is an approximation, be compared with numerical results of multidimensional quantum mechanics?

Professor Brumer responded: We have not made this comparison. Rather, we have focused on analyzing the validity of the widely adopted Markovian approximation, and also have some preliminary results on the range of validity of the second order approximation. A proper comparison with the exact results can, of course be made.

Professor de Vivie-Riedle asked: What is the most important parameter, the physical back ground, for a long-lived coherence in a system like FMO?

Professor Brumer answered: This question is in reference to the model that I introduced at the meeting, in which my postdoc Leonardo Pachon designed a simple spin-boson model that provides excellent agreement with long lived dynamics in dimers that model FMO and PC645 dynamics.

Amazingly, the model is in a regime where the relaxation rates, the decoherence rates, and the site dynamics are given analytically. Hence, a simple formula provides insight into the role of both the bath and system parameters that control the observed long lived decoherence. As to the question, parameters that are found to play a role include the exciton-exciton level spacing, the dimer coupling, the reorganization energy, and the phonon-bath cutoff frequency.

Professor de Vivie-Riedle commented: Can the electronic energy transfer between chromophores in a complex biological system like FMO be compared to the intramolecular energy transfer in a large polyatomic system, where donor and acceptor side are connected through a molecular bridge?

Professor Brumer replied: This is an excellent topic, and one which I expect to explore in the future. The stumbling block, however, is that the computational tools necessary to properly treat electronic energy transfer in a system of donor–bridge–acceptor (including decoherence by coupling to molecular vibrations) are cumbersome and computationally demanding.

As a start to this effort we have resolved the general problem of defining that operator whose average value gives the "electronic energy on a site". Surprisingly, this basic question was not previously addressed in the literature. It is certainly the first step toward the general computation implicit in your question.

Professor Miller asked: My understanding is that it is only in the limit the electronic coupling is significantly greater that the bath coupling that the excitonically coupled states are immune to decoherence from bath fluctuations. It seems Figure 3 in the paper bears this expectation out as the decay in the coherent population oscillations takes much longer to damp out with larger electronic couplings and smaller reorganization energies. It takes many more modulations for the bath (re: γ-1 parameter) to damp out the coherent oscillations. Can you define the boundary in relation to electronic coupling/reorganization energy (bath couplings) where the system passes over to the Förster picture of incoherent energy transfer in terms of typical system parameters for electronic coupling, and relaxation times for Markovian and non-Markovian limits of the bath interaction?

Professor Brumer responded: Carrying out such a correlation requires carrying out a number of numerical computations. Our method is computationally quite efficient, so this can be done, but we have not undertaken this study.

Professor Miller said: In Figure 2 of your paper (top panel for example), can you provide some physical insight into why the population decay ($\rho 11$) becomes faster with increasing reorganization energy, λ, for the non-Markovian treatment of an excitonically coupled system? The reorganization energy is related to the change in the charge distribution with the electronic excitation of the molecule and leads to localized relaxation. This process should lead to decoupling of the two states. I would have thought that the population decay would be longer lived on site 1 as the reorganization energy increases, which seems to be opposite of what you find for both Markovian and non-Markovian limits.

Professor Brumer answered: The equilibrium state here is equal population on both sites. What we are seeing here, in this parameter range, is an increased rate of relaxation to this equilibrium as the reorganization energy increases.

Professor Tannor remarked: The Ishizaki and Fleming paper referenced in your article seems to be a breakthrough. They derive a master equation that can deal with system-bath coupling of arbitrary strength[1] (not just second-order in the system-bath interaction). This allows using a master equation treatment for system-bath coupling parameters that are strong enough to accurately describe the physical situation in the photosynthetic reaction center. Recently, another approach to deriving a master equation was developed, based on the strong polaron transformation,[2] that is also valid in the regime of strong system-bath coupling (this method is also referenced in your paper). These methods didn't exist several years ago, but given this recent progress in handling the strong coupling regime, when modeling the photosynthetic reaction center is there any motivation for applying the second-order master equation?

1 A. Ishizaki and G. R. Fleming, *J. Chem. Phys.*, 2009, **130**, 234111.
2 S. Jang, Y.-C. Cheng, D. R. Reichman and J. D. Eaves, *J. Chem. Phys.*, 2008, **129**, 101104.

Professor Brumer answered: The master equation approach is, of course, applicable to a wide range of problems, with photosynthetic systems being just one of many. Given that the method that we introduce in this paper is far faster than those that you cite, it remains important to assess the range over which second order methods can be applied.

The results of this particular application, to the dimer system, make clear that higher order approaches are indeed required for typical reorganization energies in photosynthetic systems. In these cases the second order approach is clearly insufficient.

Professor Ohmori asked: To my understanding of the excitation transfer among LH2s in the photosynthetic system, this transfer of the excitation corresponds to the motion of the wavepacket which is the superposition of different k states of Frenkel exciton delocalized over the LH2s. These different k states with different eigenenergies are coherently superposed by a broadband femtosecond laser pulse in the laboratory experiments performed by Graham Fleming, Greg Scholes, and other people.[1,2] Although I admire these pioneering experiments, I wonder if this coherent superposition is really created with one photon from the sun that should be heavily decohered when it arrives on green plants on the earth. Do you believe that these laboratory experiments with femtosecond lasers really mimic the relevant biological systems under the sunlight? If not, what would be the most important motivation to investigate coherent features of such photosynthetic systems?

1 H. Lee, Y.-C. Cheng and G. R. Fleming, *Science*, 2007, **316**, 1462–65.
2 E. Collini, C. Y. Wong, K. E. Wilk, P. M. G. Curmi, P. Brumer and G. D. Scholes, *Nature*, 2010, **463**, 644–647.

Professor Brumer replied: There are significant differences between the coherence experiments that are done in the laboratory, and the situation under natural conditions. First, studies like those of FMO remove the FMO from its natural environment, where electronic excitation occurs by receiving energy from the baseplate (and not directly from solar excitation) and transferring the energy to a reaction center. This is quite different from coherent laboratory experiments where energy transfer down the FMO is studied as an isolated event. Second, even for natural processes that are activated by light, the incident light is, as you suggest, incoherent. By contrast, the laser radiation used experimentally is coherent. There is a significant difference. As I showed long ago[1] coherent light forms a quantum superposition state, whereas incoherent light prepares stationary states.

Recently, to examine the difference between these types of light sources in a biological system, we examined[2] the relevance of the measured 200 fs time scale for Photoizomerization, obtained in coherent excitation experiments, to the natural visual process. We found that this time scale was essentially irrelevant because the rate determining step was the (very low) natural incoherent photon flux.

Hence, much more work will have to be done, on a case-by-case basis, to establish the link between the results of coherent excitation experiments in the laboratory and processes that occur in vivo.

1 X-P. Jiang and P. Brumer, *J. Chem. Phys.*, 1991, **94**, 5833–5843.
2 K. Hoki and P. Brumer, *Procedia Chem.*, 22nd Solvay Meeting, 2011, in press.

Professor Engel responded: Laser excitation of an ensemble is clearly different from excitation by incoherent light. That is, after excitation by an ultrafast laser pulse, the ensemble is synchronized and in a coherent superposition of electronic states. The subsequent dynamics, however, are governed by the same Hamiltonian as in the biologically relevant environment. These experiments allow us to probe relaxation dynamics (both coherent and incoherent) that we would be unable to probe in other ways. In particular, these experiments are important to understand the underlying design principles in photosynthetic light harvesting.

Professor Brumer commented: This is indeed the case. However, one should keep in mind that even the nature of the decoherence and relaxation dynamics depends upon the initial state, which is different in coherent *vs.* incoherent excitation.

Professor de Vivie-Riedle opened the discussion of the paper by Professor van Hulst: In Fig. 4 of your paper the flipping of the relative phase between the two sub pulses from $\Phi=0$ to $\Phi=\pi$ leads to the expected change in sign, which shows very clearly in your data, but also to a significant difference in the amplitude of the relative fluorescence, which I would not expect. How can this difference in amplitude be explained?

Professor van Hulst replied: Indeed the excitation with in-phase and anti-phase delayed pulse does predominantly change the sign of the observed coherent oscillation, yet with deviations which are actually different from molecule to molecule. Here it should be realized that multiple vibrations are part of the dynamics. This gives a periodic response, however not a single frequency oscillation. Flipping the spectral shaper to anti-phase setting, changes the contribution of the frequency components in the excitation spectrum, which can cause more complex modifications than a simple inversion. In phase space, this means that keeping the delay constant while rotating the phase, (as shown in figure 5 in the paper) does not give a simple sinusoidal curve, and thus the amplitude is not expected to invert completely.

Professor Motzkus remarked: When you measured the fluorescence of the different single molecules with the shaped double pulse sequence the different transients are all shifted in phase in a irregular manner. What is the origin of this phase shift? Could you explain this in more detail? If you renormalize the transients with respect to frequency would they show all a cosine-like behavior?

Professor van Hulst answered: The distribution of phases reflects the environmental heterogeneity for the different molecules in the same sample. The subtly distinct conformation from molecule to molecule results in differences in the coupling strength of different vibrations to the electronic transition, and to the electronic dephasing rate of this system. The strength with which each of these vibrations gets probed depends indeed on the strength of the spectral component in the pulse addressing this energy level in the molecule. The observed variability in the transients is the result of the specific spectral shifts, exact coupling strengths, and the number of vibrational levels involved in quantum beating for each individual molecule studied. In that sense renormalizing the spectrum to a different shape (which we actually attempted) does not change the irregularity, but leads to another trace which is still specific for that molecule.

Professor Miller asked: I would argue that the change you see in the phase of the coherently driven fluorescence at the different single molecule sites can't be a simple change in the site energy with local nuclear configurations. This site specific effect would simply change the centre wavelength of the emission. The wavepacket motion depends on the potential energy surface and curvature. The change in site energy or separation between the ground and excited state surfaces would only give a constant offset. It would seem that both the position of the excited state surface relative to the Franck–Condon transitions and curvature are effected by the local environment. I would assume it is the excited state that is affected by environmental influences as it is most polarizable (standard argument) but the effect could also be related to small barriers between molecular conformations in the ground state. This is a pretty big molecule with numerous possible rotational conformations of the side groups. Can you expand further on this interesting observation?

Professor van Hulst answered: Indeed our observed variation in both phase response and oscillation period clearly indicates excited state potentials with both different site energy and curvature. The environmental heterogeneity is both spatial and temporal. Unfortunately some of the temporal conformational changes average out, when the polymer matrix changes on time scales faster than our experiment. Our experimental data could be expanded by simultaneously monitoring emission spectra, polarization or even fluorescence lifetime. Particularly in single molecule spectral diffusion experiments temporal shifts in central wavelength, *i.e.* changes in site energy, have been reported, which are related to conformational variations of molecular side groups and the local environment. Correlation of the spectral shape with our coherent fs excitation would give direct access to resolve changes in potential energy surface curvature. Similarly polarization data can give direct insight in fluctuations of the emission dipole moment orientation, while lifetime fluctuations are a good probe of local polymeric conformational dynamics. We plan such correlation experiments, but they are beyond the scope of the current paper.

Professor Kleiman asked: In your experiments, the same molecule is interrogated thousands of time over a period of seconds, and throughout that time there will be some slow nuclear motion that in principle could affect the molecules' spectrum. How come averaging of those slightly different initial conformations does not lower the contrast on the oscillations?

Professor van Hulst responded: The detection of single molecules based on their fluorescence of typically kcounts/s does require integration over ms or second timescale indeed. In our experiments the molecule of study is immobilized in a PMMA host. Working at room temperature and with a PMMA glass-transition temperature of 110 °C, only limited significant conformational changes occur on the timescale of the experiment, and the molecular wavepacket interferences can be observed with sufficient contrast. Occasionally jumps in fluorescence level or polarization are observed, which are clear indications of conformational changes; such molecular traces are interesting cases for further analysis. In the current paper results on molecules with stable fluorescence level and orientation, *i.e.* free of jumps, are presented.

Professor Weinacht addressed Professor van Hulst and Professor Motzkus: Initially I had thought that the shift in phase between molecules cannot correspond to a difference in their absorption spectra. However, now I understand that this can be the case. Perhaps it would be nice to address this point explicitly in the paper or with a comment (and perhaps reference to the work of N. Scherer for experiments in I2) since there were a few people in the audience for whom it was not immediately obvious. I think that one of the issues which is confusing is that there is a vibrational wave packet which is being driven in the excited state and the phase of this vibrational wave packet is separate from the optical phase for the carrier between the two pulses for which one gets optimal fluorescence. I guess that the issue is if one programs on a relative phase of $\cos(\omega\tau+\varphi)$ between the two optical pulses, but the central frequency ω doesn't match the molecule resonance frequency, then one will need a different phase, φ, at delay, τ, in order to get constructive interference as compared with hitting the molecule on resonance, for which the phase would simply be zero.

Professor van Hulst responded: Thank you for the comment, which captures the essence of the observed differences in phase response amongst individual molecules in the same sample. The intrinsic phase differences between the molecules can be attributed to changes in the coupling strength of different vibrations to the electronic transition, and to the electronic dephasing rate of this system. The choice of the experimental carrier phase difference between the delayed pulses in relation to the molecular phase dictates the effective excitation probability. In the frequency domain one can intuitively understand this as a (mis)match between the molecular resonance (absorption spectrum) and the excitation spectrum. Yet the information about the energy levels in the quantum system can equivalently described in the time domain (wave packet oscillations). In that sense the wavepacket interference model and the absorption spectrum model are equivalent in describing linear interactions between photons and molecules. The wavepacket model gets is added value in two cases: when nonlinear interactions start playing a role, and when the dynamics during interaction are of interest. For both research lines this is preparatory work.

Professor Tannor commented: As you indicated, each molecule has a different environment. This difference between individual molecules includes a difference in orientation relative to the polarization of the incoming electric field. Does this play any role in your experiments, and in particular could this account for the differences in magnitude of the coherent interference transients from different molecules?

Professor van Hulst answered: In this experiment the molecular orientation does not play a role since only one linear polarization was used for the full spectrum: this means that only the projection of the laser excitation polarization onto the molecular excitation dipole moment matters. As a result the total efficiency of interaction can be different between molecules depending on their specific orientation, but the relative interaction between different pulse shapes on the same molecule is not. For this to happen, either the different spectral components of the excitation pulse would need to have different polarizations, which is not the case, or the molecular excitation dipole would need to have different orientations for different frequencies, which we do not expect for this molecule.

Dr Offerhaus asked: I expect that you are using amplitude and phase shaping to generate the pulse-pair. If so, could it be that the modulation in the fluorescence is due to a change in the spectral intensity that falls within the window for stimulated emission?

Professor van Hulst replied: Present experiments are carried out at excitation intensity below 1 kW/cm^2, which is in the linear regime, far below saturation conditions. Thus the power level is not sufficient for stimulated emission and the presented vibrational wavepacket oscillation contrast is basically independent of the excitation intensity. It should be noted that we also performed experiments with a more narrow band higher power laser; in those experiments, published in ref. 36 in the paper,[1] we present coherent control of electronic transitions and the effect of stimulated emissions is clearly observed in the Rabi-oscillations of single molecules at room temperature.

1 R. Hildner, Daan Brinks and Niek F. van Hulst, *Nat. Phys.*, 2011, **7**, 172.

Professor Kosloff addressed Professor van Hulst: The spectral modulation that you see in your beautiful single molecule experiment can arise from both dynamics on the ground state as well as dynamics on the excited state. I expect that for these molecules the frequencies are similar. The difference can be traced to the phase of the modulation. Chirping the pump pulse may cause a phase shift which could be used to discriminate the two types of dynamics.[1]

1 E. Gershgoren, J. Vala, R. Kosloff, and S. Ruhman, *J. Phys. Chem. A*, 2001, **105**, 5081–5095.

Professor van Hulst responded: Yes exploring ground state vibrational dynamics at the level of single molecules is an important research goal. In our current experiments we operate in the linear regime, below saturation. We are relatively sure that all observed dynamics occurs in the excited state, because we do not exert the power for multiple Rabi-oscillations. As you suggest, chirping the excitation pulses is a very interesting approach, which we started to explored but still needs more systematic experiments.

Professor Kosloff addressed Professor van Hulst: The phase of the modulation of the signal in your pump probe experiment is a result of vibrational dynamics either on the excited state or on the ground state.

On the excited state the initial wavepaket will be positioned on a classical turning point therefore it will have zero initial phase. But an impulsive short pulse will generate a dynamical hole in the ground surface initial stationary wavefunction. This dynamical hole will move with a different phase depending on the position of the hole. Moreover by chirping the pulse this initial phase can be modified. momentum is generated.

It seems therefore that the signal of different individual molecules comes from a different combination of ground and excited state dynamics, see for example ref. 1, 2.

1 U. Banin, A. Bartana, S. Ruhman and R. Kosloff, *J. Chem. Phys.*, 1994, **101**, 8461–8481.
2 E. Gershgoren, J. Vala, R. Kosloff, and S. Ruhman, *J. Phys. Chem. A*, 2011, **105**, 5081–5095.

Professor van Hulst replied: This is also related to the previous questions by Professors Motzkus, Miller and Weinacht. For the large molecule studied several different molecular vibrations are excited. Thus the broad band delayed excitation pulses, with certain carrier phase difference setting, probe the effective response of the excited state wavepacket, involving different vibrations with their respective coupling strengths. As a result the observed phase is different from molecule to molecule due to the environmental heterogeneity. The suggested role of ground state dynamics is very interesting, however with excitation in the low power linear regime, far below saturation, we are confident to probe primarily excited state dynamics.

Professor Rabitz asked: Did you try chirping the pulses?

Professor van Hulst answered: So far we have concentrated on identical delayed pulses with flat spectral phase, as the observed heterogeneity of molecular responses is already quite extensive. Clearly controlled chirp would give us access to the shape of the excited state potential. So far we have not tried that systematically.

Mr Lane asked Professor van Hulst: To obtain phase control of single molecule wave packets trains of phase-locked pulses were used. Was a relationship seen between the magnitude of enhancement of fluorescence excitation and the number of pulses in the sequence? Would eight-pulse sequences be possible before decoherence occurs?

Professor van Hulst responded: In our experiments, with trains of up to 4 pulses the "control ratio" became markedly better, however for more pulses the effects where minimal. Basically all depends on the number of pulses that can be used for excitation before total dephasing occurs. A good rule of thumb is given by the FWHM of the pulses, compared to the dephasing time. In our case, with 12 fs pulses and a dephasing time of about 50 fs, the observed optimum of 4 pulses fits nicely.

Professor Weinacht opened the discussion of the paper by Professor Kleiman: It is interesting that the mechanism you propose is opposite to what has been discussed in some recent papers by Herschel Rabitz and coauthors. In their work on discrimination between similar molecules (PRL and JCP), they see a decrease in fluorescence when they pump the molecule up from the initial S1/S2 state. It seems a little counter-intuitive to think of additional pumping up leading to increased fluorescence, but I suppose this depends on the details of the potentials. Are there any calculations or additional measurements outside of the control and pump probe measurements that might test this hypothesis regarding the potentials?

Professor Kleiman responded: The papers mentioned, by Rabitz and co-workers,[1] measured a depletion ratio induced by autoionization states at the accessed energies. Indeed, those experiments include an initial excitation photon of 400 nm follow by two additional 800 nm photons. This brings the system >6 ev above the ground state. As the authors mention, these energies are consistent with ionization processes that induced the depletion of the $S-_n$ population. There are also major differences in the experimental conditions. Whereas their pump beam (400 nm) has similar fluences as the ones we use (800 nm), we are accessing the first state by two-photon absorption, thus the initial population transfer is much smaller. More importantly, the follow up interaction in our experiment arises from the same excitation pulse (as it corresponds to a low-intensity component of the shaped pulse) and a subsequent two-photon absorption is unlikely. In the work by Rabitz *et al.*,[1] the second pulse is independent, it is TL and has fluences 10x larger, making it very likely that multiphoton absorption (at least 2, as they propose) will induce depletion through ionization at higher energies. In these dendritic molecules, linear absorption in the 200–300 nm region (corresponding to the ~4.5 ev) does not show signs of ionization. While in one experiment the conditions are such to induce a non reversible loss of S_n population, in our experiments, condition are milder and no irreversible process occurs. Confirmation of this last statement are the results from two color transient absorption data using

400 nm + 800 nm. They show the full recovery of the ground state population, corroborating the reversibility of the induced excitations.

1 J. Roslund, M. Roth, L. Guyon, V. Boutou, F. Courvoisier, J.-P. Wolf and H. Rabitz, *J. Chem. Phys.*, 2011, **134**, 034511.

Professor Rabitz asked: Did you do the optimization experiments at different time delays?

Professor Kleiman responded: In a very similar system (Donor–acceptor) we performed an experiment where we generated two pulses, with either 0 or π phase difference between them. We normalized the quantum yield to the valued obtained when pulses overlap ($\Delta t=0$). As the delay between these pulses was varied, we observed that for the 0 phase, there is a maximization of the quantum yield at ∼70 fs (which coincides with the subpulse delay time obtained in the optimization experiments) while for the other conditions (pulse difference of π), the quantum yield was always lower than the value obtained at $\Delta t=0$. (See supplemental information of ref. 1).

1 D. G. Kuroda, C. P. Singh, Z. Peng and V. D. Kleiman, *Science*, 2009, **326**, 263.

Professor Whitaker addressed Professor Kleiman and Professor Baumert: In your paper you conclude that the mechanisms responsible for the control of the emission efficiency in the conjugated dendrimer and rhodamine 6G are different. In the former you propose a 2+1 mechanism (which you illustrate in Fig. 5 of you paper) but the mechanism acting in R6G is perhaps harder to understand. Here you propose that the phase modulation of the excitation pulse leads to better or worse spectral matching of the red and blue components in the pulse with the two photon transition probability. This is certainly possible and quite likely the right explanation but I wonder if you have also considered other possibilities; in particular the possibility of a non-linear interaction with the solvent rather than the solute? The reason I ask is because a few years ago we performed an experiment, rather ill conceived as it turns out, in which we excited a dilute solution of R6G with a phase only shaped pulse centred at 600 nm.[1] The pulse shaper used in these experiments cut the spectrum of the pulses sharply at 575 and 625 nm, so the short wavelength limit of the excitation spectrum lay well to the red of the one photon absorption band of the molecule, where the molar extinction coefficient falls below 10^{-2} above 570 nm. We were surprised therefore to observe a strong fluorescence signal (whose intensity we could control by modulating the spectral phase). The (mundane) explanation was that the laser pulse was generating a weak continuum spectrum in the solvent with wavelength components capable of exciting the molecule. The estimated laser fluence in our experiments (3.6×10^{13} W cm^{-2}) was a hundred times that reported in your paper but, nonetheless, I worry about possible non-linear effects induced in the solvent in these types of experiment; which brings me to address a supplementary question to Professor Baumert, where in his excellent introductory lecture he talked about strong field control experiments in solution. How is this possible without simultaneously inducing undesirable effects in the solvent?

1 A. Barman, N. T. Form and B. J. Whitaker, *Chemical Physics Letters*, 2006, **427**, 317.

Professor Baumert answered: In our experiment[1] we demonstrated efficient and robust population transfer in two photosensitizer dyes in solution being prepared in the triplet ground state. We investigated a one photon resonant process, where the measured peak intensity for the bandwidth limited pulse was 60 GW/cm^{-2}. Strong fields in our case means, that we are in the regime of Rabi cycling. Taking into account that dye molecules provide oscillator strengths in the range of unity, the pulse area reaches 8 pi. The damage threshold in nonlinear microscopy is given as 200 GW/cm^{-2}.[2] For the shaped laser pulses (GDD *versus* TOD) the peak intensity was correspondingly lower. So due to the large oscillator strength our strong fields have a much lower peak intensity as mentioned in your question. In addition we compared our experimental results with simulations on generic molecular potentials by solving the time-dependent Schrödinger equation for excitation with shaped pulses. Control landscapes with respect to population transfer confirmed the general trends from the experiments. An analysis of regions with maximum or minimum population transfer indicated that coherent processes were responsible for the outcome of our excitation process. The physical mechanisms of joint motion of ground and excited state wave packets or population of a vibrational eigenstate in the excited state permitted us to discuss the molecular dynamics in an atom-like strong field interaction picture.

1 J. Schneider, M. Wollenhaupt, A. Winzenburg, T. Bayer, J. Köhler, R. Faust, and T. Baumert, *Phys. Chem. Chem. Phys.*, 2011, **13**, 8733–8746.
2 K. E. Sheetz and J. Squier, *J. Appl. Phys.*, 2009, **105**, 051101.

Professor Kleiman replied: Under certain conditions, you can observe white light generation from the cuvette or Raman scattering from the solvent. The experiments in this paper were performed under conditions where non of those side effects were observed. We tested cuvettes with pure solvent and observe no emission. More importantly, the same modulated pulses, applied to a solution of similar concentration ethynylene perylene did not show the phase sensitive asymmetric increase in quantum yield. It was possible to optimize the excitation process (by using a TL pulse), but the quantum yield simply follows the curve of a two-photon absorption process.

Professor Miller asked: The peak power of the transform limited pulses was 700 GW/cm^2. We have done similar experiments at intensities below 100 GW/cm^2 and see clear evidence for multiphoton ionization of just the solvent.[1] Did you measure the transmitted excitation and power dependence to see how much of the excitation may be absorbed through higher than a 2-photon process? The reason I am asking is that it is generally difficult to selectively have 2-photon absorption for large molecules. Once you have significant coherence between the ground and excited state through a non-resonant 2-photon process, there is generally a resonant 1-photon process of much higher cross section that leads to higher order multiphoton processes and ionization. This problem limits the application of high intensity or strong field control of molecular systems, especially in solution phase where there are broad linewidths and no clear gaps in resonantly driven multiphoton processes. This situation is very different from atoms or small molecules in the gas phase where there is a much sparser density of optically active transitions and much smaller possibility of resonant multiphoton absorption. This multiphoton absorption channel may be affecting the fluorescence of the dendrimer re: pulse width and peak power effects. Here the effect is similar to your proposal of an alternative excited state pathway to fluorescence but involves another photoproduct that does not fluoresce that gives rise to the observed phase dependence.

1 V. I. Prokhorenko, A. Halpin, P. J. M. Johnson, L. S. Brown and R. J. D. Miller, *J. Chem. Phys.*, 2011, **134**, 085105.

Professor Kleiman answered: We have performed several "control" experiments to rule out further absorption of photons and spurious mechanisms like ionization (either of the solvent or the molecule under investigation). We measured the power dependence of the emission over a range of laser intensities and obtained 2nd order dependence. The intensities used in the experiments (same beam diameter) were chosen on the lower end of those used for the power dependence studies. Under the experimental conditions, measurements performed with pure solvent did not show any photoluminescence and moreover, experiments performed with ethynylene-perylene did not show an increase in quantum yield after excitation with equally modulated pulses. Transient absorption experiments (400 nm + 800 nm) show full recovery of the ground state population in accordance with the lifetime of the emission. All these control experiments led us to trust that the total population of the initial excited state is a consequence of the 2-hv absorption process and that the additional 3rd-photon interaction does not induce any ionization of the molecular system.

Dr Amitay remarked: Could you please clarify what is the spectral position of the variable phase step for which the results of Fig. 3 in the paper correspond to? What is the time delay and relative phase between the two sub-pulses composing the corresponding shaped pulse with a 1.4π spectral phase step, which enhances the emission efficiency of the 2G2-m-OH molecule as presented in Fig. 3 in the paper? Are both these values of the time delay and relative phase indeed consistent with the pump–probe emission results presented for the 2G2-m-OH molecule in Fig. 4 in the paper (right panel)?

Also, could you please explain your interpretation for the asymmetric structure that the results presented in Fig. 3 in the paper for the 2G2-m-OH molecule (blue line) have around a spectral phase step-size of π?

Professor Kleiman answered: The applied spectral step function is at 800 nm (the center wavelength of the fundamental excitation pulse). This spectral phase leads to subpulses separated by c.a.70 fs. This delay time is consistent with the delay time (also $\sim$70 fs) between two pulses (with 0 phase difference) from which the maximum emission is obtained in Figure 4 in the paper. The curves in Figure 3 in the paper are built from measuring the molecular response as a function of the size of the step spectral phase, ratioed to the response from a GaAsP photodiode. For the perylene sample, the response obtained from measuring the fluorescence (following 2-photon excitation) is similar to the response obtained from the photodiode and thus the curve is symmetric around π (the small shift is due to a third order phase component). In the case of 2G2-mOH, the molecular response is quite shifted from the 2-photon photodiode and thus the ratio shows the asymmetric function. This asymmetry is a consequence of the molecular response beyond the absorption process.

Professor Weinacht addressed Dr Amitay and Professor Kleiman: In response to the question from Dr Amitay regarding the asymmetry of Figure 3 in the paper, I just point out that the x axis of the graph is not the location of the Π phase flip but rather the depth of the phase jump.

Professor Kleiman responded: Indeed, the parameter we varied in this experiment is the is the size of the step function, not its position. This translates on a change in relative intensity of the two sub-pulses created by the spectral step function. The optimal pulse has a high-intensity earlier subpulse (to create excited state) and a lower-intensity later subpulse to induce further excitation.

Professor de Vivie-Riedle opened the discussion of the paper by Professor Baumert with a question for all: In theoretical optimizations you will always reach high efficiencies, provided the system is controllable. The easiest control mechanisms are achieved when resonances available in the molecule are used. Comparable high quantum yield are not reported in the experiment. Are the idealized conditions used in theory like optimal preorientation, no laser fluctuation, no coupling to the environment the reason for these large differences?

Professor Kosloff responded: Our calculations confirm that using optimal control theory we are able to obtain very high yields close to 100%. Almost all calculations performed start with a pure initial state. Thermal averaging is the main reason for low yields in experiment. To reach high yields a first step is to purify the thermal ensemble. Experiments on cold matter or that used an initial purification step show high yields of control.

Professor Miller addressed Professor Brumer: This comment is in response to Professor Brumer's suggestion during discussion that *a priori* theoretical calculations of control pulses is more informative than closed loop searches of experimental parameters. I would agree if one approach is clearly superior in generating accurate potential energy surfaces. I appreciate that there are two approaches to rational coherent control of molecular dynamics. There is the open loop control in which pulses are crafted based on theoretical analysis of the excited state dynamics convolved to a potential energy surface and closed loop control in which experimentally the key parameters are found typically using genetic learning algorithms. The process involves an exploration of a considerable amount of parameter space to see clear effects in phase or phase/amplitude dependencies. In both cases, it is exceedingly difficult to invert the discovered optimal and anti-optimal control pulses using closed loop methods, or to refine the assumed potential energy surface in open loop control. The problem is that both the creation of the excited state and the subsequent observable are both connected to the expectation value of the dipole operator (or other transitions). There is simply not enough information to invert the pulse shapes to potential energy surfaces or refine the assumed potential energy surface in either case. The only exception is when one can define a well defined reference state to effectively heterodyne detect the phase cycled pulses to map out an excited state potential as is discussed by Professor Tannor. There are other approximations one can make to help in effectively phasing the signal as Professor Shapiro states. The problem is that these approaches only work for small molecules and it could be argued we know the surfaces for such systems reasonably well within current limitations to ab initio methods. The major challenge is to be able invert the coherent control pulses to construct the wavefunction interferences along reaction pathways for molecular systems that form a basis for typical chemical processes. This will be equivalent to seeing quantum mechanics in action. We need another approach that does not use the same operator to both prepare and observe the quantum state dynamics. In this context, I would like to point out that there has been tremendous progress in the development of "ultrabright" electron and X-ray sources that now provide direct observation of the atomic motions with time resolution approaching 10 femtoseconds so that even the fastest nuclear motions are now within reach. Here I am referring to high bunch charge femtosecond electron sources and X-ray Free Electron Lasers or 4th Generation Light Sources.[1] One of the next major frontiers will the combination of coherent control methodology with these new structural probes to directly observe quantum interference effects. We will be able to observe the time dependent velocities of the various atoms and from this directly construct potential energy surfaces. We will effectively have exactly what we would like to retrieve from any potential energy surface, a means to visualize the forces on the atoms driving the chemistry. With this approach, we will be able to see it directly. This approach will solve the inversion problem.

I would like to add that the Coherent Control field is going to have a very significant impact in the future applications of these ultrabright electron and X-ray sources. There are only a small number of problems in which it is possible to optically prepare the system under barrierless conditions to induce structural changes with high quantum efficiency. Most photochemical processes involve a barrier in which case all the correlated atomic motions that direct the system into the product channel will be lost through the rate limited diffusive sampling of the reaction barrier. The impact of these new sources in terms of understanding chemistry and directly observing mode coupling along reaction coordinates will be very limited. The excitement in this emerging field will be very short lived if we can not expand the base of problems that can be studied. Here Coherent Control methods have a very important role to play to open up new methods of probing reaction dynamics and greatly extending the applications of these new structural probes. Here I am making a case to revisit strong field control along ground state potential energy surfaces. I think the IR source technology is now sufficient to drive the true reaction modes and thereby selectively drive chemistry. This of course has been a dream for some time. We are getting closer and now have new tools to renew the hunt.

1 G. Sciaini and R. J. D. Miller, *Rep. Prog. Phys.*, 2011, **74**, 096101.

Professor Brumer responded: To be precise, I did not suggest that *a priori* theoretical calculations of control pulses are more informative than closed loop searches of experimental parameters. Rather, I commented that closed loop searches often yield complex electric fields whose physical content is unclear, and that a detailed theoretical analysis of such experiments is necessary if we are to learn something about the underlying physics. Indeed, there are very few detailed analyses of adaptive feedback experiments, and two examples make clear the importance of such work. In both "Mechanisms for the Control of Two-Mode Transient Stimulated Raman Scattering in Liquids",[1] and "Mechanisms in Adaptive Feedback Control: Photoisomerization in Liquid",[2] we demonstrated that published proposals to explain the experimental results based on a coherent control scenario were incorrect, and that a non-coherent control explanation fit the data well. These serve as examples of the importance of such an analysis to extract the proper physics from experiment.

1 M. Spanner and P. Brumer, *Phys. Rev. A.*, 2006, **73**, 023809.
2 K. Hoki and P. Brumer, *Phys. Rev. Lett.*, 2005, **95**, 168305.

Professor Rabitz asked: Discerning quantum control mechanism is an important objective and a major challenge in this regard is to determine mechanism without detailed knowledge of the Hamiltonian or an ability to solve Schrödinger's equation. Can the high duty cycle of laser-probe experiments be used in this fashion?

Professor Miller answered: The use of closed loop control as you have introduced is a very important experimental approach. The challenge is always to connect the pulse shape to the mechanism. I don't think the duty cycle of the laser-probe experiment is the limitation in making this connection but one could argue we are restricting our parameter search due to time constraints in collecting data. I think the major advance will be made using femtosecond structural probes (high brightness X-ray or electron probes) to directly observe the atomic motions and thereby discern the control mechanism, rather than rely on spectral changes that are not uniquely related to molecular structure.

Professor Tannor asked Professor Miller: The recent progress on real time X-ray imaging of molecular dynamics is indeed very promising. But I would not be so fast to dismiss optical spectroscopies to reconstruct potential energy surfaces; in fact, this is the subject of my contribution to this meeting. You are fully correct, that inverting to the potential from spectroscopy, even if there is information from the microwave to the X-ray, is non-unique. The difficulty comes from the fact that the energy spectrum is essentially a 1-dimensional set of data and does not provide a unique mapping for the potential surfaces of a polyatomic with 3N-6 spatial degrees of freedom. In our approach we assume that the ground state potential is known in all 3N-6 coordinates, in which case the excited state potential can be uniquely inverted from resonant CARS spectroscopy (assuming there is only one excited state potential that is resonant). I believe this property of uniqueness applies also to the inversion method developed recently by Moshe Shapiro and coworkers.[1,2]

1 X. Li, C. Menzel-Jones, D. Avisar and M.Shapiro, *Phys. Chem. Chem. Phys.*, 2010, **12**, 15760.
2 X. Li and M. Shapiro, *J. Chem. Phys.*, 2011, **134**, 094113.

Professor Miller replied: I agree and did not mean to imply there was no solution to inverting optimal pulse shapes to real space wavepacket motion along the excited state surface. Your approach as well as that of Moshe Shapiro's (as stated above) provide theoretical methods to invert pulse shapes. The question is how large a system can be done with these methods. From a practical standpoint, these methods are currently limited to small isolated molecular systems. We need to be able to scale to more complex systems and be able to include bath interactions to probe issues relevant to most chemistry of interest. From an experimental standpoint, we need to go beyond using the expectation value of the dipole operator to both prepare and observe the state evolution and then try to invert the shaped pulses to excited state potentials. There is not enough information. One needs some form of well known reference to invert as you outlined in your paper and here lies the problem with respect to scaling to larger systems. It is a classic problem in solving the phasing issue. Here I will restate my above comment to Professor Brumer (and my hopes). The latest advances in both X-ray and electron structural probes will allow us to view the atomic motions (forces) directly. In my opinion, the prospect of using direct structural probes is exciting as this experimental approach would be general and would open up the study of even complex molecular systems. The problem right now is that the X-ray sources are far from table top and it will difficult to get sufficient beam time to conduct this class of experiments. The new electron sources are table top and I hold out hope that this approach can be used to unravel the mechanism of coherent control in complex systems. The main point I would like to make is that the field of Coherent Control can have a significant impact in this emerging field of atomically resolved structural dynamics. There are very few systems that can be optically prepared under barrierless conditions, as required to enable direct observation of the reaction forces. Most processes are rate limited by barriers and all the information on the relative atomic motions is lost. Coherent Control protocols, especially Strong Field Control in the IR, could open up a much broader class of problems for the observation of the light-matter interaction at the atomic level of inspection.

Extracting dynamics of excitonic coherences in congested spectra of photosynthetic light harvesting antenna complexes

Justin R. Caram and Gregory S. Engel*

Received 30th March 2011, Accepted 27th April 2011
DOI: 10.1039/c1fd00049g

We present an analysis of dephasing rates for multiple zero-quantum electronic coherences in the Fenna–Matthew–Olson (FMO) pigment–protein complex using two-dimensional electronic spectroscopy. We employ the linear prediction Z-transform to determine both the frequency and decay rates of 8 individually assigned exciton–exciton coherences. Despite congestion in the spectra, we can isolate multiple crosspeaks signals and analyze their dephasing rates. A non-trivial relationship exists between the excitons and the bath determining the lifetimes of different exciton–exciton coherences. We propose that the correlations that protect long-lived electronic coherence may yield microscopic knowledge regarding the structure of the protein bath surrounding the chromophores.

Introduction

Photosynthetic organisms convert sunlight into chemical energy. Typically, a photon is absorbed by a large network of chlorophylls or bacteriochlorophylls; this energy is then transferred toward a reaction center where charge separation occurs. Interestingly, the energy transfer to the reaction center proceeds with near unity quantum efficiency.[1] The Fenna–Mathews–Olson pigment–protein complex (FMO) functions as an excitonic conductor between the large peripheral chlorosome antenna and the reaction center in *Chlorobium tepidum*, a photosynthetic green sulfur bacteria.[2] FMO transmits energy through a network of 7 coupled bacteriochlorophyll *a* chromophores (BChl *a*) that collectively form 7 distinct excitonic states. These excitons are arranged such that the highest energy (7) is oriented toward the chlorosome, and the lowest (1) points toward the reaction center, facilitating unidirectional transfer toward the reaction center.[3]

FMO has a number of favorable properties as a target for nonlinear electronic spectroscopy. Its absorption spectrum spans the typical output of a regeneratively amplified Ti:Sapphire laser system (780–820 nm). The protein is water soluble decreasing scattering associated with aggregation which aids acquisition of non-linear signals. Furthermore, FMO is asymmetric, leading to 7 non-degenerate excitonic states, all of which are all optically accessible from the ground state. Also, the crystal structure of the protein pigment complex is known to a high degree of accuracy.[4] Because of these properties, FMO has been extensively studied as a model for photosynthetic electronic energy transfer (EET).[5–12]

Electronic energy transfer occurs on 200 fs–10 ps time scale in light harvesting complexes.[1] For many years, photosynthetic energy transfer was commonly treated in two regimes. If the electronic coupling between two chromophores was small in

Department of Chemistry and The James Franck Institute, The University of Chicago, 929 East 57th Street, Chicago, IL, 60637, USA. E-mail: gsengel@uchicago.edu

comparison to electron–phonon coupling, energy transfer was considered to be strictly incoherent and could be properly modeled using Forster Resonance Energy Transfer (FRET).[13] If the coupling was large, excitation was delocalized among chromophores forming excitonic states, and energy transfer could be modeled with the Redfield equation.[14] However, recent experiments show that electronic coherences between excitons (zero-quantum coherences) persist for much longer than expected, at least 660 fs.[15] Later work showed that at 77 K, that the coherence or superposition of exciton 1 and 2 has a lifetime over 1 ps and can be unambiguously observed at 2 ps.[16] Interestingly, this dephasing is slow relative to the fluctuation of the electronic energy gap (the dephasing of the 1-quantum coherence) which occurs with a lifetime of under 100 fs at 77 K (15 fs at 277 K). This observation appears general, extending both to physiological temperatures[16-18] and to many different systems including phycobillin proteins,[18] LHC II,[19] and reaction centers.[14] Because zero-quantum coherences persist beyond the time scale for energy transfer, the coherences must be relevant to energy transfer. The involvement of coherence in the energy transfer process indicates an intermediate EET regime, where the coupling energy between excitons is on the same order as the coupling to the phonon bath.

From these observations, several theories have emerged to show how quantum coherence can contribute to the remarkable quantum efficiency of energy transfer in this intermediate regime. Ishizaki and Fleming showed that by taking advantage of quantum coherence, the excitation can sample the energy landscape, without getting trapped in a low lying energetic state.[7] Cheng et al. have suggested that while FRET transfer leads to exponential increases and decreases in site population reaching an equilibrium value, coherence within the system can result in population amplitudes above those governed by equilibrium statistics, leading to highly efficient energy transfer.[10] Caruso and Rebentrost have shown that the correct ratio between thermal dephasing and coherence can actually enhance quantum efficiency.[6,11] At room temperature the coherence-dephasing rate appears optimized such that energy is quickly and efficiently transferred to the reaction center.[12] This prediction coincides with the experimental results near room temperature showing that FMO shows excitonic coherences persisting for at about 1.5 oscillations before dephasing.[16]

Quantum coherence is thought to be a delicate and fragile phenomenon. For it to persist in such a soft, wet environment, some correlation must exist between the ground-excited state transition energies of two different excitons.[14] Interestingly, the correlation mechanism appears to be quite general. Recent experiments demonstrate that coherences persist even when the phytyll tails in the BChl are modified, and when the protein is randomly deuterated.[20] Collini and Scholes have shown long-lived quantum coherence in a conjugated polymer.[21] A common feature of all these systems is some kind of semi-rigid scaffold which correlates the energy levels of two excitons on the short time scales associated with energy transfer. The nature of this correlation is one of the key unanswered questions in photosynthetic energy transfer.

Lee et al.[14] prepared a quasi two state system by reducing the primary electron donor of the reaction center, leaving a Bchl state (B), and a bacteriopheophytin (H). Long lived coherence between two states was then observed. To fit this data, they examine the B–H coherence and fit the dephasing rate with a simulation that invokes a cross-correlation term representing the extent to which nuclear modes in the bath modulate transition energies of both states. In their best simulation they observe a cross correlation constant of 0.9, indicating that chromophores embedded in the same protein matrix feel nearly the same short time Gaussian fluctuations.

Using the relaxation superoperator formalism, an expression for the dephasing rate of an electronic coherence expanded in terms of the second order perturbation in the system bath interaction Hamiltonian can be expressed as follows[22]

$$\Gamma_{ab} = -2\sum_{\alpha,\beta}(\langle B\beta|H_{s-b}|B\alpha\rangle - \langle A\beta|H_{s-b}|A\alpha\rangle)^2\,\delta(\varepsilon_\alpha - \varepsilon_\beta). \quad (1)$$

where |A> and |B> are pure states with no overlap in energy, |α> and |β> are bath eigenstates with energies ε_α and ε_β. The first (squared) term represents the difference between the energetic overlap between each exciton and the bath. When this difference term is large, fast dephasing occurs. However, if each electronic/excitonic state projects onto the same bath motions, the dephasing rate is slower because the difference is smaller. Thus the dephasing rate can be used as a proxy for measuring the correlation between two excitonic energy levels. However, for a seven exciton system, the situation is much more complicated, and microscopic insight might be gained by understanding whether fluctuations are correlated among all seven excitons or whether different excitonic coherences dephase differently. Here we present evidence that shows different coherences within the one-exciton manifold dephase at different rates. These data indicate that the correlation is not spatially homogenous in a complex system such as FMO.

Two-dimensional electronic spectroscopy of photosynthetic systems

Two dimensional Electronic Spectroscopy (2DES) can uniquely probe a variety of phenomena in multi-chromophoric systems. A four wave mixing experiment, the technique directly probes the correlation function between states of different chromophores.[23,24] This information appears in a 2D electronic spectrum as a crosspeak. The x-axis (the ω_τ-axis) effectively depicts the frequency of the excitation while the y-axis (ω_t) shows the frequency of the excitation some time, T, later. These correlation maps provide optimal specificity and permit 2DES to resolve specific exciton-exciton correlations while simultaneously maintaining precision in the time domain.[24]

Briefly, the 2D correlation spectrum arises from third order polarization as follows:[25]

$$S_{2D}(\omega_\tau, T, \omega_t) = \int_{-\infty}^{\infty} iP^{(3)}(\tau, T, \omega)\exp(i\omega_\tau\tau)d\tau \quad (2)$$

Three pulses ($\mathbf{k}_1, \mathbf{k}_2, \mathbf{k}_3$), separated in time ($\tau$ between $\mathbf{k}_1$ and $\mathbf{k}_2$ and T between $\mathbf{k}_2$ and $\mathbf{k}_3$) and arranged in a box geometry, generate a third order polarization in the sample. This polarization is heterodyned with a non-interacting fourth pulse and frequency resolved on a camera. Pulse 1 and 2 are stepped in time (τ), while pulse 3 is fixed to obtain a single 2D spectrum for different waiting times T. Using double-sided Feynman diagrams, we relate elements of the density matrix to the particular perturbation applied to the system by the three pulses.[26] By selecting only those phase matched under the following condition ($\mathbf{k}_s = -\mathbf{k}_1 + \mathbf{k}_2 + \mathbf{k}_3$) and enforcing that $\mathbf{k}_3$ is final interaction, we reduce the number of response pathways from 48 to 6. We can further limit number of terms to only 3 by only enforcing the time ordering between pulse 1 and pulse 2 (positive coherence time, τ). These diagrams, shown in Fig. 1, correspond to rephasing pathways.[10] By frequency and phase resolving that photon echo signal, we directly measure which frequencies in the system are correlated at later times to other frequencies in the system.

As Fig. 1 demonstrates, if $a = b$ in pathway 1, the spectral feature appears on the main diagonal as a stimulated emission signal. If $a \neq b$, then it appears as a crosspeak, which oscillates at a frequency $\exp(-i(E_a - E_b)T/\hbar)$. Pathway 2 corresponds to a ground state bleach signal, which will appear on the main diagonal, and pathway 3 is an excited state absorption pathway, which appears as a crosspeak red-shifted by the exciton–exciton binding energy.[27] In pathway 3, if $a \neq b$, the crosspeak oscillates.

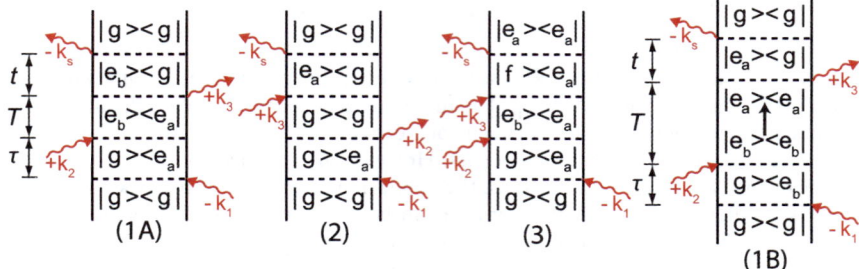

Fig. 1 Four third-order rephasing response pathways represent signals that come out in $k_s = -k_1 + k_2 + k_3$. Time increases from bottom to top and is divided into three periods: coherence time, waiting time and rephasing time or τ, T, and t respectively. These Feynman diagrams show electric field interactions with the bra and ket of a multilevel system. If $a = b$, the feature is static during waiting time T, and appears on the diagonal. If $a \neq b$, it beats with a frequency of $\exp(-i(E_a - E_b)T/\hbar)$, on a crosspeak. 1A and 1B represent stimulated emission pathways, and a positive signal. In 1B population transfer has occurred during T. Pathway 2 represents a ground state bleach signal, on the diagonal, and pathway 3 represents an excited state absorption signal, which is negative.

The oscillation on a crosspeak as we vary waiting time (T) represents coherence between two different excitonic states. Likewise, as shown in pathway 1B, an increase in the crosspeak signal can also indicate incoherent energy transfer. For this reason, several groups[18,28] have fit this signal sequentially, first to two independent exponential decays and an offset (associated with incoherent dynamics like pathway 1A), and then to the following function:

$$\sum_{\alpha=1}^{N} A_\alpha \exp(-k_\alpha T)\sin(\omega_\alpha T + \phi_\alpha) \qquad (3)$$

where N is the number of crosspeaks believed to be associated with a region, k is the decay rate, ω is the frequency, Φ is the phase. In congested spectra, cross-peaks are often incompletely separated in the 2D spectrum, leading to multiple beating signatures. With this approach, a recent study fit the beating signature for a 1–2 crosspeak with $N = 2$, seeing a clearly improved fit.[28] Interestingly, this analysis reveals statistically significant differences in the dephasing attributed to the 1–2 and 1–3 coherences. However, problems arise from applying a least square fitting procedure to a specific function, especially in regions of the 2D spectra that are poorly resolved. In the FMO spectrum, there can be as many as 21 different beating frequencies, each with 4 fitting parameters. With truncated data sets of fewer than 100 time points, the high dimensionality of the fitting space increases the likelihood of finding a false minimum. At the same time, recent work has shown that the long lived coherence in the FMO complex can be exploited to resolve even more crosspeaks than those evident in the 2D spectrum displayed in Fig. 2.[29] By Fourier transforming over waiting time, Hayes and Engel were able to isolate oscillations that beat with specified difference frequency on the 2D spectrum.[29] Here, we simultaneously extract decay rates and frequencies from those newly resolved cross peaks using a linear prediction based analysis instead of least squares fitting. This Z-transform method improves frequency resolution while allowing simultaneous extraction of decay rates thereby unambiguously assigning decay rates to crosspeaks. These dephasing rates provide clues regarding the mechanism for protection of long-lived coherences by the protein.

Methods

FMO was isolated from *Chlorobium tepidum* as described previously.[4] Samples were concentrated to OD $\sim$50 cm^{-1} at 809nm in 800mM tris/HCl buffer (pH 8.0) and then

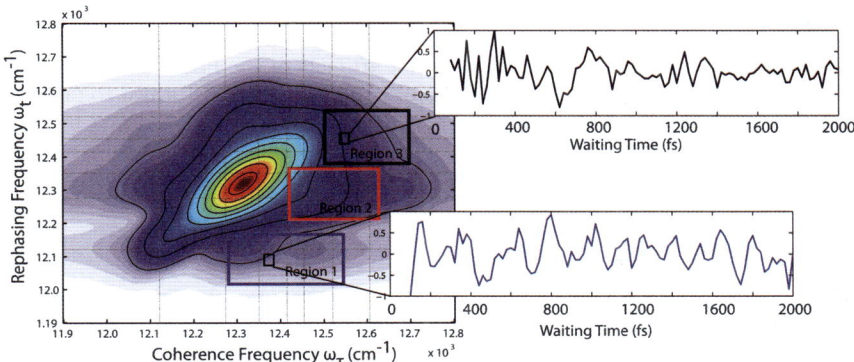

Fig. 2 Left: a representative absolute value rephasing 2D spectrum is shown at $T = 400$fs. Visible below the diagonal are two crosspeaks, arising from coupling and energy transfer. Vertical and horizontal lines represent the excitonic Hamiltonian energies determined previously.[24] The three regions where time series were extracted and Z-transforms were performed are labeled as region 1 (representing primarily coherences with exciton 1 and 2), region 2 (exciton 2 and 3) and region 3 (excitons 3 and 4). This analysis was conducted with crosspeaks below the diagonal, but beating signals are also observed above the diagonal. The boxes represent a sample region where a time series was extracted. Right: Representative time series from region 1 and 3 showing 80 fs–2 ps in steps of 20 fs, after fitting and subtracting a biexponential decay representing population transfer. Signals show complex beating structure and different decay dynamics in each region.

mixed with lauryldimethylamine oxide detergent (0.1%). Sixty μl of a mixture with 35 : 65 (v/v) sample : glycerol were put into a 200 μm quartz cuvette and cooled to 77k in a liquid nitrogen cryostat (Optistat DN, Oxford Instruments). The final OD was measured to be .36.

The experimental details of two-dimensional electronic spectroscopy have been described in detail elsewhere.[25,30] Briefly, the output of a self-mode locking Ti: Sapphire oscillator laser (Coherent Micra) was regeneratively amplified (Coherent Legend Elite) to provide a 5 kHz pulse train centered at 809 nm with 34 nm bandwidth and 38 fs duration as measured by intensity autocorrelation. These pulses had a <0.15% stability (10 Hz, SD/mean). This beam is split twice, first with a 50/50 beamsplitter (CVI), with one leg that is delayed relative to another by use of a retroreflector mounted on a motorized translation stage (Aerotech). Both beams focus onto a diffractive optic, giving two pairs of phase matched beams. Two are delayed individually using matched pairs of fused silica wedges mounted to two translation stages (Aerotech) to give precise interferometric delays (τ). One beam is attenuated by a factor of 100–1000 and serves as the local oscillator for phase matched detection. The beams are then focused onto the sample with a spot size of approximately 70 μm and pulse energy of approximately 1.8 nJ pulse^{-1}. The local oscillator and the signal are then dispersed by a spectrometer (Andor Shamrock) onto a 1600 × 5 pixel region of the CCD camera (Andor Newton). Delay calibration was performed interferometrically as described by previously.[25,31]

Each 2D spectrum is constructed from interferograms with different coherences times. Coherence times were scanned from −300 fs to 600 fs in steps of 4 fs. This sampling strategy aliases the frequency of the signal, but allows full sampling of the optical response with reasonable acquisition time. Data from waiting times (T) from 0 to 2 ps in 20 fs increments were taken, with several 0 fs replicates interspersed to ensure sample integrity. The analysis was performed as described elsewhere.[25] Briefly, the signal is transformed from wavelength to frequency space, then Fourier transformed over the ω_t dimension to produce at tau vs. time image. This signal is windowed using a moving Welch window in the t domain that follows the photon echo signal, and a Gauss-Lorentz window in the τ domain (taking only positive τ) to minimize apodization artifacts. Both the window size and type were varied to ensure spectral signatures were not artifacts of the Fourier analysis. The signal is then zero-padded in both dimensions and Fourier transformed to produce the interpolated spectra reported. The time series reported are produced by integrating a 20 cm^{-1} box from each spectrum in waiting time. Separate pump–probe spectra were also acquired, however they are not used in the current analysis.

Linear prediction Z-transform

For a discretely sampled time-domain signal, the Discrete Fourier Transform (DFT) can be used to determine its spectral components.[32] Here the finite DFT for a signal consisting of N points is expressed as follows:

$$S(z) = \sum_{n=0}^{N-1} y(\tau) z^{-n} \qquad (4)$$

where $y(\tau)$ are the values of the time series and $z = \exp(-i\omega\tau)$ with $\omega = 1/(2N\Delta\tau)$. The DFT has long been implemented in spectroscopy and other fields by use of the Fast Fourier Transform algorithm (FFT).[33] Unfortunately, the frequency resolution is limited by the reciprocal of the time spacing and the duration of the detection. As a result the frequency resolution of highly damped or truncated signals is severely limited. Strategies, such as windowing and apodization can be used to improve resolution and lessen truncation artifacts; however, a tradeoff exists between signal enhancement and noise reduction.

To avoid these drawbacks linear prediction can be used simultaneously to enhance the signal-to-noise ratio and to improve resolution. To demonstrate these benefits, we take N points made up of K exponentially damped sinusoids (free induction decays, or FIDs) sampled with sufficient time resolution to resolve the highest frequency. The direct Fourier transform of such a series is K Lorentzians centered at each frequency. The width of each Lorentzian is related to the decay of each frequency component. Using the following function, we can estimate future points of a data series as a linear function of previous points

$$y(n) = \sum_{m=1}^{M} a_m y_{n-m} \qquad (5)$$

where M is the number of linear prediction coefficients (filter coefficients) needed to model the signal and n is the index of the discrete signal. Using this approach, we can extend a limited data series to an infinite one thereby eliminating truncation errors. First, we must select M using a suitable number of linear prediction coefficients (LPCs) to filter out uncorrelated signals (white noise) and represent the signals making up the spectral response. Too many coefficients lead to spurious peaks and occasional instability. In general, for a real valued system, $M > 2K$ (or twice the number of signals expected), but less than $N - 2K$, so the system is not overdetermined.[34] There are several methods of computing linear prediction coefficients,[35,36] but we chose the autocorrelation method of autoregressive analysis.[37] Here, the signal can be estimated by the Yule–Walker equation which exists for the autocorrelation functions derived from the signal.[37] From this function it is possible to efficiently calculate the LPCs without any least squares fitting.

Several methods have emerged that operate on linear prediction coefficients in order to determine the frequency and decay components of an underlying signal. Some common ones include the Prony spectral line estimation,[38] and the Burg maximum entropy method.[38] In Prony's method, the linear prediction coefficients are used to create a characteristic polynomial, the roots of which are the frequencies and decays of the signal. Subsequent fitting can reproduce the phase and magnitude information for each signal. The drawback is that both root finding and least squares fitting procedures are unstable and neither provides a guarantee of finding the global minimum. The maximum entropy method can estimate the power spectrum effectively, but does not extract decay coefficients. We therefore employ a method described by Tang and Norris, the two dimensional linear prediction Z-transform.[39]

The Z-transform is the following relation

$$S(\hat{z}) = \sum_{n=0}^{\infty} y(n)\hat{z}^{-n} \quad (6)$$

where $\hat{z} = \exp(-i\omega\tau - k\tau)$. This infinite sum can be reduced to a finite sum using the repetitive LP relation in eqn (5). The result is the LPZ spectral function.[39]

$$S(z) = \frac{H(z)}{1 + G(z)} \quad (7)$$

$$G(z) = -\sum_{m=1}^{M} a_m z^m \quad (8)$$

$$H(z) = -\sum_{m=1}^{M} b_m z^m \quad (9)$$

$$b_m = \sum_{n=m}^{M} a_n y(n-m) \quad (10)$$

In this formulation, linear prediction coefficients can be used with the FFT to calculate the function for discrete values of the continuous function z. Here $G(z)$ and $H(z)$ are calculated using the FFT and k is varied by multiplying each exp $(-kt)$ for a discrete set of decay values. Where ω (the frequency) and k (the decay rate) are exactly equal to those roots of the polynomial equation described previously, the Z-transform diverges. By sampling k and the frequency domain, this function can be used to build a two-dimensional plot of decay rate *vs.* frequency, where the peaks represent poles in the Z-transform. This process is more stable than the root finding and fitting procedures described previously and has the added benefit of an easily interpreted graphical representation.

Applicability to two-dimensional electronic spectroscopy

Linear prediction has been widely applied in NMR,[34,36,37,39–41] FTIR,[42] Raman[43] and other spectroscopies. The linear prediction Z-transform (2D-LPZ) has been applied successfully to analyze NMR signals.[37,39,40] Here, we extend the 2D-LPZ approach for two-dimensional electronic spectroscopy. The 2D-LPZ will extract signals arising from zero quantum electronic coherences and vibrational coherences that beat and decay during the waiting time even in congested spectra. While experimental considerations make it very difficult to collect more than 100 points during a single experimental run, LPZ analysis allows for a model-free method of enhancing frequency resolution and decay rates without making any underlying assumption about the number of FID signals in a given sample, and avoiding least squares curve fitting to a high dimensional model.

This method of determining linear prediction coefficients works best for autoregressive processes to an order less than $0.5N$ (for real valued signals).[43] This condition simply means that linear prediction depends on the ability of the previous values to model the underlying behavior of the function. Functions with underlying exponential behavior accomplish this well as the Taylor expansion derivatives of an exponential resemble each other. However, higher power terms in the dephasing rate (such as Gaussian terms) are autoregressive to infinite order and thus are unstable to linear prediction.[43] In spectroscopy, Gaussian lineshapes are common due to inhomogeneous broadening that comes from static inhomogeneity in electronic environments. However, because these signals have successfully been fit to exponential decay previously, it is likely that a) we lack the resolution to see higher order dephasing processes and b) this source of broadening is relatively minimal in comparison to homogenous broadening.

Dephasing rates were determined in the following way. Twenty time series were extracted from each of 3 different regions of the 2D spectra corresponding to $T = 80$ to 2000 fs (early times were eliminated pulse overlap effects below $T = 80$). A biexponential fit was performed to remove from each time series population dynamics (sample results are shown in Fig. 2). Each region corresponds to different exciton coherences. Region 1 primarily consists of crosspeaks associated with exciton 1 and 2 coherences, region 2 coherences are primarily associated with excitons 2 and 3, and region 3 with excitons 3 and 4. Representative time series from regions 1 and 3 are shown on the figure. A Z-transform was performed for each time series with a linear prediction basis set of 48, determined by estimating the variance on several Z-transforms with different size filters. Linear prediction coefficients were determined using the built in algorithm in Matlab (MathWorks).[45] A typical Z-transform from region 1 is shown in Fig. 3. Changing the filter size to several other values did not significantly affect the results. Peaks from the Z-transform corresponding to the primary frequencies shown in the Fourier transform were sorted by decay rate and frequency. We report the mean and standard deviation of 20 timeseries for each peak measured. Assignment of each zero quantum coherence was made using the Hamiltonian provided in Hayes and Engel.[29] The numbers agree and are internally self consistent.

Results

As can be seen in Fig. 3, the Z-transform produces a clear improvement in the specificity of the frequency resolution for certain peaks. The improvements come from two sources. First, a decrease in the white-noise spectrum eliminates the underlying flat spectral response. This filtering is inherent in the linear prediction. Second, apodization artifacts associated with the truncation in the time domain are markedly diminished. A cut in the k-domain reveals the principle components that decay near $k = 20$ cm^{-1}. The lineshapes are far narrower near the exact decay rate of the underlying function. From this spectrum we identify the 1–2, 1–3 and 2–6 coherence frequencies, consistent both with the region that was sampled, as well as the Fourier analysis done previously.[29] Decay rates separate clearly in the k domain. Consistent with previous reports, the 1–2 coherence decays more slowly than the 1–3 coherence.[28] We did not consider coherences with energy differences under 100 cm^{-1} out of concern that the number of periods sampled in the time domain was not sufficient for unambiguous assignment to specific coherences. The presence of 3 significant frequencies also demonstrates that in 2D spectra, crosspeaks still overlap significantly.

Fig. 4 shows the results of our analysis of the principle frequencies across the three different regions of the 2D spectrum. Eight individual crosspeaks were identified and decay rates were extracted. The crosspeaks are 1–2 ($\omega = 163 \pm 1, k = 18 \pm 2$), 1–3 ($\omega = 203 \pm 1, k = 36 \pm 6$), 2–4 ($\omega = 147 \pm 1, k = 40 \pm 3$), 2–6 ($\omega = 263 \pm 2, k = 26 \pm 4$), 3–5 ($\omega = 104 \pm 1, k = 66 \pm 7$), 3–6 ($\omega = 159 \pm 3, k = 45 \pm 9$), 3–7($\omega = 264 \pm 2, k = 36 \pm 5$), and 4–6 ($\omega = 107 \pm 2, k = 47 \pm 9$) (cm^{-1}). The error reported comes from sampling different time series from within a region of a single experiment, and thus represents the consistency of the results of the linear prediction. The frequencies reported are consistent with those reported in the experimental Hamiltonian presented by Hayes and Engel,[29] despite arriving from a distinct data set. The frequencies that

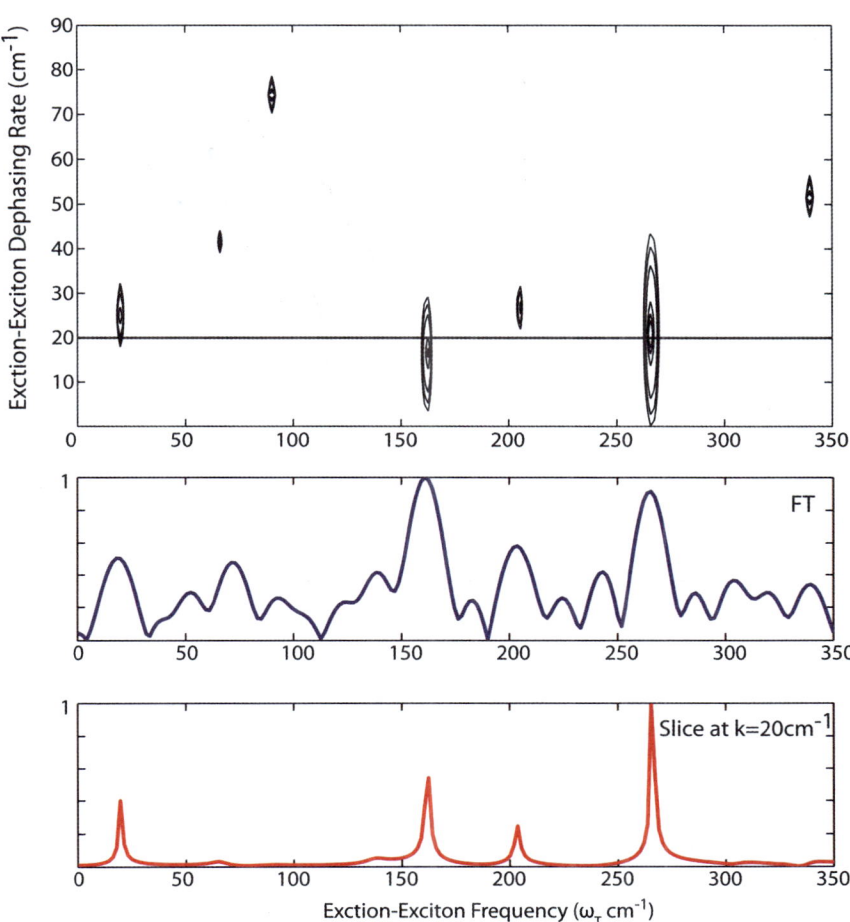

Fig. 3 Top: A two dimensional Z-transform spectrum of the time series from region 1 depicted in Fig. 2. The peaks represent poles of the Z-transform where the transform diverges. The contour map has lines representing evenly spaced contours from .1 to 1.0, the middle range of total intensity of the spectra. Middle: The Fourier transform for the region 1 time series. While two frequencies are dominant, much of the signal is difficult to distinguish from apodization artifacts and noise. Bottom: A cut at $k = 20 \text{cm}^{-1}$, of the top spectrum showing improved resolution of frequencies that decay, at or near that rate.

are within a similar frequency range (*i.e.* 150–170 cm^{-1}) arise from different regions in the 2D spectrum and represent different response pathways.

Conclusion

Our analysis allows us to create a frequency and decay rate representation of the coherences within the one-exciton manifold. We find that the dephasing rate follows a complex pattern, which cannot be described simply by variations in the exciton energy difference. For example, the 3–5 exciton coherence and the 4–6 exciton coherence decay at nearly 3 times the rate that the 1–2 exciton coherence does despite its higher frequency. If these excitonic dephasing rates followed a similar pattern to vibrational dephasing rates, the higher frequency would decay faster than the lower frequency, due to inertial interactions with the environment. Even for coherences

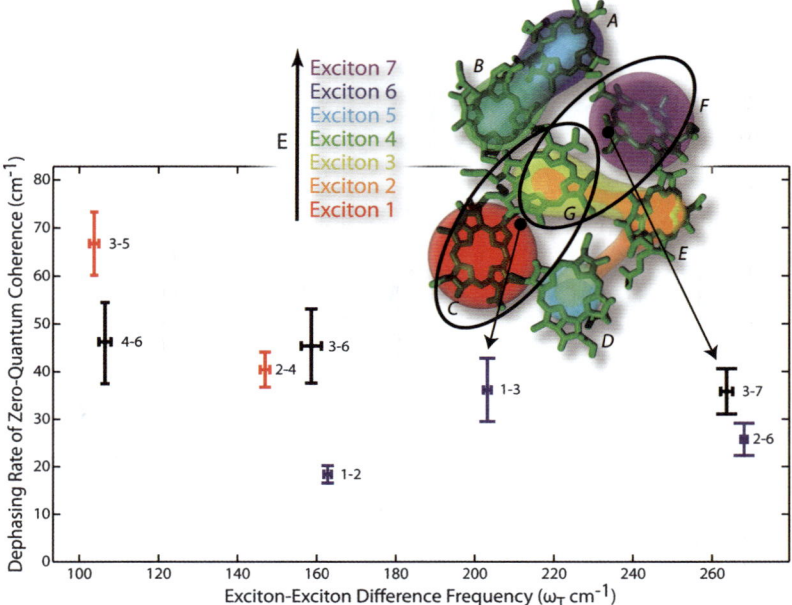

Fig. 4 The frequencies and decay rates extracted for signals above 95 cm^{-1}. Blue crosses represent signals taken from region 1, red from region 2 and black from region 3 (shown in Fig. 2). The error bars represent the variation in the linear prediction (the standard deviation over 20 different sampling boxes in each region). In the upper right, the excitonic structure of FMO is shown based on site participation calculated from prior studies.[24] The colormap represents the degree to which each site (lettered) participates in each exciton (numbered). Bchl sites are labeled in italics according to the original site numbering by Fenna and Mathews, with site 1 corresponding to site A and site 7 corresponding to site G.[44] Coherences between excitons 1 & 3 and 3 & 7 are highlighted to show spatial correlation between the excitons pictured.

with similar energy differences such as the 1–2, 2–4 and 3–6 coherences, we see significant variation in decay rates. This difference implies that the spectral overlap of the energy gap between excitons and the power spectrum of the protein is complex, and that the mechanism of correlation must have spatial heterogeneity.

In Fig. 4 we include a picture of the excitonic structure of FMO as derived from the site energies and couplings determined by Hayes and Engel.[29] The brightness the colors represents the degree to which each BChl site participates in forming a particular exciton. Highlighted are two exciton–exciton coherences, specifically those between excitons 1 & 3 and 3 & 7. Suggestively, all the most prevalent frequencies found correspond to coherences between excitons that are spatially adjacent, with the exception of those connected to exciton 2. However, exciton 2 is highly delocalized, participating significantly in sites D, E, and G, which are adjacent to nearly all the other excitons in the system. The data suggests some kind of spatial dependence of the dephasing rate, instead of one that is dependent either on resonance to a specific energy gap (a peak in the spectral density of the protein), or one that is simply based on the energy difference between two excitons. Physically, this model suggests coupling to a somewhat delocalized phonon bath.

In light of other recent results, long-lived electronic coherence appears to be a wide-ranging phenomenon, fundamentally a function of a generally correlated environment, such as a protein or a rigid polymer. Our results do not contradict this finding, but they do suggest that the specific arrangement of the chromophores (or the bath surrounding them) matters when examining the lifetime of the coherence. Light harvesting complexes have evolved to transfer energy efficiently, and therefore, coherent energy transfer may be better interpreted as a consequence of arranging the chromophores such that transfer is efficient and directional, not as an optimized result in and of itself. However, despite its generality, the exact mechanism of protection of coherences remains elusive. Based on these results and others, we hypothesize that while the spectral density of the bath may be general, the projection of the individual excitons onto that phonon bath is state-specific, and an understanding of the spatial relationship between exciton and bath is necessary to predict the correlation parameter.

Linear predictive coding allows deconvolution of spectral responses. When applied carefully, it can extract relevant quantities such as decay rates and frequencies from limited or noisy datasets. However, it must be applied with caution, as it works best with data sets that are inherently autoregressive. This analysis however cannot distinguish between homogeneous sources of dephasing (arising from the same random fluctuations inside different proteins leading to a loss of correlated transition energies), or inhomogeneous sources (arising from different environmental starting points). Homogenous broadening leads to Lorentzian line shapes as they arise from natural lifetimes of correlations, while inhomogeneous broadening tends to be Gaussian (or more complex) in nature, as it arises from statistical distributions in a solvent environment. Interestingly, inhomogeneous broadening is significantly less temperature dependent than homogenous broadening, and FMO shows significant changes in the dephasing rate of the 1–2 coherence as a function of temperature.[16] Nevertheless, this assignment is tenuous, and the dephasing rate measured will likely be a combination of both sources of broadening.

With this uncertainty, we cannot relate the time scale of dephasing observed directly to the correlation factor discussed previously. However, the source of inhomogeneous broadening is likely similar for all coherences observed, and thus, the trend in the dephasing rates should be preserved in any explanation of coherent energy transfer within the FMO complex. What remains to be understood is the source of these correlations. Is it a passive process coming from underlying dynamics of the protein? Or do these correlations only appear upon electronic excitation? Further experiments and simulation both on photosynthetic and model systems are needed to elucidate the mechanism of the correlation.

Many studies and simulations show that coherent energy transfer may play an important role in enhancing the efficiency and directionality of energy transfer. Photosynthetic systems provide an excellent system to study highly evolved energy transfer machinery. However, even the simplest photosynthetic antennae are extremely complex, consisting of multiple chromophores, complex excitonic structures and non-trivial solvent environments. While tools are being developed to dissect these complex systems, novel analysis techniques must also be employed to filter the relevant information out of the immense number of signals. Much of the algorithmic machinery has already been developed, especially in the NMR community. Just as the formalism of 2DES grew as an extension of 2D NMR, the analytical techniques used in NMR can be applied to dissect the complex spectra of light harvesting complexes.

Acknowledgements

The authors thank Robert E. Blankenship and Jianzhong Wen for sample and Dugan Hayes for experimental data and discussions. This work has been supported in part by the Searle Foundation, AFOSR (Grant No. FA9550-09-1-0117) and DARPA (Grant No. N66001-10-1-4060). JRC was supported by the NSF GRF program.

References

1 H. V. Amerongen, L. Valkunas and R. V. Grondelle, *Photosynthetic Excitons*, World Scientific Publishing Company, 2000.

2. R. E. Blankenship, *Molecular Mechanisms of Photosynthesis*, Wiley-Blackwell, 2002.
3. J. Wen, H. Zhang, M. L. Gross and R. E. Blankenship, *Proc. Natl. Acad. Sci. U. S. A.*, 2009, **106**, 6134–6139.
4. A. Camara-Artigas, R. E. Blankenship and J. P. Allen, *Photosynth. Res.*, 2003, **75**, 49–55.
5. S. I. E. Vulto, S. Neerken, R. J. W. Louwe, M. A. de Baat, J. Amesz and T. J. Aartsma, *J. Phys. Chem. B*, 1998, **102**, 10630–10635.
6. F. Caruso, A. W. Chin, A. Datta, S. F. Huelga and M. B. Plenio, *Phys. Rev. A: At., Mol., Opt. Phys.*, 2010, **81**, 062346.
7. A. Ishizaki and G. R. Fleming, *Proc. Natl. Acad. Sci. U. S. A.*, 2009, **106**, 17255–17260.
8. M. Cho, H. M. Vaswani, T. Brixner, J. Stenger and G. R. Fleming, *J. Phys. Chem. B*, 2005, **109**, 10542–10556.
9. M. Sarovar, A. Ishizaki, G. R. Fleming and K. B. Whaley, *Nat. Phys.*, 2010, **6**, 462–467.
10. Y.-C. Cheng, G. S. Engel and G. R. Fleming, *Chem. Phys.*, 2007, **341**, 285–295.
11. P. Rebentrost, M. Mohseni, I. Kassal, S. Lloyd and A. Aspuru-Guzik, *New J. Phys.*, 2009, **11**, 033003.
12. J. Cao and R. Silbey, *J. Phys. Chem. A*, 2009, 13825–13838.
13. T. Förster, *Ann. Phys.*, 1948, **437**, 55–75.
14. H. Lee, Y.-C. Cheng and G. R. Fleming, *Science*, 2007, **316**, 1462–1465.
15. G. Engel, T. Calhoun, E. Read, T. Ahn, T. Mancal, Y. Cheng, R. Blankenship and G. Fleming, *Nature*, 2007, **446**, 782–786.
16. G. Panitchayangkoon, D. Hayes, K. A. Fransted, J. R. Caram, E. Harel, J. Wen, R. E. Blankenship and G. S. Engel, *Proc. Natl. Acad. Sci. U. S. A.*, 2010, **107**, 12766–12770.
17. E. Collini and G. D. Scholes, *Science*, 2009, **323**, 369–373.
18. E. Collini, C. Y. Wong, K. E. Wilk, P. M. G. Curmi, P. Brumer and G. D. Scholes, *Nature*, 2010, **463**, 644–647.
19. T. R. Calhoun, N. S. Ginsberg, G. S. Schlau-Cohen, Y.-C. Cheng, M. Ballottari, R. Bassi and G. R. Fleming, *J. Phys. Chem. B*, 2009, **113**, 16291–16295.
20. D. Hayes, J. Wen, G. Panitchayangkoon, R. E. Blankenship and G. S. Engel, *Faraday Discuss.*, 2011, DOI: 10.1039/c0fd00030b.
21. E. Collini and G. D. Scholes, *J. Phys. Chem. A*, 2009, **113**, 4223–4241.
22. S. Mukamel, *Nonlinear Optical Spectroscopy*, Oxford, New York, 1995.
23. M. Cho, *Chem. Rev.*, 2008, 1331–1418.
24. D. Jonas, *Annu. Rev. Phys. Chem.*, 2003, 425–463.
25. T. Brixner, T. Mancal, I. V. Stiopkin and G. R. Fleming, *J. Chem. Phys.*, 2004, **121**, 4221.
26. S. Mukamel, *Principles of Nonlinear Optical Spectroscopy*, Oxford University Press, USA 1st, 1995.
27. K. Stone, K. Gundogdu, D. Turner, X. Li, S. Cundiff and K. Nelson, *Science*, 2009, 1169–1173.
28. D. Hayes, G. Panitchayangkoon, K. A. Fransted, J. R. Caram, J. Wen, K. F. Freed and G. S. Engel, *New J. Phys.*, 2010, **12**, 065042.
29. D. Hayes and G. S. Engel, *Biophys. J.*, 2011, **100**, 2043–2052.
30. J. D. Hybl, A. A. Ferro and D. M. Jonas, *J. Chem. Phys.*, 2001, **115**, 6606–6622.
31. L. Lepetit, G. Chiriaux and M. Joffre, *J. Opt. Soc. Am. B*, 1995, **12**, 2467–2474.
32. L. B. Jackson, *Digital Filters and Signal Processing: With MATLAB Exercises*, 3rd Edition, Kluwer Academic, 1995.
33. J. Cooley and J. Tukey, *Math. Comput.*, 1965, 297.
34. G. L. Millhauser and J. H. Freed, *J. Chem. Phys.*, 1986, **85**, 63.
35. D. Tufts and R. Kumaresen, *Proc. IEEE*, 1982, 975–989.
36. J. Tang, C. Lin, M. Bowman and J. Norris, *J. Magn. Reson.*, 1985, 167–171.
37. J. Tang and J. R. Norris, *Chem. Phys. Lett.*, 1986, **131**, 252–255.
38. S. Kay and S. Marple, *Proc. IEEE*, 1981, 1380–1419.
39. J. Tang and J. R. Norris, *J. Magn. Reson.*, 1986, (69), 180–186.
40. J. Tang and J. R. Norris, *J. Magn. Reson.*, 1969, **1988**(78), 23–30.
41. J. J. Led and H. Gesmar, *Chem. Rev.*, 1991, **91**, 1413–1426.
42. Y. P. Lee and D. S. Chen, *Microchim. Acta*, 1988, **94**, 85–87.
43. O. P. Sievanen, *Appl. Spectrosc.*, 1999, **53**, 144–149.
44. R. E. Fenna and B. W. Matthews, *Nature*, 1975, **258**, 573–577.
45. *Matlab 2008b*, Mathworks, Natick, MA.

PAPER

Multiconfigurational Ehrenfest approach to quantum coherent dynamics in large molecular systems

Dmitrii V. Shalashilin*

Received 4th March 2011, Accepted 24th May 2011
DOI: 10.1039/c1fd00034a

This article briefly describes recently developed Multiconfigurational Ehrenfest dynamics method to simulate quantum dynamics in systems with many degrees of freedom. The central idea is to guide the trajectories of basis wave functions by means of the Ehrenfest trajectories. The amplitudes of guided basis functions are coupled through a system of linear equations. The approach has been applied to simulations of nonadiabatic dynamics in Spin–Boson model and in pyrazine molecule. A new application to nonadiabatic dynamics in 24D model of pyrazine, where good spectrum for is obtained with the basis of only 34 basis Ehrenfest configurations is reported. This application provides the ground for future fully quantum direct dynamics. Another new application to the model of sticking to the surface described by the System-Bath Hamiltonian is presented to demonstrate the broadness of the approach, which can be applied to both electronically adiabatic and nonadiabatic dynamics. For all applications the results are in good agreement with those of MCTDH, which is very difficult to achieve with other trajectory-based methods. Therefore MCE can serve as a starting point for future use with "on the fly" direct dynamics. MCE provides an efficient fully quantum method capable of catching coherent dynamics in multidimentional systems, which is a necessary step for developing and understanding coherent control in realistic quantum systems.

1 Introduction

1.1 Why quantum dynamics?

Nuclei are heavy particles and usually their motion can be treated classically, but there are significant exceptions. For example light absorption always creates a coherent initial wave packet which on a time scale slower than the decoherence time retains its quantum nature. Therefore ultrafast photochemistry on the subpicosecond timescale is an essentially quantum process. Many mechanisms of light harvesting in plants, fluorescence of living organisms and vision are due to photochemical reactions which include excitation of electronic states and subsequent electronically nonadiabatic dynamics. Recently significant progress has been made in experimental ultrafast time resolved spectroscopy studies of various photochemical reactions, focused on biologically related molecules. There is growing experimental evidence that coherent quantum nuclear dynamics in many cases plays an important role in such processes[1–5] on short time scale. Other examples of the situation when quantum coherent effects are important are in the dynamics of hydrogen and at low temperatures.

School of chemistry, Univertsity of Leeds, Leeds, LS2 9JT, UK. E-mail: d.shalashilin@leeds.ac.uk

1.2 Exponential curse

Theoretical study of molecular quantum dynamics is a difficult task. Potential energy surfaces (PES) can be quite complex and the reactions of interest can occur through multiple avoided crossings and conical intersections. Many vibrational modes can be involved in quantum motion. The greatest challenge comes from the fact that quantum wave packet dynamics in complex systems with many degrees of freedom (DOF) is prohibitively expensive to simulate numerically with traditional methods due to the exponential scaling of the quantum basis sets with the number of DOF. A problem in M-dimensions (or degrees of freedom) for example M vibrational modes of a molecule requires a regular basis of the size

$$N = l^M \tag{1}$$

where l is the number of basis functions needed for a similar 1D case. The grid (or basis) size N grows so fast with M that the eqn(1) is often called the "*exponential curse*" of quantum mechanics.

1.3 Existing methods of highdimentional quantum dynamics

Only recently, new approaches, which push the limits of quantum dynamics, have started to emerge. Multiconfigurational Time Dependent Hartree (MCTDH)[6–8] is by far the most well known technique capable of treating many nuclear degrees of freedom. The important limitation of MCTDH for the current proposal is that MCTDH requires a certain analytical form of the PES.

Another class of methods relies on randomly selected trajectory guided basis functions, which are usually chosen to be Frozen Gaussian wave packets; also known as Coherent States (CS). This approach has several advantages.

(1) Trajectory guided basis functions evolve following the wave function thus staying within the dynamically important region and minimising the basis set size.

(2) A randomly selected basis set does not have to scale exponentially with the number of DOF and therefore avoids the "exponential curse".

(3) Mechanisms of various physical processes can often be associated with the trajectories of basis functions and visualised

(4) An important advantage of the methods which exploit trajectory guided grids is that they can be interfaced with an *ab initio* electronic structure code for calculation of the Potential Energy Surface "on the fly" Therefore simulations can proceed without preconditions and assumptions about the PES.

Fig.1 gives a sketch of the main idea of quantum methods based on trajectory guided grids. The wave function is represented on a basis (or grid) of multidimentional Gaussian wave packets (Coherent States). Quantum evolution is described by the motion of the basis and exchange of the amplitudes between the CS. Several related methods exist.

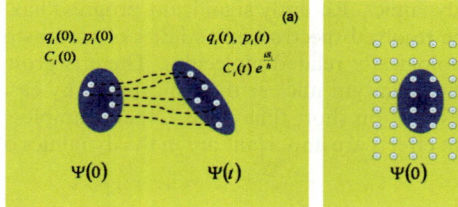

Fig. 1 Wave function evolution on a small trajectory guided grid (grid points are shown by white circles) compared to that on a large regular grid.

(i) The *ab initio* Full Multiple Spawning (FMS) method[9–14] was the first method of this sort. MS has been developed particularly for simulating electronically nonadiabatic reactions. MS uses a basis of Gaussian Coherent States guided by their classical trajectories. The dynamics on a single PES is described by the motion of the basis functions and exchange of their amplitudes. The central feature of the method is spawning. When a "parent" Gaussian approaches a region of strong nonadiabatic coupling with another PES a new Gaussian (a "child") is generated on the other PES. The coupling between "parent" and "child" basis Gaussians ensures transfer of amplitude and population between the two electronic states. Being a trajectory-based technique MS allows combination of quantum dynamics with "on the fly" calculations of the PES, thus resulting in full scale *ab initio* dynamics, similar to the direct classical molecular dynamics (MD).

(ii) Similarly to MS the method of variational Multiconfigurational Gaussians (vMCG) and related method of Gaussian MCTDH[15–17] represents the wave function as a superposition of Gaussian Coherent States. The difference is that their trajectories are determined from the full variational principle applied simultaneously to all parameters of the wave function (*i.e.* phase space positions of the Coherent States and their amplitudes). The trajectories are therefore nonclassical and can go into classically forbidden regions. However finding variational vMCG trajectories can be rather expensive even without "on the fly" calculations of the PES.

(iii) The method of Coupled Coherent States (CCS)[18–22] is somewhat in between MS and vMCG. It uses trajectories with quantum corrections to guide the basis, but the CCS trajectories are not much more expensive that those of classical mechanics. Recently a new method termed as Multi-Configurational Ehrenfest (MCE) dynamics, which is a generalisation of the CCS, has been proposed in 23,24. Similarly to MS approach the MCE method focuses on nonadiabatic dynamics. Converged results were obtained for the Spin-Boson model with a broad range of parameters[23] and for nonadiabtic dynamics near a conical intersection in pyrazine.[24] Both systems[23,24] are important case studies of nonadiabatic dynamics. For the Spin–Boson model the number of quantum degrees of freedom treated by MCE reaches several thousand[23] but remarkably the number of trajectory guided basis functions used was no greater than a hundred. MCE is capable of treating problems previously accessible only by MCTDH, and the efficiency of MCE is unique among the trajectory based methods. Therefore, MCE has a prospect of becoming a quantum counterpart of classical Molecular Dynamics (MD) capable of producing well converged results in simulations of nonadiabatic dynamics. This paper gives a brief sketch of the MCE theory and its previous applications to the models of electronically nonadiabatic dynamics. Also a new case study of the nonadiabatic dynamics in pyrazine molecule is presented in which main features of the dynamics are reproduced with the basis of only 34 Ehrenfest trajectory guided configurations. This example demonstrates that MCE method can yeild good results with the basis sets sufficiently small for direct dynamics. Another new case study of the dynamics of hydrogen atom sticking to the surface is presented in order to demonstrate the broadness of the MCE approach, which can also be used for electronically adiabatic processes.

2. Theory

Ehrenfest approximation in quantum mechanics introduced by Billing[25] and Mayer and Miller[26] has proven to be a very useful tool of simulations in chemistry and physics. In its standard form the method assumes that the full system can be split into quantum "subsystem" and classical "bath". Then the trajectory of classical subsystem is assumed to be driven by the Hamiltonian averaged with the quantum subsystem represented as a superposition of several basis functions with coefficients follwing coupled equations. Unlike standard Ehrenfest approach the following MCE theory aims at treating all degrees of freedom including those of the "bath" at a fully quantum level. Let us first represent the wave function of the whole system as a product of quantum wave function and a frozen Gaussian Coherent State wave packet which "dresses" trajectory of the classical "bath"

$$|\Psi(t)\rangle = |\Psi^{sstm}(t)\rangle |\mathbf{z}^{bth}(t)\rangle = (a_1(t)|1\rangle + a_2(t)|2\rangle + ...) |\mathbf{z}^{bth}(t)\rangle \qquad (2)$$

Here we use the Klauder's z-notations for the Coherent States which labels a 1D Gaussian Coherent State

$$\langle x|z\rangle = \left(\frac{\gamma}{\pi}\right)^{\frac{1}{4}} \exp\left(-\frac{\gamma}{2}(x-q)^2 + \frac{i}{\hbar}p(x-q) + \frac{ipq}{2\hbar}\right) \qquad (3)$$

characterised by its position q and momentum p with a single complex number z

$$z = \frac{\gamma^{\frac{1}{2}}q + i\hbar^{-1}\gamma^{-\frac{1}{2}}p}{\sqrt{2}} \qquad (4)$$

and multidimensional wave packet $|\mathbf{z}^{bth}(t)\rangle = |z^{(1)}(t)\rangle|z^{(2)}(t)\rangle...|z^{(M)}(t)\rangle$ is a product of M 1D CSs. Similarly to 25 and 26 the evolution of the "classical bath" in the wave function (2) is described by a single trajectory $|\mathbf{z}^{bth}(t)\rangle$.

Then variational principle $\delta S = 0$ where $S = \int \langle \Psi | i\frac{\hat{\partial}}{\partial t} - \widehat{H} | \Psi \rangle dt$ can be used to obtain the equations for the coefficients in the expansion of the quantum subsystem wave function

$$\dot{d}_1 = -iH_{12}d_2 \exp(i(S_2 - S_1))$$
$$\dot{d}_2 = -iH_{21}d_1 \exp(i(S_1 - S_2)) \qquad (5)$$

and the trajectories of the classical bath modes

$$i\dot{z} = -\frac{\partial H^{Ehr}}{\partial z*} \qquad (6)$$

In eqn (6)

$$H^{Ehr} = \frac{\langle \Psi^{sstm} | \widehat{H} | \Psi^{sstm} \rangle}{\langle \Psi^{sstm} | \Psi^{sstm} \rangle}$$
$$= \frac{H_{11}\ a_1{}^*a_1 + H_{22}\ a_2{}^*a_2 + H_{12}\ a_1{}^*a_2 + H_{21}\ a_2{}^*a_1}{a_1{}^*a_1 + a_2{}^*a_2} \qquad (7)$$

and the equations for the amplitude are written for the preexponential factor

$$a_{1,2} = d_{1,2}\exp(iS_{1,2}) \qquad (8)$$

rather then for the amplitudes a themselves. In (8)

$$S_{l=1,2} = \int \left[i\frac{\dot{z}z^* - \dot{z}^*z}{2} - H_{ll}\right]dt \qquad (9)$$

is the action.

The eqn (5,6) constitutes the semiclassical Ehrenfest approximation to the dynamics of quantum-classical systems known from a number of previous works.[25-28] The eqn (5,6) are not more expensive than those of classical mechanics and very easy to solve. However, the wave function (2) is not flexible enough to represent accurately the quantum dynamics. In Refs 23 and 24 it was proposed to use a set of the wave functions (2) as a basis for exact quantum propagation. The exact wave function is represented as a superposition of Ehrenfest configurations as follows

$$|\Psi(t)\rangle = \sum_{k=1,N} D_k(t)|\phi_k(t)\rangle = \sum_{k=1,N} D_k(t)(a_{1k}(t)|1\rangle + a_{2k}(t)|2\rangle + ...)\left|\tilde{z}_k(t)\right\rangle \quad (10)$$

where now the wave function is described not by a single configuration (2), but by a linear combination of several configurations $|\phi_k(t)\rangle = (a_{1k}(t)|1\rangle + a_{2k}(t)|2\rangle + ...)|\tilde{z}_k(t)\rangle$. The basis CSs $|\tilde{z}_k(t)\rangle$ describing the dynamics of the bath differ by their initial conditions. Unlike (2) the wave function (10) can (at least in principle) be converged to the numerically exact result. Assuming the time dependence of a configuration $|\phi_k(t)\rangle$ to be given by the eqn(5,6) variational principle can be used again to derive the equations for their amplitudes $D_k(t)$.

$$\sum_i \langle \varphi_j(t)|\varphi_i(t)\rangle \frac{dD_i(t)}{dt} = -i \sum_i \Delta^2 \langle H \rangle_{ji} D_i(t) \quad (11)$$

where

$$\langle \varphi_j(t)|\varphi_i(t)\rangle = \langle \tilde{z}_j(t)|\tilde{z}_i(t)\rangle(a_{1j}^* a_{1i} + a_{2j}^* a_{2i}) \quad (12)$$

is the overlap matrix of the Ehrenfest basis wave functions and the elements of the matrix $\Delta^2\langle H\rangle$ are

$$\Delta^2\langle H\rangle_{ij} = \langle\varphi_j(t)|\hat{H}|\varphi_i(t)\rangle - \left\langle\tilde{z}_j(t)\Big|\dot{\tilde{z}}_i(t)\right\rangle H_{ji} - i\langle\varphi_j(t)|\varphi_i(t)\rangle(\tilde{z}_j^* - \tilde{z}_i^*)\dot{\tilde{z}}_i \quad (13)$$

In (13)

$$\langle\varphi_j(t)|\hat{H}|\varphi_i(t)\rangle =$$
$$\langle\tilde{z}_j|\hat{H}_{11}|\tilde{z}_i\rangle a_{1j}^* a_{1i} + \langle\tilde{z}_j|\hat{H}_{22}|\tilde{z}_i\rangle a_{2j}^* a_{2i} + ... \quad (14)$$
$$+ \langle\tilde{z}_j|\hat{H}_{12}|\tilde{z}_i\rangle a_{1j}^* a_{2i} + \langle\tilde{z}_j|\hat{H}_{21}|\tilde{z}_i\rangle a_{2j}^* a_{1i} + ...$$

and the matrix $\langle\tilde{z}_j(t)|\dot{\tilde{z}}_i(t)\rangle H_{ji}$ is given as

$$\left\langle\tilde{z}_j(t)\Big|\dot{\tilde{z}}_i(t)\right\rangle H_{ji} =$$
$$a_{1j}^* \langle\tilde{z}_i|\hat{H}_{12}|\tilde{z}_i\rangle a_{2i} + a_{2j}^* \langle\tilde{z}_i|\hat{H}_{21}|\tilde{z}_i\rangle a_{1i} + ... \quad (15)$$
$$+ a_{1j}^* \langle\tilde{z}_i|\hat{H}_{11}|\tilde{z}_i\rangle a_{1i} + a_{2j}^* \langle\tilde{z}_i|\hat{H}_{22}|\tilde{z}_i\rangle a_{2i} + ...$$

The elements of coupling matrix $\Delta^2\langle H\rangle$ are always small. The matrix is sparse, and has zero diagonal. In the case of single PES the eqn (11)–(15) for the amplitudes become those of the CCS theory.[18] Therefore the current Multiconfigutrational Ehrenfest approach represents an extension of the CCS technique to nonadiabatic multisurface dynamics. See Ref. 24 for more details.

Another version of the MCE method uses slightly different anzatc for the wave function

$$|\Psi(t)\rangle = \sum_{k=1,N} (a_{1k}(t)|1\rangle + a_{2k}(t)|2\rangle) |\tilde{z}_k(t)\rangle \quad (16)$$

It still uses the eqn(6) for $\tilde{z}_k(t)$ but the equations for the coefficients a in (16) are also obtained from variational principle. They are different from (5) and (11). See Refs 23 and 24 for more details.

In both variations of the MCE method the quantum subsystem is treated with a regular basis $|1\rangle$, $|2\rangle$,... and the basis of trajectory guided coherent states $|\bar{\mathbf{z}}_k(t)\rangle$ is used to describe "classical" bath.

3 Applications and results

This section summarises applications of MCE to several model systems previously treated by MCTDH method and demonstrates that MCE can treat multidimensional problems on a fully quantum level. In all cases the quality of the results is comparable with that of MCTDH, which is very difficult to achieve with a technique based on randomly selected trajectory guided basis.

3.1 Spin–Boson model of nonadiabtic dynamics

The efficiency of MCE has been first demonstrated by simulating nonadiabatic dynamics in the Spin–Boson model with up to 2000 nuclear vibrational modes.[23] Spin–Boson model describes two quantum electronic states in many dimensions coupled to each other. It is used in many areas of physics to simulate such physical objects as light harvesting complexes, quantum dots, decoherence and many others. A good MCTDH benchmark is available and the full study of SB model with a broad range of parameter by MCE can be found in. For instance Fig.2 shows a complicated case of the thermally averaged population difference between two electronic states described by asymmetric Spin–Boson model obtained with MSE converges to the MCTDH result.

Each electronic state included 70 vibrational modes so that the problem is really multidimensional. The quantum subsystem basis included simply two electronic states $|1\rangle$ and $|2\rangle$ and the classical "bath" of vibrational modes was described by a small basis of multidimensional coherent states $|\mathbf{z}^{bth}(t)\rangle = |z^{(1)}(t)\rangle|z^{(2)}(t)\rangle\ldots|z^{(M)}(t)\rangle$. The basis was biased to the initial position of propagating bath state, which was selected from quantum Boltzmann distribution. Thermally averaged populations of the two electronic states were then obtained from several such propagations

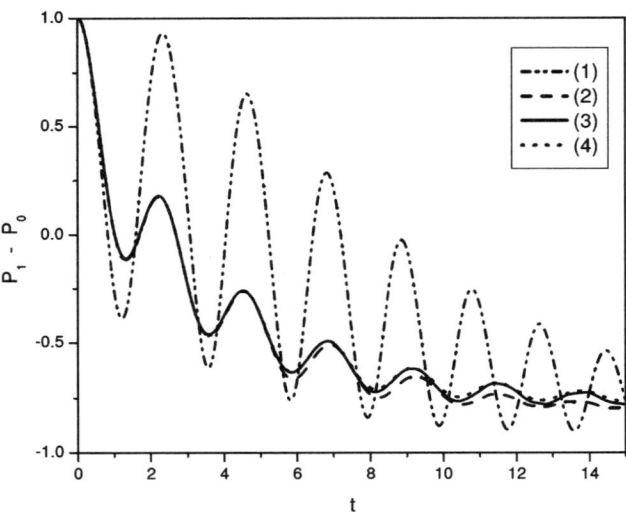

Fig. 2 Convergence of population difference between two states described by asymmetric Spin–Boson model obtained with Multi-Configurational Ehrenfest approach (curves 2 and 3 obtained with two different basis set sizes) to MCTDH result (curve 4). Curve 1 is produced by standard Ehrenfest approximation (single configuration). See Ref. 23 for more details.

with initial position of the bath selected from quantum Boltzmann destribution. Obtaining well converged result for the challenging case of multidimensional asymmetrical Spin–Boson model proves the efficiency of the MCE approach. The oscillations of the populations are due to coherent quantum dynamic, which is efficiently reproduced by the MCE method.

3.2 Efficient calculation of the Franck–Condon spectrum of pyrazine

Thus MCE approach is capable of treating accurately nonadiabatic dynamics. One of the advantages of trajectory-based techniques such as MCE is that they can be used in direct dynamics *i.e.* with "on the fly" calculation of interaction. Electronic structure calculations are expensive and it is important to mimnimise the number of MCE trajectories, which can be done by using efficient sampling of their initial conditions. This section reports a new study of 24D multidimensional vibronic-coupling dynamics at the S_2-S_1 conical intersection of the pyrazine molecule, and shows that a good result can be obtained with very small basis. Only 34 MCE trajectory guided MCE configurations can reproduce the main features of the spectrum. This system has been the subject of many previous investigations using the MCTDH method[8] for a second-order vibronic-coupling model, which will be used here as well. The aim was to illustrate the performance of the MCE method for this model system.

The parameters of the 24D model potential energy surfaces (PES) were supplied by Burghardt and they are similar to those used previously in Ref. 8. The Franck–Condon absorption spectrum was calculated as a real part of the Fourier transform of the autocorrelation function

$$I(E) = \mathrm{Re} \int e^{iEt} \langle \Psi(0)|e^{-iHt}|\Psi(0)\rangle \quad (17)$$

The autocorrelation function is obtained by propagating the wave packet initially moved vertically from the ground state S_0 to the surface S_2 and evolving on two coupled surfaces (S_2 and S_1). Previously the spectrum of pyrazine has already been calculated by the MCE method.[24] Following the understanding developed previously[8] that 4 vibrational modes are most important the wave function was first expanded in the basis of N Ehrenfest configurations using the so called pancake sampling[20] with broad random distribution of the basis Coherent States in the most important modes and narrow distribution in the rest 20 modes treated as a bath. The calculation was then repeated several times with different initial positions of less important part of the wave packet. In Ref. 24 a good spectrum was obtained with the basis of $N = 500$ and $N = 1000$ Ehrenfest configurations. In the present work special efforts were made to minimise the size of the basis set by introducing some regularity in the initial sampling. Two algorithms to choose initial positions of the Ehrenfest basis set have been used. In both samplings the "main" Coherent State was positioned in the centre of the wave packet. In the sampling 1 sixteen more basis CS were generated by shifting the central Coherent State by $\pm\delta$ and $\pm i\delta$ for the 4 most important modes. For each of the 17 CS their amplitude at S_2 was set equal to one and zero respectively ($a_{1k}(0) = 0$, $a_{2k}(0) = 1$). The amplitudes $D_k(0)$ were set by projecting the initial wave packet on the 17 configurations. After that another 17 configurations were created on S_1 by setting $a_{1k}(0) = 1$, $a_{2k}(0) = 0$. For these configurations the coefficient $D_k(0) = 0$ was initially set to zero. The initial position of the basis CS in 20 less important modes was the same but calculation was repeated to average over these 20 DOF. As has been shown in Ref. 24 such averaging does not introduce any approximations. Therefore the calculation with the sampling 1 required the basis of 34 configurations and several repetitions. In sampling 2 the basis Coherent States were obtained by allowing no more than two shifts $\pm\delta$, $\pm 2\delta$ $\pm i\delta$ and $\pm 2i\delta$ from the central CS in each of the 4 main modes or one shift in each pair of the 4 modes. The sampling 2 basis set included $N = 290$ configurations.

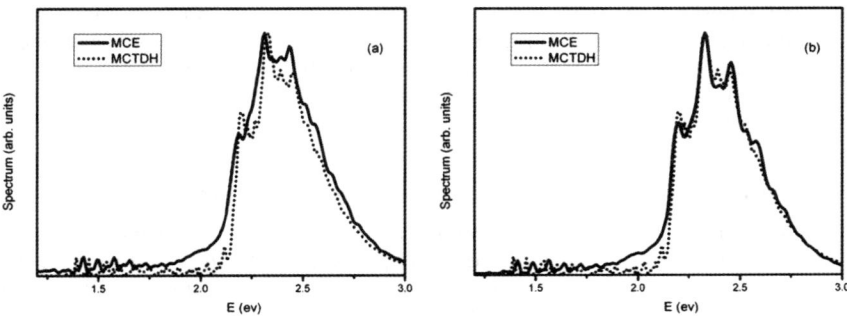

Fig. 3 The absorption spectrum for the 24D model of pyrazine obtained with the basis of 24-D Gaussian Coherent States forming Ehrenfest configurations (solid line) compared with MCTDH calculation (dashed line). Frame (a) shows the spectrum obtained with only 24 configurations. Frame (b) is for 290 configurations.

Fig. 3 shows the pyrazine absorption spectrum obtained with the sampling 1 (frame (a)) and sampling 2 (frame (b)) with $\delta = 0.75$. The CPU time required by propagation with the sampling 1 is 9 s, which is very fast even though the propagation had to be repeated 200–300 times with different initial positions of the 20 less important modes. The basis of only 34 Ehrenfest configurations reproduces the main features of the spectrum. The propagation with 290 configurations is in good agreement with MCTDH benchmark. Variation of the "spacing" δ does not significantly affect the result for the sampling 2 providing that δ is small enough to ensure good initial overlap between the basis configurations.

Thus the conclusion of this section is that if a good sampling of initial positions of the trajectory guided Ehrenfest configurations is found accurate multidimensional propagation can be achieved at very low cost. Very small number of Ehrenfest trajectories is sufficient to obtain main features of the dynamics.

3.3 System-bath model of sticking to the surface

The MCE technique can be applied to the processes other than electronically nonadiabatic dynamics. An example of such process is sticking of hydrogen to the surface, which can be described by the so called system-bath model. The interaction of H with the surface is given by Morse oscillator coupled to a harmonic bath of phonons.

$$\hat{H} = \frac{\hat{P}^2}{2M} + D\left(e^{-2\alpha \hat{X}} - 2e^{-\alpha \hat{X}}\right) + \sum_{k=1}^{f}\left(\frac{\hat{p}_k^2}{2m} + \frac{m\omega_k^2 \hat{x}_k^2}{2}\right) - \sum_{k=1}^{f} c_k \frac{1 - e^{-\alpha \hat{X}}}{\alpha} \hat{x}_k \quad (18)$$

The parameters of the model [29] $D = 1.55$ eV $= 0.0569614$ E_h, $\alpha = 1.238$ a_0, $M = 1837.361916$ m_e, $m = 10^4 m_e$ have been supplied by M. Nest.[29] The frequencies of the $f = 50$ phonon modes were chosen as $\omega_k = k\Delta\omega = k\frac{\omega_f}{f}$ where $\omega_f = 0.826827$ fs^{-1} = 0.02 au. The coupling constants between phonon oscillators and the hydrogen mode were chosen as $c_k = \left(\frac{2Mm\gamma\Delta\omega^3}{\pi}\right)^{\frac{1}{2}}$ with the fixed relaxation rate $\gamma = 10^{-3}$ fs^{-1}.

In the most recent MCE calculation the wave function of hydrogen considered as quantum subsystem (X coordinate in (18)) was represented on a grid of 512 Discrete Variable Representation (DVR) points covering the interval 0.2–20 au of X, the distance of H from the surface. Thus in eqn(17) each Ehrenfest configuration is $a_{1k}|1\rangle + a_{2k}|2\rangle + \ldots + a_{512k}|512\rangle$ contains 512 terms as opposed to 2 terms in the previous examples. The H wave packet was initially positioned at the distance *8 au* from the surface and the wave function of phonons was represented by Gaussian Coherent States at Equilibrium. An important technicality is that due to a large number of DVR points Fast Fourier Transform method has been employed to perform all matrix multiplications. The sticking probability was calculated as

$$P_{stick} = Tr\left\{\left(\sum_n |n\rangle\langle n|\right)|\Psi\rangle\langle\Psi|\right\} \quad (19)$$

where the sum is over the bound states of hydrogen on the surface and Ψ is the total wave function. Similarly to the Spin–Boson case the phonon part of the density matrix $|\Psi\rangle\langle\Psi|$ was first expanded as a superposition of coherent states and later each of them was propagated independently using the MCE technique. The initial state of the bath at $T = 0$ was taken to be the ground state $|\Psi\rangle = |z_0\rangle$. Then at time $t = 0$ the density matrix was written in the diagonal form as

$$|\Psi\rangle\langle\Psi| \approx |z\rangle\frac{|\langle z|z_0\rangle|^2}{A}\langle z| \quad (20)$$

Then each initial state $|\Psi^{sstm}\rangle|z\rangle$ was expanded on a small basis of Ehrenfest configurations (10) and propagated independently. The sticking probability was then found by averaging several propagations (about a 100).

Fig. 4 shows that the sticking coefficient of the wave packet initially located 8 Å from the surface at $T = 0$ K as a function of incident energy is close to that of obtained by MCTDH. The MCE sticking probability is 15% below that of MCTDH. The only approximation used in this calculation is the expression (20) for the diagonal form of the density matrix. The basis of coherent states is nonorthogonal and overcomplete still operators can be written in the diagonal form, although the exact form of the density operator (20) is more complex.[30]

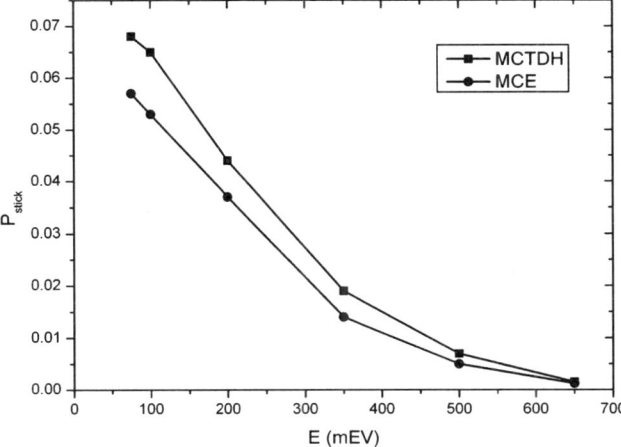

Fig. 4 Dependence of sticking coefficient on the incident energy of H atom wave packet.

4 Comparison with other trajectory-based methods. Why MCE?

As described in the introduction above, short time quantum dynamics can play an important role even in large molecular systems and a theoretical method is needed to simulate such dynamics. One of the advantages of the methods based on trajectory guided grids (or basis sets) of Gaussian coherent states is that they can be coupled with electronic structure theories and implemented "on the fly" just like direct classical dynamics. In principle this allows simulations without preconditions. The trajectory-guided methods also allow strategies which "grow" potential energy surfaces in the dynamically important regions accumulating the data about PES in the areas visited by trajectories. On the fly methodology has already been implemented in the FMS technique,[9,14] which has been used to simulate electronically dynamics in a variety of practical systems. Fig. 5(a) sketches the trajectories of Ehrenfest configurations driven by the eqn (6,7) and compares them with those of the FMS method (Fig.5(b)). In FMS when a Gaussian approaches the region of PESs intersection another Gaussian ("child") is spawned. The parent and the child move on two very different PESs and run away from each other quickly so that quantum coupling between them is quickly lost. In this case FMS still works produsing good results, but rather as a semiclassical technique. The Ehrenfest trajectories guided by the average Hamiltonian (7) stay close to each other for longer than those of FMS which are moving on two different potential energy surfaces and therefore diverge very quickly.

Thus we might expect that MCE can provide faster convergence and the above examples (section 3.1 and 3.2) show that indeed good quality fully quantum results can be achieved. In addition, as shown in the section 3.3, the MCE can treat quantum systems which require large bases sets. It would be difficult to define good spawning algorithm and use FMS for such problems. On the other hand FMS algorithm of spawning is advantageously based on a physically motivated picture of surface hopping. Perhaps one can combine the advantages of the two techniques using MCE in the region of intersection and later reprojecting the wave function on the individual surfaces in the spirit of FMS thus recovering the intuitive surface-hopping-like picture.

Currently we are also working on comparison of the MCE technique with fully variational methods [15–17] aiming to understand whether the use of fully variational trajectories would result in gains in the basis set size sufficient to justify the extra cost of the fully variational methodology.[30]

In summary this paper briefly describes the recently developed method of Multiconfigurational Ehrenfest Dynamics which has been successfully tested on a number of model problems, yielding well converged results for electronically nonadiabatic dynamics. Just like multiple spawning the MCE method is well suited for

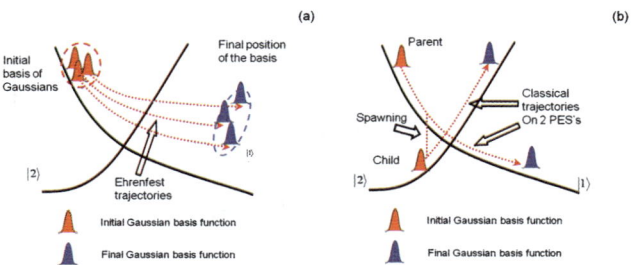

Fig. 5 A basis evolving along Ehrenfest trajectories stays compact for a long time ensuring coupling between the basis configurations. In MS approach the "child" basis function runs away from its "parent" very quickly. Initial and final positions of the basis Gaussian are shown by red and blue, respectively.

simulations "on the fly". The section 3.1 reviews previous calculation of the Spin–Boson model, which is a prototype for light harvesting and many other physical systems currently treated with approximate techniques.[31,32] Section 3.2 shows that the nonadiabatic dynamics and the main features of the pyrazine absorbtion spectrum were obtained on the basis of 34 Ehrenfest configurations only. Such small basis set is certainly within the reach of "on the fly" methodology. In addition the section 3.3 shows that MCE works for quantum dynamics of hydrogen coupled with the bath of surface phonons, where quantum subsystem must be described by a large number of basis functions as opposed to few electronic states in nonadiabatic dynamics. This new example shows that the MCE technique can be applied to various dynamics of hydrogen, which is difficult for a spawning technique like FMS. In all examples presented in this article all degrees of freedom including those of the nuclei are treated on a fully quantum level. A number of projects are under way, which are related to photodynamics and quantum dynamics of hydrogen. The MCE method should become a useful tool for simulating coherent quantum dynamics in large multidimensional systems, which is a necessary step towards the development of coherent control techniques.

Acknowledgements

The author is grateful to M. Nest for sending the parameters of the system-bath model of sticking and the results of MCTDH simulations.

References

1 Y.-Ch. Cheng and G. R. Flemming, *Annu. Rev. Phys. Chem.*, 2009, **60**, 241.
2 E. Collini, C. Y. Wong, K. E. Wilk, P. M. G. Curmi, P. Brumer and G. D. Scholes, *Nature*, 2010, **463**, 644.
3 J. L. Herek, W. Wohlleben, R. J. Cogdell, D. Zeidler and M. Motzkus, *Nature*, 2002, **417**, 533.
4 V. Blanchet, M. Z. Zgierski, T. Seideman and A. Stolow, *Nature*, 1999, **401**, 2.
5 D. G. Kuroda, C. P. Singh, Z. Peng and V. D. Kleiman1, *Science*, 2009, **326**, 263.
6 H.-D. Meyer and G. A. Worth, *Theor. Chem. Acc.*, 2003, **109**, 251.
7 M. H. Beck, A. Jackle, G. A. Worth and H.-D. Mayer, *Phys. Rep.*, 2000, **324**, 1.
8 A. Raab, G. A. Worth, H.-D. Meyer and L. S. Cederbaum, *J. Chem. Phys.*, 1999, **110**, 936.
9 M. Ben-Nun and T. J. Martinez, *Adv. Chem. Phys.*, ed. By I. Prigogineand S. Rice, Wiley, 121 (2002) 439.
10 T. J. Martinez, *Acc. Chem. Res.*, 2006, **39**, 119.
11 B. G. Levine and T. J. Martinez, *Annu. Rev. Phys. Chem.*, 2007, **58**, 613.
12 S. Yang, J. D. Coe, B. Kaduk and T. J. Martinez, *J. Chem. Phys.*, 2009, **130**, 134113.
13 B. G. Levine, J. D. Coe, A. M. Virshup and T. J. Martınez, *Chem. Phys.*, 2008, **347**, 3.
14 M. Ben-Nun and T. J. Martinez, *Isr. J. Chem.*, 2007, **47**, 75.
15 I. Burghardt, H.-D. Meyer and L. S. Cederbaum, *J. Chem. Phys.*, 1999, **111**, 2927; I. Burghardt, K. Giri and G. A. Worth, *J. Chem. Phys.*, 2008, **129**, 174104.
16 G. A. Worth and I. Burghardt, *Chem. Phys. Lett.*, 2003, **368**, 502; I. Burghardt, M. Nest and G. A. Worth, *J. Chem. Phys.*, 2003, **119**, 5364.
17 G. A. Worth, M. A. Robb and I. Burghardt, *Faraday Discuss.*, 2004, **127**, 307; M. Araújo, B. Lasorne, A. L. Magalhães, G. A. Worth, M. J. Bearpark and M. A. Robb, *J. Chem. Phys.*, 2009, **131**, 144301.
18 D. V. Shalashilin and M. S. Child, *Chem. Phys.*, 2004, **304**, 103.
19 D. V. Shalashilin and M. S. Child, *J. Chem. Phys.*, 2004, **121**, 3563.
20 D. V. Shalashilin and M. S. Child, *J. Chem. Phys.*, 2008, **128**, 054102.
21 D. V. Shalashilin, M. S. Child and A. Kirrander, *Chem. Phys.*, 2008, **347**, 257.
22 D. R. Glowacki, S. K. Reed, M. J. Pilling, D. V. Shalashilin and Emilio Martınez-Nunez, *Phys. Chem. Chem. Phys.*, 2009, **11**, 963.
23 D. V. Shalashilin, *J. Chem. Phys.*, 2009, **130**, 244101.
24 D. V. Shalashilin, *J. Chem. Phys.*, 2010, **132**, 244111.
25 G. D. Billing, *J. Chem. Phys.*, 1976, **65**, 1; G. D. Billing, *Chem. Phys. Lett.*, 1983, **100**, 535; G. D. Billing, *The quantum classical theory*, Oxford University Press, 2003; A. García-Vela, R. B. Gerber and D. G. Imre, *J.Chem.Phys.*, 1992, **97**, 7442.
26 H.-D. Mayer and W. H. Miller, *J. Chem. Phys.*, 1979, **70**, 3214.

27 S. Y. Kim and S. Hammes-Schiffer, *J. Chem. Phys.*, 2003, **119**, 4389.
28 S. Hammes-Schiffer and S. R. Billeter, *Int. Rev. Phys. Chem.*, 2001, **20**, 591.
29 M. Nest and H.-D. Mayer, *J. Chem. Phys.*, 2003, **119**, 24.
30 D. Shalashilin (in preparation).
31 A. Olaya-Castro, C. F. Lee, F. Fassioli Olsen and N. F. Johnson, *Phys. Rev. B: Condens. Matter Mater. Phys.*, 2008, **78**, 085115.
32 P. Huo and D. F. Coker, *J. Chem. Phys.*, 2010, **133**, 184108.

PAPER

The influence of the optical pulse shape on excited state dynamics in provitamin D_3

Kuo-Chun Tang and Roseanne J. Sension*

Received 7th March 2011, Accepted 1st April 2011
DOI: 10.1039/c1fd00035g

Broadband visible transient absorption spectroscopy was used to characterize the excited state population of 7-dehydrocholesterol (provitamin D_3, DHC) following excitation by UV pulses with systematically varied linear chirp. These experiments demonstrate that the phase of the excitation pulse can modify the observed excited state decay. The results suggest that coherent mechanisms involving multiple interfering pathways may be exploited to control branching between excited state pathways and manipulate product formation.

1 Introduction

Ultrafast photochemical isomerization reactions of simple polyene chromophores play an essential role in the function of many key biological systems.[1] The electrocyclic ring-opening reaction of 7-dehydrocholesterol (provitamin D_3, DHC) is one important example of such a reaction.[2–6] The photochemical scheme for the formation of vitamin D_3 from DHC *in vivo* involves UV photoexcitation of a 1,3-cyclohexadiene chromophore embedded in the much larger molecule.[7] Excitation of this chromophore results in an electrocyclic ring-opening reaction as illustrated in Fig. 1. The hexatriene containing previtamin D_3 molecule undergoes a thermally activated hydrogen migration to form vitamin D_3. Production of vitamin D_3 from DHC is complicated by competing photochemical reactions producing byproducts containing conjugated dienes or trienes that absorb in the same general region between 300 nm and 250 nm.[7] This system presents an interesting opportunity for optical control of complex molecular dynamics. The goal is selective formation of the previtamin and concurrent suppression of the competing side-reactions. Control of such a complex, chemically important transformation remains a serious challenge to the field of coherent optical control.

Theoretical and experimental studies of the isolated 1,3-cyclohexadiene (CHD) molecule have demonstrated that the ring opening reaction begins on the initially excited 1^1B state surface. Distortion from C_2 symmetry is accompanied by rapid movement of the excited state population to the "dark" 2^1A state.[8–14] The ring-opening reaction is completed following internal conversion to the ground state surface. The entire process is completed within 150 fs and the reaction is described as "more or less ballistic" with negligible barriers along the reaction path.[15,16] The Franck–Condon region of the excited state surface accessed by the original excitation pulse has a lifetime of 10–20 fs in both gas phase and solution phase environments. Fluorescence measurements by Trulson *et al.* are consistent with a ~10 fs lifetime in cyclohexane.[17] Multiphoton ionization experiments on the isolated gas phase molecule place the lifetime of the initially excited state at 21 fs.[16,18] The short

Department of Chemistry, Department of Physics, University of Michigan, 930 N. University Ave., Ann Arbor, Michigan, USA. E-mail: rsension@umich.edu; Fax: +001-734-647-4865; Tel: +001-734-763-6074

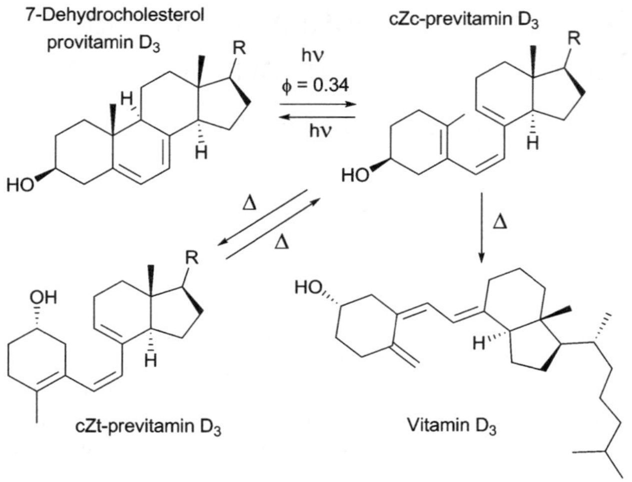

Fig. 1 Photochemical formation of vitamin D_3. The R group is shown explicitly for vitamin D_3 and is the same for all for isomers. The electrocyclic ring-opening and ring-closure reactions are photochemical.[19] The conformational relaxation of the previtamin and the formation of vitamin D_3 are thermally activated reactions.

lifetime of the optically allowed excited electronic state limits the available mechanisms for coherent optical control of the CHD ring-opening reaction.

In contrast, the ring-opening reaction of the cyclohexadiene chromophore embedded in the DHC molecule requires 1–2 ps at room temperature, with evidence for both intramolecular and intermolecular barriers to internal conversion as the molecule isomerizes.[2,3,6] Ultrafast transient absorption measurements have demonstrated that excitation of DHC in the strongly allowed UV absorption band results in formation of an excited state population with a strong visible absorption spectrum as shown in Fig. 2.[2–4,6] There is also evidence for excited state absorption in the UV region of the spectrum, but this is difficult to separate from coherent two-photon absorption of the solvent and the absorption of the hot photoproduct formed following return to the ground state. The disappearance of the visible excited state absorption is well modelled using a biexponential decay of ~0.4–0.65 ps and a long component of 1–1.8 ps around room temperature. The relative amplitudes of the fast and slow components depend on solvent.[6]

Integrated fluorescence measurements have demonstrated that the optically allowed excited state of DHC has a nanosecond lifetime at cryogenic temperatures.[20] A barrier of *ca.* 11 kJ mole^{-1} for the ring-opening reaction in the related ergosterol provitamin D_2 molecule is deduced from the temperature dependence. At room temperature the fluorescence quantum yield of DHC is $2-3 \times 10^{-4}$, consistent with an excited state lifetime of 1–2 ps given the oscillator strength of the allowed transition.[6] From this it is concluded that the excited state absorption monitors the population of the optically allowed state of DHC. The picosecond lifetime of the cyclohexadiene chromophore in DHC extends the range of possible control mechanisms allowing time for interaction between the optical pulse and the electronically excited molecule. The visible absorption band provides a probe for the influence of the optical pulse on excited state dynamics.

Optical control, coherent or incoherent, of the CHD ring-opening reaction has been explored experimentally[21–23] and theoretically.[8,12,24–27] The theoretical calculations have emphasized pump–dump mechanisms manipulating the excited state wave packet. The experimental efforts have used learning algorithms to determine optimal pulses and have attempted to deduce the mechanism from the resulting

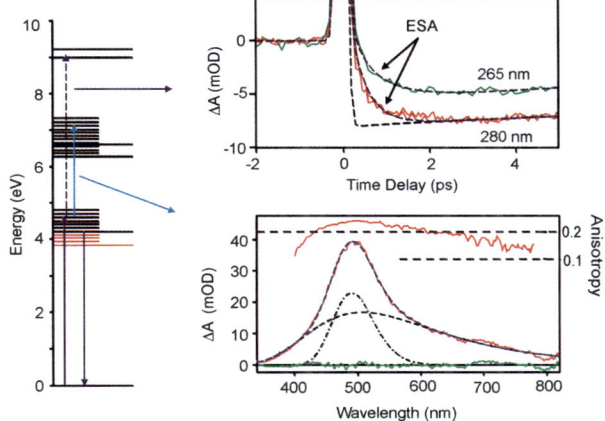

Fig. 2 Excitation of DHC at 266 nm produces an excited state population with a *ca.* 1 ps lifetime. The excited state population can be probed through a UV excited state absorption (top panel) or a visible excited state absorption (bottom panel). The kinetic traces at 265 nm and 280 nm for DHC in n-heptane are dominated by ground state bleaching and recovery, but there is a clear contribution from excited state absorption (ESA) at early times. The visible absorption signal probes transitions to two overlapping higher excited state as indicated by the breadth and shape of the absorption and the absorption anisotropy as a function of wavelength. In the sketch on the left the red states overlapping the optically allowed state at 4 eV indicate that the optically "dark" 2 ^{1}A state may also play a role in the dynamics as observed for cyclohexadiene and other small polyenes.

optimal pulse shapes. Multiphoton excitation with shaped 800 nm pulses was used to manipulate the excited state population and modify the yield of the ring-opening reaction in cyclohexane and hexane solution.[21,22] The photoproduct yield was determined using the absolute absorption of the permanent hexatriene photoproduct nanoseconds after the excitation pulse to define a fitness function. These experiments were able to enhance the formation of hexatriene, but the mechanism of control was unclear. Possible mechanisms include ground state vibrational excitation of CHD, strong-field 'dressing' of the excited electronic state surfaces, and recycling of population after return to the ground state.

An optical control experiment to manipulate the ring-opening reaction of gas phase CHD was performed by Weinacht and coworkers using strong-field UV excitation.[23] The UV pulse was shaped using an acoustooptic modulator and the optimal pulse shape determined by a closed-loop learning algorithm. The hexatriene photoproduct was identified and quantified by the photofragmentation pattern. The UV pulses were ≤ 10 TW cm^{-2} with control observed only in a regime where the UV pulse intensity was sufficient to ionize the CHD molecule. These experiments were able to enhance the formation of hexatriene by 37% over unshaped pulses. The optimal pulses were typically over 100 fs and often displayed subpulses spaced by 80 to 150 fs. Possible control mechanisms suggested include manipulation of the excited state wave packet launched on the optically allowed state and recycling of population after return to the ground state.

The strong-field experiments probing and controlling the ring-opening reaction of CHD are interesting, but ultimately limited in usefulness for three significant reasons. First, the requirement of a strong-field limits the applicability of this approach in any practical sense. For coherent optical control to be useful in any significant chemical way it must use pump-pulses with modest intensities to avoid collateral damage, photofragmentation or photoionization, of either the reactant or the surroundings. Control of the photoisomerization of bacteriorhodopsin has demonstrated that it is possible to modify photochemical processes in the regime where the excitation energy is far below saturation of the one-photon process.[28] These results motivate the attempt to use modest excitation energies to control the ring-opening reaction of a cyclohexadiene chromophore. Second, the absence of an accessible probe of the excited state behavior inhibits analysis of the mechanisms operating in the control process. The excited state absorption of DHC provides one useful probe of the excited state population. Finally, the complex pulse shapes obtained using a learning algorithm to search the available parameter space tend to hinder the detailed understanding of mechanism. Learning algorithms will continue to play an important role in the search for optimal pulses, but there is important work to be done using systematic pulse variation to explore the influence of pulse parameters on molecular dynamics.

In the work reported here we have used the systematic variation of the linear chirp of a UV excitation pulse as a means to manipulate the excited state population in DHC. The visible transient absorption signal is used to monitor the influence of the chirp on the excited state dynamics. It is well established that linear chirp can modify the population of an excited state surface through an intrapulse pump–dump mechanism.[29–35] The pump–dump mechanism itself only operates within the duration of the excitation pulse. Wave packet shaping, however, may persist to influence population dynamics beyond the direct interaction with the pulse. The chirp of a pulse, even in the limit of one-photon interactions, can focus or disperse an excited state wave packet and modify photoproduct formation.[36] The transient absorption probe of DHC allows monitoring of the excited state population over the entire lifetime of the excited state and offers the potential to shed light on the reaction dynamics. These experiments will provide groundwork for future studies combining measures of photoproduct yield, more complex excitation pulses, and more sophisticated probes of the excited state molecules.

2. Experimental

Femtosecond pump–probe transient absorption measurements were performed using a Ti:sapphire oscillator and a home-built 1 kHz multipass ultrafast amplifier to generate femtosecond laser pulses. The laser system yields compressed pulses with a central wavelength of 797 nm, pulse duration of ∼60 fs and pulse energy of ∼700 µJ. A beamsplitter was used to generate pump and probe pulses separately. The major portion (650 µJ) was frequency-doubled and then frequency-summed with residual fundamental to generate an ultraviolet femtosecond pulse with pulse energy of ∼17 µJ at 266 nm. This UV pulse was then passed through an acousto-optic programmable dispersive filter (AOPDF, Dazzler, Fastlite) to output a pulse with programmed linear chirp as measured at the sample position.

The minor portion of the fundamental beam (50 µJ) was delayed by a computer-controlled optical delay line with respect to the pump and then focused into a 1 mm quartz flow cell containing ethylene glycol to produce a broadband white-light continuum pulse (400 nm to 780 nm). Alternatively, a continuum was generated in a CaF_2 window which extended the available spectrum from 340 nm to 810 nm.

Pump and probe pulses were both focused and overlapped with an angle of ∼11° into a 0.5 mm quartz flow cell containing DHC sample solutions. Some experiments were performed using a wire-guided flow arrangement to avoid the cross phase modulation introduced by the cell walls. A mechanical optical chopper (MC1000A, THORLabs) was used to modulate pump pulses. Probe pulses with pump on and pump off at each time delay point were recorded by a multichannel spectrometer (AvaSpec-2048-USB2, Avantes) and used to calculate the difference spectrum. The relative polarization of the pump and probe was set to 54.7° (the magic angle) to eliminate contributions from reorientation of the transition dipole.

Transient absorption measurements were performed for DHC dissolved in n-heptane, n-hexadecane, methanol, and 2-butanol. The sample concentration was 2.6 to 2.8 mM and the sample temperature was 30 °C except in methanol where the temperature was 27 °C. The beam diameter of the pump in the sample cell was about 140 µm measured by a knife-edge scan. DHC samples were pumped with a pulse energy of 0.25 to 0.26 µJ which provides 3.4×10^{11} photons per pulse. From the pump energy, the spot size, and the sample concentration, it is estimated that ~2.5% of the molecules in the sample volume were excited by the pump pulse.

In the experiments reported here the linear chirp of the excitation pulse was varied from -10^4 fs^2 to $+10^4$ fs^2. In order to avoid the possibility that systematic drift of the laser system could perturb the trends these measurements were made using an alternating pattern of chirps: $+12, -10, +1, +4, -2, +8, -6, +2, 0, +6, -4, +10, -8 \times 10^3$ fs^2. The extremes of this range stretched the pulse to *ca.* 0.8 ps, still within the lifetime of the excited state absorption, but longer than the intrinsic lifetime of the fast decay component. For the shortest pulses the peak energy flux was $\sim 2 \times 10^{10}$ W cm^{-2}, almost three orders of magnitude smaller than the UV pulses used in control experiments on CHD described above;[23] at the extremes of the chirp range this dropped by an order of magnitude. The pulse parameters are sufficient to enable non-linear pathways for control, but the pulses do not saturate the linear transition or produce strong non-linear contributions to the signal.

3. Results

The excited state spectrum of DHC in n-heptane is plotted in Fig. 2. Analysis of the decay of the excited state absorption as a function of wavelength is consistent with a biexponential decay of the excited state population. Fig. 3 shows a range of wavelengths for typical data obtained with near-transform-limited excitation pulses. The slow component is ~1.3 ps, and the fast component is ~0.6 ps. The time constants are independent of wavelength; the relative amplitudes vary slightly with wavelength as the spectrum of the slow component is blue-shifted about 4 nm with respect to the spectrum of the fast component.[6] There is no evidence for a coherent vibrational oscillation in the excited state absorption measured in heptane at any wavelength within the available signal-to-noise ratio of the data. When the absorption intensity is integrated across the entire spectrum the fast component accounts for 66% of the decay of the absorption while the slow component accounts for 34% of the decay. Because there is no significant wavelength dependence to the dynamics of the excited state absorption other than the small frequency shift of the two components, we have chosen to analyze the influence of chirp on the excited state dynamics by

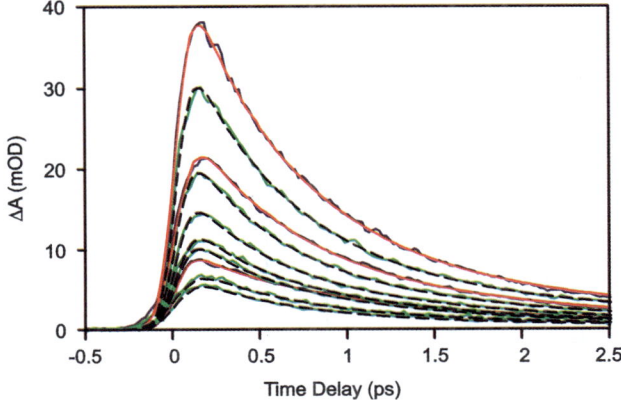

Fig. 3 Selected traces from the excited state absorption of DHC in n-heptane. The wavelengths are 410 nm, 450 nm, 490 nm (blue: data traces, red: fits) and 530 nm, 570 nm 610nm, 650 nm, 690 nm, 730 nm and 770 nm (green: data traces and black dashed lines: fits).

integrating across the spectrum. If the data are analyzed using only the wavelengths near the peak absorption (470 nm–510 nm) or on the red edge of the spectrum (600 nm–770 nm) the results are similar, but the signal/noise ratio is lower.

The decay of the integrated excited state absorption is compared for positive and negative chirps in Fig. 4. The difference between positive and negative chirp for $\pm 2 \times 10^3$ fs^2 could be accounted for by a slight difference in the pump intensity. For the larger chirps there is a clear reduction of the signal at early times for a negatively chirped excitation pulse compared with the signal observed for a positively chirped excitation pulse. The difference cannot be accounted for by a simple scale factor. Integrated over time the decrease in intensity is 2–5% of the total intensity.

In order to further characterize the behaviour of the excited state population the traces for each value of the linear chirp were fit to a single exponential decay of the absorption. The single exponential fit provides an estimate of the excited state behaviour without the ambiguity that may be introduced when the data are fit to a biexponential decay. The time constants, plotted in Fig. 5, show that the average excited state lifetime increases from ca. 0.78 ps with positively chirped excitation pulses to 0.85 ps with negatively chirped excitation pulses. The difference is small, but the trend is clear. The excited state decay is faster for positively chirped pulses than for negatively chirped pulses. This difference, as also observed qualitatively in the comparisons in Fig. 3, is consistent and outside of the error of the data.

A more complete analysis of the excited state dynamics requires fitting the data to a biexponential decay as described above. Such a fit yields time constants of 0.56 $\pm$ 0.06 ps and 1.3 $\pm$ 0.1 ps, in reasonable agreement with the results obtained for more extensive measurement using transform-limited pulses.[6] The time constants do not depend on the chirp of the excitation pulses, but the relative amplitudes of the two components are influenced by the phase of the excitation pulses. The chirp dependence of the amplitudes is plotted in Fig. 5 as fraction slow $F_{slow} = A_{slow}/(A_{fast} + A_{slow})$, where A_{slow} is the amplitude of the 1.3 ps decay component and A_{fast} is the amplitude of the 0.56 ps component. The fraction of the slow component increases from 0.3 when the excitation pulse is positively chirped to 0.39 when the excitation pulse is negatively chirped.

Less extensive data sets were obtained for DHC in three additional solvents, n-hexadecane, methanol, and 2-butanol. Again the data were fit to a biexponential decay of the excited state absorption. The ratio of the amplitude of the fast

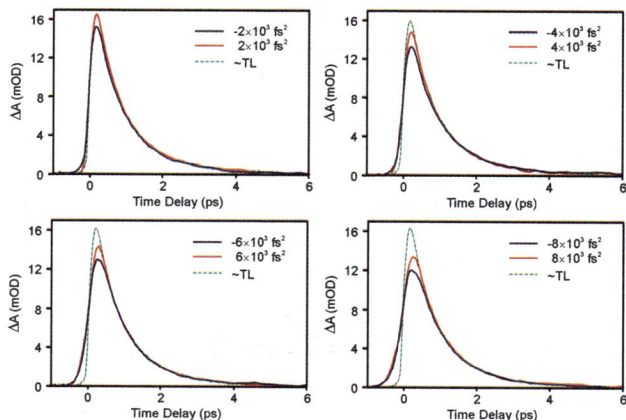

Fig. 4 The influence of the linear chirp of the excitation pulse on the excited state decay. The plots compare positive and negative chirps where the effect of chirp on the integrated pulse duration is similar. The trace obtained with transform-limited (TL) excitation is also shown for comparison.

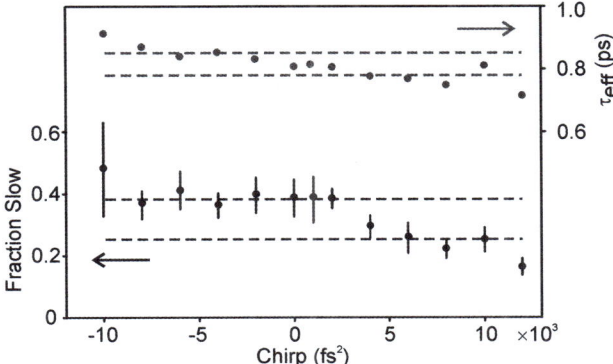

Fig. 5 The influence of chirp on the excited state decay in n-heptane. Top: The gray circles plot the effective lifetime when the data is fit to a single exponential decay of the excited state absorption. Bottom: The traces were fit to a biexponential decay of the excited state absorption. The ratio of the amplitude of the slow component to the sum of the amplitudes of the fast and slow components is plotted (black circles) as a function of the chirp of the excitation pulse.

component to the amplitude of the slow component depends on solvent as reported previously. All three of these solvents have an intrinsically larger contribution from the slow component; F_{slow} = 0.44 in 2-butanol, 0.55 in hexadecane and 0.7 in methanol for transform-limited pulses.[6] The data plotted in Fig. 6 exhibit a dependence of F_{slow} on the chirp of the excitation pulse that is consistent with the trend observed in n-heptane. Positively chirped excitation pulses give rise to a faster decay of the excited state that can be attributed to a lower fractional contribution from the slow component.

The analysis of the influence of chirp on the excited state dynamics of DHC in n-heptane can be considered in more detail by plotting the total amplitude of the excited state absorption ($A_{fast} + A_{slow}$) and the amplitude of each component as a function of the chirp of the excitation pulse. This is shown in Fig. 7. The absolute amplitude is constant for negative chirp, equal to the amplitude of the near transform-limited pulses, but increases ~10% for positive chirp. The increase in the total amplitude is decomposed into an increase in the amplitude of the fast component accompanied by a somewhat smaller decrease in the slow component.

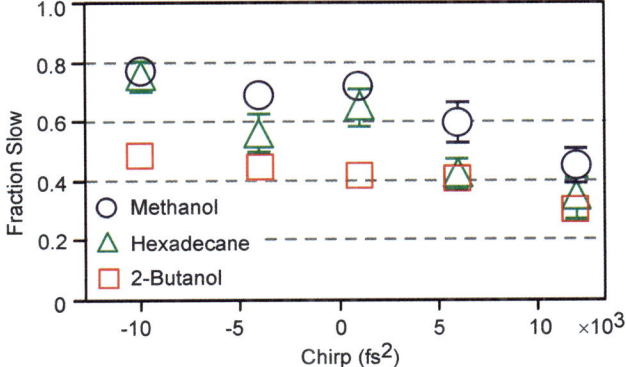

Fig. 6 The influence of chirp on the biexponential fit to the excited state decay of DHC in methanol, 2-butanol, and hexadecane. Error bars are shown when they are larger than the symbols representing the data, for most other point the error is approximately the same as the symbol size.

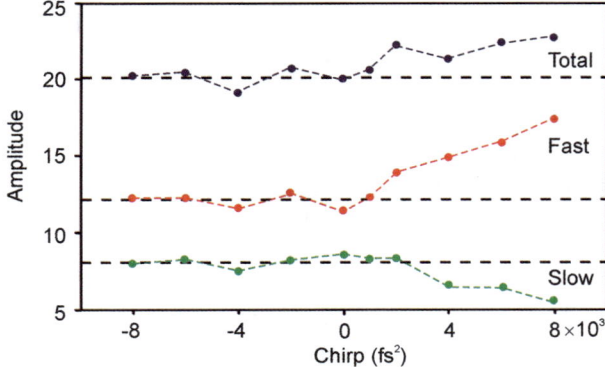

Fig. 7 Amplitude change in the total signal (blue) for DHC in n-heptane. The change is also broken down into the contributions from the slow component (green) and the fast component (red).

4. Discussion

The results presented above demonstrate that the linear chirp of the excitation pulse has an influence on the excited state population of DHC as probed through the visible excited state absorption. In order to put this observation into context it is useful to first summarize the potential explanations for the biexponential excited state decay in DHC.[6]

The biexponential decay of the excited state population in DHC has two alternative potential explanations. The initial excitation process will populate the optically allowed 1^1B state. This state decays on an ultrafast time scale, ~10 fs, in the simple isolated cyclohexadiene molecule, but persists for a much longer time in DHC. The fluorescence spectrum and quantum yield suggest that the 1^1B state has a lifetime of ca. 1 ps. Thus it is concluded that the excited state absorption probes this optically allowed state. The anisotropy of the excited state absorption is constant in time, leading to the conclusion that the excited state absorption arises from one excited electronic state and the decay of the absorption reflects the population decay of this excited state.

Decay of the optically allowed 1^1B state of cyclohexadiene leads to population of a "dark" 2^1A with a lifetime of ca. 80 fs. If the 1^1B and 2^1A states of the cyclohexadiene chromophore are isoenergetic or nearly isoenergetic in DHC, the biexponential decay of the excited state absorption may represent equilibration between these two states accompanied by internal conversion of the equilibrated population to the ground state. The general scheme is $1^1B \to (1^1B \rightleftarrows 2^1A) \to$ products. The ground state product is either previtamin D_3 or recovery of the original DHC configuration.

Alternatively, the biexponential decay of the excited state population may reflect parallel reaction pathways on the excited state surface. The CHD chromophore is helical, but carries C_2 symmetry. Theoretical calculations of CHD predict that the initial motion on the optically populated 1^1B state preserves the C_2 symmetry. Distortion from the initial C_2 symmetry provides two symmetry equivalent pathways for internal conversion from the 1^1B state to the 2^1A state; two symmetry equivalent minima on the 2^1A state are populated prior to internal conversion to the ground state.[16] In DHC the symmetry is lowered and the analogs of these two pathways will no longer be equivalent. The rapid bifurcation of population on the excited state surface will populate more than one distinct conformation. Excited state absorption from these two conformations will have similar spectral form and similar rotation of the transition dipole with respect to the transition dipole direction excited by the pump pulse.

We have presented arguments previously for preferring the parallel pathway model to account for the biexponential decay of the excited state of DHC.[6] Such a hypothesis is consistent with the observed data, but the argument based on transform-limited pulses alone is not conclusive. The chirp dependent data presented here helps to distinguish between these possible mechanisms for the excited state dynamics of DHC. The chirp *independence* of the excited state lifetimes along with the chirp *dependence* of the amplitudes provides an additional argument against the equilibration model and in favor of parallel excited state pathways. If the chirp of the excitation pulse modifies an equilibration process it will do so by modifying the apparent rate of equilibration. The relative amplitudes are determined by the rate constants and by the relative degeneracy or density of states in the 1^1B and 2^1A manifolds.[6] The excitation pulse cannot modify the density of states and thus cannot modify the amplitudes without modifying the dynamics.

In contrast there are several mechanisms by which the excitation pulse may modify the branching of an initial population on the excited state surface. Assuming parallel excited state pathways, the changes in the amplitudes of the fast and slow components are directly related to changes in the populations of the distinct excited state conformations. The subsequent dynamics are independent of the chirp of the excitation pulse, inherent properties of the distinct incoherent excited state populations.

Both linear one-photon interactions and nonlinear multiphoton interactions provide possible mechanisms for coherent optical control of excited state population in DHC under the experimental conditions employed here. An energy flux of 0.2 to 2 $\times 10^{10}$ W cm^{-2} is sufficient for measurable two-photon absorption at 266 nm. One signature of two-photon absorption and ionization of the solvent, with possible contributions from the solute, is seen in the small persistent absorption observed in all solvents investigated here. The signal is weak, with an absorbance of <1 mOD across the visible spectrum, but it is present in all of the transient absorption data including transient absorption data measured on the pure solvents. On the other hand, these experiments were carried out in an intensity realm where linear one-photon interactions dominate. There is no evidence for saturation of the absorption.

Nonlinear mechanisms for control

Nonlinear mechanisms for coherent or incoherent optical control involve multiphoton interactions through multiple pathways. One of the simplest mechanisms is the intrapulse pump–dump manipulation of excited state population. In such a mechanism the initial population excited by the leading high frequency portion of a negatively chirped pulse can be dumped back to the ground state by the trailing lower frequency portion of the pulse. In contrast, for a positively chirped pulse the entire pulse acts to place population in the excited state. The initial population cannot be dumped back to the ground state by the trailing higher frequency portion of the pulse. Such mechanisms have been demonstrated to modify the population of the excited state in a number of different systems, including organic dyes and proteins.[29–35]

The data plotted in Fig. 7 illustrates that the total amplitude of the excited state absorption of DHC increases for positive chirp, consistent with the prediction of an intrapulse pump–dump mechanism. This increase, however, cannot be correlated directly with the excited state population apart from an assumption about the intrinsic oscillator strength of the transitions giving rise to each component. If the oscillator strength is constant, the change in total intensity points to a nonlinear control mechanism contributing to the observed dynamics. If, however, the intrinsic oscillator strength of the state probed by the slow decay component is approximately a factor of two smaller than the oscillator strength of the fast component, the change in total signal will correlate with a change in the branching between excited state pathways without a concomitant change in the total excitation probability. Thus the amplitude of the excited state absorption alone cannot be taken as an indication of coherent or incoherent manipulation of total excited state population.

The intrapulse pump–dump mechanism is also rendered improbable by the detuning of the pump pulse from the red edge of the ground state absorption spectrum of DHC. The influence exercised in the pump–dump mechanism is maximized when the pump pulse falls near the red edge of the spectrum, overlapping both the absorption spectrum and the fluorescence spectrum. When the excitation pulse is detuned from the red edge of the absorption spectrum the transition will saturate as the pump intensity is increased, but the dependence of the population on the phase of the excitation pulse diminishes or disappears.[34] The 266 nm excitation pulse used to excite DHC in the experiments presented here falls $\sim$4000 cm^{-1} above the red edge of the absorption band, well away from the observed fluorescence. Thus it is unlikely that the pump–dump mechanism plays a significant role in the influence of chirp of a 266 nm pulse on the total amplitude of the excited state population.

Alternative nonlinear mechanisms that are potentially operational in the excitation of DHC involve multiphoton interactions and higher excited electronic states. The excitation pulse at 266 nm can access an excited state absorption from the initially excited state. The signature of this excited state absorption is seen in the transient absorption traces plotted in the top panel of Fig. 2.[2,3] The very strong positive signal at zero time delay, truncated in these plots, arises from coherent two-photon absorption by the solvent. This signal is observed throughout the UV and is present for pure solvent. It provides a measure of the cross correlation of the pump and probe pulses. The subsequent dynamics reflect the behavior of the DHC solute in the excited electronic state. Although the signal at 265 nm and 280 nm is dominated by a net bleaching of the ground state absorption followed by ground state recovery, there is a clear indication of an excited state absorption with a *ca.* 1 ps lifetime on top of the ground state bleaching signal.

Mechanisms for manipulating the population of the 1^1B state through interaction with higher excited states rely on absorption and emission moving population between electronically excited states. Two-photon absorption at 266 nm will either deplete the population of the excited state accessed following one-photon excitation leading to new channels for photochemistry unobserved in the current measurements or will result in an additional channel for population of the one-photon allowed state through internal conversion. There is no evidence in the data for significant population returning to the lowest excited electronic state *via* internal conversion.

Two-photon excitation will place $\sim$37600 cm^{-1} of excess energy into the molecule in addition to the $\sim$3500 cm^{-1} supplied in the initial one-photon excitation process. If the vibrations are assumed to be harmonic with frequencies comparable to the ground state frequencies of DHC the internal vibrational temperature of the molecule will be $\sim$870 K. This is 500 K hotter than the molecules produced *via* one-photon excitation. If the energy is not dispersed throughout the molecule, the effective temperature at the chromophore will be even higher. The additional energy placed in the molecule would be expected to modify the observed rate constants and to introduce new dynamical components into the decay of the excited state absorption. Yet the rate constants extracted from the data are independent of the chirp, only the amplitudes appear to vary. As a result two-photon population transfer seems unlikely to contribute to the results of these experiments on DHC.

While 'trivial' nonlinear mechanisms for manipulating the excited state population of DHC can be eliminated from consideration, more complex coherent interactions cannot be eliminated from consideration *a priori* under the pulse parameters employed in the experiments described here. The evaluation of the likelihood of coherent multiphoton interactions, with multiple interfering pathways exploited to manipulate the excited state wave packet and the branching between excited state pathways, will require more detailed theoretical simulation.

Control mechanisms in the linear regime

Nonlinear interactions between the optical pulse and the material system do not exhaust the realm of control possibility. Multiple pathway interference can influence wave packet formation and product yields in the regime where only linear one-photon absorption occurs. In a pivotal experiment Prokhorenko *et al.* demonstrated phase-only optical control of the bacteriorhodopsin isomerization reaction with pulse intensities in the weak-field limit.[28] These results, with the claim of phase only control of product yield in the limit of one-photon excitation, countered the widely accepted idea that one-photon control was impossible. With the impetus provided by these experimental observations, Katz, Ratner, and Kosloff demonstrated coupling of a system to a bath can change the picture and permit weak-field control.[36] Dissipation of energy into the bath provides an effective dump pulse and leads to the control of the population. The time scale of energy relaxation becomes an important parameter in the process.

Spanner, Arango, and Brumer considered the conditions for one-photon phase control systematically. They demonstrated that one-photon control is possible for isolated molecules if the measured observable does not commute with the system Hamiltonian.[37] This could be true, for example, when a photochemical process is an isomerization involving substantive molecular rearrangement. Open systems, with the quantum system being measured embedded in a bath, constitute a more important class of systems where one-photon control is theoretically possible. The conclusion reached by Spanner *et al.* is that virtually every property of the open quantum system is in principle controllable. Practical coherent control, produced by available pulses and observable in experiment, is another issue yet to be investigated completely.

The ring-opening reaction of DHC provides an example of an open quantum system embedded in an environment provided both by the vibrational modes of the molecule apart from the cyclohexadiene chromophore and the external interaction with the surrounding solvent. Excited state conformations are differentiable when coupling to the bath allows dissipation of the energy trapping the molecule in a specific potential well. Thus multiple pathway interference as described by Spanner *et al.* could in principle manipulate product yields, even in the limit of linear one-photon excitation.[37]

With or without a change in the total excitation probability, there is a clear change in the branching between two excited state pathways in the data plotted in Fig. 7. For positively chirped pulses there is increased population in the species responsible for the fast decay component and decreased population in the species responsible for the slow decay component. A full interpretation of the excited state dynamics of DHC and the role of pulse parameters in manipulating this population awaits accurate theoretical calculations of the relevant DHC potential energy surfaces.

5. Conclusions

The experiments presented here explore the influence of linear chirp in an excitation pulse on the excited state dynamics of provitamin D_3. The experimental results demonstrate that positive chirp on the excitation pulse can influence the branching between excited state pathways. In n-heptane positive chirp increases the population of a pathway characterized by fast, ~0.6 ps, decay of the excited state and decreases the population of a pathway characterized by slower, ~1.3 ps decay of the excited state. Preliminary measurements in other solvents show a similar influence of chirp on the excited state dynamics. Positive chirp decreases the fractional contribution of the slow, 1.0–1.8 ps component and increases the fractional contribution of the fast 0.4–0.6 ps component. The influence of chirp on the excited state population is not explained by "trivial" incoherent population mechanisms and points to coherent manipulation of population through nonlinear or through linear interactions.

More detailed theoretical simulations, including quantum chemical calculations of the potential energy surfaces in DHC and simulation of the interaction of optical pulses with the material system, will be required for a full interpretation of the results.

Acknowledgements

This work has been supported by the National Science Foundation through Grant No. CHE-0718219 and through the FOCUS Center at the University of Michigan. We would also like to thank Prof. Ken Spears for useful advice and conversation.

References

1. V. Sundstrom, *Annu. Rev. Phys. Chem.*, 2008, **59**, 53–77.
2. N. A. Anderson and R. J. Sension, in *Liquid Dynamics: Experiment, Simulation, and Theory*, ed. J. T. Fourkas, American Chemical Society, Washington, D.C., 2002, vol. 820, pp. 148–158.
3. N. A. Anderson, J. J. Shiang and R. J. Sension, *J. Phys. Chem. A*, 1999, **103**, 10730–10736.
4. W. Fuss, T. Hofer, P. Hering, K. L. Kompa, S. Lochbrunner, T. Schikarski and W. E. Schmid, *J. Phys. Chem.*, 1996, **100**, 921–927.
5. W. Fuss and S. Lochbrunner, *J. Photochem. Photobiol., A*, 1997, **105**, 159–164.
6. K.-C. Tang, A. Rury, M. B. Orozco, J. Egendorf, K. G. Spears and R. J. Sension, *J. Chem. Phys.*, 2011, **134**, 104503.
7. I. P. Terenetskaya, *Theor. Exp. Chem.*, 2008, **44**, 286–291.
8. J. B. Schonborn, J. Sielk and B. Hartke, *J. Phys. Chem. A*, 2010, **114**, 4036–4044.
9. A. Nenov, P. Kolle, M. A. Robb and R. de Vivie-Riedle, *J. Org. Chem.*, 2010, **75**, 123–129.
10. T. Mori and S. Kato, *Chem. Phys. Lett.*, 2009, **476**, 97–100.
11. H. Tamura, S. Nanbu, T. Ishida and H. Nakamura, *J. Chem. Phys.*, 2006, **124**, 084313.
12. M. Garavelli, C. S. Page, P. Celani, M. Olivucci, W. E. Schmid, S. A. Trushin and W. Fuss, *J. Phys. Chem. A*, 2001, **105**, 4458–4469.
13. A. Y. Li, S. A. Yuan, Y. S. Dou, Y. B. Wang and Z. Y. Wen, *Chem. Phys. Lett.*, 2009, **478**, 28–32.
14. H. Tamura, S. Nanbu, H. Nakamura and T. Ishida, *Chem. Phys. Lett.*, 2005, **401**, 487–491.
15. F. Rudakov and P. M. Weber, *Chem. Phys. Lett.*, 2009, **470**, 187–190.
16. K. Kosma, S. A. Trushin, W. Fuss and W. E. Schmid, *Phys. Chem. Chem. Phys.*, 2009, **11**, 172–181.
17. M. O. Trulson, G. D. Dollinger and R. A. Mathies, *J. Chem. Phys.*, 1989, **90**, 4274–4281.
18. W. Fuss, W. E. Schmid and S. A. Trushin, *J. Chem. Phys.*, 2000, **112**, 8347–8362.
19. H. J. C. Jacobs, J. W. J. Gielen and E. Havinga, *Tetrahedron Lett.*, 1981, **22**, 4013–4016.
20. N. Nakashima, S. R. Meech, A. R. Auty, A. C. Jones and D. Phillips, *J. Photochem.*, 1985, **30**, 207–214.
21. E. C. Carroll, J. L. White, A. C. Florean, P. H. Bucksbaum and R. J. Sension, *J. Phys. Chem. A*, 2008, **112**, 6811–6822.
22. E. C. Carroll, B. J. Pearson, A. C. Florean, P. H. Bucksbaum and R. J. Sension, *J. Chem. Phys.*, 2006, **124**, 114506.
23. K. Kotur, T. Weinacht, B. J. Pearson and S. Matsika, *J. Chem. Phys.*, 2009, **130**, 134311.
24. D. Geppert and R. de Vivie-Riedle, *J. Photochem. Photobiol., A*, 2006, **180**, 282–288.
25. D. Geppert and R. de Vivie-Riedle, *Chem. Phys. Lett.*, 2005, **404**, 289–295.
26. A. Hofmann and R. de Vivie-Riedle, *Chem. Phys. Lett.*, 2001, **346**, 299–304.
27. A. Hofmann and R. de Vivie-Riedle, *J. Chem. Phys.*, 2000, **112**, 5054–5059.
28. V. I. Prokhorenko, A. M. Nagy, S. A. Waschuk, L. S. Brown, R. R. Birge and R. J. D. Miller, *Science*, 2006, **313**, 1257–1261.
29. V. I. Prokhorenko, A. M. Nagy and R. J. D. Miller, *J. Chem. Phys.*, 2005, **122**, 184502.
30. O. Nahmias, O. Bismuth, O. Shoshana and S. Ruhman, *J. Phys. Chem. A*, 2005, **109**, 8246–8253.
31. C. J. Bardeen, V. V. Yakovlev, J. A. Squier and K. R. Wilson, *J. Am. Chem. Soc.*, 1998, **120**, 13023–13027.
32. C. J. Bardeen, V. V. Yakovlev, K. R. Wilson, S. D. Carpenter, P. M. Weber and W. S. Warren, *Chem. Phys. Lett.*, 1997, **580**, 151–158.
33. G. Cerullo, C. J. Bardeen, Q. Wang and C. V. Shank, *Chem. Phys. Lett.*, 1996, **262**, 362–368.
34. E. C. Carroll, A. C. Florean, P. H. Bucksbaum, K. G. Spears and R. J. Sension, *Chem. Phys.*, 2008, **350**, 75–86.

35 A. C. Florean, E. C. Carroll, K. G. Spears, R. J. Sension and P. H. Bucksbaum, *J. Phys. Chem. B*, 2006, **110**, 20023–20031.
36 G. Katz, M. A. Ratner and R. Kosloff, *New J. Phys.*, 2010, **12**, 015003.
37 M. Spanner, C. A. Arango and P. Brumer, *J. Chem. Phys.*, 2010, **133**, 151101.

PAPER

Wavepacket and potential reconstruction by four-wave mixing spectroscopy: preliminary application to polyatomic molecules

David Avisar and David J. Tannor*

Received 27th March 2011, Accepted 3rd May 2011
DOI: 10.1039/c1fd00048a

We have recently shown how the excited-state wavepacket of a polyatomic molecule can be completely reconstructed from resonant coherent anti-Stokes Raman spectroscopy [Avisar and Tannor, *Phys. Rev. Lett.*, 2011, **106**, 170405]. The method assumes knowledge of the ground-state potential but not of any excited-state potential, however the latter can be computed once the excited-state wavepacket is known. The formulation applies to dissociative as well as bound excited potentials. We demonstrate the method on the Li_2 molecule with its bound first excited-state as well as with a model dissociative excited state potential. Preliminary results are shown for a model two-dimensional molecular system. The calculations assume constant transition dipole moment (Condon approximation), δ-pulse excitation and a single excited-state potential, but we discuss the implications of removing these assumptions.

1. Introduction

Probing the real time dynamics of reacting molecules is a long-standing aim in chemical research. For several decades now, femtosecond pump–probe spectroscopies have been employed to study transition states of molecules reacting on excited potential surfaces.[1-5] Although these studies have shed a tremendous amount of light on excited-state dynamics, none of the methods in use provides complete information on the excited-state wavefunction. The need for an experimental method that will provide this information is compounded by the fact that theoretical *ab initio* calculations for excited states are difficult and generally of limited accuracy except in small systems.[6]

Several methods have been proposed for reconstructing excited-state wave functions from spectroscopic signals. Shapiro has suggested wave function imaging using the excited vibrational eigenstates as an expansion basis.[7] The coefficients are obtained from spectroscopic information and later used to reconstruct the excited potential, from which the basis is obtained. Cina has suggested a method of wave function reconstruction that assumes that one of the excited-state potentials is known (not necessarily the potential on which the propagation occurs).[8-10] There have also been various proposals for reconstructing excited-state potentials from spectroscopic data.[11-17] Experimental work has focused on wavepacket interferometry of vibrational wavepackets[18-20] as well as electronic Rydberg wavepackets.[21,22]

The approach we present here assumes knowledge of the ground-state potential but not of any excited potential. Our strategy is to express the reacting-molecule wavefunction, $|\Psi(t)\rangle$, as a superposition of the vibrational eigenstates of the ground-state Hamiltonian, $\{|\psi_g\rangle\}$:

Department of Chemical Physics, The Weizmann Institute of Science, PO Box 26, Rehovot, 76100, Israel. E-mail: david.tannor@weizmann.ac.il; Tel: +97 2-8-934-2094

$$|\Psi(t)\rangle = \sum_g |\psi_g\rangle\langle\psi_g|\Psi(t)\rangle \equiv \sum_g C_g(t)|\psi_g\rangle. \qquad (1)$$

Since the vibrational eigenstates $\{|\psi_g\rangle\}$ are assumed known, the challenge is to find the time-dependent superposition coefficients $C_g(t)$. Note that in principle the approach is completely general for polyatomics.

2. Theoretical methodology

2.1 Preliminaries

Consider a two-state molecular system within the Born–Oppenheimer approximation. The nuclear Hamiltonians H_g and H_e correspond, respectively, to the (known) ground- and (unknown) excited-state potentials, which can be of any dimension. Applying first-order time-dependent perturbation theory, the wavepacket that we want to reconstruct is[23]

$$|\Psi(t)\rangle = -i\int_{-\infty}^{t} dt_1 e^{-iH_e(t-t_1)}\{-\mu\varepsilon_1(t_1)\}|\psi_0\rangle, \qquad (2)$$

where $\varepsilon(t)$ is the incident laser pulse that promotes ground-state amplitude to the excited state and μ is the electronic transition dipole moment. The initial state, $|\psi_0\rangle$, is the vibrational ground-state of H_g with the eigenfrequency ω_0. (Here and henceforth we take $\hbar = 1$.) For simplicity, we consider a δ-pulse excitation as well as a coordinate-independent electronic transition dipole, μ (Condon approximation). Thus, the wavepacket takes the simplified form

$$|\Psi(t)\rangle = i\mu\varepsilon_1 e^{-iH_e t}|\psi_0\rangle \equiv i\mu\varepsilon_1|\psi(t)\rangle, \qquad (3)$$

where ε_1 is the amplitude of the pulse and t is the propagation time on the excited state measured from the time of pulse excitation. Note that within a proportionality constant the excited-state wavepacket $|\Psi(t)\rangle$ is equal to $|\psi(t)\rangle = e^{-iH_e t}|\psi_0\rangle$, the vibrational ground-state of H_g propagated on H_e.

A physical interpretation of the superposition coefficients, $C_g(t)$, may be obtained by substituting eqn (3) into eqn (1). We find that

$$C_g(t) = i\mu\varepsilon_1 c_g(t), \qquad (4)$$

where

$$c_g(t) = \langle\psi_g|\psi(t)\rangle = \langle\psi_g|e^{-iH_e t}|\psi_0\rangle. \qquad (5)$$

Hence, the central quantities required for reconstructing $|\Psi(t)\rangle$ are the overlaps $\langle\psi_g|\psi(t)\rangle$. These overlaps have a physical interpretation as the projections of $|\psi(t)\rangle$ onto the basis of ground vibrational eigenstates: as the wavepacket moves on the excited-state potential its *shadow* on the ground-state potential is completely recorded in these time-dependent projections. The rightmost expression in eqn (5) indicates that $c_g(t)$ has the form of a time correlation function between $|\psi_0\rangle$ and $|\psi_g\rangle$. It has long been recognized that such correlation functions appear in the time-dependent formulation of resonance Raman scattering (RRS);[24–27] however, the experimental RRS signal involves the absolute-value-squared of the half-Fourier transform of the correlation function, hence the latter cannot be recovered from that signal.

Fully resonant coherent anti-Stokes Raman scattering (CARS) has been shown to be a useful probe of ground and excited electronic states properties[28,29] but its power has not been fully exploited. We now show that the correlation functions $\{c_g(t)\}$ may be completely recovered from femtosecond resonant CARS spectroscopy, allowing complete reconstruction of the excited-state wavepacket. The formula for the CARS signal produced by a three-pulse pump-dump-pump sequence is[30]

$$P^{(3)}(t) = \langle \psi^{(0)}(t)|\mu|\psi^{(3)}(t)\rangle + \text{c.c.}, \tag{6}$$

where $\psi^{(3)}(t)$ is the third-order wavefunction and $\psi^{(0)}(t) = e^{-iH_g t}\psi_0$. Writing out eqn (6) explicitly we obtain

$$P^{(3)}(t) = (-i)^3 \int_{-\infty}^{t} dt_3 \int_{-\infty}^{t_3} dt_2 \int_{-\infty}^{t_2} dt_1 \langle \psi_0|\mu|e^{-iH_e(t-t_3)}\{-\mu\varepsilon_3(t_3)\} \\ \times e^{-iH_g(t_3-t_2)}\{-\mu\varepsilon_2(t_2)\}e^{-iH_e(t_2-t_1)}\{-\mu\varepsilon_1(t_1)\}|\psi_0\rangle + \text{c.c.}, \tag{7}$$

where the **k**-vector dependence of the pulses (with the sequence $\mathbf{k_1} - \mathbf{k_2} + \mathbf{k_3}$) is omitted. Within the δ-pulse and Condon approximations, $P^{(3)}(t)$ takes the form

$$P^{(3)}(\tau) = \tilde{\varepsilon}\langle\psi_0|e^{-iH_e\tau_{43}}e^{-i\tilde{H}_g\tau_{32}}e^{-iH_e\tau_{21}}|\psi_0\rangle, \tag{8}$$

where $\tau_{ij} = \tau_i - \tau_j$ is the (positive) time-delay between the centers of the ith and jth pulses and $\tau_{43} = \tau - \tau_3$ with τ being the time of signal measurement. We have denoted $\tilde{H}_g = H_g - \omega_0$, $\tilde{\varepsilon} = i^3\mu^4\varepsilon_1\varepsilon_2\varepsilon_3 e^{i\omega_0(\tau_{21}+\tau_{43})}$ with $\varepsilon_{1,2,3}$ as the first, second and third pulse amplitudes, respectively, and $\tau \equiv [\tau_{21}, \tau_{32}, \tau_{43}]$. In writing $P^{(3)}(\tau)$ as a complex quantity we have assumed the signal is measured in a heterodyne fashion.

Fig. 1 illustrates the physical interpretation of eqn (8): a first laser pulse, the 'pump' pulse, transfers amplitude to the excited potential surface creating the wavepacket whose time-dependence we are interested in reconstructing. After evolving on the excited state for some time, τ_{21}, a second pulse, the 'dump' pulse, transfers part of this amplitude back to the ground state where it evolves for a second interval of time, τ_{32}. Finally, a third laser pulse, a second 'pump' pulse, excites part of the second-order amplitude to the excited state, generating the third-order polarization that produces the CARS signal, measured after a time τ_{43}.

The wavepacket $|\psi(t)\rangle$ (eqn (3)) may already be recognized in the rightmost factors in eqn (8); the question is how to extract it without knowing H_e. Since, as described above, all we need to reconstruct $|\Psi(t)\rangle$ are the correlation functions $\langle\psi_g|e^{-iH_e t}|\psi_0\rangle$, the problem reduces to extracting these correlation functions from eqn (8). Introducing a complete set of ground vibrational states, $\Sigma_g|\psi_g\rangle\langle\psi_g| = \hat{1}$, into eqn (8), we obtain the following suggestive formula for the signal:

$$P^{(3)}(\tau) = \tilde{\varepsilon}\sum_g e^{-i\tilde{\omega}_g\tau_{32}} P_g^{(3)}(\tau_{43}, \tau_{21}), \tag{9}$$

where $P_g^{(3)}(\tau_{43},\tau_{21}) = \langle\psi_0|e^{-iH_e\tau_{43}}|\psi_g\rangle\langle\psi_g|e^{-iH_e\tau_{21}}|\psi_0\rangle$, and $\tilde{\omega}_g = \omega_g - \omega_0$. Examining the form of $P_g^{(3)}$ we see that the desired correlation functions are closely related to the

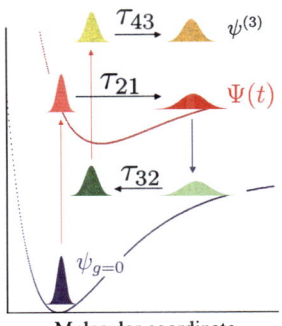

Fig. 1 The pump-dump-pump CARS scheme. $\Psi(t)$ is the desired wavefunction.

square roots of the $P_g^{(3)}$'s. Thus, a general strategy for extracting the overlaps is clear. First, the signal $P^{(3)}(\tau)$ is Fourier-transformed along τ_{32} to resolve the individual $P_g^{(3)}$'s. Then, we take the square-root of each $P_g^{(3)}$ to obtain $c_g(t)$ and reconstruct the wavefunction $|\Psi(t)\rangle$ by superposing the basis functions $\{|\psi_g\rangle\}$ with the coefficients $\{c_g(t)\}$. We now proceed to describe the steps for reconstructing $|\Psi(t)\rangle$ from $P^{(3)}(\tau)$ in more detail.

2.2 Procedure for reconstruction the wavefunction from the CARS signal

1. *Fourier-transform $P^{(3)}(\tau)$ with respect to τ_{32}.* The transformation is designed to resolve $P^{(3)}(\tau)$ into individual ground-state components, $\{P_g^{(3)}\}$. Since τ_{32} is defined to be positive, we multiply eqn (9), prior to the transformation, by the rectangular function, $\Pi(\tau_{32})$, that takes the value 1 for the τ_{32} domain and 0 elsewhere. Using the Fourier convolution theorem we obtain a sinc-type of spectrum with peaks at the frequencies $\omega = \tilde{\omega}_g$:

$$\tilde{P}^{(3)}(\tau_{43}, \omega, \tau_{21}) = \tilde{\varepsilon} \sum_{g=0}^{N} P_g^{(3)}(\tau_{43}, \tau_{21}) \int_{-\infty}^{\infty} d\tau_{32} \Pi(\tau_{32} - (\bar{\tau}_{32} + T)) e^{i(\omega - \tilde{\omega}_g)\tau_{32}}$$

$$= 2T\tilde{\varepsilon} \sum_{g=0}^{N} P_g^{(3)}(\tau_{43}, \tau_{21}) e^{i(\omega - \tilde{\omega}_g)(\bar{\tau}_{32} + T)} \operatorname{sinc}[(\omega - \tilde{\omega}_g)T] \qquad (10)$$

$$\equiv \sum_{g=0}^{N} S(\omega, \tilde{\omega}_g) P_g^{(3)}(\tau_{43}, \tau_{21}),$$

where $S(\omega, \tilde{\omega}_g) = 2T\tilde{\varepsilon} e^{i(\omega - \tilde{\omega}_g)(\bar{\tau}_{32} + T)} \operatorname{sinc}\left[(\omega - \tilde{\omega}_g)T\right]$, $2T = \hat{\tau}_{32} - \check{\tau}_{32}$, and $\check{\tau}_{32}$ ($\hat{\tau}_{32}$) is the minimal (maximal) value of τ_{32}. Fixing (τ_{43}, τ_{21}), eqn (10) can be written as a matrix equation:

$$\tilde{\mathbf{P}}^{(3)} = \mathbf{S}\mathbf{P}_g^{(3)}, \qquad (11)$$

where $\tilde{\mathbf{P}}^{(3)}$ is a vector that runs over the continuous index ω and $\mathbf{S}$ is a two dimensional matrix with one continuous and one discrete frequency index, $(\omega, \tilde{\omega}_g)$.

2. *Invert Eq. (11) to obtain $P_g^{(3)}(\tau_{43}, \tau_{21})$.* To do this we need the matrix $\mathbf{S}$ to be square; we therefore choose the number of frequency elements ω equal to the number of $\tilde{\omega}_g$ elements and calculate $\mathbf{P}_g^{(3)} = \mathbf{S}^{-1}\tilde{\mathbf{P}}^{(3)}$. For numerical accuracy, the inversion is implemented separately around each of the peaks $\tilde{\omega}_g$.

3. *Take the square-root of $P_g^{(3)}$.* Assuming the functions $\{\psi_g(x)\}$ are real, we can rewrite $P_g^{(3)}$ as

$$P_g^{(3)}(\tau_{43}, \tau_{21}) = \langle \psi_g | e^{-iH_e\tau_{43}} | \psi_0 \rangle \langle \psi_g | e^{-iH_e\tau_{21}} | \psi_0 \rangle. \qquad (12)$$

Taking the square-root of the diagonal of $P_g^{(3)}(\tau_{43}, \tau_{21})$ (*i.e.* $\tau_{43} = \tau_{21} = t$), we recover the $c_g(t)$ up to a sign:

$$\sqrt{P_g^{(3)}(t)} = a_g \langle \psi_g | e^{-iH_e t} | \psi_0 \rangle \equiv \langle \tilde{\psi}_g | e^{-iH_e t} | \psi_0 \rangle, \qquad (13)$$

where $a_g = \pm 1$ and the sign of $\tilde{\psi}_g(x)$ is as yet undetermined. By demanding continuity of the cross-correlation functions (and their derivatives), the coefficients a_g can be regarded as time-independent. Substituting eqn (13) instead of $c_g(t)$ into the expression $C_g(t) = i\mu\varepsilon_1 c_g(t)$ and using the resulting $C_g(t)$ in eqn (1) yields

$$|\tilde{\Psi}(t)\rangle = i\mu\varepsilon_1 \sum_{g=0}^{N} |\psi_g\rangle \langle \tilde{\psi}_g | e^{-iH_e t} | \psi_0 \rangle. \qquad (14)$$

The different sign combinations of $\tilde{\psi}_g(x)$ generate 2^{N+1} possible superpositions. (In fact, only 2^N are physically meaningful since we are free to set the sign of one of the g-components.) Only one out of the 2^N $|\tilde{\Psi}(t)\rangle$ coincides with $|\Psi(t)\rangle$: the $|\tilde{\Psi}(t)\rangle$ for which the sign combination satisfies $\Sigma_g|\psi_g\rangle\langle\tilde{\psi}_g| = \hat{1}$.

4. *Discriminating* $|\Psi(t)\rangle$ *from the set* $\{|\tilde{\Psi}(t)\rangle\}$. The set of wavefunctions $\{|\tilde{\Psi}(t)\rangle\}$ are all consistent with the CARS signal at a specific value of $\tau_{43} = \tau_{21}$†. However, only one $|\tilde{\Psi}(t)\rangle$ is consistent with the signal *derivatives*. To see this, consider the nth derivative of the experimental signal, eqn (8), with respect to τ_{21}:

$$\frac{\partial^n P^{(3)}(\tau)}{\partial \tau_{21}^n} = \varepsilon^\dagger \langle \Psi^*(\tau_{43})|e^{-iH_g\tau_{32}}\tilde{H}_e^n|\Psi(\tau_{21})\rangle$$

$$= \varepsilon^\dagger \sum_{g,g'} e^{-i\omega_g\tau_{32}} C_g(\tau_{43}) C_{g'}(\tau_{21}) \tilde{H}^n_{e,gg'}, \quad (15)$$

where $\varepsilon^\dagger = (-i)^{n-1}\mu^2\varepsilon_1^{-1}\varepsilon_2\varepsilon_3 e^{i\omega_0\tau_{41}}$, $\tau_{41} = \tau - \tau_1$, $\tilde{H}_e^n = (H_e - \omega_0)^n$, and $\tilde{H}^n_{e,gg'} = \langle \psi_g|\tilde{H}_e^n|\psi_{g'}\rangle$. Substituting $|\tilde{\Psi}(t)\rangle$ instead of $|\Psi(t)\rangle$, into eqn (15) gives

$$\frac{\partial^n \tilde{P}^{(3)}(\tau)}{\partial \tau_{21}^n} = \varepsilon^\dagger \sum_{g,g'} e^{-i\omega_g\tau_{32}} a_g a_{g'} C_g(\tau_{43}) C_{g'}(\tau_{21}) \tilde{H}^n_{e,gg'}. \quad (16)$$

Accordingly, the $|\tilde{\Psi}(t)\rangle$ for which $\frac{\partial^n \tilde{P}^{(3)}(\tau)}{\partial \tau_{21}^n} = \frac{\partial^n P^{(3)}(\tau)}{\partial \tau_{21}^n}$ for all n is the wavefunction that coincides with $|\Psi(t)\rangle$ of eqn (3), and hence, is the reconstruction solution. An alternative and somewhat more practical analysis of the sign determination is described in the next Section.

2.3 Alternative analysis of the sign determination

The idea of using the signal time-derivative for discriminating $\Psi(t)$ can be applied in a somewhat more convenient way. Let us look at a fictitious time-dependent Schrödinger equation for each $|\tilde{\Psi}(t)\rangle$, inverted for a corresponding potential $\tilde{V}_e$

$$\tilde{V}_e(x) = \frac{1}{\tilde{\Psi}(x,t)}\left[i\frac{\partial}{\partial t} + \frac{1}{2m}\frac{\partial^2}{\partial x^2}\right]\tilde{\Psi}(x,t), \quad (17)$$

where m is the system's reduced mass. As we show below, the potentials $\tilde{V}_e$ calculated by the $|\tilde{\Psi}(t)\rangle$ are generally time-*dependent*; only the potential calculated with $|\tilde{\Psi}(t)\rangle = |\Psi(t)\rangle$ is time-independent and hence corresponds to the excited-state Hamiltonian H_e of the measured system. Thus, in order to find the correct wavefunction we use the set of calculated potentials—as if they were time-independent—to propagate the corresponding $\{|\tilde{\Psi}(t)\rangle\}$ back to time zero. Of all the potentials, only the truly time-independent one will propagate the corresponding $|\tilde{\Psi}(t)\rangle$ correctly back to $|\Psi_0\rangle$, and therefore this $|\tilde{\Psi}(t)\rangle$ is the correct wavefunction. Note that the above procedure requires knowing the signal as a function only of τ_{32} and $\tau_{21} = \tau_{43}$.

We prove now that the potentials calculated by the functions $\{\tilde{\Psi}(t)\}$ that do not coincide with $\Psi(t)$ are time-*dependent*. Let us look once again at the wavefunctions that can be constructed using the information obtained from the CARS signal:

$$|\tilde{\Psi}(t)\rangle = i\mu\varepsilon_1 \sum_g |\psi_g\rangle\langle\tilde{\psi}_g|e^{-iH_e t}|\psi_0\rangle$$

$$= \sum_g |\psi_g\rangle\langle\tilde{\psi}_g|\Psi(t)\rangle \equiv \tilde{1}|\Psi(t)\rangle, \quad (18)$$

---

† In fact, the set of wavefunctions given by eqn (14) are consistent with the CARS signal for any pair (τ_{21}, τ_{43}).

where we have denoted $\sum_g |\psi_g\rangle\langle\tilde{\psi}_g| \equiv \tilde{1}$. Recall that $\hat{1} \equiv \sum_g |\psi_g\rangle\langle\tilde{\psi}_g| \equiv \sum_g |\psi_g\rangle a_g \langle\psi_g|$ where a_g may take one out of two possible values: ± 1. A useful property of the operator $\tilde{1}$ is that its square equals the identity operator $\hat{1}$:

$$\tilde{1}\tilde{1} = \sum_{gg'} a_g a_{g'} |\psi_g\rangle\langle\psi_g|\psi_{g'}\rangle\langle\psi_{g'}|$$
$$= \sum_g a_g^2 |\psi_g\rangle\langle\psi_g| = \sum_g |\psi_g\rangle\langle\psi_g| = \hat{1}. \quad (19)$$

We can derive an equation of motion for $\tilde{\Psi}(t)$:

$$\frac{\partial}{\partial t}|\tilde{\Psi}(t)\rangle = \frac{\partial}{\partial t}\tilde{1}|\Psi(t)\rangle = \tilde{1}\frac{\partial}{\partial t}|\Psi(t)\rangle = -i\tilde{1}H_e|\Psi(t)\rangle$$
$$= -i\tilde{1}H_e\tilde{1}\tilde{1}|\Psi(t)\rangle = -i\tilde{1}H_e\tilde{1}|\tilde{\Psi}(t)\rangle \quad (20)$$
$$\equiv -i\tilde{H}_e|\tilde{\Psi}(t)\rangle,$$

where, we have used the fact that $\tilde{1}$ is time-independent and therefore commutes with $\frac{\partial}{\partial t}$. Eqn (20) shows that $\tilde{\Psi}(t)$ obeys a time-dependent Schrödinger equation with the effective Hamiltonian $\tilde{H}_e = \tilde{1}H_e\tilde{1}$.

The Hamiltonian H_e has the conventional form of $H_e = V_e + T$, where T is the kinetic-energy operator. The Hamiltonian $\tilde{H}$ therefore takes the form:

$$\tilde{H}_e \equiv \tilde{1}H_e\tilde{1} = \tilde{1}V_e\tilde{1} + \tilde{1}T\tilde{1} \equiv \tilde{V}_e + \tilde{T}, \quad (21)$$

Note that the operator $\tilde{1}$ does not commute with V_e, T or H_e since it does not share a common basis of eigenvectors with the last three operators. Note also that the operators $\tilde{V}_e$, $\tilde{T}$ and $\tilde{H}_e$ are all time-independent.

Rearranging eqn (20), we obtain:

$$\tilde{V}_e = \frac{1}{\tilde{\Psi}(t)}\left[i\frac{\partial}{\partial t} - \tilde{T}\right]\tilde{\Psi}(t). \quad (22)$$

where we emphasize that $\tilde{V}_e$ is time-independent. Let us now define the related quantity

$$\tilde{\tilde{V}}_e = \frac{1}{\tilde{\Psi}(t)}\left[i\frac{\partial}{\partial t} - T\right]\tilde{\Psi}(t), \quad (23)$$

where T is the usual kinetic energy operator. Obviously, for $\tilde{\Psi}(t) \equiv \Psi(t)$ eqn (23) is equivalent to the usual time-dependent Schrödinger equation for $\Psi(t)$ and therefore $\tilde{\tilde{V}}_e \equiv V_e$ is time-independent. For any other, incorrect, wavefunction $\tilde{\Psi}(t)$, eqn (23) results in a time-*dependent* potential $\tilde{\tilde{V}}_e$.

In order to show this we substitute $\tilde{T} = T + \Delta T$ in eqn (22), where $\Delta T = \tilde{T} - T$, and obtain:

$$\tilde{V}_e = \frac{1}{\tilde{\Psi}(t)}\left[i\frac{\partial}{\partial t} - (T + \Delta T)\right]\tilde{\Psi}(t)$$
$$= \tilde{\tilde{V}}_e - \frac{1}{\tilde{\Psi}(t)}[\Delta T]\tilde{\Psi}(t). \quad (24)$$

The term $\frac{1}{\tilde{\Psi}(t)}[\Delta T]\tilde{\Psi}(t)$ is time-*dependent* (unless $\tilde{\Psi}(t)$ is an eigenfunction of ΔT, which has no general reason to hold). Also, $\Delta T \neq 0$ in general‡. Therefore, in order to preserve the time-independence of $\tilde{V}_e \equiv \tilde{1}V_e\tilde{1}$, $\tilde{V}_e$ must also be time-*dependent*.

2.4 Practical determination of the reconstruction solution $\Psi(t)$

Each wavefunction $\tilde{\Psi}(t)$ of the set $\{\tilde{\Psi}(t)\}$, including the correct wavefunction $\Psi(t)$, is characterized by a specific sign-combination $\{a_g\}$. This sign-combination assigns to each $|\psi_g\rangle$ the consistent coefficient $\langle\psi_g|e^{-iH_e t}|\psi_0\rangle$; see eqn (13)–(14). Thus, finding the correct wavefunction $\Psi(t)$, means finding the corresponding combination of signs $\{a_g\}$.

Due to limited numerical accuracy, applying the above potential-reconstruction and backward-propagation procedure at time t allows us to find the signs only of those superposition components that contribute significantly at this instant in time. For similar reasons, the potential $\tilde{V}_e$ that is reconstructed at time t is expected to be accurate only in the region where the wavefunction has considerable amplitude. There are two implications to these statements. The first is that, if applied in a single iteration, the potential-reconstruction and backwards-propagation procedure is expected to be accurate only at times relatively close to zero. The second, which follows from the first, is that in order to find the signs at any desired (future) time, we have to apply the potential-reconstruction and backward-propagation procedure at several points in time.

In practice, we have used the following algorithm to find the signs of the coefficients of the correct wavefunction $\Psi(t)$:

1. Choose a time, t_1, close enough to time zero.
2. Construct a set of wavefunctions $\{\tilde{\Psi}(t_1)\}$, using eqn (14), each characterized by a specific combination of signs $\{a_g\}$. Note that only a relatively small number, n, of superposition components are required to construct the wavefunctions at this point in time; the number is determined by the modulus of the $c_g(t_1)$'s.
3. Construct a set of potentials $\{\tilde{V}_e\}_{t_1}$ from the set of wavefunctions $\{\tilde{\Psi}(t_1)\}$.
4. Propagate each $\tilde{\Psi}(t_1)$ of the set backward to time zero using its corresponding potential $\tilde{V}_e$, as if the latter were time-independent.
5. For each wavefunction propagated from t_1 back to time zero, $\tilde{\Psi}_{t_1}(0)$, calculate $|\langle\psi_0|\tilde{\Psi}_{t_1}(0)\rangle|$. Note that ψ_0 is a fixed initial condition.
6. Collect the wavefunctions $\tilde{\Psi}_{t_1}(0)$ for which $|\langle\psi_0|\tilde{\Psi}_{t_1}(0)\rangle|$ is the closest to being equal 1 (subject to numerical accuracy considerations), and inspect the signs, a_g, of the first several (n_1) superposition components. This subset of n_1 signs correctly characterize the actual wavefunction.
7. Construct a new initial condition, $\Psi^{\{n_1\}}(t_1)$, using the n_1 signs determined in stage 6: $\Psi^{\{n_1\}}(t_1) = \sum_{g=0}^{n_1}|\psi_g\rangle a_g \langle\psi_g|e^{-iH_e t_1}|\psi_0\rangle$. Note that the upper limit of the sum is $n_1 < N$: the number n_1 will grow at later iterations, but at this point, n_1 determines the excited-state wavepacket to within the limits of our numerical accuracy.
8. Choose a time $t_2 > t_1$.
9. Construct a set of wavefunctions $\{\tilde{\Psi}(t_2)\}$, using eqn (14), each characterized by a specific combination of signs $\{a_g\}$. Note, the first n_1 signs are already determined and are the same for each $\{\tilde{\Psi}(t_2)\}$ of the set. The total number of superposition components required to construct $\{\tilde{\Psi}(t_2)\}$ is determined, again, by the modulus of the $c_g(t_2)$'s.
10. Construct a set of potentials $\{\tilde{V}_e\}_{t_2}$ from the set of wavefunctions $\{\tilde{\Psi}(t_2)\}$.

---

‡ $\Delta T = (\tilde{T} - T) = \tilde{1}\tilde{1}(\tilde{1}T\tilde{1} - T) = \tilde{1}(T\tilde{1} - \tilde{1}T) = \tilde{1}[T,\tilde{1}]$. The commutator $[T,\tilde{1}]$ is not identically zero. Therefore, $\tilde{1}[T,\tilde{1}] \equiv \Delta T$ is not identically zero as well.

11. Propagate each $\tilde{\Psi}(t_2)$ of the set backward to time t_1 by its corresponding potential $\tilde{V}_e$, as if the latter were time-independent.

12. For each wavefunction propagated from t_2 back to time t_1, $\tilde{\Psi}_{t_2}(t_1)$, calculate $|\langle \Psi^{\{n_1\}}(t_1)|\tilde{\Psi}_{t_2}(t_1)\rangle|$. Note that $\Psi^{\{n_1\}}(t_1)$ is the initial condition constructed at stage 7.

13. Collect the wavefunctions $\tilde{\Psi}_{t_2}(t_1)$ for which $|\langle \Psi^{\{n_1\}}(t_1)|\tilde{\Psi}_{t_2}(t_1)\rangle|$ is closest to being equal 1 (subject to numerical accuracy considerations), and inspect the next n_2 common signs, a_g. At this point $n_1 + n_2$ signs are determined as correctly characterizing the actual wavefunction.

14. Proceed by going back to stage 7, constructing a new initial condition, $\Psi^{\{N_2\}}(t_2)$ using the $N_2 = n_1 + n_2$ determined signs, and follow the subsequent stages to determine $\Psi(t)$ until a desired time t_j.

3. Results

3.1 One-dimensional system

3.1.1 Wavefunction reconstruction.
To test the above reconstruction methodology, we simulated the CARS signal by calculating $\langle \psi^{(0)}(\tau)|\hat{\mu}|\psi^{(3)}(\tau)\rangle$ as a function of the three time-delays, for two diatomic systems. The first is the Li_2 molecule, with its ground (X) and first-excited (A) electronic states as Morse-type potentials with the form, $V(x) = D(1 - e^{-b(x-x_0)})^2 + T$. The second system, henceforth denoted d-Li_2, has the Li_2 ground state (X) but a dissociative excited potential of the form $V(x) = De^{-b(x-x_0)} + T$ (denoted $\tilde{A}$). Table 1 gives the potential parameters in atomic units used for the simulations. The parameters for the Morse-type potentials are based on data published in ref. 31.

The simulation of $P^{(3)}$ was performed using the split-operator method[32] on a spatial grid of 256 points in the range of 2–12a.u. with time spacing of $\Delta t = 0.1$ fs. A constant transition-dipole of 2 a.u. was used, and the pulse amplitudes $\varepsilon_{1,2,3}$ were taken to be 10^{-4} a.u. The range of time-delay for the Li_2 system was $\tau_{21,\,43} = 0$–200 fs and for the d-Li_2 system was 0–80 fs, with spacing of 0.2 fs for both. For both systems, we took $\tau_{32} = 3$–6000 fs with 1 fs spacing. For Li_2, we inverted eqn (11) for each of the first 25 peaks of $\tilde{P}^{(3)}(\omega)$ using the matrices **S** with 25 frequency grid points centered around the peaks at $\tilde{\omega}_g$. This produced 25 two-dimensional functions $P_g^{(3)}$, $g = 0, ..., 24$. For d-Li_2 the procedure was performed for the first 40 peaks, producing 40 two-dimensional functions $P_g^{(3)}$, $g = 0, ..., 39$.

In Fig. 2 and 3 we present snapshots of the real part of the reconstructed first-order wavefunction for the Li_2 and the d-Li_2 molecules, respectively. For Li_2 (d-Li_2) we superpose the first 25 (40) eigenfunctions $\psi_g(x)$ using the cross-correlation functions $\{c_g(t)\}$ obtained by the CARS analysis and maintaining $\Sigma_g |\psi_g\rangle\langle \tilde{\psi}_g| = \hat{1}$. The reconstructed wavefunctions are seen to be in excellent agreement with the exact ones, obtained by direct calculation of the first-order population, for all propagation times. For the Li_2 system, a high quality reconstruction is already obtained by superposing just 20 basis functions.

Table 2 shows the details of successive application of the potential-reconstruction and back-propagation procedure applied to determine the (correct) wavefunction Ψ

Table 1 The parameters, in atomic units, for the X, A and $\tilde{A}$ potentials used in simulating the CARS signals

	X	A	$\tilde{A}$
D	0.0378492	0.0426108	9.11267×10^{-5}
b	0.4730844	0.3175063	1.5875317
x_0	5.0493478	5.8713786	7.3699313
T	0	0.0640074	0.0640074

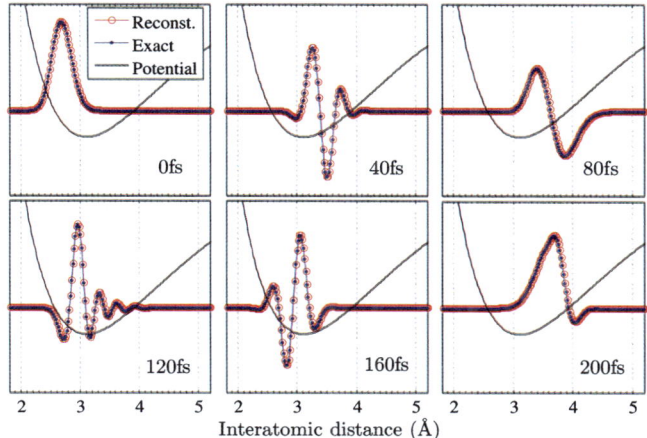

Fig. 2 Snapshots of the real part of the reconstructed (circles, red) vs. the exact (dots, blue) wavefunction, at various times on the excited (A) potential (solid line) of Li_2.

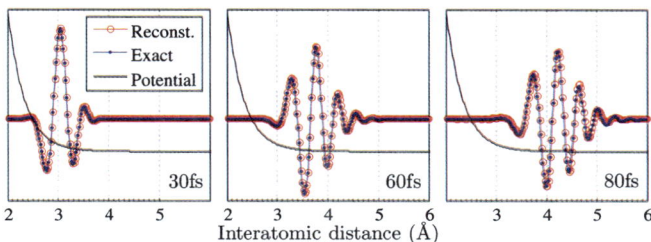

Fig. 3 Snapshots of the real part of the reconstructed (circles, red) vs. the exact (dots, blue) wavefunction, at various times on the excited ($\bar{A}$) potential (solid line) of d-Li_2.

for the Li_2 system. Accordingly, in the first step, we used $t_1 = 10$ fs and constructed the set $\{\tilde{\Psi}(10)\}$ using 12 basis functions ψ_g. By back propagation to time zero (using the obtained potentials $\{\tilde{V}_e\}_{10}$) and calculating $|\langle\psi_0|\tilde{\Psi}_{10}(0)\rangle|$, we determined the first seven signs. In the second step, we used the first seven g-components, with the corresponding determined signs, to determine a new initial condition $\Psi^{\{7\}}(10)$. We

Table 2 Details of the eight successive stages in determining the wavefunction Ψ for Li_2. Refer to the text for details

Procedure step index, i	1	2	3	4	5	6	7	8
Reconstruction time, t_i, fs	10	20	25	30	40	50	60	70
Total number of basis functions used in constructing $\{\tilde{\Psi}(t_i)\}$	12	18	23	25	25	25	25	25
Number of basis functions with undetermined signs used in constructing $\{\tilde{\Psi}(t_i)\}$	12	11	11	9	7	5	3	2
Number of (new) sign-coefficients determined at step i, n_i	7	5	4	2	2	2	1	0
Total number of sign-coefficients determined after step i, N_i	7	12	16	18	20	22	23	23

chose $t_2 = 20$ fs and constructed the set $\{\tilde{\Psi}(20)\}$ using 18 basis functions ψ_g. Note that the signs of the first seven g-components were already fixed. We constructed the set $\{\tilde{V}_e\}_{20}$ and propagated the wavefunctions $\{\tilde{\Psi}(20)\}$ back to $t_1 = 10$ fs. We calculated $|\langle \Psi^{\{7\}}(10)|\tilde{\Psi}_{20}(10)\rangle|$ for each $\tilde{\Psi}_{20}(10)$, and then set the next 5 signs. We proceeded until we could not accurately determine any more coefficients, ending up with 24 signs—enough to provide an excellent reconstruction, as seen in Fig. 2. Fig. 4 shows the measures $|\langle \Psi^{\{N_i\}}(t_{i-1})|\tilde{\Psi}_{t_i}(t_{i-1})\rangle|$ obtained for the different sign-combinations for each t_i used. Table 3 gives the details of successive application of the potential-reconstruction and back-propagation procedure applied to determine the correct wavefunction for the d-Li$_2$ system. Fig. 5 shows the measured $|\langle \Psi^{\{N_i\}}(t_{i-1})|\tilde{\Psi}_{t_i}(t_{i-1})\rangle|$ obtained for the different sign combinations for each t_i used, for the d-Li$_2$ system. There is one main difference between the procedures for the Li$_2$ and the d-Li$_2$ systems that stems from the fact that Li$_2$ is bound while d-Li$_2$ is dissociative. For the Li$_2$ system, only a finite number of ψ_g basis functions are required for a good reconstruction of the wavepacket at any time, due to the quasi periodic nature of the excited-state wavepacket. Thus, at each step of the procedure described in Table 2, fewer and fewer new sign-coefficients are determined. On the other hand, for the d-Li$_2$ system, the excited-state wavepacket visits new regions as time progresses. Therefore, a good reconstruction of the wavepaket at different times requires more and more (new) basis functions as time increases, as indicated in Table 3.

3.1.2 Potential reconstruction.
Having determined the correct wavefunction at different times, we calculate the corresponding excited potential surfaces from eqn (17) using central finite-differencing with eight-points (three-points) for the time (spatial) derivatives. The time-step used was 0.2 fs but very good results were also

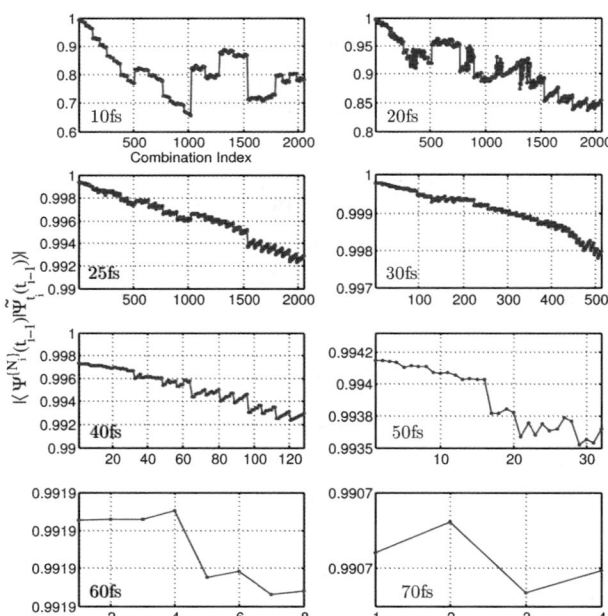

Fig. 4 The measured $|\langle \Psi^{\{N_i\}}(t_{i-1})|\tilde{\Psi}_{t_i}(t_{i-1})\rangle|$ for the eight stages of the Li$_2$ wavefunction determination, as a function of the sign combinations. Next to each plot is written the time, t_i, at which the potentials are reconstructed. Note the change in both the horizontal and vertical scale as time progresses, indicating that the signs of more and more coefficients in Ψ have been determined with certainty.

Table 3 Details of the six successive stages in determining the wavefunction Ψ for d-Li$_2$

Procedure step index, i	1	2	3	4	5	6
Reconstruction time, t_i, fs	10	20	30	40	50	60
Total number of basis functions used in constructing $\{\bar{\Psi}(t_i)\}$	12	15	19	24	27	32
Number of basis funcions with undetermined signs used in constructing $\{\tilde{\Psi}(t_i)\}$	12	11	11	11	11	11
Number of (new) sign-coefficients determined at step i, n_i	4	4	5	3	5	4
Total number of sign-coefficients determined after step i, N_i	4	8	13	16	21	25

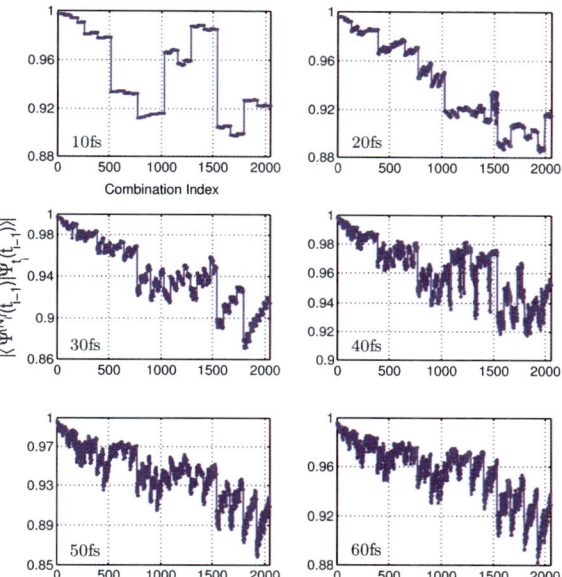

Fig. 5 The measure $|\langle \Psi^{\{N_i\}}(t_{i-1})|\tilde{\Psi}_t(t_{i-1})\rangle|$ for the six stages of the d-Li$_2$ wavefunction determination, as a function of the sign combinations. Next to each plot is written the time, t_i, at which the potentials are reconstructed.

obtained using 0.5 fs. Fig. 6 and 7 compare the reconstructed vs. the exact potentials. The wavefunction (absolute value) used in calculating the potential is shown by a black solid line. Note from Fig. 6 and 7 that combining the reconstructed potential from two points in time (e.g. 5 and 70 fs for Li$_2$ and 5 and 79 fs for d-Li$_2$) is sufficient to reconstruct the potential over the full range of interest (2–5 Å). Once the potential is known one can calculate the excited-state wavefunction as a function of time for any excitation pulse sequence without the need for any additional laboratory experiments.

3.2 Two-dimensional system

The ultimate goal of this work is the reconstruction of excited-state wavefunctions and potentials for polyatomic molecules. Perhaps the biggest advantage of the

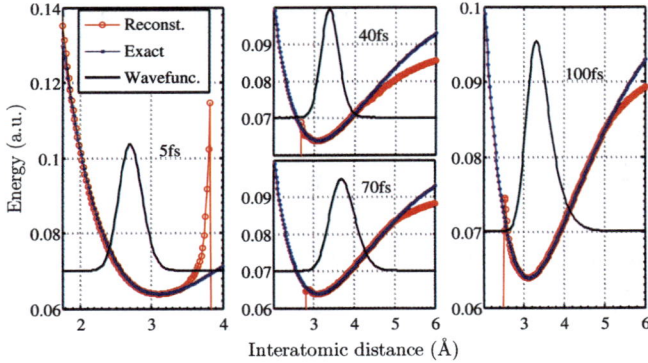

Fig. 6 The reconstructed (circles, red) vs. the exact (dots, blue) A potential of Li_2.

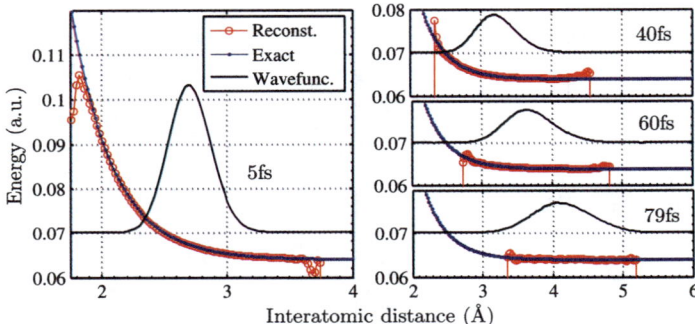

Fig. 7 The reconstructed (circles, red) vs. the exact (dots, blue) $\tilde{A}$ potential of d-Li_2.

approach is that it is completely general with respect to the molecular system dimensionality, provided the ground-state potential is known.

In this section we present preliminary results on two-dimensional wavefunction and potential reconstruction. We chose a model, non-symmetric, two dimensional Morse-type ground- and excited-state potentials with the general form $V_i(x,y) = D_x(1 - e^{-b_x(x-x_0)})^2 + D_y(1 - e^{-b_y(y-y_0)})^2 + T; i = g, e$. The potential parameters used are specified in Table 4. The results shown here were obtained not by simulating the CARS signal itself but rather simulating directly the correlation functions $\{c_g\}$. However, we do not believe this to be a serious limitation: in

Table 4 The parameters, in atomic units, for the two-dimensional ground and excited potentials

	V_g	V_e
D_x	0.0283869	0.0468719
D_y	0.0378492	0.0426108
b_x	0.5203929	0.2698804
b_y	0.4730844	0.3175063
x_0	5.0493478	6.1649475
y_0	5.0493478	5.8713786
T	0	0.0640074

the one-dimensional case the CARS signal analysis gives the correlation functions $\{c_g\}$ with very high precision, and we expect the same will be true in multi-dimensions, although the ground-state spectrum will be significantly more congested.

We calculate the cross-correlation functions $c_g = \langle \psi_g | e^{-iH_e t} | \psi_0 \rangle$ by projecting $|\psi(t)\rangle = e^{-iH_e t}|\psi_0\rangle$ onto the basis set $\{|\psi_g\rangle\}$. We then superpose these correlation functions and reconstruct the excited-state wavepacket $|\Psi(t)\rangle = \Sigma_g |\psi_g\rangle \langle \psi_g | e^{-iH_e t} | \psi_0 \rangle$, using 700 basis functions ψ_g; the wavepacket is shown at different times in the rightmost column of Fig. 8. Just as in the one-dimensional case, we use these wavepackets to calculate the excited-state potentials; these are seen in the leftmost column of Fig. 8 (compared to the exact potential in the middle column). Although we did not base the reconstruction on a full CARS signal simulation, these preliminary results show the potential of our reconstruction methodology for multi-dimensional systems; *i.e.* polyatomics.

In Fig. 9(a) we present spectra (eqn (10)) obtained from the CARS signal of the two-dimensional system. The spectra correspond to two different pairs of time delays $\tau_{21} = \tau_{43}$ and clearly illustrate the large number of ground-vibrational

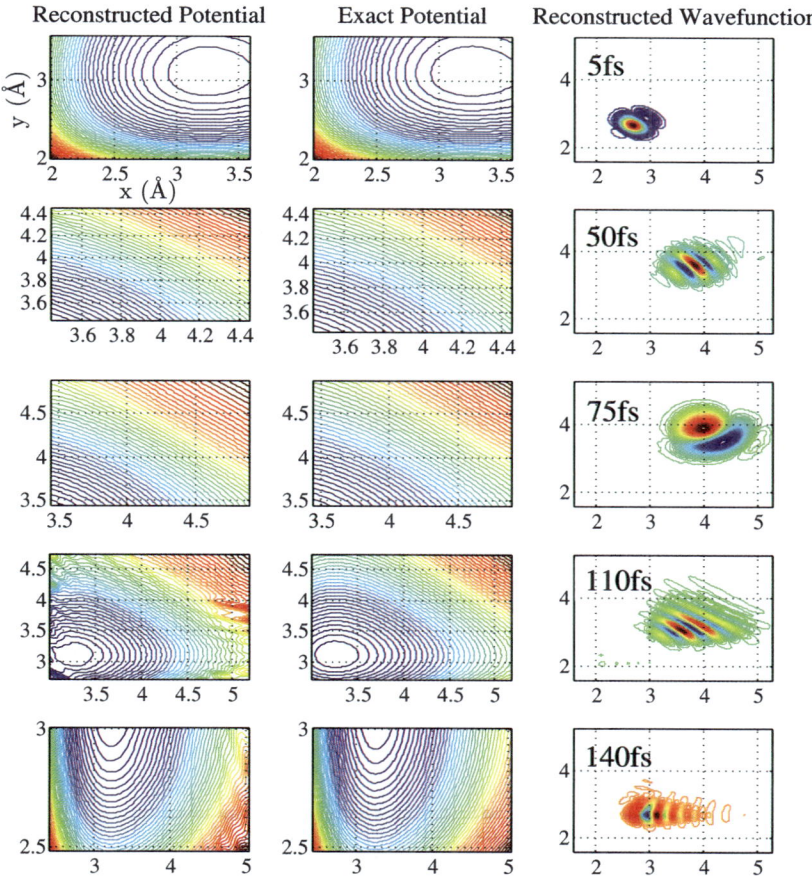

Fig. 8 Excited-state wavefunctions and potentials for the two-dimensional system at different times. The rightmost column shows the real part of the reconstructed wavefunction. The middle column shows the exact excited-state potential, $V_e(x, y)$, and the leftmost column shown the reconstructed excited-state potential calculated from the reconstructed wavefunction.

components that contribute to the signal. In comparison, in Fig. 9(b) we present an illustrative spectrum obtained from the CARS signal of the one-dimensional Li_2 system. Note the difference in the amount of information contained in the spectrum of the two-dimensional system relative to that of the one-dimensional system.

4. Discussion of the approximations in the method

The methodology presented above relies on several simplifying approximations. Here we discuss the consequences of removing these approximations.

4.1 Coordinate-dependent transition dipole moment

Considering a coordinate-dependent transition dipole operator, $\hat{\mu}$, the wavefunction of interest is

$$|\Psi(t)\rangle = i\varepsilon_1 e^{-iH_e t}\hat{\mu}|\psi_0\rangle = i\varepsilon_1 \sum_g |\psi_g\rangle\langle\psi_g|e^{-iH_e t}\hat{\mu}|\psi_0\rangle. \qquad (25)$$

According to Eq. (25), the coefficients required for the reconstruction are

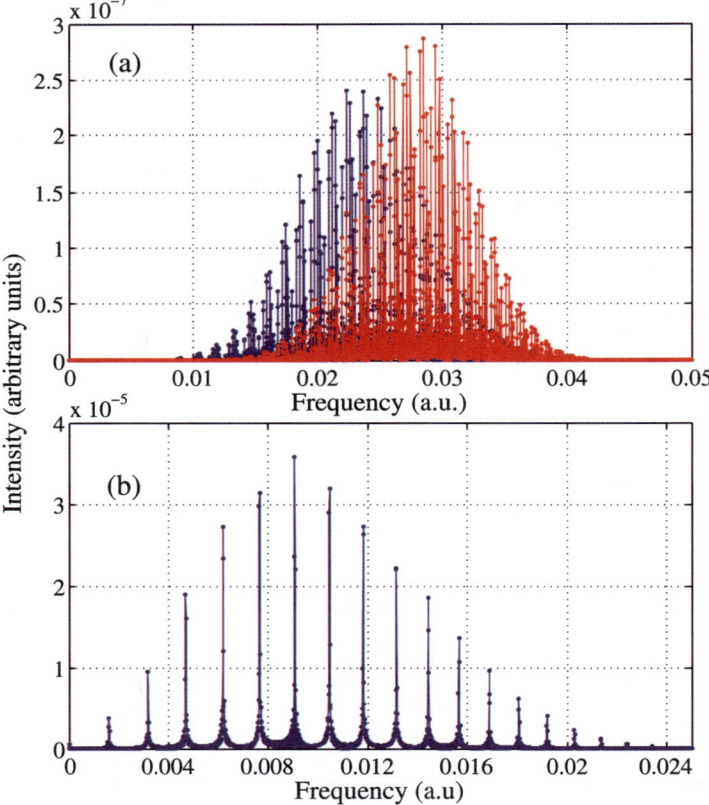

Fig. 9 (a) Ground vibrational spectrum of the two-dimensional model-system for two different values of (τ_{21}, τ_{43}) (red and blue) obtained from the CARS signal; see (eqn (10)). (b) Ground vibrational spectrum of the one-dimensional Li_2 system for fixed (τ_{21}, τ_{43}) obtained from the CARS signal; see eqn (10).

$$c_g(t) = \langle \psi_g | e^{-iH_e t} \hat{\mu} | \psi_0 \rangle = \langle \psi_g | e^{-iH_e t} | \phi_0 \rangle, \tag{26}$$

where we have defined $|\phi_0\rangle = \hat{\mu}|\psi_0\rangle$. The CARS signal is now given by

$$\begin{aligned} P^{(3)}(\tau) &= \tilde{\varepsilon} \langle \psi_0 | \hat{\mu} e^{-iH_e \tau_{43}} \hat{\mu} e^{-i\tilde{H}_g \tau_{32}} \hat{\mu} e^{-iH_e \tau_{21}} \hat{\mu} | \psi_0 \rangle \\ &= \tilde{\varepsilon} \langle \phi_0 | e^{-iH_e \tau_{43}} \hat{\mu} e^{-i\tilde{H}_g \tau_{32}} \hat{\mu} e^{-iH_e \tau_{21}} | \phi_0 \rangle, \end{aligned} \tag{27}$$

where $\tilde{\varepsilon} = i^3 \varepsilon_1 \varepsilon_2 \varepsilon_3 e^{i\omega_0(\tau_{21}+\tau_{43})}$. Following the above analysis of the signal, we can recover the coefficients

$$\begin{aligned} \sqrt{P_g^{(3)}(t)} &= a_g \langle \psi_g | \hat{\mu} e^{-iH_e t} | \phi_0 \rangle \equiv \langle \tilde{\psi}_g | \hat{\mu} e^{-iH_e t} | \phi_0 \rangle \\ &= \langle \tilde{\psi}_g | \hat{\mu} e^{-iH_e t} \hat{\mu} | \psi_0 \rangle. \end{aligned} \tag{28}$$

Using these coefficients (and omitting the coefficients $i\varepsilon_1$ for convenience) we may construct a set of wavefunctions $\{|\tilde{\Phi}(t)\rangle\}$ of the form:

$$\begin{aligned} |\tilde{\Phi}(t)\rangle &= \sum_g |\psi_g\rangle \langle \tilde{\psi}_g | \hat{\mu} e^{-iH_e t} \hat{\mu} | \psi_0 \rangle = \tilde{1} \hat{\mu} e^{-iH_e t} \hat{\mu} | \psi_0 \rangle \\ &= \tilde{1} \hat{\mu} | \Psi(t) \rangle. \end{aligned} \tag{29}$$

Note that for the particular case where $\tilde{1} = \hat{1}$, the reconstructed wavefunction is

$$\begin{aligned} |\Phi(t)\rangle &= \sum_g |\psi_g\rangle \langle \psi_g | \hat{\mu} e^{-iH_e t} \hat{\mu} | \psi_0 \rangle = \hat{\mu} e^{-iH_e t} \hat{\mu} | \psi_0 \rangle \\ &= \hat{\mu} | \Psi(t) \rangle. \end{aligned} \tag{30}$$

Let us expand the transition dipole function in a power series:

$$\mu(\mathbf{r}) = \mu_0 + \mu_1 \mathbf{r} + \mu_2 \mathbf{r}^2 + \dots . \tag{31}$$

Now, by considering the zeroth-order of the expansion, Eqs. (29) and (30) become $|\tilde{\Phi}(t)\rangle \approx \mu_0 |\tilde{\Psi}(t)\rangle$ and $|\Phi(t)\rangle \approx \mu_0 |\Psi(t)\rangle$, respectively. This means that to zeroth order $|\tilde{\Phi}(t)\rangle$ and $|\Phi(t)\rangle$ have behavior similar to $|\tilde{\Psi}(t)\rangle$ and $|\Psi(t)\rangle$ when plugged into Eq. (23). The result is that we may discriminate $|\Phi(t)\rangle = \hat{\mu}|\Psi(t)\rangle$ from the set $\{|\tilde{\Phi}(t)\rangle\}$ in a way similar to the way that we discriminate $|\Psi(t)\rangle$ from the set $|\tilde{\Psi}(t)\rangle$. However, we are left in this stage with $\mu(\mathbf{r})\Psi(\mathbf{r},t)$ rather than $\Psi(\mathbf{r},t)$. Obviously, if the transition dipole function is known we can immediately recover $\Psi(\mathbf{r},t)$.

4.2 Finite pulse width

Considering a finite pulse-width, the excited-state wavefunction we want to reconstruct is given by eqn (2), where $\varepsilon_n(t_n) = \varepsilon_n e^{-\alpha_n(t_n-\tau_n)^2 - i(-1)^{n+1}\omega_n(t_n-\tau_n)}$ with τ_n as the pulse center, ε_n as a real-positive coefficient, and ω_n as the central frequency of the pulse. Substituting $t_1 - \tau_1 = T_1$, we get

$$|\Psi(t)\rangle = ie^{-iH_e(t-\tau_1)}\Pi e^{-i\omega_0(t_1-t_0)}|\Psi_0\rangle, \tag{32}$$

where $\Pi = \int_{-\infty}^{t-\tau_1} dT_1 e^{+iH_e T_1} \mu \bar{\varepsilon}_1(T_1) e^{-i\omega_1 T_1} e^{-iH_g T_1}$ is a 'pulse propagator', and $\bar{\varepsilon}_n(T_n) = \varepsilon_n e^{-\alpha_n T_n^2}$. The third-order polarization is given by

$$P^{(3)}(\tau) = \tilde{\varepsilon} \langle \Psi_0 | \mu e^{-iH_e \tau_{43}} \Pi_3 e^{-i\tilde{H}_g \tau_{32}} \Pi_2^\dagger e^{-iH_e \tau_{21}} \Pi_1 | \Psi_0 \rangle, \tag{33}$$

where $\tilde{\varepsilon} = i^3 e^{i\omega_0(\tau_{21}+\tau_{43})}$, and the pulse propagators are defined by

$$\Pi_1 = \int_{-\infty}^{T_2+\tau_{21}} dT_1 e^{iH_e T_1} \mu \bar{\varepsilon}_1(T_1) e^{-i\omega_1 T_1} e^{-iH_g T_1}$$

$$\Pi_2^\dagger = \int_{-\infty}^{T_3+\tau_{32}} dT_2 e^{iH_g T_2} \mu \bar{\varepsilon}_2(T_2) e^{i\omega_1 T_2} e^{-iH_e T_2} \quad (34)$$

$$\Pi_3 = \int_{-\infty}^{\tau_{34}} dT_3 e^{iH_e T_3} \mu \bar{\varepsilon}_3(T_3) e^{-i\omega_3 T_3} e^{-iH_g T_3}.$$

In the limit that the pulses are short enough we can neglect Hamiltonian dynamics in the pulse propagators. Assuming, moreover, that the time delays, τ_{ij}, are long enough, we can take the upper limit of the integral to $+\infty$, and the pulse propagators can be expressed in a simplified form:

$$\Pi_1 = \int_{-\infty}^{\infty} dT_1 \mu \bar{\varepsilon}_1(T_1) e^{-i\omega_1 T_1} = \mu \varepsilon_1 \sqrt{\frac{\pi}{\alpha_1}} e^{\frac{-\omega_1^2}{4\alpha_1}}$$

$$\Pi_2^\dagger = \int_{-\infty}^{\infty} dT_2 \mu \bar{\varepsilon}_2(T_2) e^{i\omega_1 T_2} = \mu \varepsilon_2 \sqrt{\frac{\pi}{\alpha_2}} e^{\frac{-\omega_2^2}{4\alpha_2}} \quad (35)$$

$$\Pi_3 = \int_{-\infty}^{\infty} dT_3 \mu \bar{\varepsilon}_3(T_3) e^{-i\omega_3 T_3} = \mu \varepsilon_3 \sqrt{\frac{\pi}{\alpha_3}} e^{\frac{-\omega_3^2}{4\alpha_3}}.$$

Thus, provided the assumptions above eqn (35) hold, the reconstruction methodology may be performed in a manner similar to that described above. Perhaps the most critical point concerning finite pulse width has to do with the second pulse; the pulse that projects the excited-state wavepacket on the ground-state potential. As the bandwidth of the (second) pulse is finite, it may not project effectively the excited-state wavefunction onto the ground-vibrational set $\{|\psi_g\rangle\}$. In order to compensate for this limitation, more than one pulse may be needed in order to produce the entire information required for a complete reconstruction.

The finite bandwidth may actually turn out to be an advantage, since the use of different frequency windows may discriminate non-CARS processes that also produce a signal at the direction $\mathbf{k} = \mathbf{k}_1 - \mathbf{k}_2 + \mathbf{k}_3$.

4.3 Several excited potentials

In principle, the 'pump' pulse of the CARS scheme might transfer amplitude to several excited-state potential surfaces, and not to only one as we assumed above. Still, the formulation for the excited-state wavepacket and the CARS signal remain correct: the excited wavepacket is now an "effective" wavepacket corresponding to the sum of the wavepackets on all the excited electronic states. However, a question remains on how to perform the potential inversion, since several potentials are now involved. Since we require local inversion of the potential in our sign determination procedure, more work will be required to solve this problem.

4.4 Above the dissociation limit of the ground state

Naturally, the most interesting regions for the excited-state wavepacket to be reconstructed are those far from the Frank-Condon region, but before the complete dissociation. This region on the excited-state may lie above the dissociation limit of the ground-state; in such a case a question arises how the CARS signal should be analyzed. Formally, eqn (9) should include another term that corresponds to vibrational eigenstates of the ground-state potential continuum, as these states might contribute to the excited-state wavepacket. This term takes the form

$$P^{(3)}(\tau) = \tilde{\varepsilon} \int d\omega_g e^{-i\bar{\omega}_g \tau_{32}} P^{(3)}_g(\tau_{43}, \tau_{21}). \quad (36)$$

Further analysis of such a term, in order to extract the corresponding ground-vibrational components $P_g^{(3)}$, requires mathematical manipulations that would allow the replacement of the integral with a discrete summation over ground vibrational states.

5. Existing methods for polyatomic potential reconstruction

In this section we give an overview of methods that have been proposed to obtain potential energy surfaces for polyatomics from spectroscopic information.

Gerber, Roth and Ratner have proposed a method for inverting spectra to obtain potential energy surfaces using semiclassical SCF.[11,12] The methodology combines a 1-d RKR methodology for each degree of freedom with an SCF treatment to include the effect of mode coupling. As stated by the authors,[12] the method should work in the low state-density energy range and requires a preliminary assignment of quantum numbers to the spectroscopic frequencies used in the inversion.

Bernstein and Zewail have proposed inverting pump–probe measured signals to obtain excited-state potentials.[13] Their reconstruction approach is classical, rather than quantum mechanical, but more relevant to the present discussion it seems to be limited to one-dimensional systems and to require some *a priori* assumptions about the shape of the potential.

Ho and Rabitz,[14] and Baer and Kosloff[15] have suggested iterative procedures for reconstructing excited potential surfaced, based on sensitivity analysis. In particular, Baer and Kosloff demonstrated their approach on a multi-dimensional molecular system. The approach requires a good initial guess for the excited-state potential and as much experimental information as possible. The method results in a potential surface that reproduces the experimental results, however, as stated by the authors the obtained potential is not unique.

Shapiro *et al.*[16] have proposed inverting measured fluorescence signals to reconstruct excited potentials of multidimensional systems. The method is based on an iterative procedure to determine the correct phases of the transition dipole matrix-elements between the ground and excited vibrational eigenstates. The method requires an initial guess for the excited state potential and proceeds by extrapolating the potential further in an iterative fashion. The method is promising but it is not clear whether it can be used for dissociating systems. More work will be required to gauge its general applicability.

In summary, it would seem that all of the existing methods for polyatomic potential reconstruction have their limitations. In particular, we would argue that all the schemes seem to be based on either a quasi-one-dimensional inversion procedure or to require a good initial guess for the desired potential. Although the method we propose here also has its shortcomings, it differs significantly from the existing methodologies in that it does not require an initial guess for the excited-state potential, nor is it quasi-one-dimensional. As such, we believe that it provides a welcome addition to the existing methodologies for polyatomic potential reconstruction.

6. Conclusions

To conclude, we have presented a methodology for the complete reconstruction of the excited-state wavefunction of a reacting molecule by analyzing a multi-dimensional resonant CARS signal. The methodology is general for polyatomics; it assumes that the ground-state potential is known but that nothing is known about the excited-state potential. Highly accurate reconstruction is obtained even far from the Franck–Condon region. In fact, in practice the method may be more accurate far from the Franck–Condon region, since the frequency shift between the pump and dump pulses will be more effective in discriminating unwanted processes that may contribute to the measured signal at $\mathbf{k} = \mathbf{k}_1 - \mathbf{k}_2 + \mathbf{k}_3$. In our numerical work we made several simplifying assumptions: δ-function pulse excitations, a coordinate-independent transition dipole moment and only one excited-state potential. However, we discussed ideas for the removal of all these assumptions.

We have shown that once the time-dependent wavefunction is found, the excited potential can be reconstructed with quite high accuracy. We presented preliminary results for polyatomics, where obtaining multidimensional potential surfaces from spectroscopic data has been one of the long standing challenges of molecular spectroscopy. An important application of excited-state potential reconstruction will be the *ab initio* simulations of laser control of chemical bond breaking. Experimental laser control has been greatly hindered by the lack of detailed theoretical guidance, which in turn is due to the lack of accurate excited-state potentials. The present methodology could have a significant impact in this field by providing the necessary information about excited-state potentials.

Acknowledgements

This research was supported by the Minerva Foundation and in part by the National Science Foundation under Grant No. PHY05-51164, KITP preprint number: NSF-KITP-11-038. This research is made possible by the historic generosity of the Harold Perlman family.

References

1 A. H. Zewail, *Science*, 1988, **242**, 1645.
2 J. C. Polanyi and A. H. Zewail, *Acc. Chem. Res.*, 1995, **28**, 119.
3 P. Kukura, D. W. McCamant, S. Yoon, D. B. Wandschneider and R. A. Mathies, *Science*, 2005, **310**, 1006.
4 S. Takeuchi, S. Ruhman, T. Tsuneda, M. Chiba, T. Taketsugu and T. Tahara, *Science*, 2008, **322**, 1073.
5 U. Banin and S. Ruhman, *J. Chem. Phys.*, 1993, **99**, 9318.
6 L. Serrano-Andres and M. Merchan, *THEOCHEM*, 2005, **729**, 99.
7 M. Shapiro, *J. Chem. Phys.*, 1995, **103**, 1748, see alsoC. Leichtle, W. P. Schleich, I. Sh. Averbukh and M. Shapiro, *Phys. Rev. Lett.*, 1998, **80**, 1418.
8 J. A. Cina, *J. Chem. Phys.*, 2000, **113**, 9488.
9 T. S. Humble and J. A. Cina, *Phys. Rev. Lett.*, 2004, **93**, 060402.
10 J. A. Cina, *Annu. Rev. Phys. Chem.*, 2008, **59**, 319.
11 R. M. Roth, M. A. Ratner and R. B. Gerber, *Phys. Rev. Lett.*, 1984, **52**, 1288.
12 R. B. Gerber, R. M. Roth and M. A. Ratner, *Mol. Phys.*, 1981, **44**, 1335.
13 R. B. Bernstein and A. H. Zewail, *J. Chem. Phys.*, 1989, **90**, 829.
14 T. Ho and H. Rabitz, *J. Chem. Phys.*, 1988, **89**, 5614.
15 R. Baer and R. Kosloff, *J. Phys. Chem.*, 1995, **99**, 2534.
16 X. Li, C. Menzel-Jones, D. Avisar and M. Shapiro, *Phys. Chem. Chem. Phys.*, 2010, **12**, 15760.
17 X. Li and M. Shapiro, *J. Chem. Phys.*, 2011, **134**, 094113.
18 N. F. Scherer, R. J. Carlson, A. Matro, M. Du, A. J. Ruggiero, V. RomeroRochin, J. A. Cina, G. R. Fleming and S. A. Rice, *J. Chem. Phys.*, 1991, **95**, 1487.
19 K. Ohmori, H. Katsuki, H. Chiba, M. Honda, Y. Hagihara, K. Fujiwara, Y. Sato and K. Ueda, *Phys. Rev. Lett.*, 2006, **96**, 093002.
20 K. Ohmori, *Annu. Rev. Phys. Chem.*, 2009, **60**, 487.
21 T. C. Weinacht, J. Ahn and P. H. Bucksbaum, *Phys. Rev. Lett.*, 1998, **80**, 5508.
22 A. Monmayrant, B. Chatel and B. Girard, *Phys. Rev. Lett.*, 2006, **96**, 103002.
23 D. J. Tannor, *Introduction to Quantum Mechanics: A Time-Dependent Perspectives*, University Science Books Sausalito, 2007, Eq. (13.8).
24 Y. Soo-Lee and E. J. Heller, *J. Chem. Phys.*, 1979, **71**, 4777.
25 E. J. Heller, R. L. Sundberg and D. Tannor, *J. Phys. Chem.*, 1982, **86**, 1822.
26 A. B. Myers, R. A. Mathies, D. J. Tannor and E. J. Heller, *J. Chem. Phys.*, 1982, **77**, 3857.
27 D. Imre, J. L. Kinsey, A. Sinha and J. Krenos, *J. Phys. Chem.*, 1983, **88**, 3956.
28 P. L. Decola, J. R. Andrews and R. M. Hochstrasser, *J. Chem. Phys.*, 1980, **73**, 4695.
29 N. A. Mathew, L. A. Yurs, S. B. Block, A. V. Pakoulev, K. M. Kornau, E. L. SibertIII and J. C. Wright, *J. Phys. Chem.*, 2010, **114**, 817.
30 J. Faeder, I. Pinkas, G. Knopp, Y. Prior and D. J. Tannor, *J. Phys. Chem.*, 2001, **115**, 8440.
31 G. Herzberg, *Molecular Spectra and Molecular Structure; I. Spectra of Diatomic Molecules*, Krieger Publishing Company, Malabar, Florida, 1950.
32 J. M. D. Feit, J. A. Fleck and A. Steinger, *J. Comput. Phys.*, 1982, **47**, 412.

Entanglement in interference-based quantum control: the wave function is not enough

Moshe Shapiro*[ab] and Paul Brumer[c]

Received 24th March 2011, Accepted 14th April 2011
DOI: 10.1039/c1fd00046b

We analyze the way entanglement affects features of interference-based quantum control. We show that quantum interferences vanishes in several cases in the process of extracting probabilities from wave functions. As an example, we discuss the loss of quantum interferences when tracing over *number* states of the radiation field. We also consider the way in which controllability is reduced when tracing over an entire manifold of states whose cumulative probability we wish to control. Finally, we show that it is impossible to control the relative populations of degenerate states (occurring, *e.g.*, in dissociation and chemical reactions) when the relevant transition amplitude is factorizeable, *i.e.*, when it can be written as a product of a purely classical field-dependent part and a purely material-dependent part. Differences between entanglement and non-factorizability of amplitudes are emphasized.

I. Introduction

It is well understood[1,2] that quantum interferences are responsible for active control over molecular processes, and that they are the single most important factor that distinguishes quantum control from classical control. For example, quantum interferences allow one to *completely* block off an undesired process,[3] thereby channelling the outcome to orthogonal subspaces with 100% probability.

It is often assumed that it is sufficient, in efforts to gain control, to attain final wave functions that are as close as possible to preset target states, a feat which is strongly dependent on quantum interferences.[4,5] In spite of this extremely useful strategy, achieving such a target wavefunction does not guarantee success in achieving desired product cross sections. This is because, in accord with quantum mechanics, predicting the probability of a given outcome entails squaring the wave function and summing (*tracing*) the result over all the events that contribute to the measured outcome. This tracing operation may well destroy interferences, thereby rendering the control-objective unattained, even though the final system wave function has been optimized to be in a very closely proximity of the target state.

In this paper we discuss several examples that show how tracing can nullify control. Of particular interest is the outcome of properly accounting for the *quantum* nature of the light source(s) used to navigate the system towards the target state. One finds that certain radiative states (such as number states) become entangled with a material system with which they interact. As a result, tracing over the final radiation states eliminates all the quantum interferences within the system, and control is lost.

[a]*Department of Chemistry, The University of British Columbia, Vancouver, Canada*
[b]*Department of Chemical Physics, The Weizmann Institute, Rehovot, Israel. E-mail: mshapiro@chem.ubc.ca; Tel: +1 604 8228449*
[c]*Chemical Physics Theory Group, Department of Chemistry, University of Toronto, Toronto, Ontario, M5S 3H6, Canada. E-mail: pbrumer@chem.utoronto.ca; Tel: +1 416 9783569*

Another example where control is lost deals with situations, commonly encountered in *non-linear* optics, in which the relevant transition amplitudes are products of purely material ("response") entities, such as non-linear polarizabilities, and field dependent entities, such as some power of the radiation field. We show that under these circumstances, control over degenerate states (occurring for example in multi-channel photodissociation or branching chemical reactions, or indeed in any scattering process) cannot be attained.

II. Loss of control due to entanglement

A. The role of the radiative states

To illustrate the issues involved we focus on the 1-photon *vs.* 2-photon symmetry-breaking control scenario[6] and consider exciting a material system, initially in bound state $|E_g\rangle$, to the final dissociative state $|E, \mathbf{m}^-\rangle$, where E denotes the total energy and $\mathbf{m}$ all other quantum numbers ("channels"). For this example we use a radiation source composed of a two mode *number* state, $|N_\omega, N_{\omega/2}\rangle$, where $|N_\omega, N_{\omega/2}\rangle$ denotes $|0,\ldots, 0, N_\omega, 0,\ldots, 0, N_{\omega/2}, 0,\ldots\rangle$, a number state containing N_ω photons in the ω mode; $N_{\omega/2}$ photons in the $\omega/2$ mode; and zero photons in all the other modes. The initial state of the matter + radiation is then,

$$|\Psi^i\rangle = |N_\omega, N_{\omega/2}\rangle|E_g\rangle. \tag{1}$$

The final matter + radiation state is given (in second-order perturbation theory) as,

$$|\Psi^f\rangle = (2\pi i)^{\frac{1}{2}}\sum_{\mathbf{m}} |E, \mathbf{m}^-\rangle \Big\{ \epsilon(\omega) \langle E, \mathbf{m}^- |d|E_g\rangle |N_\omega - 1, N_{\omega/2}\rangle - \epsilon^2(\omega/2)$$
$$\times \langle E, \mathbf{m}^- |D|E_g\rangle |N_\omega, N_{\omega/2} - 2\rangle \Big\}, \tag{2}$$

where $\langle E, \mathbf{m}^- |d|E_g\rangle$ is the one-photon transition-dipole matrix element, and $\langle E, \mathbf{m}^- |D|E_g\rangle$ is the two-photon matrix element, given as

$$\langle E, \mathbf{m}^- |D|E_g\rangle = \sum_j \langle E, \mathbf{m}^- |d|E_j\rangle\langle E_j |d|E_g\rangle [\hbar\omega/2 + E_g - i\epsilon - E_j]^{-1}. \tag{3}$$

The field amplitude $\epsilon(\omega)$ is related to N_ω by:

$$\epsilon(\omega) = i\left(\frac{\hbar\omega N_\omega}{\varepsilon_0 V}\right)^{\frac{1}{2}} \exp(i\omega z/c - i\phi_\omega), \tag{4}$$

with z denoting the axis of propagation of the light beams, ε_0 the permittivity of the vacuum, and V the cavity volume.

The ϕ_ω phase in $\epsilon(\omega)$ is an added (controllable) phase that we can vary by, *e.g.*, modifying the optical path transversed by the field. As discussed in the context of the classical-field version of this scenario,[6] one controls the process by varying the $\phi_\omega - 2\phi_{\omega/2}$ phase difference and the $|\epsilon(\omega)/\epsilon(\omega/2)|$ amplitude ratio. Of interest is the probability $P_\mathbf{m}(E)$ of observing the dissociation products in a particular state $\mathbf{m}$ in the distant future, given by

$$P_\mathbf{m}(E) \equiv |\langle E, \mathbf{m}^- |\psi^f\rangle|^2 = 2\pi|\epsilon(\omega)\langle E, \mathbf{m}^- |d|E_g\rangle|N_\omega - 1, N_{\omega/2}\rangle$$
$$- \epsilon^2(\omega/2)\langle E, \mathbf{m}^- |D|E_g\rangle |N_\omega, N_{\omega/2} - 2\rangle|^2. \tag{5}$$

Varying the phase difference allows controls of the interference term arising in this expression.

The difference between the above formula and that obtained when using classical light[6] is that here the radiative states become entangled with the material states. A result of the orthogonality between the radiative states and the entanglement is that if we *do* measure either of the radiative states then we need to compute either

$$|\langle N_\omega - 1, N_{\omega/2}|\langle E, \mathbf{m}^-|\Psi^f\rangle|^2 = 2\pi|\epsilon(\omega)\langle E, \mathbf{m}^-|d|E_g\rangle|^2, \quad (6)$$

or

$$|\langle N_\omega, N_{\omega/2} - 2|\langle E, \mathbf{m}^-|\Psi^f\rangle|^2 = 2\pi|\epsilon^2(\omega)\langle E, \mathbf{m}^-|D|E_g\rangle|^2. \quad (7)$$

Note that neither probability expressions contain any interference contribution. Furthermore, if we *do not* measure the radiative state explicitly then the observed result is obtained by summing (tracing over) these probabilities to obtain that

$$\overline{P}_\mathbf{m}(E) \equiv |\langle N_\omega - 1, N_{\omega/2}|\langle E, \mathbf{m}^-|\Psi^f\rangle|^2 + |\langle N_\omega, N_{\omega/2} - 2||\langle E, \mathbf{m}^-|\Psi^f\rangle|^2$$
$$= |\epsilon(\omega)\langle E, \mathbf{m}^-|d|E_g\rangle|^2 + |\epsilon^2(\omega/2)\langle E, \mathbf{m}^-|D|E_g\rangle|^2 \quad (8)$$

in which there are no interference terms either. We therefore conclude that coherent control cannot take place when the initial radiative states are *number states*. This property is not, as one might pre-suppose, the result of the indeterminacy of the electric field phase associated with number state. Rather, it is a result of the entanglement between the number states and matter states and the fact that the number states are orthogonal to one another.

Note that these results are not restricted to the one *vs.* two control scenario. As shown by Gong *et al.*,[7] similar conclusions hold for other coherent control scenarios as well.

B. Entanglement in one photon dissociation

The above results, pertaining to the way entanglement eliminates interferences, are not restricted to quantum fields. Another example is ordinary multichannel photodissociation in the weak field domain. In this case we irradiate a molecule with weak cw light and the excited molecule breaks apart into a number of energetically accessible (open) internal fragment states **m** ("channels"). Such a process creates a superposition of all the open channels:

$$|\Psi^f\rangle \xrightarrow{t\to\infty} (2\pi i)^{1/2} \sum_\mathbf{m} |e_\mathbf{m}\rangle|k_\mathbf{m}\rangle \epsilon(\omega)\langle E, \mathbf{m}^-|d|E_g\rangle, \quad (9)$$

where $e_\mathbf{m}$ is the internal energy of the photofragments and $k_\mathbf{m}$ is the momentum associated with their relative translational motion. The latter quantity is fixed by energy conservation

$$E = E_g + \hbar\omega = \hbar^2 k_\mathbf{m}^2/(2\mu) + e_\mathbf{m}. \quad (10)$$

Hence, a measurement of $e_\mathbf{m}$ fixes $k_\mathbf{m}$, and *vice versa*. Thus, energy conservation *entangles* the translational motion with the internal states of the fragments. A result of this entanglement is that although Ψ^f is a superposition of many dissociation channels, no interferences between these channels can be observed.[7,8] It is only when the bandwidth of the dissociating laser *exceeds* an internal energy difference $|e_\mathbf{m} - e_\mathbf{n}|$ that we can begin to see interferences between the $|k_\mathbf{m}\rangle$ and $|k_\mathbf{n}\rangle$ translational states.[8] This argument solidifies the well known prescription of scattering theory that the *total* cross section is the sum of the individual scattering amplitudes-squared, rather than the square of the sum of the scattering amplitudes.

C. Regaining control *via* the use of coherent states

The situation is dramatically different than that in Sect. II A if the radiation field is comprised of a product of coherent states, $|\alpha_\omega\rangle$, defined as,

$$\hat{a}_\omega |\alpha_\omega\rangle = \alpha_\omega |\alpha_\omega\rangle. \tag{11}$$

In this case, the initial matter + radiation state associated with the one *vs.* two photons control scenario is

$$|\Psi^i\rangle = |\alpha_\omega, \alpha_{\omega/2}\rangle |E_g\rangle. \tag{12}$$

Because $|\alpha_\omega\rangle$ is an eigenstate of the annihilation operator, which is the only part of the quantum field operator that contributes in the rotating wave approximation, the absorption of one photon from an $|\alpha_\omega\rangle$ state results in the production of $\epsilon(\omega)|\alpha_\omega\rangle$, *i.e.* the initial coherent state multiplied by the field amplitude:

$$\epsilon(\omega) = i\left(\frac{\hbar\omega}{\epsilon_0 V}\right)^{\frac{1}{2}} \alpha_\omega \exp(i\omega z/c - i\phi_\omega). \tag{13}$$

Hence, the final matter + radiation state can be written as,

$$|\Psi^f\rangle = (2\pi i)^{\frac{1}{2}} \sum_{\mathbf{m}} |E, \mathbf{m}^-\rangle \{\epsilon(\omega)\langle E, \mathbf{m}^-|\mathrm{d}|E_g\rangle - \epsilon^2(\omega/2)\langle E, \mathbf{m}^-|D|E_g\rangle\} |\alpha_\omega, \alpha_{\omega/2}\rangle. \tag{14}$$

The resultant wave function $|\Psi^f\rangle$ is seen to be completely unentangled. That is, contrary to the number state situation discussed above, the coherent field states do not get entangled with the material system. Hence the interference terms do not vanish when we trace-over the radiative states. In other words, the measurement of the final radiation states (or lack thereof) collapses this final state to the material superposition of eqn (14), allowing for quantum control. Hence, coherent control with coherent states is seen to be essentially identical to the classical-field scenario, in keeping with the fact that coherent states are known to be the closest quantum states to classical electromagnetic fields.[9]

III. Reduced interference in the control of a *manifold* of states

Another example where the control objective cannot be expressed as that of attaining a particular final *wave function* is the case analyzed early on[3] in which one attempts to achieve control over a *manifold* of states. In this case as well we need to trace over states, but here it is not done with respect to unobserved states of the bath or radiation field. Rather the trace arises as the sum over all final states of interest. As a result, the control objective in this case is to maximize or minimize a *cumulative probability* given as the sum over many states:

$$P_q(E) = \sum_{\mathbf{m}} P_{q,\mathbf{m}}(E), \tag{15}$$

where q denotes the multi dimensional manifold to which we wish to navigate the system.

As an example of this situation, consider bichromatic control attained by irradiating a system with two cw fields given as:

$$\varepsilon(t) = 2\sum_{i=1,2} \hat{\epsilon}_i |\epsilon(\omega_i)| \cos[\omega_i t - \phi(\omega_i)] = 2\sum_{i=1,2} \hat{\epsilon}_i Re[\epsilon(\omega_i)\exp(-i\omega_i t)]. \tag{16}$$

Here the system is initially in a superposition of two bound states,

$$\chi = a_1|E_1\rangle + a_2|E_2\rangle. \tag{17}$$

Tuning the frequencies ω_1 and ω_2 such that $\omega_2 - \omega_1 = (E_1 - E_2)/\hbar$, we have that $P_q(E)$, the cumulative probability at energy $E = E_1 + \hbar\omega_1 = E_2 + \hbar\omega_2$, is given as

$$P_q(E) = 2\pi\left\{|a_1|^2|\epsilon(\omega_1)|^2 d_q(22) + 2Re\left[a_1 a_2{}^*\epsilon(\omega_1)\epsilon^*(\omega_2) d_q(12)\right]\right\}, \tag{18}$$

where

$$d_q(ij) \equiv \sum_\mathbf{m} \langle E_i | d | E, q, \mathbf{m}^-\rangle\langle E, q, \mathbf{m}^- | d | E_j\rangle, \quad i,j = 1,2. \tag{19}$$

Expressing each complex dipole amplitude as its absolute value times a phase factor,

$$\langle E_i|d|E, q, \mathbf{m}^-\rangle \equiv |\langle E_i|d|E, q, \mathbf{m}^-\rangle|\exp(i\alpha^{(1)}_{q,\mathbf{m}}), \tag{20}$$

exposes the $d_q(12)$ interference term as

$$d_q(12) = \sum_\mathbf{m} |\langle E_1 | d | E, q, \mathbf{m}^-\rangle\langle E, q, \mathbf{m}^- | d | E_2\rangle| \exp(i\alpha^{(1)}_{q,\mathbf{m}} - i\alpha^{(2)}_{q,\mathbf{m}}). \tag{21}$$

As more and more **m** terms are added to the sum we expect the oscillatory $\exp(i\alpha^{(1)}_{q,\mathbf{m}} - i\alpha^{(2)}_{q,\mathbf{m}})$ terms to average out, thereby reducing the magnitude of the interference term. In the classical limit the size of the q manifold, and hence the number of **m** terms, becomes very large. In that limit, if the $\alpha^{(i)}_{q,\mathbf{m}}$ phases vary in an essentially random fashion, the interference term vanishes. Thus, the desire to control the population of an *entire* manifold, q, can result in a highly reduced interference term and the reduction of our ability to control the process, even when optimization procedures are nonetheless capable of attaining a single target wave function.

IV. Non-factorizability as a prerequisite for degenerate state control

Although, as discussed above, entanglement between the radiation field states and matter states can result in loss of control, the inability to factorize the wavefunction into a product of matter and radiation terms is sometimes necessary for control.

To appreciate the difference[10] between "non-factorizable" and "entangled", consider eqn (14). This wavefunction is an *unentangled product* of a matter state $|E, \mathbf{m}^-\rangle$ and a state of the radiation field $|\alpha_\omega, \alpha_{\omega/2}\rangle$. However, the field and matter dependence is given by

$$\{\epsilon(\omega)\langle E, \mathbf{m}^-|d|\rangle E_g - \epsilon^2(\omega/2)\langle E, \mathbf{m}^-|D|E_g\rangle\}, \tag{22}$$

which is a *non-factorizable* product of matter and field components. It is this last feature that will be shown necessary for control in the cases discussed below. Specifically, we now show that in one such case, the control of the branching ratios between *degenerate* states, a situation common to all multichannel dissociation and scattering phenomena in general, non-factorizability is essential for interference-based control.[11]

This statement is an extension of the well known result [12] that it is impossible to control the branching ratio between two dissociation channels, and indeed all processes involving degenerate final states, using weak field one-photon transitions. It is also an extension of the (possibly less well known) result [1,2] that it is impossible to control the branching ratio between degenerate final states even in the strong field domain, if the process involves the net absorption of one photon from a *single* initial ("precursor") state.

The prevailing opinion[13,14] was, in fact, that the above conclusions do not hold in the (non-linear) N-photon absorption case. This perspective was supported by a series of seminal experiments[15–19] that showed that the use of coherent control techniques in the short pulse realization of N-photon processes, whether performed *via* intermediate resonances or not, allows one to control the transitions to *non-degenerate* states.

Here we show that the situation concerning the control over *degenerate* final states is different: If the relevant amplitudes can be factorized into radiative and material parts, no control over branching ratios of degenerate states can be attained, even in the non-linear N-photon absorption case. That is, the mere existence of two or more interfering pathways to the same final states,[1,2] though *necessary* for quantum control, is not *sufficient*. Rather, in addition to this necessary condition, one must also satisfy certain conditions on the radiative states, as discussed above, and the non-factorizability condition, to be discussed below.

In order to see why non-factorizability is necessary, we utilize N-order perturbation theory to write the N-photon transition amplitude between the initial state $|E_g\rangle$ and a final state $|E_f\rangle$, where the former is subject to the action of a broad-band pulse of light as,

$$b_{E_f}^{(N)} = \int \cdots \int (\Pi_{j=1}^{N-1} d\omega_j) \epsilon(\omega_N)$$
$$\times \sum_{i_1,\ldots,i_{N-1}} \langle E_f|d|E_{i_{N-1}}\rangle \Pi_{k=1}^{N-1} \frac{\epsilon(\omega_k)\langle E_{i_k}|d|E_{i_{k-1}}\rangle}{\sum_{j=1}^{k}\hbar\omega_j - (E_{i_k} + \Delta_{i_k} - i\Gamma_{i_k}/2 - E_g)}, \quad (23)$$

where $\omega_N = \omega_{f,g} - \sum_{j=1}^{N-1}\omega_j$, $i_0 = g$, with each i_k, $k = 1,\ldots, N-1$, ranges over all the intermediate states. In the absence of intermediate resonances, *e.g.* when $|\omega_{f,g}| \ll |E_i - E_g|/\hbar$ for all $i \ne f$ or g, we can replace the term $\sum_{j=1}^{k}\hbar\omega_j - (E_{i_k} + \Delta_{i_k} - i\Gamma_{i_k}/2 - E_g)$ in the denominator with a constant term $\hbar\omega\bar{\omega} - (E_{i_k} + \Delta_{i_k} - i\Gamma_{ik}/2 - E_g)$, and write that

$$b_{E_f}^{(N)} = \langle E_f|M^{(N)}|E_g\rangle \int \cdots \int (\Pi_{j=1}^{N-1} d\omega_j) \Pi_{k=1}^{N} \epsilon(\omega_k) \quad (24)$$

where

$$\langle E_f|M^{(N)}|E_g\rangle \equiv \sum_{i_1,\ldots,i_{N-1}} \langle E_f|d|E_{i_{N-1}}\rangle \Pi_{k=1}^{N-1} \frac{\langle E_{i_k}|d|E_{i_{k-1}}\rangle}{\hbar\bar{\omega} - (E_{i_k} + \Delta_{i_k} - i\Gamma_{i_k}/2 - E_g)}.$$

Using the fact that the Fourier transform of a product of functions is the convolution integral of their respective Fourier transforms,

$$\int_{-\infty}^{\infty} dt \varepsilon'(t)\varepsilon''(t)\exp(-i\omega t) = 2\pi \int_{-\infty}^{\infty} d\omega' \epsilon'(\omega')\epsilon''(\omega - \omega'), \quad (25)$$

we obtain, by repeated application of eqn (25), that $b^{(N)}{}_{E_f}$ in eqn (24) can be written as,

$$b_{E_f}^{(N)} = \frac{\langle E_f|M^{(N)}|E_g\rangle}{(2\pi)^{N-1}} \int dt \varepsilon^N(t) \exp(-i\omega_{f,g}t). \quad (26)$$

It follows from eqn (26) that it is possible to control the strength of a general N-photon transition between *non-degenerate* states, by essentially using variants of the pulse modulations of the $N = 2$ case.[20] However, it also follows from eqn (26) that if the final state is energetically *degenerate* it is not possible to use N-photon processes to optically control the *relative* populations of its components in the absence of intermediate resonances. In the continuum case this means that such N-photon processes cannot be used to control *branching ratios* to different scattering channels of the same total energy.

This extension of the *one-photon* result of ref. 12 follows because in the absence of intermediate resonances the transition amplitude $b^{(N)}_{E_f}$ factorizes into a product of a material part $\langle E_f|M^{(N)}|E_g\rangle$ and a radiative part $\int dt \varepsilon^N(t)\exp(-i\omega_{f,g}t)$. Since the only information regarding the final state appearing in the radiative part is its energy (contained in the $\omega_{f,g}$ factor), there is no way that the shape of the pulse can affect the populations of different states with the *same* energy.

We can extend this result to the strong field domain where N-order perturbation theory is not valid by including the time dependent initial coefficient in our expression. In particular, in a multi-channel photodissociation process, in the absence of an intermediate resonance, we can write the non-perturbative amplitude at time t for observing channel $\mathbf{n}$ in the far future as,

$$b^{(N)}_{E,\mathbf{n}}(t) = \frac{\langle E,\mathbf{n}^-|M^{(N)}|E_g\rangle}{(2\pi)^{N-1}}\int_{-\infty}^{t} dt'\varepsilon^N(t')\exp(-i\omega_{E,g}t')b_{E_g}(t'), \qquad (27)$$

where $b_{E_g}(t)$ is the amplitude of the initial state, and

$$\langle E,\mathbf{n}^-|M^{(N)}|E_g\rangle = \sum_{i_1,\ldots,i_{N-1}} \langle E,\mathbf{n}^-|d|E_{i_{N-1}}\rangle \Pi_{k=1}^{N-1}\frac{\langle E_{i_k}|d|E_{i_{k-1}}\rangle}{\hbar\omega - (E_{i_k} + \Delta_{i_k} - i\Gamma_{i_k}/2 - E_g)}, \qquad (28)$$

where $|E_{i_0}\rangle \equiv |E_g\rangle$. We see that the branching ratio between different final channels is given as

$$R_{\mathbf{n},\mathbf{m}}(t) \equiv \left|\frac{b^{(N)}_{E,\mathbf{n}}(t)}{b^{(N)}_{E,\mathbf{m}}(t)}\right|^2 = \left|\frac{\langle E,\mathbf{n}^-|M^{(N)}|E_g\rangle}{\langle E,\mathbf{m}^-|M^{(N)}|E_g\rangle}\right|^2, \qquad (29)$$

and the pulse attributes, as well as the detailed evolution history of the initial state coefficient $b_{E_g}(t')$, have completely disappeared! Thus it is not possible, by manipulating the laser field(s), to control the branching ratio into different dissociation channels in a fixed-N multi-multi-photon process which starts with only one precursor state.

As discussed in ref. 1, 2, the remedy in this case is to interfere N photon processes with $L \neq N$ photon processes, both leading to the same final state $|E,\mathbf{n}^-\rangle$. In this case the transition amplitude is given in a field-matter *non-factorizable* form,

$$b_{E,\mathbf{n}}(t) = b^{(N)}_{E,\mathbf{n}}(t) + b^{(L)}_{E,\mathbf{n}}(t) =$$

$$\left\{\frac{\langle E,\mathbf{n}^-|M^{(N)}|E_g\rangle}{(2\pi)^{N-1}}\int_{-\infty}^{t}dt'\varepsilon^N(t') + \frac{\langle E,\mathbf{n}^-|M^{(L)}|E_g\rangle}{(2\pi)^{L-1}}\int_{-\infty}^{t}dt'\varepsilon^L(t')\right\}e^{-i\omega_{E,g}t'}b_{E_g}(t'). \qquad (30)$$

Alternatively we can start from two ($|E_1\rangle$ and $|E_2\rangle$) or more precursor states, in which case

$$b_{E,\mathbf{n}}(t) = \left(\frac{1}{2\pi}\right)^{N-1}\{\langle E,\mathbf{n}^-|M^{(N)}|E_1\rangle\int_{-\infty}^{t}dt'\varepsilon_1^N(t')e^{-i\omega_{E,1}t'}b_{E_1}(t') +$$

$$\langle E,\mathbf{n}^-|M^{(N)}|E_2\rangle\int_{-\infty}^{t}dt'\varepsilon_2^N(t')e^{-i\omega_{E,2}t'}b_{E_2}(t')\}. \qquad (31)$$

Controllability over degenerate states can also be attained if there exists an intermediate (M-photon) resonance, because in that case the photon variables and the material factors are also non-factorizable. Specifically, in the case of an intermediate resonance we have that

$$b_{E_f}^{(N)} = \left\langle E, \mathbf{n}^- | M'^{(N)} | E_g \right\rangle \int \cdots \int (\Pi_{j=1}^{N-1} d\omega_j) \Pi_{k=1}^{N} \epsilon(\omega_k) +$$

$$\int \cdots \int (\Pi_{j=1}^{N-1} d\omega_j) \epsilon(\omega_N) \sum_{i_1,\ldots,i_{M-1},i_{M+1},\ldots,i_{N-1}} \langle E_f | d | E_{i_{N-1}} \rangle$$

$$\frac{\epsilon(\omega_M) \langle E_{i_M} | d | E_{i_{M-1}} \rangle}{\sum_{j=1}^{M} \hbar\omega_j - (E_{i_M} + \Delta_{i_M} - i\Gamma_{i_M}/2 - E_g)} \cdot \Pi_{k \neq M}^{N-1} \frac{\epsilon(\omega_k) \langle E_{i_k} | d | E_{i_{k-1}} \rangle}{\sum_{j=1}^{k} \hbar\omega_j - (E_{i_k} + \Delta_{i_k} - i\Gamma_{i_k}/2 - E_g)},$$

(32)

where

$$\langle E, \mathbf{n}^- | M'^{(N)} | E_g \rangle = \sum_{i_1,\ldots,i_{M-1},i_M \neq i'_M, i_{M+1},\ldots,i_{N-1}} \langle E, \mathbf{n}^- | d | E_{i_{N-1}} \rangle \Pi_{k=1}^{N-1}$$

$$\frac{\langle E_{i_k} | d | E_{i_{k-1}} \rangle}{\hbar\bar{\omega} - (E_{i_k} + \Delta_{i_k} - i\Gamma_{i_k}/2 - E_g)}$$

V. Conclusions

In this paper we have shown that, in certain cases, attaining a particular target wave function is insufficient for control. For example, when entanglement is involved, the need to sum probabilities over all events that contribute to a given outcome, whether such events are measured explicitly or not, can destroy all interferences, making interference-based control impossible. Several examples were given: the vanishing of quantum interferences when material states are entangled with number states, and the reduction in controllability when control of an entire manifold of states is desired. An alternate example, where field-matter non-factorizability (as distinct from entanglement) is essential for exercising control over degenerate states was also discussed.

Acknowledgements

NSERC Canada support for this research is gratefully acknowledged.

References

1 M. Shapiro and P. Brumer, *Principles of the Quantum Control of Molecular Processes*(Wiley, New York, 2003).
2 M. Shapiro and P. Brumer, Quantum Control of Molecular Processes, 2nd ed. (Wiley-VCH, Wertheim, in press).
3 P. Brumer and M. Shapiro, *Chem. Phys. Lett.*, 1986, **126**, 541.
4 D. J. Tannor and S. A. Rice, *J. Chem. Phys.*, 1985, **83**, 5013.
5 S. A. Rice and M. Zhao, *Optical Control of Molecular Dynamics*(Wiley, New York, 2000).
6 G. Kurizki, M. Shapiro and P. Brumer, *Phys. Rev. B*, 1989, **39**, 3435.
7 J. B. Gong and P. Brumer, *J. Chem. Phys.*, 2010, **132**, 054306.
8 M. Shapiro and J. Phys, *Chem. A*, 2006, **110**, 8580.
9 R. Loudon, *The Quantum Theory of Light*, 2nd ed. (Clarendon Press, Oxford, 1983).
10 This discussion corrects that in ref. 11 in which "entangled" was used to mean "non-factorizable".
11 M. Shapiro and P. Brumer, *J. Chem. Phys.*, 2010, **132**, 186101.
12 P. Brumer and M. Shapiro, *Chem. Phys.*, 1989, **139**, 221.
13 M. Joffre, *Science*, 2007, **317**, 453.

14 V. I. Prokhorenko, A. M. Nagy, S. A. Waschuk, L. S. Brown, R. R. Birge and R. J. D. Miller, *Science*, 2007, **317**, 453.
15 D. Meshulach and Y. Silberberg, *Nature*, 1998, **396**, 239–242.
16 D. Meshulach, D. Yelin and Y. Silberberg, *J. Opt. Soc. Am. B*, 1998, **15**, 1615.
17 D. Meshulach and Y. Silberberg, *Phys. Rev. A: At., Mol., Opt. Phys.*, 1999, **60**, 1287.
18 Y. Silberberg, *Nature*, 2001, **414**, 494.
19 N. Dudovich, B. Dayan, S. M. Gallagher Faeder and Y. Silberberg, *Phys. Rev. Lett.*, 2001, **86**, 47.
20 B. Dayan, A. Pe'er, A. A. Friesem and Y. Silberberg, *Phys. Rev. Lett.*, 2004, **93**, 023005.

Searching for pathways involving dressed states in optimal control theory

Philipp von den Hoff,* Markus Kowalewski and Regina de Vivie-Riedle

Received 3rd March 2011, Accepted 5th April 2011
DOI: 10.1039/c1fd00031d

Selective population of dressed states has been proposed as an alternative control pathway in molecular reaction dynamics [Wollenhaupt *et al.*, *J. Photochem. Photobiol. A*: *Chem.*, 2006, **180**, 248]. In this article we investigate if, and under which conditions, this strong field pathway is included in the search space of optimal control theory. For our calculations we used the proposed example of the potassium dimer, in which the different target states can be reached *via* dressed states by resonant transition. Especially, we investigate whether the optimization algorithm is able to find the route involving the dressed states although the target state lies out of resonance in the bare state picture.

1 Introduction

Recent developments in ultrashort laser pulse generation and shaping technology opened the door to new strategies for the control of ultrafast molecular photoreactions.[1,2] One of the novel routes uses the phase of the electric field with respect to the envelope as control parameter.[3] Another approach utilizes strong electric fields to shift electronic states in energy in order to steer the molecular reactions.[4] In this work we focus on the selective population of dressed states (SPODS), a strategy which nicely combines both routes of phase and strong field control. SPODS can be implemented *via* the photon locking technique,[5] the optical counterpart to the spin locking technique, originally developed in NMR.[6] The photon locking technique in combination with light field control of molecular reactions was theoretically exemplified for ground state dynamics.[7] Selective population of dressed states is also the key in adiabatic radid passage (ARP)[8] or stimulated Raman adiabatic passage (STIRAP).[9,10] Both are adiabatic processes characterized by a smooth change in the temporal phase and enevlope of the electric field. SPODS *via* photon locking, on the other hand, is characterized by a rapid phase jump and is therefore of non-adiabatic nature.

Strong field quantum control *via* SPODS using pulse shaping techniques was experimentally demonstrated for the potassium atom by M. Wollenhaupt and T. Baumert.[11] Within their investigations, they used sinusoidal spectral phase modulation,[12,13] chirped excitation[14] and adaptive optimization of the spectral phase[11] to realize the SPODS scheme. They already extended the SPODS mechanism theoretically to the molecular system K_2[15] to control the final population in the $4^1\Sigma_g^+$ and $5^1\Sigma_g^+$ target states.

In a previous work we performed calculations on the SPODS in K_2 to analyze the interaction of the ultrashort light field with the initialized oscillating electronic dipole moment[16] as the key element for the control mechanism. As in ref. 15, we

Department Chemie, Ludwig-Maximilians-Universität, München, 81377, München, Germany. E-mail: Philipp.vondenhoff@cup.uni-muenchen.de; Fax: +49 89 218077133; Tel: +49 89 218077535

used a simple double pulse sequence. Depending on the parameters pulse delay and intensity, we reached a maximum efficiency of about 66% for both target states and showed the correlation between the optimal pulse delay and the lifetime of the induced electronic coherence.

In this work we shortly revisit the basics of the SPODS mechanism. With optimal control theory (OCT) we investigate whether the SPODS mechanism is an optimal solution for the given control task and—if yes—whether its efficiency can be further improved by OCT. In addition, we outline a strategy to include the SPODS mechanism in the search space of the OCT—if needed—by selecting special starting conditions.

2 SPODS excitation scheme

In this section we revisit the excitation scheme for the non-adiabatic SPODS mechanism for the example of K_2 and summarize the previous findings.[16] In the ideal case a weak and resonant pulse first creates a state of maximum coherence, *i.e.* a 50 : 50 superposition between the bare electronic states $X^1\Sigma_g^+$ and $A^1\Sigma_u^+$ (see Fig. 1 dash dotted arrow). This process simultaneously launches an oscillating dipole moment following the driving field with a phase shift of $\frac{\pi}{2}$. The formation of the superposition and the rise of the induced dipole moment following the driving laser field is schematically illustrated in Fig. 2 (a) and (b) (first 40 fs).

In the second step the pre-pulse is followed by an intense pulse with the same frequency. This pulse is shifted in phase by $\pm\frac{\pi}{2}$ relative to the first one. Thus the electric field of the second pulse is either exactly in phase with the prepared oscillating molecular dipole or exactly shifted by π (see Fig. 2 (b). The in-phase situation selectively populates the lower dressed state (DS) (see Fig. 1 lower dotted curve), the π shift leads to a selective population of the upper DS state (see Fig. 1 upper dotted curve). During the second pulse the bare state populations are locked due to the phase relation, preventing population transfer between the $X^1\Sigma_g^+$ and $A^1\Sigma_u^+$, although

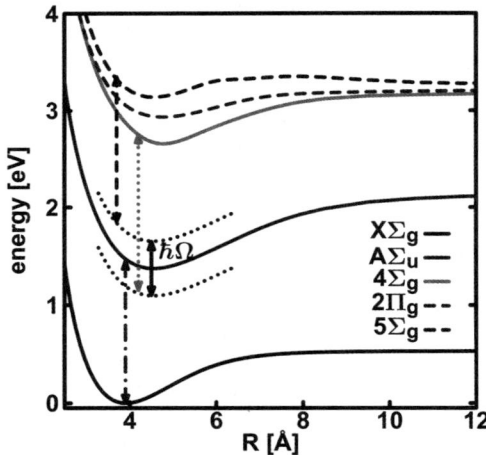

Fig. 1 SPODS scheme of potassium dimer. The first pulse in the sequence creates a superposition between the $X^1\Sigma_g^+$ and $A^1\Sigma_u^+$ states (dash dotted arrow). During the second pulse the $X^1\Sigma_g^+$ and the $A^1\Sigma_u^+$ states are 'photon locked'. The optical phase controls which of the dressed states (indicated as dotted lines) energetically separated by $\hbar\Omega$ is selectively populated. Absorption of another photon leads to population transfer to either the $4^1\Sigma_g^+$ (dotted arrow) or $5^1\Sigma_g^+$ (dashed arrow).

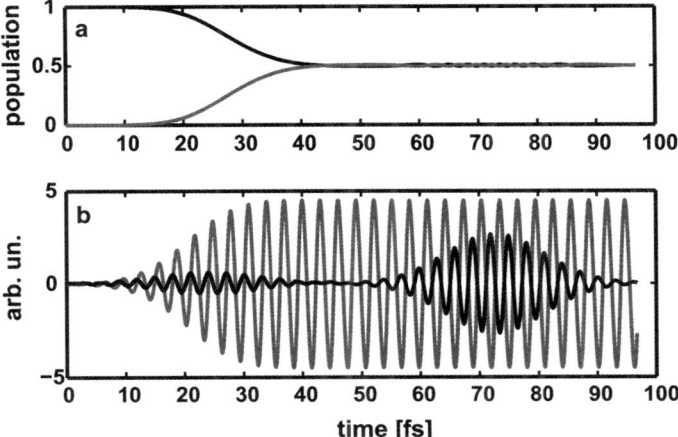

Fig. 2 Schematic illustration of the SPODS pulse sequence populating the upper dressed state acting on two electronic states. (a) Temporal evolution of the populations of the two electronic states. (b) Temporal evolution of the electric field (black) and the induced dipole moment (gray).

the frequency is resonant on this transition (see Fig. 2 (a)).[7] Due to the laser intensity dependent energy splitting of the DS in the order of $\hbar\Omega$ (solid black arrow in Fig. 1) resonance is reached either with the $4^1\Sigma_g^+$ or the $5^1\Sigma_g^+$ target state. With the described pulse sequence it is in principle possible to control the final populations in the $4^1\Sigma_g^+$ and $5^1\Sigma_g^+$ state by switching the relative phase between the two sub-pulses.

In the case of the potassium atom it is possible to separate the two involved sub-pulses well in time.[12,13] But in the case of the potassium dimer it turned out that the two sub-pulses have to overlap partially in time to ensure an efficient SPODS.[15] To explain their results, we used our approach for the coupled electron and nuclear quantum dynamics[17–19] and followed the time-dependent expectation value of the induced dipole moment along the molecular z-axis $\langle\mu_z\rangle(t)$. The value of $\langle\mu_z\rangle(t)$ is related to the time-dependent electron density $\varphi_{tot}(t)$[16] through:

$$\langle\mu_z\rangle(t) = \langle\varphi_{tot}(t)|\hat{\mu}_z|\varphi_{tot}(t)\rangle = \int\rho_{tot}(t)\hat{\mu}_z dr. \quad ((1))$$

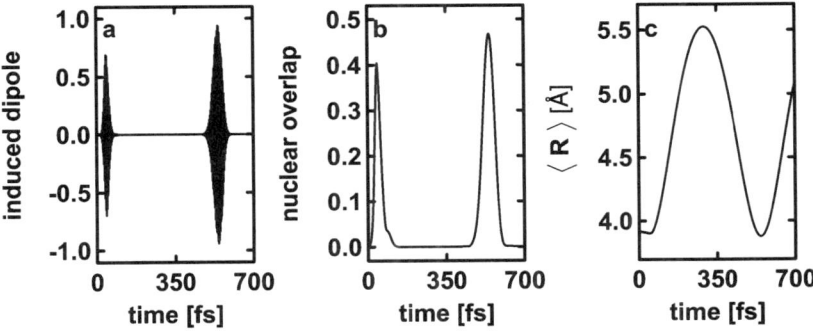

Fig. 3 Temporal evolution in the presence of the pre-pulse of (a): the expectation value of the induced dipole moment $\langle\mu_z\rangle$; (b): the overlap of the nuclear wavefunctions on the $X^1\Sigma_g^+$ and $A^1\Sigma_u^+$ state; (c): the expectation value of the internuclear distance of the nuclear wavefunction on the $A^1\Sigma_u^+$ state ($\langle R\rangle$).

$\hat{\mu}_z$ is the dipole operator along the intramolecular axis. Fig. 3 (a) shows the expectation value of the induced dipole moment in the presence of the pre-pulse alone. The corresponding overlap of the nuclear wavefunctions on the $X^1\Sigma_g^+$ and the $A^1\Sigma_u^+$ surface is plotted in Fig. 3 (b). Both observables reflect the formation of the superposition between the $X^1\Sigma_g^+$ and $A^1\Sigma_u^+$ state within the first 40 fs. Right after the end of the pulse both values begin to decrease in the same manner, although the populations of the involved sates stay constant. The reason for this decay lies in the nuclear wavepacket dynamics. After the pre-pulse, the two nuclear wavepackets start to evolve differently, due to the shift of the $A^1\Sigma_u^+$ state towards larger internuclear distances and the difference in potential shape (see Fig. 1). In addition only large amplitude motion is induced in the upper potential, while the wavepacket in the ground state nearly stays fixed in space. This extended nuclear motion can be seen in the time-dependent expectation value of the internuclear distance $\langle R \rangle (t)$ for the nuclear wavepacket in the $A^1\Sigma_u^+$ state (see Fig. 3 (c)).

Now the reason for the partial temporal overlap of the two sub-pulses becomes obvious. As the nuclear wavepackets start to evolve asynchronously, the induced dipole moment is damped by the reducing overlap of the two nuclear wavepackets. As a consequence the interaction between the induced dipole moment and the second pulse (i.e. the photon locking) gets less effective and leads to the loss of control. In the potassium atom this effect is absent due to the non-existing nuclear dynamics. In Fig. 3 (a) a revival of the expectation value of the induced dipole moment with even an increase in amplitude can be seen after one oscillation period of the nuclear wavepacket in the upper potential (after approx. 500 fs). From the theory, this revival can also be used to continue the SPODS mechanism with a long pulse delay of about 550 fs.

3 Optimal control theory

OCT is established as a powerful method for various control tasks and for the interpretation of the induced control mechanisms. Different schemes for quantum control investigations were developed based on the calculus of variations.[20–22] The OCT algorithm is used to find the optimal laser pulse, driving a quantum system from a defined initial state $\psi(t=0)$ to the wanted target state $\Phi(t=T)$.[23,24] For the optimization of the SPODS mechanism, the final population of the target state needs to be optimized. Therefore we used the expectation value of a positive definite operator as control aim[21] in the following form:[25]

$$J[\psi(t),\Phi(t),\varepsilon(t)] = |\langle\psi(T)|\hat{O}|\psi(T)\rangle|^2 - \int_0^T \alpha(t)|\varepsilon(t) - \varepsilon_{ref}(t)|^2 dt$$
$$- 4\,\Re\left[\langle\psi(T)|\hat{O}|\psi(T)\rangle\int_0^T \langle\Phi(t)|\left[\frac{i}{\hbar}(\hat{H}_0 - \hat{\mu}\,\varepsilon(t)) + \frac{\partial}{\partial t}\right]|\psi(t)\rangle dt\right]$$

(2)

It includes three terms, the optimization target, an integral over the laser field, penalizing the pulse fluence and the time dependent Schrödinger equation as a side condition. The optimization target is to maximize the absolute square of the expectation value of any positive definite operator $|\langle\psi(T)|\hat{O}|\psi(T)\rangle|^2$ using the wavefunction ψ after the laser excitation time T. The initial states can be chosen as eigenstates or arbitrary superpositions of eigenstates. The second term of eqn (2) is an integral over the laser field $\varepsilon(t)$ and a reference field $\varepsilon_{ref}(t)$ with a time-dependent factor $\alpha(t)$. Depending on the implementation of $\varepsilon_{ref}(t)$, it is known as the penalty factor or Krotov change parameter. With the choice of $\alpha(t) = \alpha_0/s(t)$ and e.g. a gaussian shaped function $s(t)$, an field envelope function can be imprinted on the laser field.[26,27] This envelope guarantees smooth on and off switching of the pulse for the times $t=0$ and $t=T$. The last term of the functional (eqn (2)) comprises the time-dependent Schrödinger equation as an additional constraint.

The time independent part of the molecular Hamiltonian is denoted by H_0. The laser field couples to the dipole moment with the interaction term $-\hat{\mu}\varepsilon(t)$.

The calculation of optimal laser fields now relies on finding the extremal value of the functional J with respect to the functions $\psi(t)$, $\Phi(t)$ and $\varepsilon(t)$. Separable differential equations can be derived[28] which involve the propagation of the initial state $\psi(0)$, the target state $\Phi(T) = \hat{O}\psi(T)$ and an equation for the electric field. The equations are solved iteratively with the Krotov method.[28,29] The laser field for the next iteration $n+1$ can be formulated as:

$$\varepsilon^{n+1}(t) = \varepsilon^n(t) + \frac{s(t)}{2\alpha}\Im[\langle\psi(T,\varepsilon^{n+1})|\hat{O}|\psi(T,\varepsilon^n)\rangle\langle\psi(t,\varepsilon^n)|\hat{\mu}|\psi(t,\varepsilon^{n+1})\rangle].$$

with $\varepsilon^n = \varepsilon_{\text{ref}}$ (3)

4 The SPODS benchmark for the OCT calculations

To set a benchmark for the OCT algorithm, we first performed calculations using a time dependent electric field $\varepsilon(t)$ representing a double pulse of the form:

$$\varepsilon(t) = E_1 e^{-2\ln 2\left(\frac{t-t_1}{\sigma}\right)^2}\cos(\omega\cdot(t-t_1))$$
$$+E_2 e^{-2\ln 2\left(\frac{t-t_2}{\sigma}\right)^2}\cos(\omega\cdot(t-t_2)+\phi). \quad (4)$$

Here $E_{1,2}$ denote the maximum electric fields of the two sub-pulses centered in time at $t_{1,2}$. Both pulses have the same frequency ω and full width at half maximum (FWHM) $\sigma = 20$ fs. The second pulse holds a phase ϕ. By setting the parameters $\Delta T = t_2 - t_1$ and ϕ to specific values, one can obtain phase shifts of $\pm\frac{\pi}{2}$. For the technical details on the calculation of the potential energy surfaces and on the wavepacket propagation see ref. 16.

First we optimized the maximum electric field of the pre-pulse E_1 to reach the state of maximum coherence. Subsequently we systematically varied the maximum electric field E_2 and ΔT. The final population in the individual target state is monitored. In order to conserve phase shifts of $\pm\frac{\pi}{2}$ we used a constant $\phi = \pm\frac{\pi}{2}$ and varied the time delay in integer multiples of the oscillation period of the driving field ($\omega = 0.05392$ a.u. $\hat{=}$ 845 nm; 1 period $\hat{=}$ 2.8 fs). For both target states we scanned E_2 values ranging from 0.005 to 0.055 GV cm^{-1} and ΔT values between 1 and 15 periods.

A maximum efficiency for the SPODS scheme of about 66% for both target states was achieved. The optimized parameters are given in Table 1. The resulting population dynamics, together with the optimized laser pulses and the pulse characterizations are summarized in Fig. 4 and 5. The upper panels show the population dynamics of the electric states involved for the $4^1\Sigma_g^+$ (Fig. 4) and $5^1\Sigma_g^+$ (Fig. 5) target

Table 1 Optimized laser parameters for population of the $4^1\Sigma_g^+$ and $5^1\Sigma_g^+$ target state

target state	$4^1\Sigma_g^+$	$5^1\Sigma_g^+$
E_1 [GV cm^{-1}]	0.002	0.002
E_2 [GV cm^{-1}]	0.048	0.036
Δ T [periods^a]	5	5

^{a} 5 periods correspond to 14 fs.

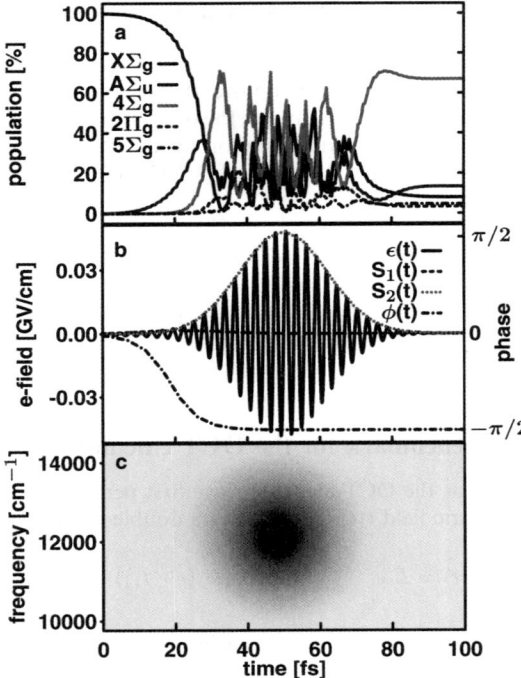

Fig. 4 Population dynamics, laser pulses and the pulse characterizations for optimized SPODS mechanisms. (a): Temporal evolution of the population in the electronic states involved. (b): Pulse sequence ($\epsilon(t)$; solid line) for the population of the $4^1\Sigma_g^+$ target state is plotted together with the envelopes of the pre-pulse (S_1; dashed line) and the second pulse (S_2; dotted line). The phase shift between the two pulses can be seen in the temporal phase $\phi(t)$ (dashed dotted line; right ordinate). (c): Cross-correlated frequency-resolved optical gating (XFROG) diagrams of the pulse sequence.

state. Both graphs show the build up of the superposition between the $X^1\Sigma_g^+$ and the $A^1\Sigma_u^+$ within the first 30 fs. After 30 fs the temporal phase $\phi(t)$ of the laser field has changed (phase shift $-\frac{\pi}{2}$; see Fig. 4 dash dotted line; phase shift $\frac{\pi}{2}$; see Fig. 4 dash dotted line) and both population dynamics proceed differently. In the case of the $-\frac{\pi}{2}$ phase shift (Fig. 4 (a)) the population is transferred form the superposition *via* the lower DS to the $4^1\Sigma_g^+$ target state (solid light gray line), while in the case of the $\frac{\pi}{2}$ phase shift (Fig. 5 (a)) the population is transferred from the superposition *via* the upper DS to the $5^1\Sigma_g^+$ target state (black dashed dotted line). The results in Fig. 4 and 5 demonstrate, that the control task is realized by only changing the sign of the interpulse phase shift and by adjusting the intensity of the second pulse. The overall structure of the two optimized pulses stays identical, as shown in the cross-correlated frequency-resolved optical gating (XFROG) representations (Fig. 4 (c) and 5 (c)).

5 Optimization of the SPODS mechanism using OCT

As target definition for OCT we use the absolute square of the expectation value of an operator $|\langle\psi(T)|\hat{O}|\psi(T)\rangle|^2$. To give the algorithm as much flexibility as possible we used projection operators $\hat{O}$, projecting the nuclear wavefunction

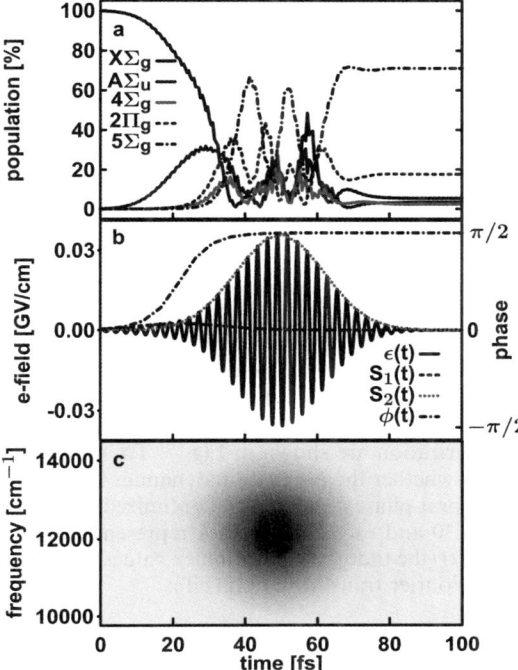

Fig. 5 Population dynamics, laser pulses and the pulse characterizations for optimized SPODS mechanisms. (a): Temporal evolution of the population in the electronic states involved. (b): Pulse sequence ($\varepsilon(t)$; solid line) for the population of the $5^1\Sigma_g^+$ target state is plotted together with the envelopes of the pre-pulse (S_1; dashed line) and the second pulse (S_2; dotted line). The phase shift between the two pulses can be seen in the temporal phase $\phi(t)$ (dashed dotted line; right ordinate). (c): Cross-correlated frequency-resolved optical gating (XFROG) diagrams of the pulse sequence.

$$\psi(t) = \begin{pmatrix} \chi_{X^1\Sigma_g^+}(t) \\ \chi_{A^1\Sigma_u^+}(t) \\ \chi_{4^1\Sigma_g^+}(t) \\ \chi_{2^1\Pi_g}(t) \\ \chi_{5^1\Sigma_g^+}(t) \end{pmatrix} \quad (5)$$

on the target electronic state *e.g.* the $4^1\Sigma_g^+$ ($\hat{O}_{4^1\Sigma_g^+}$) or $5^1\Sigma_g^+$ ($\hat{O}_{5^1\Sigma_g^+}$) state:

$$\hat{O}_{4^1\Sigma_g^+} = \begin{pmatrix} 0 & 0 & 0 & 0 & 0 \\ 0 & 0 & 0 & 0 & 0 \\ 0 & 0 & 1 & 0 & 0 \\ 0 & 0 & 0 & 0 & 0 \\ 0 & 0 & 0 & 0 & 0 \end{pmatrix} \quad \text{and} \quad \hat{O}_{5^1\Sigma_g^+} = \begin{pmatrix} 0 & 0 & 0 & 0 & 0 \\ 0 & 0 & 0 & 0 & 0 \\ 0 & 0 & 0 & 0 & 0 \\ 0 & 0 & 0 & 0 & 0 \\ 0 & 0 & 0 & 0 & 1 \end{pmatrix}. \quad (6)$$

These projection operators (eqn (6)) make the target independent from the spatial shape of the wavefunction. In agreement with SPODS only the final population of the target state is decisive. Thus additional constraints on the final wavefunction are avoided.

5.1 Optimization of the $4^1\Sigma_g^+$ target using OCT

We started the OCT calculations with optimization for the $4^1\Sigma_g^+$ state. For the initial laser field $\varepsilon^0(t)$ we used a gaussian shaped pulse with a central frequency $\omega = 911$ nm, a full width at half maximum FWHM = 20 fs and a maximum electric field $E_{max} = 0.0026$ GV cm^{-1}. The field strength was chosen in order to start the algorithm in the weak field regime. The frequency and the FWHM was chosen in order to include the $X^1\Sigma_g^+$ to $A^1\Sigma_u^+$ and simultaneously the $A^1\Sigma_u^+$ to $4^1\Sigma_g^+$ transition within the frequency spectrum of the pulse but to exclude the $A^1\Sigma_u^+$ to $5^1\Sigma_g^+$ transition. The initial laser field $\varepsilon^0(t)$ and the corresponding temporal evolution of the population in the electronic states involved are shown in Fig. 6 (a) and (b). From panel (b) it becomes obvious, that the initial field already populates the $4^1\Sigma_g^+$ target state up to 4%.

The optimization is performed using the OCT algorithm with a given time span of $T = 75$ fs, a Krotov change parameter $\alpha_0 = 1$ a.u. and the initial Gaussian shaped pulse described above (see Fig. 6 (a)). The algorithm yields a highly efficient laser field which transfers about 98.3% from the electronic and vibrational ground state to the $4^1\Sigma_g^+$ target state. The induced population dynamics, the optimized laser pulse and the pulse characterization are shown in Fig. 7. To analyze the resulting pulse sequence and to verify whether the population dynamics follow the SPODS scheme we extracted the temporal phase $\phi(t)$ from the optimized electric field $\varepsilon(t)$. We followed the idea of ref. 30 and used the complex representation of the electric field in the time domain $\tilde{\varepsilon}^+(t)$ (the tilde denotes complex values) which delivers only positive frequencies after Fourier transformation (FT):

$$\tilde{\varepsilon}^+(t) = FT^{-1}\{\tilde{\varepsilon}^+(\Omega)\} = \frac{1}{2\pi}\int_{-\infty}^{\infty}\tilde{\varepsilon}^+(\Omega)e^{i\Omega t}d\Omega$$

$$\text{with } \tilde{\varepsilon}^+(\Omega) = |\tilde{\varepsilon}(\Omega)|e^{i\Phi(\Omega)} = \begin{cases} \tilde{\varepsilon}(\Omega) & \text{for } \Omega \geq 0 \\ 0 & \text{for } \Omega < 0 \end{cases} \quad (7)$$

$$\text{and } \tilde{\varepsilon}(\Omega) = FT\{\varepsilon(t)\} = \int_{-\infty}^{\infty}\varepsilon(t)e^{-i\Omega t}dt.$$

From $\tilde{\varepsilon}+(t)$ the temporal phase can be calculated as:

$$\tilde{\varepsilon}^+(t) = \frac{1}{2}S(t)e^{i\Gamma(t)} = \frac{1}{2}S(t)e^{i\phi(t)}e^{i\omega_l t}. \quad (8)$$

Here $S(t)$ is the field envelope, ω_l the central frequency of the optimized electric field $\hat{\varepsilon}(t)$ (in this case 11600 cm^{-1} (860 nm) see Fig. 7 (c)). The extracted temporal phase $\phi(t)$ (dash dotted line in Fig. 7 (b); right ordinate) clearly shows a jump of $-\frac{\pi}{2}$ in the time interval between 15 and 20 fs. Right after this phase jump the $4^1\Sigma_g^+$ target state starts to be populated form the initially prepared superposition between the $X^1\Sigma_g^+$ and the $A^1\Sigma_u^+$ state.

This population dynamics, together with the temporal phase and the shape of the pulse sequence, is very similar to those observed in Sec. 4 but with markedly higher efficiency. Thus we conclude, that the OCT algorithm has found the SPODS scheme as the optimal path to selectively populate the $4^1\Sigma_g^+$ target state within the given boundary and starting conditions.

5.2 Optimization of the $5^1\Sigma_g^+$ target using OCT

To optimize the selective population of the $5^1\Sigma_g^+$ target state by OCT, we cannot use the same initial laser field as in Sec. 5.1 because the target state is not accessible by this electric field. As a consequence the optimization fails or gets at least extremely inefficient. But from SPODS we know, that changing the phase shift from $-\frac{\pi}{2}$ to $\frac{\pi}{2}$

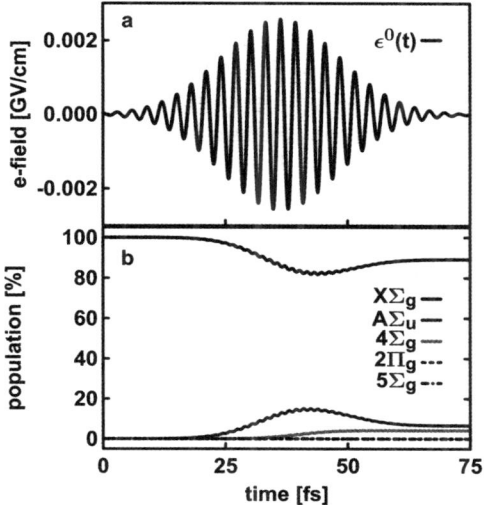

Fig. 6 Population dynamics using the initial laser field, starting the OCT algorithm. (a) Initial electric field $\varepsilon^0(t)$. (b) Temporal evolution of the population in the electronic states involved.

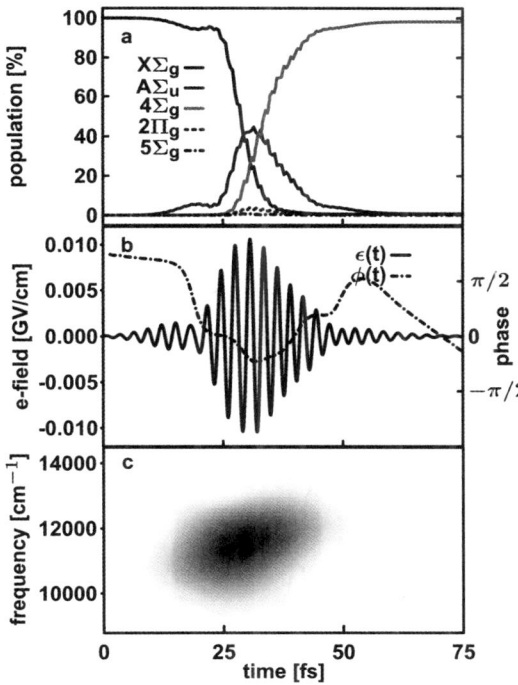

Fig. 7 Population dynamics, laser pulse and pulse characterization for the optimized SPODS mechanism. (a) Temporal evolution of the population in the electronic states involved. (b) Optimized pulse sequence for the selective population of the $4^1\Sigma_g^+$ target state $\phi(t)$ (solid line). The phase shift between the two pulses can be seen in the temporal phase ($\phi(t)$; dashed dotted line; right ordinate). (c) XFROG diagrams of the pulse sequence.

leads to a selective population of the upper DS, making the $5^1\Sigma_g^+$ state accessible. Thus we inverted the sign of the temporal phase of the optimized electric field (from Sec. 5.1) and reconstructed the corresponding laser field, now exhibiting a phase shift of $\frac{\pi}{2}$. The reconstruction was done by reversing the steps used to extract the temporal phase, but with the inverted sign of $\phi(t)$. Both pulses, the optimized ($\varepsilon(t)$) and the reconstructed pulse ($\varepsilon_r(t)$), together with their corresponding temporal phase ($\phi(t)$ and $\phi_r(t)$) are shown in Fig. 8 (a). Here the overall phase shift of π between the two pulses can be seen directly in the electric fields. The population dynamics, using the reconstructed electric field, is plotted in Fig. 8 (b) and clearly shows the switching between the DS. After the phase shift of $\frac{\pi}{2}$ (after approx. 25 fs) now all three electronic states (i.e. the $4^1\Sigma_g^+$, the $2^1\Pi^+_g$ and the $5^1\Sigma_g^+$) are accessible from the upper DS within the spectral width of the laser pulse. As a result we now found a population of 6.4% in the $5^1\Sigma_g^+$ state and a significantly reduced population of 20.8% in the $4^1\Sigma_g^+$. This behavior emphasizes, that the OCT algorithm has found the SPODS mechanisms.

The further optimization for the $5^1\Sigma_g^+$ target state is performed using the OCT algorithm with the same time span as in Sec. 5.1 ($T = 75$ fs), a Krotov change parameter $\alpha_0 = 1$ a.u. and the reconstructed electric field $\varepsilon_r(t)$ exhibiting a phase shift of $\frac{\pi}{2}$ (see Fig. 8 (a)). Again the OCT algorithm yields a highly efficient laser field which transfers of about 96.7% from the electronic and vibrational ground state to the $5^1\Sigma_g^+$ target state. The resulting population dynamics, the optimized laser pulse and the pulse characterization are summarized in Fig. 9. The temporal evolution of the population in the states involved (panel (a)) again show the build up of the superposition between the $X^1\Sigma_g^+$ state and the $A^1\Sigma_u^+$ in the first 25 fs. Right after the phase jump of $\frac{\pi}{2}$ (Fig. 9 (b) dash dotted line; 11800 cm^{-1} (844 nm)) in the time interval between 15 and 25 fs, the $5^1\Sigma_g^+$ target state gets populated up to the final value.

The comparison between the OCT solutions and the manually optimized SPODS sequences (see Sec. 4) show that the OCT algorithm uses additional control knobs

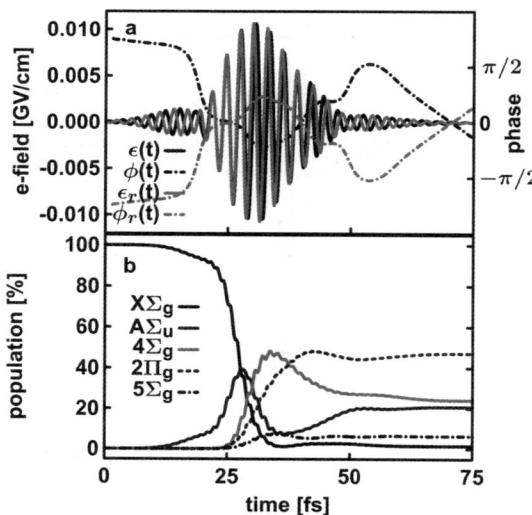

Fig. 8 (a) Optimized pulse sequence for the selective population of the $4^1\Sigma_g^+$ target state ($\varepsilon(t)$; gray solid line) and the corresponding temporal phase ($\phi(t)$; gray dash dotted line). Pulse sequence ($\varepsilon_r(t)$; black solid line) with reversed temporal phase ($\phi_r(t)$; black dash dotted line). (b) Temporal evolution of the population in the electronic states involved propagated with $\varepsilon_r(t)$.

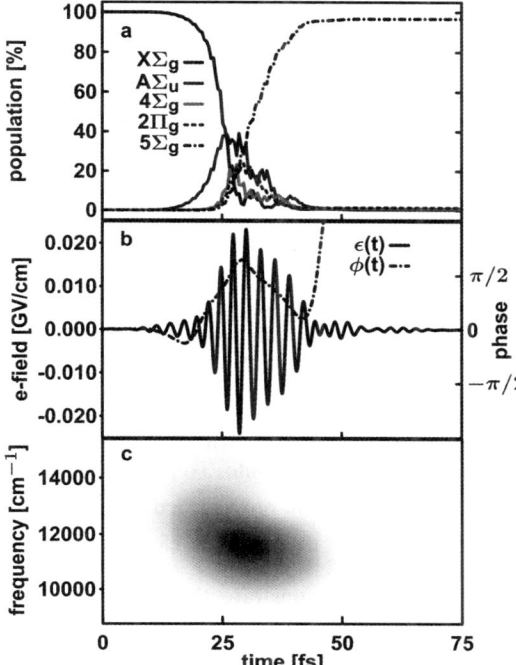

Fig. 9 Population dynamics, laser pulses and the pulse characterizations for optimized SPODS mechanisms. (a) Temporal evolution of the population in the electronic states involved. (b) Pulse sequence for the selective population of the $5^1\Sigma_g^+$ target state ($\varepsilon(t)$; solid line). The phase shift between the two pulses can be seen in the temporal phase ($\phi(t)$; dashed dotted line; right ordinate). (c) XFROG diagrams of the pulse sequence.

like chirp to allow efficient adiabatic transitions to the target states by keeping the laser intensity as low as possible (see Fig. 4 (c), and 5 (c) , 7 (c) and 9 (c)). The direct comparison between the two optimized laser fields ($\varepsilon^{4^1\Sigma_g^+}(t)$ and $\varepsilon^{5^1\Sigma_g^+}(t)$) is shown in Fig. 10. Here the two most prominent differences become obvious. The first is the π phase shift between the pulses in the time interval from 0 to 25 fs, while the two pulses are nearly in phase in the subsequent interval between 25 and 45 fs. This phase shift proves, that the control of both target states achieved by the OCT follow the SPODS scheme. The short pulse duration of the optimized pulses with the fast varying envelope underlines the non-adiabacity of the process. The second difference

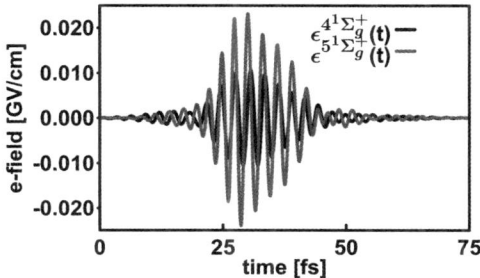

Fig. 10 Comparison of the optimized pulse sequences selectively populating the $4^1\Sigma_g^+$ ($\varepsilon^{4^1\Sigma_g^+}(t)$) and the $5^1\Sigma_g^+$ ($\varepsilon^{5^1\Sigma_g^+}(t)$).

between the two pulses is the intensity. The lower intensity is found for the $4^1\Sigma_g^+$ target state in order to suppress the competing transitions in the frequency spectrum of the pulse. The higher intensity for the $5^1\Sigma_g^+$ target state is needed, because the energies of the two target states are not symmetric around twice the energy difference between the $X^1\Sigma_g^+$ and $A^1\Sigma_u^+$ (which would be the ideal case for the SPODS scheme). Thus a higher intensity for the $5^1\Sigma_g^+$ target state is needed in order to suppress the other competing transitions in the frequency spectrum.

6 Conclusion

In this work we performed OCT calculations to investigate whether the SPODS mechanism working in the strong field regime is included in the search space of the OCT algorithm. Based on these calculations, we can conclude that the SPODS mechanism is included in the search space, and that OCT is able to remarkably increase the efficiency compared to the double pulse sequence presented in Sec. 4. For the optimization of the selective population of the lower lying $4^1\Sigma_g^+$ target state, the algorithm found the SPODS mechanism without any further constraint or additional starting condition as the optimal route. To optimize the selective population of the higher lying $5^1\Sigma_g^+$ target state we only needed the $\frac{\pi}{2}$ phase jump as an additional starting condition, in order to make the electronic state accessible within the initial guess.

In addition our calculations demonstrate that the SPODS mechanism is an optimal solution in the OCT search space. From the properties of the OCT algorithm it is known that high quality control and robust solutions are found even for complex quantum systems including a large number of control variables.[31] In this sense the SPODS can be regarded as a robust way to control the selective population of higher lying electronic states, opening a wide spectrum of application ranging form reaction control within molecules up to discrimination between different molecules in a mixture. For larger molecules, holding a more complex electronic structure, the frequency shaped OCT algorithm[32] might be helpful to optimize the SPODS mechanisms in order to avoid competing resonant transition.

Acknowledgements

The authors would like to thank Thomas Baumert and Matthias Wollenhaupt for fruitful discussions. We are also grateful for support by the DFG *via* the Cluster of Excellence: Munich Centre of Advanced Photonics, the SFB749 and the Normalverfahren.

References

1 T. Brixner and G. Gerber, *ChemPhysChem*, 2003, **4**, 418.
2 F. Krausz and M. Ivanov, *Rev. Mod. Phys.*, 2009, **1**, 163.
3 M. F. Kling, C. Siedschlag, A. J. Verhoef, J. I. Khan, M. Schultze, T. Uphues, Y. Ni, M. Uiberacker, M. Drescher, F. Krausz and M. J. J. Vrakking, *Science*, 2006, **312**, 246–248.
4 M. Y. I. B. J. Sussman, D. Townsend and A. Stolow, *Science*, 2006, **314**, 278.
5 E. T. Sleva, I. M. Xavier Jr. and A. H. Zewail, *J. Opt. Soc. Am. B*, 1986, **3**, 483.
6 S. R. Hartmann and E. L. Hahn, *Phys. Rev.*, 1962, **128**, 2042.
7 A. D. H. R. Kosloff and D. Tannor, *Phys. Rev. Lett.*, 1992, **69**, 2172.
8 R. Netz, T. Feurer, G. Roberts and R. Sauerbrey, *Phys. Rev. A: At., Mol., Opt. Phys.*, 2002, **65**, 043406.
9 U. Gaubatz, P. Rudecki, S. Schiemann and K. Bergmann, *J. Chem. Phys.*, 1990, **92**, 5363.
10 K. Bergmann, H. Theuer and B. W. Shore, *Rev. Mod. Phys.*, 1998, **70**, 1003.
11 M. Wollenhaupt, A. Präkelt, C. Sarpe-Tudoran, D. Liese and T. Baumert, *J. Opt. B: Quantum Semiclassical Opt.*, 2005, **7**, S270.
12 M. Wollenhaupt, A. Präkelt, C. Sarpe-Tudoran, D. Liese and T. Baumert, *J. Mod. Opt.*, 2005, **52**, 2187.

13 M. Wollenhaupt, D. Liese, A. Präkelt, C. Sarpe-Tudoran and T. Baumert, *Chem. Phys. Lett.*, 2006, **419**, 184.
14 M. Wollenhaupt, A. Präkelt, C. Sarpe-Tudoran, D. Liese and T. Baumert, *Appl. Phys. B: Lasers Opt.*, 2006, **82**, 183.
15 M. Wollenhaupt and T. Baumert, *J. Photochem. Photobiol., A*, 2006, **180**, 248.
16 P. von den Hoff, R. Siemering, M. Kowalewski and R. de Vivie-Riedle, *IEEE Journal of Selected Topics in Quantum Electronics*, 2011, (in press).
17 D. Geppert, P. von den Hoff and R. de Vivie-Riedle, *J. Phys. B: At., Mol. Opt. Phys.*, 2008, **41**, 074006.
18 I. Znakovskaya, P. von den Hoff, S. Zherebtsov, A. Wirth, O. Herrwerth, M. Vrakking, R. de Vivie-Riedle and M. Kling, *Phys. Rev. Lett.*, 2009, **103**, 103002.
19 P. von den Hoff, I. Znakovskaya, M. Kling and R. de Vivie-Riedle, *Chem. Phys.*, 2009, **366**, 139.
20 W. Zhu, J. Botina and H. Rabitz, *J. Chem. Phys.*, 1998, **108**, 1953–1963.
21 W. Zhu and H. Rabitz, *J. Chem. Phys.*, 1998, **109**, 385–391.
22 D. J. Tannor and S. A. Rice, *Adv. Chem. Phys.*, 1988, **70**, 441.
23 C. M. Tesch and R. de Vivie Riedle, *Phys. Rev. Lett.*, 2002, **89**, 157901.
24 C. M. Tesch and R. de Vivie-Riedle, *J. Chem. Phys.*, 2004, **121**, 12158.
25 R. de Vivie-Riedle, K. Sundermann and M. Motzkus, *Faraday Discuss.*, 1999, **113**, 303.
26 K. Sundermann and R. de Vivie-Riedle, *J. Chem. Phys.*, 1999, **110**, 1896.
27 J. Manz, K. Sundermann and R. de Vivie-Riedle, *Chem. Phys. Lett.*, 1998, **290**, 415–422.
28 J. Somloi, V. A. Kazakov and D. J. Tannor, *Chem. Phys.*, 1993, **172**, 85–98.
29 J. P. Palao and R. Kosloff, *Phys. Rev. A: At., Mol., Opt. Phys.*, 2003, **68**, 062308.
30 J. C. Diels and W. Rudolph, *Ultrashort Laser Pulse Phenomena*, Academic Press, Inc., 1996.
31 H. Rabitz, *J. Mod. Opt.*, 2004, **51**, 2469.
32 C. Gollub, M. Kowalewski and R. de Vivie-Riedle, *Phys. Rev. Lett.*, 2008, **41**, 073002.

PAPER

Photoelectron photoion coincidence imaging of ultrafast control in multichannel molecular dynamics

C. Stefan Lehmann,[†] N. Bhargava Ram,[†] Daniel Irimia[†] and Maurice H. M. Janssen*

Received 24th March 2011, Accepted 6th May 2011
DOI: 10.1039/c1fd00047k

The control of multichannel ionic fragmentation dynamics in CF_3I is studied by femtosecond pulse shaping and velocity map photoelectron photoion coincidence imaging. When CF_3I is photoexcited with femtosecond laser pulses around 540 nm there are two major ions observed in the time-of-flight mass spectrum, the parent CF_3I^+ ion and the CF_3^+ fragment ion. In this first study we focussed on the influence of LCD-shaped laser pulses on the molecular dynamics. The three-dimensional recoil distribution of electrons and ions were imaged in coincidence using a single time-of-flight delay line detector. By fast switching of the voltages on the various velocity map ion lenses after detection of the electron, both the electron and the coincident ion are measured with the same imaging detector. These results demonstrate that a significant simplification of a photoelectron-photoion coincidence imaging apparatus is in principle possible using switched lens voltages. It is observed that shaped laser fields like chirped pulses, double pulses, and multiple pulses can enhance the CF_3^+/CF_3I^+ ratio by up to 100%. The total energetics of the dynamics is revealed by analysis of the coincident photoelectron spectra and the kinetic energy of the CF_3^+ and I fragments. Both the parent CF_3I^+ and the CF_3^+ fragment result from a five-photon excitation process. The fragments are formed with very low kinetic energy. The photoelectron spectra and CF_3^+/CF_3I^+ ratio vary with the center wavelength of the shaped laser pulses. An optimal enhancement of the CF_3^+/CF_3I^+ ratio by about 60% is observed for the double pulse excitation when the pulses are spaced 60 fs apart. We propose that the control mechanism is determined by dynamics on neutral excited states and we discuss the results in relation to the location of electronically excited (Rydberg) states of CF_3I.

I. Introduction

The rapid development of ultrafast laser technology and single particle detection techniques during the first decade of the 21st century are providing novel opportunities to advance our insight and understanding of the control of molecular dynamics by ultrafast pulse shaping. Since the first pioneering theoretical studies[1–3] in the mid '80s and early '90s of the 20th century the field of optimal control has expanded very rapidly during the last decade as demonstrated by a series of recent special issues like *J. Photochem. Photobiol. A*,[4] *J. Phys. B: At. Mol. Opt. Phys.*,[5] *New*

LaserLaB Amsterdam and Department of Chemistry, Vrije Universiteit, de Boelelaan 1083, 1081 HV Amsterdam, The Netherlands. E-mail: mhmj@chem.vu.nl

[†] These authors contributed equally to this research.

J. Phys.,[6] and *Phys. Chem. Chem. Phys.*[7] A very extensive review paper on the control of quantum phenomena was published recently by Rabitz and coworkers.[8]

The control of multichannel fragmentation in strong field excitation of polyatomic molecules in the gas phase was pioneered by the experimental groups of Gerber[9] and Levis.[10] Even though these experiments were performed more than a decade ago, little is known today about the mechanism that favors certain pulse shapes over other ones in promoting targeted ionic fragmentation channels in those experiments. Theoretical calculations on these molecular systems consisting of some 10–20 atoms are quite challenging. In some cases high level one dimensional *ab initio* calculations were performed discussing the wavepacket mechanism in pulse shaping experiments of a similarly large organometallic molecule by Woste, Manz and coworkers.[11] Dantus and coworkers[12] reported a very extensive experimental study on some 16 different organic molecules studying the strong field control of multichannel ionic fragmentation. They concluded in this study that coherence does not play a role in the observed changes in yield as a function of pulse shape, only the average pulse duration. Furthermore, very recently this group debated the interpretation of some of the experimental observations in strong field optimal control experiments of larger polyatomic molecules.[13,14]

However, also in a much simpler five atomic molecule like CH_2BrCl the molecular mechanism behind the control of fragmentation in pulse shaping experiments[15] at 800 nm, that favored the breaking of the strong (C–Cl) bond over the weaker (C–Br) bond, has not been revealed. Extensive *ab initio* electronic structure and dynamical calculations have been performed on CH_2BrCl by Gonzalez and coworkers[16,17] discussing the role of the B and C excited states and nonadiabatic transitions in the neutral CH_2BrCl molecule. However, still little insight was obtained on the connection between these theoretical calculations of low lying neutral surfaces and the experimental pulse shaping data[15] employing strong field pulses near 800 nm. More recently, several multichannel fragmentation studies were reported using pump–probe spectroscopy to study wavepacket dynamics and control in similar halomethanes like CH_2BrI.[18–21] From these experiments at 800 nm in combination with theoretical calculations it was concluded that resonances between ionic surfaces of the parent ion facilitate and control the formation of ionic fragments.

Until now most of these experimental studies employing adaptive pulse shaping to control multichannel (ionic) fragmentation dynamics in polyatomic molecules have used time-of-flight mass spectrometric detection. Mass spectrometry provides only information on the masses of the various ionic species produced and no or very limited information on the mechanism of the pulse shaping dynamics. The introduction in 1987 of single particle imaging techniques by Chandler and Houston[22] has led to the development of very advanced and powerful spectroscopic imaging techniques revealing many novel and fundamental aspects of chemical dynamics.[23–26] Ionic multichannel fragmentation leads to the ejection of electrons and ions. Both particles contain information on the dynamics and therefore it is very valuable to measure both particles in coincidence. The combination of the photoelectron photoion coincidence imaging technique with time-resolved (femtosecond) pump–probe spectroscopy was pioneered in 1999 by Hayden and coworkers.[27–30] In 2002 we proposed[31] to use the coincidence imaging technique in combination with femtosecond pulse shaping to study the mechanism in pulse shaping control of multichannel ionic fragmentation dynamics. A new photoelectron-photoion coincidence imaging instrument was constructed at LaserLaB Amsterdam and has become operational for femtosecond time-resolved experiments.[26,32–34] Velocity map imaging of electrons and ions was recently[35] combined with chirped pulse shaping techniques in our group employing a first generation femtosecond imaging machine.[36–38] In these non-coincidence imaging experiments on CH_2BrCl a strong effect of up-chirp was observed enhancing the formation of the CH_2Cl^+ fragment over the CH_2BrCl^+.[35]

In this paper we present experimental results combining for the first time velocity map coincidence imaging of correlated electrons and ions with full ultrafast pulse shaping using LCD pulse shaping technology. In the experiments reported here we studied the photodynamics of CF_3I interacting with chirped laser pulses with central wavelengths around 540 nm. The photochemistry of CF_3I has attracted a lot of attention during the past two decades partially because of the involvement of CF_3I in atmospheric problems such as global warming and ozone depletion. Furthermore, CF_3I has served as a model system for photodynamical studies of polyatomic molecules, both in the time and the frequency domain. A widely varying number of experimental methods and techniques were used, like nanosecond and femtosecond lasers, single photon or multiphoton absorption, resonance-enhanced-multi-photon-ionization (REMPI), one-color and multi-color excitation, time-of-flight (TOF) detection, photoelectron-photoion coincidence (PEPICO) detection, and femtosecond velocity map photoelectron-photoion coincidence imaging.[30,34,37–52] Following photolysis with UV solar radiation, CF_3I has a short atmospheric lifetime because of the dissociation along the C–I bond. Extensive one-photon dissociation studies of the A-band were done with nanosecond lasers to obtain further insight on the influence of the conical intersection on the branching ratio between $CF_3 + I(^2P_{1/2})$ and $CF_3 + I(^2P_{3/2})$ channels.[41,42,53–55] The A-band consists of a broad absorption band centered around 4.7 eV (265 nm) and the excitation is followed by rapid dissociation in less than 500 fs leading to the formation of CF_3 and $I(^2P_{1/2},^2P_{3/2})$ fragments. The ionization of CF_3I in the nanosecond regime *via* multiphoton absorption is usually realized through higher lying Rydberg states which serve as resonant intermediate levels.[42–44,50] The spectroscopy of the ionized CF_3I^+ cation was investigated by Aguirre and Pratt[50,56] in the spectral range around 300 nm using two-photon resonant, three-photon ionization to prepare the ground state CF_3I^+, followed by the absorption of an additional (forth) photon leading to fragmentation of the parent ion. Recently, such fragmentation dynamics in CF_3I^+, due to further absorption of the parent ion, was also studied using the novel Amsterdam coincidence photoelectron-photoion imaging apparatus.[32,34]

The present paper is organized as follows, in Section 2 we present the experimental apparatus. In Section 3 we report the experimental data which are discussed in Section 4. We summarize our conclusions in Section V.

II. Experimental

The coincidence imaging apparatus used in the present work has been described in detail before.[26,32] The machine consists of three differentially pumped UHV chambers, the source chamber, a buffer chamber, and the imaging chamber. Contrary to the first series of experiments, which were done using a continuous molecular beam source, in the present setup we use a pulsed molecular beam expansion. We have replaced the continuous molecular beam source by a homebuilt high-repetition rate cantilever piezo valve.[57,58] A pulsed molecular beam source matches much better the duty cycle of the experiment because of the pulsed femtosecond laser source which operates at a repetition rate that is selectable between 1–5 kHz. The pulsed cantilever valve makes it possible to produce stronger beams and to use higher backing pressures. The piezo valve can operate both in pulsed mode and continuous mode and the gas was expanded through a nozzle with a diameter of 200 μm. In the present experiments we operated the regenerative femtosecond laser (Spectra Physics Spitfire Pro) and the pulsed piezo valve at 1 kHz. In most cases low seeding ratios (about 1% or less) of CF_3I in Neon were used. Pulsed expansions were checked for not having clusters in the mass spectrum. The typical speed ratio with 3 Bar backing pressure is about $S = 25-30$.[26]

A home-built two-stage noncollinearly phase-matched optical parametric amplifier (NOPA) (following the design of ref. 59, 60), seeded by a white light continuum and pumped by the 400 nm light of the frequency-doubled regen laser, generates ultrafast tunable laser pulses with a duration as short as 30 fs when properly compressed. For the present experimental study the pulses were set with the peak wavelength varied within the range of about 520–550 nm. The NOPA-pulses are subsequently shaped in a LCD pulse shaper (CRI, type SLM-640-D-VN) which is a dual-layer array of 640 pixels with 100 µm width per pixel. Each pixel is irradiated by a specific wavelength with a spectral width of approximately 0.15 nm. In case of phase-only shaping, the applied phases are identical for both layers. The temporal profile of the pulses was mostly characterized with an interferometric autocorrelator, and in some cases with a FROG set-up. The spectrum of the laser pulses was measured with a fiber spectrometer (Ocean Optics). The typical output after the LCD shaper is about 15–20 µJ and the typical energy used in the coincidence machine was about 10 µJ. The NOPA light is focussed into the interaction region of the reaction chamber (lens with a focal length of 30 cm) and the polarization is kept parallel to the plane of the imaging detectors. From an estimate of the beam waist at the crossing with the molecular beam we estimate the intensity to be about 10^{11}–10^{12} W cm^{-2}, for a 10 µJ pulse with 30 fs duration.

The coincidence machine has two opposite Time-of-Flight (TOF) delay line detectors.[32] However, in the present experiments we operated the machine in a novel single detector set-up. We use only the short-length TOF for both electron and ion coincidence detection. The schematic of the operation is shown in Fig. 1. The repeller (R), extractor (E) and extra lens (L) are first set with appropriate negative voltages for electron velocity map imaging. Then about 200 ns after the laser

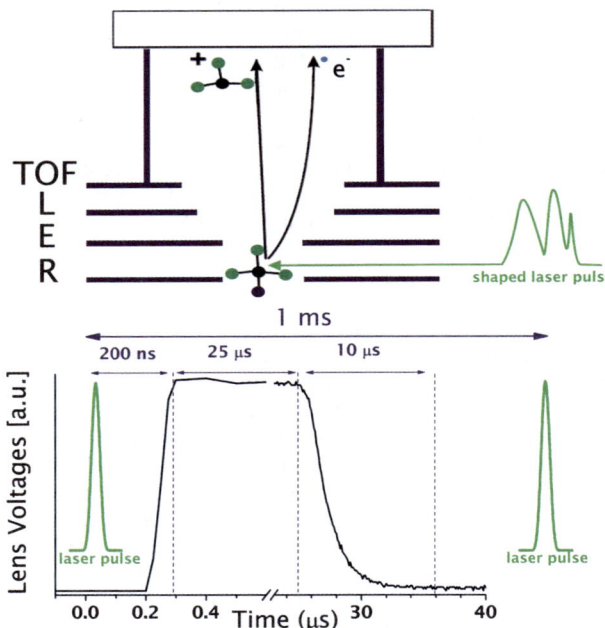

Fig. 1 Schematic of the single detector based coincidence imaging setup (top) and voltage conditions (bottom). A pulsed molecular beam, produced by a cantilever piezo valve (57; 58), is intersected by the shaped laser pulses in the interaction region between Repeller (R) and Extractor (E). Photoelectrons and photoions produced are extracted by switching the voltages on the Repeller (R), Extractor (E) and Lens (L). Electrons are detected in the first 200 ns after the laser interaction. Thereafter, the voltages are switched in about 50 ns (Behlke HV switches) to detect ionic fragments in the next 25 µs. Voltages used for electron detection are −520 V (R), −385 V (E) & −270 V (L) and for ion detection +2550 V (R), +1550 V (E) & + 500 V(L). Long after all masses of interest have reached the detector, the voltages are switched back in about 10 µs to the values for electron detection. All switching is at the repetition rate of the laser system which was 1 kHz in the present experiment.

interaction (note the typical flight time of the electron is only about 15 ns) the voltages on all 3 lenses are quickly switched (in about 50–70 ns using Behlke HV switches) to a geometry for ion extraction. These positive voltages are kept for about 25 μs so that all ions of interest have reached the same detector. Subsequently, the voltages are switched back again in about 10 μs to the optimum voltages for electron detection, so that when the next laser pulse is fired (1 ms inter pulse duration) the set-up is ready again for the detection of electrons. All voltages are switched at the repetition rate of the laser system (1 kHz). Our HV-switching set-up can also operate at 5 kHz which was used in the coincidence detection using two separate TOF delay line detectors for ions and electrons on opposite sides of the laser interaction region.[26,32–34] Even though the HV switching causes some ringing on the delay line, these false events are easily rejected in the recorded TDC events of ions, as the masses of interest under our conditions typically arrive after about 2 μs. Furthermore, the ringing occurs after the fast arrival of the electrons so the switching does not interfere with our correlated (e,ion) events. The short length electron TOF is not ideal for high-resolution 3D detection of ions,[61] however, in the present experiment the ionic fragments have low kinetic energy without any resolvable structure. In the future we will report a more extensive analysis of the single short-TOF delay line velocity map (e,ion) coincidence characteristics. We will also compare the performance with single long-TOF delay line (e,ion) detection and the dual short/long-TOF delay line (e,ion) coincidence detection.

III. Results

In Fig. 2 the TOF ion spectra obtained from multiphoton excitation of CF_3I with a femtosecond laser pulse centered at 520 nm, 530 nm, 540 nm and 550 nm is shown. The spectral width of the laser pulses was measured to be typically around 20 nm FHWM. The LCD-shaper was set to compress the pulses as short as possible to be close to the transform-limit. The pulse duration was measured to be about 35 fs, which results in a time-bandwidth-product (TBP) of about TBP = 0.72. The TOF-spectra in Fig. 2 show that there are predominantly two ions produced, the CF_3I^+ parent ion and the CF_3^+ fragment ion. In addition to these dominating ions

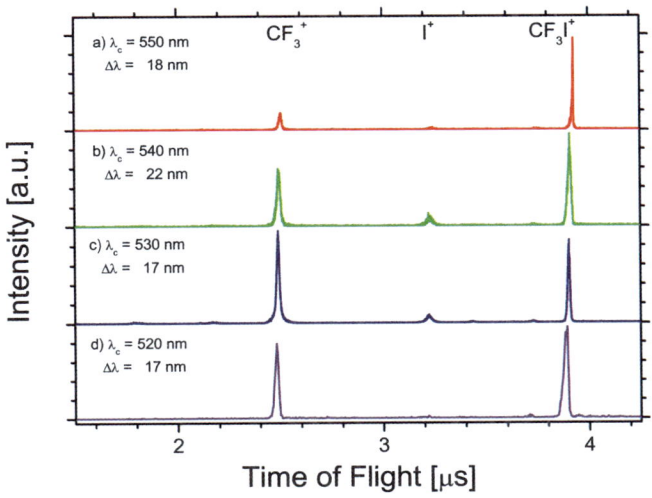

Fig. 2 Time-of-Flight (TOF) spectra of CF_3^+, I^+ and CF_3I^+ obtained from multiphoton excitation of CF_3I with femtosecond laser pulses (close to transform limited) centered at 520 nm, 530 nm, 540 nm and 550 nm.

a much smaller third peak is observed at the mass of I^+. The ratio of the two dominant ions is plotted as a function of the excitation wavelength in Fig. 3 for short pulses of about 35 fs.

It is seen that the CF_3^+/CF_3I^+ ratio changes substantially with change in the center wavelength. For excitation with pulses centered at 520 nm, 530 nm, 540 nm and 550 nm, the ratios are 0.71, 1.50, 0.73 and 0.46 respectively. Note that in calculating these ratios we not just calculated the ratio of the peak intensity in the TOF-spectrum but summed all ions within the TOF-region of the particular mass peak, as the CF_3^+ peak has somewhat broader tails due to some kinetic energy of the fragment. The absolute ion yield is mentioned in Table 1. It can be clearly seen from Table 1 that near 540 nm the absolute ion yield (consisting of mainly CF_3^+ and CF_3I^+ ions, I^+ ion yield is very small) per laser shot is strongly enhanced.

The CF_3^+/CF_3I^+ molecular fragmentation ratio recorded at two different wavelengths as a function of linear chirp is depicted in Fig. 4. At 530 nm there appears to be a strong effect of chirp on the fragmentation ratio, for both up-chirped and down-chirped pulses the CF_3^+/CF_3I^+ ratio increases from about 1.5 for short pulses to about 3.5–4 when the chirp increases to about $\pm$ 2000 fs^2. For pulses centered near 540 nm a similar behavior is seen but with lower ratios. The CF_3^+/CF_3I^+ ratio for short near transform-limit pulses is about 0.63 and the mass yield ratio increases to about 1.4 for chirped pulses of $\pm$ 3500 fs^2. It is to be noted that for both wavelengths the absolute yield of ions decreases strongly with increasing chirp. For instance, the total number of CF_3^+ or CF_3I^+ ions decreases by about a factor of five when the pulse changes from short (no chirp) to -1600 fs^2 down-chirped. A chirped pulse of -1600 fs^2 represents a laser pulse with a pulse duration of about 150 fs.

In Fig. 5 we show the change in fragmentation ratio when a phase mask is applied on the LCD-pulse shaper such that either a double pulse is created, panel (a), or an equally spaced pulse sequence, panel (b). The time duration between the pulses was varied and increased up to a spacing of about 140 fs. The center wavelength was 540 nm. It appears that both the double pulse and the pulse sequence increases the CF_3^+/CF_3I^+ ratio most when the time duration is about 60 fs. For the double pulse, around 120 fs, again the ratio increases.

In Fig. 6 the kinetic energy distributions of the CF_3^+ + I fragmentation channel is shown for excitation at 540 nm, while in the inset of Fig. 6 an example of a CF_3^+ ion

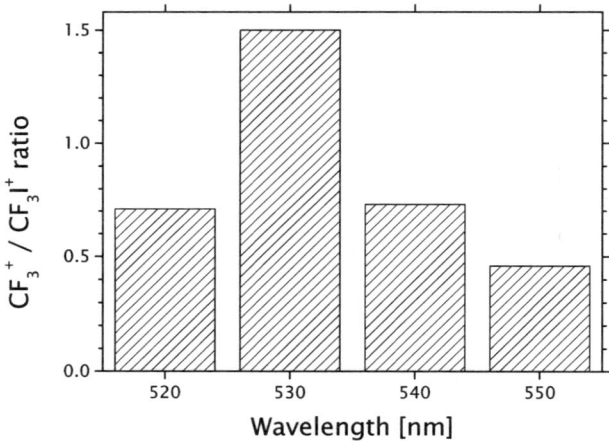

Fig. 3 CF_3^+/CF_3I^+ molecular fragmentation ratio as a function of excitation wavelength. The pulses centered at these wavelengths were near transform-limited and had a bandwidth of $\sim$ 20 nm. The corresponding absolute ion yield for these pulses is given in Table I.

Table 1 Absolute ion yield of CF_3^+, CF_3I^+ and I^+ ions at different wavelengths for short (near transform-limited) pulses of about 35 fs

λ_c [nm]	Int [μJ]	Ions	Laser Shots [LS]	Abs. Yield [Ions LS^{-1}] [per mill]	Abs. Yield per μJ [Ions LS^{-1} μJ^{-1}][per mill]
550	7.6	6640	3.6×10^6	2	0.24
540	11	16979	65321	260	24
530	8	193346	3.6×10^6	54	6.7
520	15	2105	448492	5	0.31

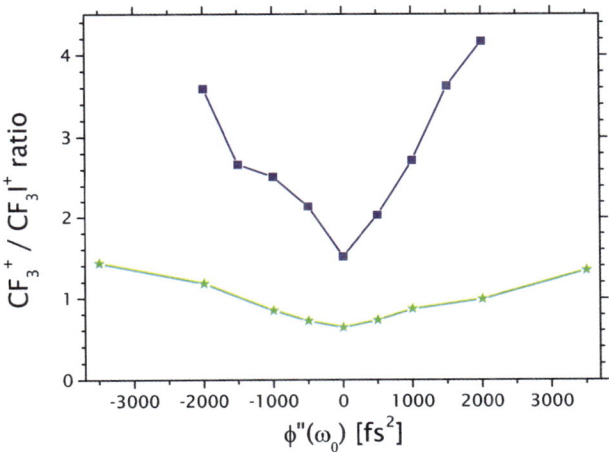

Fig. 4 CF_3^+/CF_3I^+ molecular fragmentation ratio as a function of linear chirp at 530 nm (blue squares) and 540 nm (green stars). The linear chirp is introduced using the phase function $\phi(\omega) = \frac{1}{2}\phi''(\omega_o)\delta\omega^2$ on the LCD shaper. The chirp parameter $\phi''(\omega_o)$ is plotted on the x-axis.

image recorded at 540 nm is shown. By using the conservation of linear momentum between CF_3^+ and I fragments, the total kinetic energy released into the $CF_3^+ + I$ channel is calculated according to

$$E_{kin}(total) = E_{kin}(CF_3^+)(1 + m_{CF_3}/m_I), \tag{1}$$

The ion images are quite isotropic and reveal that relatively low energy (less than 0.1 eV) is released into translational energy of the $CF_3^+ + I$ fragmentation channel.

Fig. 7 presents an overview of the photoelectron spectra resulting from the photoionization of CF_3I with near transform-limited laser pulses at three different center wavelengths, panel (a) is at 540 nm, panel (b) is at 520 nm, and panel (c) is at 400 nm. Each panel gives the photoelectron spectrum measured in coincidence with either CF_3^+ or CF_3I^+. The data in panel (c) at 400 nm is taken with longer pulses of about 120 fs and was published before.[34] At all three wavelengths basically two peaks are observed in the photoelectron spectrum in coincidence with CF_3I^+, but the relative intensity of the two peaks change with excitation wavelength. At 540 nm and 400 nm the intensity of the peak with largest photoelectron energy is smallest, whereas at 520 nm this peak is strongest. In the photoelectron spectrum in coincidence with the CF_3^+-fragment a new peak is visible with low kinetic energy.

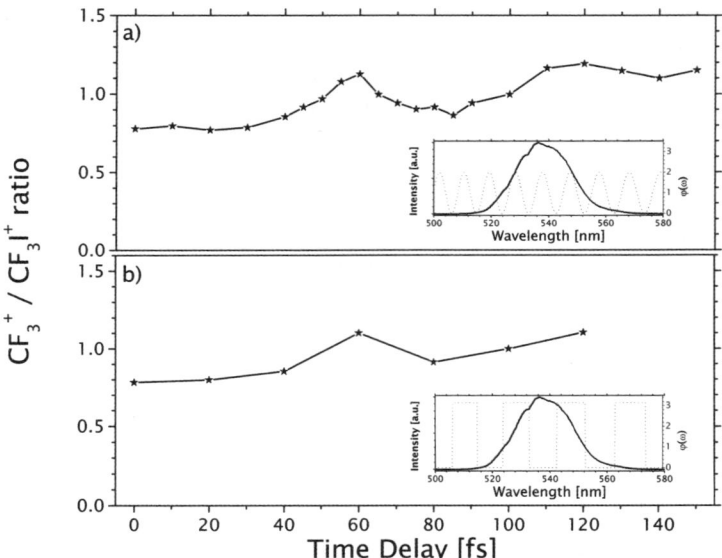

Fig. 5 CF_3^+/CF_3I^+ molecular fragmentation ratio for excitation with a double pulse (panel (a)) and an equally spaced pulse sequence (panel (b)) with wavelength centered at 540 nm. Note the increase in the ion yield ratio for pulse separations of 60 fs and 120 fs. The double pulse and the pulse sequence are generated with a periodic binary phase profile $\phi(\omega) = 0$ or π over spectral intervals of $\frac{1}{\Delta t}$; where Δt is the separation between two pulses (see inset in panel (a)) and a sinusoidal phase profile $\phi(\omega) = a \sin(b\delta\omega)$ (see inset in panel (b); parameter b is plotted on the x-axis keeping $a = 1$) on the LCD-pulse shaper respectively. The phase profiles plotted in the insets correspond to pulse separation of 100 fs for binary phase and $a = 1$, $b = 100$ fs for the sinusoidal phase profile.

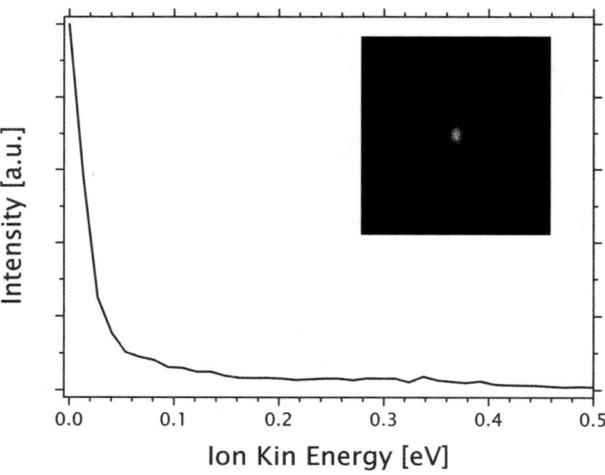

Fig. 6 Kinetic energy distribution of the $CF_3^+ + I$ fragmentation channel excited at center wavelength 540 nm (near transform limit). The inset shows the velocity image of CF_3^+ ions at 540 nm.

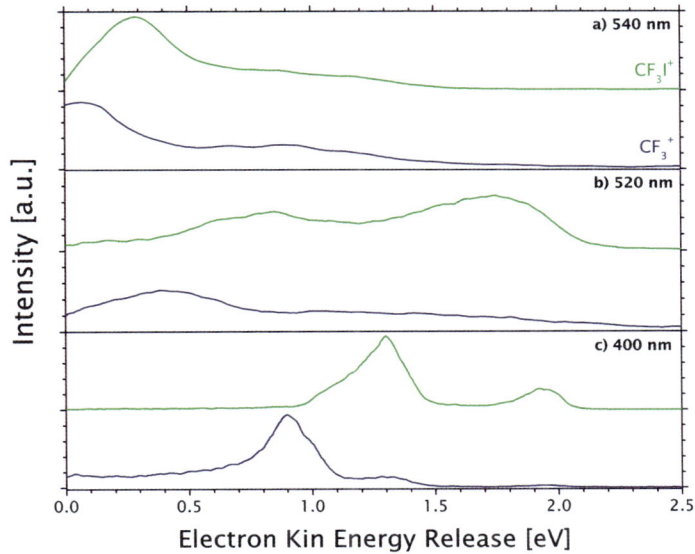

Fig. 7 Photoelectron spectrum (PES) resulting from photoionization of CF$_3$I with near transform limit pulses at (a) 540 nm, (b) 520 nm, and (c) 400 nm. The pulses at 540 nm and 520 nm were 35 fs long whereas the 400 nm pulse had a pulse duration of 120 fs. Each panel gives PES measured in coincidence with CF$_3^+$ (blue solid line) and CF$_3$I$^+$ (green solid line) respectively.

In Fig. 8 we present the laboratory angular distribution of photoelectrons in coincidence with either CF$_3$I$^+$ or CF$_3^+$ after excitation with laser pulses at 520 nm or 540 nm. The distributions were fitted to a Legendre polynomial expansion,

$$I(\theta) = \frac{\sigma}{4\pi} \times [1 + \beta_{2,e}P_2(\cos\theta) + \beta_{4,e}P_4(\cos\theta) + ...], \quad (2)$$

where θ is the polar angle between the recoil direction of the ejected electrons and the direction of the laser polarization, σ the total photoionization cross section,

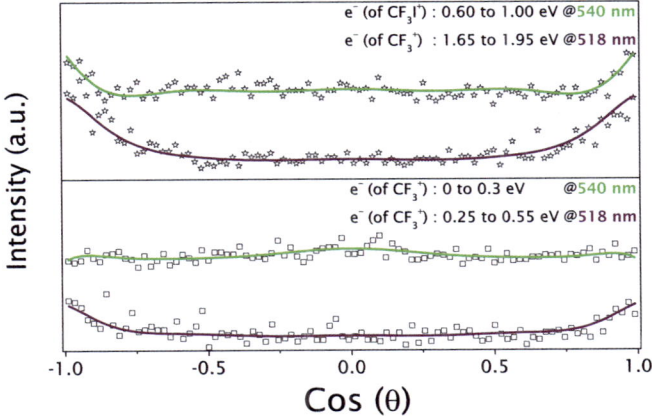

Fig. 8 Photoelectron angular distributions (PAD) measured in coincidence with CF$_3^+$ (open squares) and and CF$_3$I$^+$ (open stars) at 540 nm (green) and 520 nm (purple). The panels show PAD along with angular distribution fits of (*top*) CF$_3$I$^+$ corresponding to the high energy photoelectron peak and (*bottom*) CF$_3^+$ ions corresponding to the low energy photoelectron peak. The anisotropy parameters obtained from these fits are listed in Table II.

Table 2 Anisotropy parameters ($\beta_{2,e}$ and $\beta_{4,e}$) for photoelectron angular distributions in selected kinetic energy regions of the photoelectron measured in coincidence with CF_3^+ and CF_3I^+ ions at 540 and 520 nm shown in Fig. 8

	540 nm			520 nm	
	$\beta_{2,e}$	$\beta_{4,e}$		$\beta_{2,e}$	$\beta_{4,e}$
(e,CF_3^+) (0 to 0.3 eV)	−0.17	0.26	(e,CF_3^+) (0.25 to 0.55 eV)	0.67	0.44
(e,CF_3^+) (0.15 to 0.35 eV)	0.50	0.33	(e,CF_3^+) (0.65 to 0.95 eV)	1.07	0.61
(e,CF_3I^+) (0.6 to 1.0 eV)	0.44	0.69	(e,CF_3I^+) (1.65 to 1.95 eV)	1.49	0.81

$P_2(\cos\theta)$ and $P_4(\cos\theta)$ the second and the fourth order Legendre polynomials and $\beta_{2,e}$ and $\beta_{4,e}$ the anisotropy parameters. In Table 2 we summarize the results of the laboratory frame photoelectron angular distribution in selected energy regions in coincidence with either the CF_3I^+ or CF_3^+ ion channel.

IV. Discussion

A. Energetics and photoelectron spectra

In the present study CF_3I is excited in a single color experiment with femtosecond laser pulses with center wavelengths around 540 nm. In Table 3 we give the total excitation energy at these wavelengths for multiphoton excitation up to five photons. Furthermore, in Fig. 9 we present the most relevant energy levels of CF_3I, CF_3I^+ and the location of various neutral and ionic fragmentation channels. The ground state of CF_3I^+ is split into $X^2E_{3/2}$ and $X^2E_{1/2}$ components due to spin–orbit interaction.

Various methods and techniques such as photoelectron spectroscopy, photoionization mass spectrometry and electron impact, have been used for determining the adiabatic (AIE) and the vertical (VIE) ionization energy of CF_3I. Lias[62] recommends AIE = 10.28 ± 0.07 eV after a general review of all ionization experiments found in the scientific literature up to around 1984. This NIST value was used also by Eden et al.[51] in their recent study on the VUV absorption of fluoro-alkanes. However, Macleod et al.[46] used the more accurate technique of ZEKE photoelectron spectroscopy in 1998 and reported the adiabatic ionization energy corresponding to the $X^2E_{3/2}$ ground state of CF_3I^+, AIE = 83652 ± 2 cm^{-1} = 10.3715 ± 0.0003 eV. We adopted this latter value for the present study. We use the value of 0.64 eV^{34} for the $X^2E_{1/2}$–$X^2E_{3/2}$ spin–orbit splitting in the ionic ground state of CF_3I^+, which is slightly smaller than the value of 0.73 reported by Cvitas et al.[63] from VIE experiments. The vertical ionization energy (VIE) is slightly higher due to molecular geometry changes upon ionization. Cvitas et al. used He(I) and He(II) radiation to photoionize CF_3 and reported VIE = 10.45 eV and 11.18 eV for the ionization onsets

Table 3 Total excitation energy (eV) at various wavelengths for multiphoton excitation up to five photons

wavelength [nm]	540	520	400
1-photon	2.3	2.4	3.1
2-photon	4.6	4.8	6.2
3-photon	6.9	7.2	9.3
4-photon	9.2	9.5	12.4
5-photon	11.5	11.9	15.5

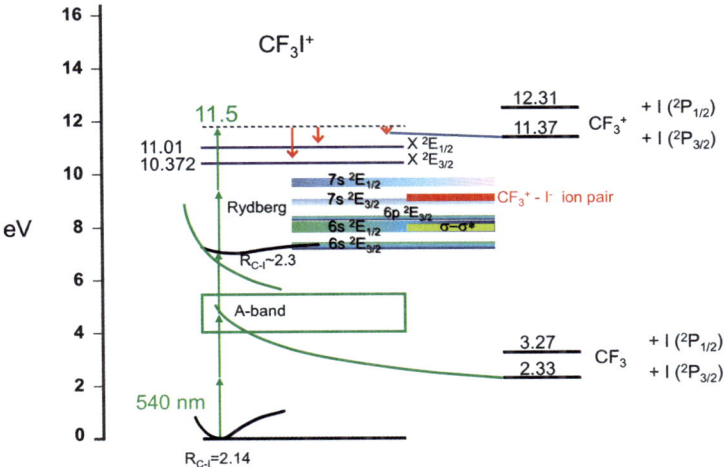

Fig. 9 Relevant energy levels of CF$_3$I and various neutral and ionic fragmentation channels.

of the X^{2}E$_{3/2}$ and X^{2}E$_{1/2}$ states respectively. Sutcliffe and Walsh[39] obtained 10.40 and 11.02 eV for the first and second ionization energies using Rydberg state assignments. The lowest fragmentation channel leading to CF$_3^+$(X^{1}A$_1$) + I(^{2}P$_{3/2}$) is located at 11.37 ± 0.05 eV.[45,46,64] From the discussion above and the energetics of Table 3 it can be directly concluded that at the wavelengths around 540 nm at least five photons are required to ionize the parent molecule. Five photons of 540 nm provide a total excitation energy of 11.5 eV. For excitation at 400 nm, as reported before,[34] a 4 photon process is energetic enough (total energy is 12.4 eV) to open the CF$_3^+$(X^{1}A$_1$) + I(^{2}P$_{3/2}$) channel.

In the (e,CF$_3$I$^+$) channel and for excitation with short pulses at 540 nm, panel (a) in Fig. 7, we basically see a single photoelectron peak near 0.25 eV, and a weaker tail to higher electron kinetic energy up to 0.8–1.0 eV. We assign the peak at 0.25 eV to the formation of the spin–orbit excited ground state X^{2}E$_{1/2}$ of the CF$_3$I$^+$ ion. The weak tail is assigned to the CF$_3$I$^+$ X^{2}E$_{3/2}$ ground state, and is apparently less favored at this excitation wavelength. For excitation with short pulses at 520 nm, panel (b) in Fig. 7 we see strong changes in the photoelectron spectra. The electron spectrum in coincidence with the CF$_3$I$^+$ parent ion now shows two clear peaks that (again) are assigned to the formation of the two spin–orbit split ground states X^{2}E$_{3/2}$ and X^{2}E$_{1/2}$ of the CF$_3$I$^+$ ion. At 520 nm the high energy photoelectron peak near 1.7 eV is now the strongest peak. The (e,CF$_3$I$^+$) photoelectron data near 400 nm were discussed before[34] and is the result of a 4 photon excitation. We proposed[34] for excitation at 400 nm that the reason that the lower photoelectron peak near 1.3 eV, correlating with the spin–orbit excited X^{2}E$_{1/2}$ ground state of the CF$_3$I$^+$ ion, is stronger than the photoelectron peak near 1.95 eV, is due to a near resonance at three photon excitation (about 9.3 eV) with (vibrational) bands of the 7s^{2}E$_{1/2}$(Ω = 0) Rydberg state and the propensity for core conservation in the subsequent one-photon further ionization.

We now discuss the photoelectron spectra in coincidence with the CF$_3^+$ fragment. Both spectra at 540 nm and 520 nm show one peak at rather low energy. For excitation at 540 nm the electron peak is near 0.1 eV, at 520 nm the peak is near 0.4 eV. The appearance energy CF$_3^+$(X^{1}A$_1$) + I(^{2}P$_{3/2}$) fragmentation channel is 11.37 eV. As shown in Fig. 6 the total kinetic energy of the CF$_3^+$, I fragments at the excitation wavelength of 540 nm is at most about 0.1 eV. This low kinetic energy of the fragments agrees well with the total energy available for the electron and kinetic energy of only about 0.15–0.2 eV for a five photon excitation at 540 nm.

In summary, all electron peaks are assigned to a five photon excitation process. The (e,CF$_3$I$^+$) coincident electron peaks correlate to the production of the two spin–orbit (X^{2}E$_{3/2}$, X^{2}E$_{1/2}$) ground states of the parent CF$_3$I$^+$. The (e,CF$_3$$^+$) electron peak at low energy of about 0.1–0.4 eV, depending on the excitation wavelength, correlates to the fragmentation channel CF$_3$$^+$(X^{1}A$_1$) + I(^{2}P$_{3/2}$). As can be seen in Fig. 9 these photon wavelengths are resonant at two-photon excitation with the fast dissociative A-band, at three-photon energy the excitation is close to the onset of the 6s Rydberg state and at four-photon level with the 7s^{2}E$_{3/2,\ 1/2}$ Rydberg states and CF$_3$$^+$–I$^-$ ion pair state.[34,39,46,51,65] We will discuss possible implications of these near resonant intermediate states on the chirped pulse control mechanism in the following section.

B. Control mechanism

The role of chirp in multi-photon excitation in ionization and fragmentation of molecules has been reported before.[9,12,66–70] Most of those experiments used either TOF-mass detection or fluorescence detection which means that very little information is obtained about the energetics of the mechanism.

Very recently,[71] we reported on the first experiments combining velocity map photoelectron and ion imaging to study the mechanism in pulse shaping control of photofragmentation and ionization in CH$_2$BrCl. In those experiments an effect of the chirp was observed in only a rather narrow wavelength region around 520 nm, and the CH$_2$Cl$^+$/CH$_2$BrCl$^+$ ratio could be enhanced by about a factor of five with up-chirped pulses. We proposed that a wavepacket following or a time-delay resonance between the fast dissociative A-band and a higher lying state (possible of Rydberg character) at the two-three photon excitation was responsible for the up-chirp enhancement of fragmentation.

In the experiments reported here, again the effect appears to be dependent on the excitation wavelength, although we are currently performing more extensive and detailed experiments at different wavelengths. For CF$_3$I we can both enhance the fragmentation to CF$_3$$^+$ at certain wavelengths with down-chirped pulses, or enhance the fragmentation with up-chirped pulses. Furthermore, it also appears that the fragmentation can be enhanced with double pulse excitation with a inter pulse separation of about 60 fs.

Because the mass ratio is so strongly dependent on the wavelength we believe that intermediate resonances with excited state surfaces in the neutral molecule play a role in the control mechanism. Two photons of 540 nm result in a total excitation energy of 4.6 eV, which is exactly at the peak of the fast dissociative A-band (see Fig. 9). Furthermore, there are various higher lying Rydberg states of 6s,6p and 7s,7p character at energies close to three and four photon excitation. Very recently, VUV negative photoion spectroscopy by Simpson et al.[72] have also provided experimental evidence of the existence of a CF$_3$$^+$–I$^-$ ion pair state located at 9 eV.

We propose here that coupling between the fast dissociative A-band at two photon excitation energy and a perturbed (6s) Rydberg state at three photon excitation energy may be responsible for the pulse shaping dependence of the fragmentation dynamics, see Fig. 9.

The perturbation of the Rydberg state is believed to be by a dissociative state of valence character and a bound ion-pair state. Recently, Lepetit and coworkers reported on an *ab initio* study of valence and Rydberg states of CH$_3$Br.[73] In CH$_3$Br the lowest Rydberg states are formed by excitation of the e-type lone pair orbitals of Br-character to the 5s, 5p levels. The resulting Rydberg state of totally symmetric ^{1}A$_1$ character is heavily perturbed by a repulsive valence state and a bound CH$_3$$^+$–Br$^-$ ion-pair state (see Fig. 2 in ref. 73). The potential outer minimum of ion-pair character at a C–Br distance of 6.5 a.u. in this double-minimum ^{1}A$_1$ state is almost 2 eV lower in energy than the inner minimum near 3.6 a.u. which is of

Rydberg character. We propose that in CF_3I a similarly perturbed Rydberg state, the analogous 6s excitation of the I-lone pair orbital, may exhibit a similarly shaped excited state that changes character from Rydberg to valence to $CF_3^+–I^-$ ion pair character when the C–I distance increases. In general the equilibrium distance of ion pair states is at a larger C–I distance than the equilibrium C–I distance of the ground state or Rydberg states.[74] Cheng et al.[75] reported a theoretical study on the dissociation energy and ionization potentials of perfluoroalkyl iodides and they calculated the C–I distance in the CF_3I^+ cation to be $R_{C–I}(CF_3I^+) = 2.3184$ Å, whereas the distance in the electronic ground state of CF_3I $R_{C–I}(CF_3I) = 2.138$ Å. When we excite the CF_3I molecule via a two-photon transition to the A-band the system will subsequently quickly evolve leading to a stretched C–I bond. If we now further excite the system with a third photon, depending on the initial excitation and the speed of the wavepacket on the A-band, this further excitation may be more favorable with down-chirped pulses, or be favored by up-chirped pulses. The difference in the slopes of the A-band and the strongly perturbed Rydberg-valence-ion pair state will have a crucial effect on the efficiency of coupling with a chirped pulse. It is further interesting to note that for the CH_3Br system the parallel electric dipole transition moment between the ground electronic state and the excited $^1A_1$ state increases dramatically once the C–Br distance has stretched to 4 a.u. (see Fig. 4 in ref. 73). We anticipate that a similar strong increase may be present in the CF_3I system. Furthermore, if we correlate the observed time spacing of 60 fs to an energy spacing we find $\Delta E = 560$ cm^{-1} = 0.069 eV. If we compare this energy with vibrational level spacing in excited Rydberg states, we see from Table 6 in ref. 51 that the ν_2 (CF_3 umbrella) vibration has an energy of 0.073 eV in the 6s $^2E_{1/2}$ Rydberg state. We further note that the CF_3^+ fragment ion is planar, whereas the CF_3 in the neutral parent is pyramidal in structure. Perhaps the controlled excitation via double pulses spaced some 60 fs apart affects the enhancement of the fragmentation in the CF_3^+ + I channel via coherent involvement of the umbrella vibration.

It is clear that high level ab initio calculations of the Rydberg, valence and ion-pair states of CF_3I are needed to further substantiate the control mechanism. At present such calculations are performed by Leininger[76] and we hope to discuss the possible implications of these calculations in relation with the more extended coincidence imaging pulse shaping experiments in the near future.

V. Conclusions

In this paper we have reported on the combination of full LCD-pulse shaping and photoelectron-photoion coincidence imaging to study the control of ionic fragmentation in the multichannel photodynamics of CF_3I. An experimentally simplified set-up was used employing a single TOF delay line detector. By switching the HV on the velocity map ion optics both the electron and the coincident ion are measured on the same detector. Photodynamics of CF_3I was studied with shaped NOPA light pulses between 520–550 nm. The dominant species formed are the parent ion CF_3I^+ either in the ($X^2E_{3/2}$) ground state or the spin–orbit excited state ($X^2E_{1/2}$) state and the $CF_3^+(X^1A_1)$ fragment ion in coincidence with ground state iodine I($^2P_{3/2}$). The energetics of the process can be fully determined by the correlated (e,ion) photoelectron spectra and the CF_3^+ ion kinetic energy distribution. The fragmentation involves multiphoton absorption mainly at the five photon level. Shaped femtosecond pulses were observed to be capable of controlling the multiphoton photofragmentation of CF_3I. Strong changes up to 60–100% in the CF_3^+/CF_3I^+ ratio are observed with chirped pulses and double or multiple pulse sequences. When excitation was performed with a double pulse an increase of the CF_3^+/CF_3I^+ ratio by about 60% was observed for a double pulse spacing of 60 fs. From comparison of the energetics and the known location of electronically excited states in the CF_3I we propose that neutral surfaces are involved in the control dynamics. At present we are further studying in more detail the wavelength dependence of the fragmentation control in combination with selected pulse shapes. Furthermore, high level *ab initio* electronic structure calculations are performed by Leininger and coworkers[76] of in particular the 6s, 7s Rydberg state region. We expect that these electronic structure and dynamics calculations in combination with further detailed coincidence imaging experiments will help to reveal the mechanism in pulse shaping control of fragmentation dynamics in this polyatomic molecule.

Acknowledgements

This research has been financially supported by the council for Chemical Sciences of the Dutch Organization for Scientific Research (NWO–CW). The authors also acknowledge support by the European Union through the Integrated Infrastructure Initiative LaserLabEurope and the Marie-Curie Initial Training Network ICONIC. The authors gratefully acknowledge theoretical collaborations on CF_3I with prof. Leininger. The authors gratefully acknowledge the excellent technical support by Mr. R. Kortekaas.

References

1 D. J. Tannor and S. A. Rice, *J. Chem. Phys.*, 1985, **83**, 5013.
2 P. Brumer and M. Shapiro, *Chem. Phys. Lett.*, 1986, **126**, 541.
3 R. S. Judson and H. Rabitz, *Phys. Rev. Lett.*, 1992, **68**, 1500.
4 J. L. Herek, *J. Photochem. Photobiol., A*, 2006, **180**, 225.
5 H. Fielding, M. Shapiro and T. Baumert, *J. Phys. B: At., Mol. Opt. Phys.*, 2008, **41**, 070201.
6 H. Rabitz, *New J. Phys.*, 2009, **11**, 105030.
7 H. H. Fielding and M. A. Rob, *Phys. Chem. Chem. Phys.*, 2010, **12**, 15569.
8 C. Brif, R. Chakrabarti and H. Rabitz, *New J. Phys.*, 2010, **12**, 075008.
9 A. Assion, T. Baumert, M. Bergt, T. Brixner, B. Kiefer, V. Seyfried, M. Strehle and G. Gerber, *Science*, 1998, **282**, 919.
10 R. J. Levis, G. M. Menkir and H. Rabitz, *Science*, 2001, **292**, 709.
11 C. Daniel, J. Full, L. Gonzalez, C. Lupulescu, J. Manz, A. Merli, S. Vajda and L. Woste, *Science*, 2003, **299**, 5606.
12 V. V. Lozovoy, X. Zhu, T. C. Gunaratne, D. A. Harris, J. C. Shane and M. Dantus, *J. Phys. Chem. A*, 2008, **112**, 3789.
13 X. Zhu, T. C. Gunaratne, V. V. Lozovoy and M. Dantus, *J. Phys. Chem. A*, 2009, **113**, 5264.
14 R. J. Levis, *J. Phys. Chem. A*, 2009, **113**, 5267.
15 N. H. Damrauer, C. Dietl, G. Krampert, S. H. Lee, K. H. Jung and G. Gerber, *Eur. Phys. J. D*, 2002, **20**, 71.
16 T. Rozgonyi and L. Gonzalez, *J. Phys. Chem. A*, 2008, **112**, 5573.
17 T. Rozgonyi and L. Gonzalez, *J. Mod. Opt.*, 2009, **56**, 790.
18 D. Geissler, B. J. Pearson and T. Weinacht, *J. Chem. Phys.*, 2007, **127**, 204305.
19 B. J. Pearson, S. R. Nichols and T. Weinacht, *J. Chem. Phys.*, 2007, **127**, 131101.
20 S. R. Nichols, T. C. Weinacht, T. Rozgonyi and B. J. Pearson, *Phys. Rev. A: At., Mol., Opt. Phys.*, 2009, **79**, 043407.
21 J. Gonzalez-Vazquez, L. Gonzalez, S. R. Nichols, T. C. Weinacht and T. Rozgonyi, *Phys. Chem. Chem. Phys.*, 2010, **12**, 14203.
22 D. W. Chandler and P. L. Houston, *J. Chem. Phys.*, 1987, **87**, 1445.
23 M. N. R. Ashfold, N. H. Nahler, A. J. Orr-Ewing, O. P. J. Vieuxmaire, R. L. Toomes, T. N. Kitsopoulos, I. A. Garcia, D. A. Chestakov, S.-M. Wu and D. H. Parker, *Phys. Chem. Chem. Phys.*, 2006, **8**, 26.
24 A. I. Chichinin, K.-H. Gericke, S. Kauczok and C. Maul, *Int. Rev. Phys. Chem.*, 2009, **28**, 607.
25 S. J. Greaves, R. A. Rose and A. J. Orr-Ewing, *Phys. Chem. Chem. Phys.*, 2010, **12**, 9129.
26 A. Vredenborg, C. S. Lehmann, D. Irimia, W. G. Roeterdink and M. H. M. Janssen, *ChemPhysChem*, 2011, **12**, 1459.
27 J. A. Davies, J. E. LeClaire, R. E. Continetti and C. C. Hayden, *J. Chem. Phys.*, 1999, **111**, 1.
28 J. A. Davies, R. E. Continetti, D. W. Chandler and C. C. Hayden, *Phys. Rev. Lett.*, 2000, **84**, 5983.
29 R. E. Continetti and C. C. Hayden, Coincidence imaging techniques, in *Modern Trends in Reaction Dynamics*, ed. X. Yang and K. Liu, World Scientific (Singapore), 2004, pp 475–528.

30 A. M. Rijs, M. H. M. Janssen, E. T. H. Chrysostom and C. C. Hayden, *Phys. Rev. Lett.*, 2004, **92**, 123002.
31 M. H. M. Janssen, www.nwo.nl/projecten.nsf/pages/1700115289.
32 A. Vredenborg, W. G. Roeterdink and M. H. M. Janssen, *Rev. Sci. Instrum.*, 2008, **79**, 063108.
33 A. Vredenborg, W. G. Roeterdink and M. H. M. Janssen, *J. Chem. Phys.*, 2008, **128**, 204311.
34 A. Vredenborg, W. G. Roeterdink, C. A. de Lange and M. H. M. Janssen, *Chem. Phys. Lett.*, 2009, **478**, 20.
35 D. Irimia, I. D. Petsalakis, G. Theodorakopoulos and M. H. M. Janssen, *J. Phys. Chem. A*, 2010, **114**, 3157.
36 W. Roeterdink, A. M. Rijs, G. Bazalgette, P. Wasylczyk, A. Wiskerke, S. Stolte, M. Drabbels, M. H. M. Janssen, Chapter III-8 in *Atomic and Molecular Beams: The State of the Art 2000*, Editor R. C. Campargue, Springer-Verlag, 2001, Berlin Heidelberg.
37 W. G. Roeterdink and M. H. M. Janssen, *Chem. Phys. Lett.*, 2001, **345**, 72.
38 W. G. Roeterdink and M. H. M. Janssen, *Phys. Chem. Chem. Phys.*, 2002, **4**, 601.
39 L. H. Sutcliffe and A. D. Walsh, *Trans. Faraday Soc.*, 1961, **57**, 873.
40 G. N. A. van Veen, T. Baller, A. E. de Vries and M. Shapiro, *Chem. Phys.*, 1985, **93**, 277.
41 P. Felder, *Chem. Phys.*, 1991, **155**, 435.
42 L. D. Waits, R. J. Horwitz, R. G. Daniel, J. A. Guest and J. R. Appling, *J. Chem. Phys.*, 1992, **97**, 7263.
43 C. A. Taatjes, G. van den Hoek, M. Jonker, D. H. Parker and S. Stolte, *Chem. Phys. Lett.*, 1993, **215**, 461.
44 C. A. Taatjes, J. W. G. Mastenbroek and S. Stolte, *Chem. Phys. Lett.*, 1993, **216**, 100.
45 R. L. Asher and B. Ruscic, *J. Chem. Phys.*, 1997, **106**, 210.
46 N. A. MacLeod, S. Wang, J. Hennessy, T. Ridley, K. P. Lawley and R. J. Donovan, *J. Chem. Soc., Faraday Trans.*, 1998, **94**, 2689.
47 P. Downie and I. Powis, *Phys. Rev. Lett.*, 1999, **82**, 2864.
48 P. Downie and I. Powis, *J. Comp. Phys.*, 1999, **111**, 4535.
49 P. Downie and I. Powis, *Faraday Discuss.*, 2000, **115**, 103.
50 F. Aguirre and S. T. Pratt, *J. Chem. Phys.*, 2003, **118**, 6318.
51 S. Eden, P. Limao-Vieira, S. V. Hoffmann and N. J. Mason, *Chem. Phys.*, 2006, **323**, 313.
52 V. N. Lokhman, D. D. Ogurok and E. A. Ryabov, *Eur. Phys. J. D*, 2008, **46**, 59.
53 G. N. A. van Veen, T. Baller, A. E. de Vries and M. Shapiro, *Chem. Phys.*, 1985, **93**, 277.
54 M. D. Person, P. W. Kash and L. J. Butler, *J. Chem. Phys.*, 1991, **94**, 2557.
55 Y. S. Kim, W. K. Kang and K.-H. Jung, *J. Chem. Phys.*, 1996, **105**, 551.
56 F. Aguirre and S. T. Pratt, *J. Chem. Phys.*, 2003, **119**, 9476.
57 D. Irimia, R. Kortekaas and M. H. M. Janssen, *Phys. Chem. Chem. Phys.*, 2009, **11**, 3958.
58 D. Irimia, D. Dobrikov, R. Kortekaas, H. Voet, D. A. van den Ende, W. A. Groen and M. H. M. Janssen, *Rev. Sci. Instrum.*, 2009, **80**, 113303.
59 T. Wilhelm, J. Piel and E. Riedle, *Opt. Lett.*, 1997, **22**, 1494.
60 S. De Silvestri and G. Cerullo, *Rev. Sci. Instrum.*, 2003, **74**, 1.
61 M. L. Lipciuc, J. B. Buijs and M. H. M. Janssen, *Phys. Chem. Chem. Phys.*, 2006, **8**, 219.
62 S. G. Lias, Ionization Energy Evaluation (NIST Chemistry Web-Book, NIST Standard Reference Database, vol. 69, p. 20899, National Intitude of Standards and Technology, Gaithersburg, MD).
63 T. Cvitas, H. Gusten, L. Klasinc, I. Novak and H. Vancik, *Z. Naturforsc*, 1978, **33a**, 1528.
64 B. Ruscic, J. V. Michael, P. C. Redfern and L. A. Curtiss, *J. Phys. Chem. A*, 1998, **102**, 10889.
65 C. A. Taatjes, J. W. G. Mastenbroek, G. van den Hoek, J. G. Snijder and S. Stolte, *J. Chem. Phys.*, 1993, **98**, 4355.
66 V. V. Yakovlev, C. J. Bardeen, J. Che, J. Cao and K. R. Wilson, *J. Chem. Phys.*, 1998, **108**, 2309.
67 I. Pastirk, E. J. Brown, Q. Zhang and M. Dantus, *J. Chem. Phys.*, 1998, **108**, 4375.
68 N. T. Form, B. J. Whitaker and C. Meier, *J. Phys. B: At., Mol. Opt. Phys.*, 2008, **41**, 074011.
69 T. Goswami, S. K. K. Kumar, A. Dutta and D. Goswami, *Chem. Phys.*, 2009, **360**, 47.
70 J. Plenge, A. Wirsing, C. Raschpichler, M. Meyer and E. Ruhl, *J. Chem. Phys.*, 2009, **130**, 244313.
71 D. Irimia and M. H. M. Janssen, *J. Chem. Phys.*, June (2010), **132**, 7.
72 M. J. Simpson, R. P. Tuckett, K. F. Dunn, C. A. Hunniford and C. J. Latimer, *J. Chem. Phys.*, 2009, **130**, 194302.
73 C. Escure, T. Leininger and B. Lepetit, *J. Chem. Phys.*, 2009, **130**, 244306.
74 K. P. Lawley and R. J. Donovan, *J. Chem. Soc., Faraday Trans.*, 1993, **89**, 1885.
75 L. Cheng, Z. Shen, J. Lu, H. Gao and Z. Lu, *Chem. Phys. Lett.*, 2005, **416**, 160.
76 T. Leininger, Personal communication.

General discussion

Professor Miller opened the discussion of the paper by Professor Engel: There has been a lot work using different methods to try to isolate different Fourier components with different decay components from time resolved data. There are generally issues with correcting for baseline, and aliasing based on the parameters used to restrict the sampling rate or time window of observation. In all cases, the analysis depends critically on knowing your baseline. As a simple example, if there is an offset that is not taken into account, the amplitudes of the decay components will be incorrect. It is quite difficult to independently determine background contributions to the signal. It seems the 2D spectrum shows strong modulations at numerous off diagonal locations, in addition to the region of interest, where you have restricted your analysis. The reason I am making this point is that we see strong off diagonal modulations on just blank controls.[1] We introduced diffractive optics nonlinear spectroscopy for 2D spectroscopy and recently fully generalized the approach to enable coherent control as discussed in our Faraday Discussion paper.[2] We were concerned about possible nonresonant contributions to the signal. The aforementioned modulation of the background signal arises from phase modulation in the diffractive optic used to create the phase locked pulse pairs. It is more than an order of magnitude down from the resonant signal from the sample but these features will show up as cross terms with your resonant signal to amplify them. I think it is difficult to properly correct for nonresonant background contributions.

Basically how do you experimentally correct for the background modulations in the signal and use of the Yuke-Walker equation that is used for isolating Fourier components that are time invariant? Is there are criterion in signal to noise ratio that is needed to extract n different Fourier components to response function?

1 V. I. Prokhorenko, A. Halpin, R. J. D. Miller, *Optics Express*, 2009, **17**(12), 9764–9779.
2 V. I. Prokhorenko, A. Halpin and R. J. D. Miller, *Faraday Discussions*, 2001, DOI: 10.1039/C1FD00095K.

Professor Engel responded: If I may summarize your comment, you are suggesting that isolating oscillatory signals in the presence of oscillatory "noise" is difficult. Perhaps, even more strongly, you are pointing out that baseline subtraction of nonlinear functions (exponential decays) can introduce oscillatory residuals. We agree with your analysis. As you also point out, there is no way to completely address this issue. One approach is to consider signal to noise ratios for thresholding. Another is to note that knowledge of the baseline character can be helpful. For example, if the baseline is a sum of exponential decays, oscillatory residuals will tend to be extremely low frequency (0–40 wavenumbers for example). We have run blank samples and bacteriochlorophyll in solvent, and we have not observed nonresonant oscillations as you have in your experiments. This might be because your have a lower noise experiment or simply because resonant bacteriochlorophyll signals are strong enough that our dynamic range is insufficient. We will reanalyze these data sets to search for such oscillations to try to establish signal to noise ratios, but in the meantime, we remain confident that the observed oscillations at the excitonic energy differences (150–300 wavenumbers) are orders of magnitude larger than the nonresonant signal. Further, we note that our beat frequencies are captured well in the real part of our data so we do not suspect nonresonant contributions which would appear in the imaginary portion of the data. Your question does raise an excellent suggestion for future studies. In particular, we chose in this paper to analyze the absolute value of the signal to eliminate phasing errors but you have made us aware that this benefit comes at the cost of perhaps introducing nonresonant oscillations. We will begin to work to quantify the benefits of this tradeoff.

Professor Kosloff asked: You showed experimental evidence of coherent transient oscillatory motion in your signal. How can you be sure that what you observe is the result of electronic coherence between the chromophores? Another possibility is vibrational coherence between low frequency modes of the molecules.

Professor Engel replied: You bring up an excellent point. We have tried many different ways to ensure that the dynamics that we attributed to electronic coherence are in fact electronic and not vibrational. I do note explicitly that this separation is somewhat artificial, but it is extremely useful conceptually and quite natural within the Born–Oppenheimer approximation.

First, the rephasing and non-rephasing pathways will show beating from vibrational wavepacket dynamics and beating electronic superpositions in different patterns.[1] For example, we expect electronic beats to appear only off of the main diagonal in the rephasing pathways, but vibrational beats should appear everywhere. The data matches this prediction.

Second, we have performed isotopic substitution experiments with deuterium.[2] We randomly deuterated the sample at 40% deuteration (by growing it in 40% deuterated water). Of course, we know that one usually thinks about deuteration affecting vibrations in the 2000–3000 wavenumber range, but the low frequency modes are also affected albeit by only 10-15 wavenumbers. One can attribute this change largely to changes in the directionality of the normal modes. By introducing such a change of 10–15 wavenumbers of inhomogeneous "noise" across the ensemble, we would expect rapid dephasing of the obsevable beating signal. We do not see such behavior. The beating remains unchanged to within experimental error.

Finally, we are currently performing experiments on bacteriochlorophyll in solvent to search for beats. We have found only one very weak beating mode near 260 wavenumbers. We have also isolated this signal in our FMO spectrum; it's about an order of magnitude weaker than the beats discussed in this paper. The work described in this final point will be the subject of a future publication.

1 Y.-C. Cheng and G. R. Fleming, *J. Phys. Chem. A*, 2008, **112**, 4254–4260.
2 D. Hayes and G. S. Engel, *Faraday Discussions*, 2011, DOI: 10.1039/C0FD00030B.

Professor Miller asked: It is interesting to postulate on what microscopic motions lead to decoherence. For these large molecular systems, the intramolecular bath is sufficiently large in terms of density of states and coupling of vibronic states to contribute to decoherence. If the intramolecular bath dominates the decoherence it is difficult to understand how spatially distinct molecules could maintain their electronic mixing/coupling or zero quantum coherence. If the intermolecular bath fluctuations dominate, then there would have to be spatially extended correlations of the bath motions over the lengths scale of separation of the excitonic coupled system. The times scales as you point out for electronic decoherence of such systems are typically on the order of 10–100 fs. Any correlated motions over the distance separating the excitonically coupled sites would involve much slower motions. The only motions fast enough in the intermolecular bath to create this fast decoherence are inertial modes of the surrounding bath. In the solution phase, these motions are quite localized librational motions. In discussing spatial correlations as a mechanism for long lived zero quantum coherences, we need to also consider the electronic coupling. As Prof. Brumer's paper showed if the coupling is stronger than the bath coupling the coherence can be preserved longer and is intimately related to the relative ratio of the electronic coupling between sites and the bath coupling. We have studied a model dimer system in which the molecules share a significant fraction of the same bath fluctuations via these correlated librational motions and have a pronounced exciton splitting so the system should be in the strong electronic coupling regime. This is preliminary information but we don't see any long lived beats in the off diagonal terms that would indicate long lived zero quantum coherences. There is a recent paper by Klaus Schulten's group[1] that indicates there is no spatial correlations in FMO complexes. Is there another mechanism beside through space coupling of excitonic sites to explain long lived coherences?

1 C. Olbrich *et al.*, *J. Phys. Chem. B*, 2011, **115**(26), 8609–8621.

Professor Engel responded: This is a fabulous question. In light of Prof. Schulten's work[1] and similar attempts by Alan Aspuru-Guzik,[2] I agree that a new explanation for the long-lived coherences is likely necessary.

I anticipate that the key to understanding the coherent dynamics in photosynthetic complexes will come through experiments on synthetic systems. As you have done, we are attempting to design dimer systems to test some of our hypotheses for such mechanisms. This work is too preliminary for me to speculate on the outcome at this time.

1 C. Olbrich *et al.*, *J. Phys. Chem. B*, 2011, **115**(26), 8609–8621.
2 S. Shim *et al.*, arXiv:1104.2943v1.

Professor Miller said: This question expands further on the question of Professor Kosloff. The long lived oscillations in the off diagonal features of the 2D spectrum is the key observation. The new analysis you present helps isolate potentially different structural connections that give rise to this intriguing observation. In this context, it is useful to note that all transitions connected to the same ground state will give rise to quantum beats if the probe spectrum is sufficiently broad to cover the different transitions. This statement is true for any intramolecular transition or excitonically coupled transition. The resulting polarization components from the excited coherences between the ground and excited states, giving rise to the signal field, are correlated through the common ground state coherence. One of the signatures of long lived electronic coherences between excitonically coupled sites or zero quantum coherences was the occurrence of anticorrelated off diagonal features in the 2D spectrum. The recent paper of Kauffmann *et al.*[1] find similar anticorrelated features in the off diagonal components that are due to vibronic states. These states are well known to live for several hundred femtoseconds, consistent with the time scales observed. Is the observation of anticorrelated beats in the off diagonals a sufficient condition for assigning the feature to long lived zero quantum coherences? As I mentioned previously, we have explored a dimer model to look at this problem selectively for the monomer and excitonically coupled dimer system so that we have a good control for isolating vibronic from excitonic effects. Have you done a control in which you decouple the excitonic mixing in the FMO complex?

1 N. Christensson *et al.*, *J. Phys. Chem. B*, 2011, **115**(18), 5383.

Professor Engel answered: I do not feel that anticorrelation is a sufficient condition to assign a beating signal to electronic coherence. I would suggest that isotopic substitution and electronic decoupling experiments are the best ways to test the origin of the signal. We have performed isotopic substitution experiments, but we have no way to execute a decoupling experiment. However, if genetic manipulations of our organism become possible, we will run such an experiment.

Professor Tannor commented: In the plots you showed of the z-transform, there was only a single decay constant for each of the frequencies. Have you seen any examples where you get multiple decay constants for a single frequency? If so, would it be useful to think in terms of generalizing the traditional 2-d spectroscopic representation of ω_1 *versus* ω_2 to (ω_1, γ_1) *versus* (ω_2, γ_2), *i.e.* in principle there could be correlations that depend not only on the frequencies but on which decay component of which frequency? A second question: instead of the z-transform, have you tried the harmonic inversion method of Neuhauser and Mandelstam?

Professor Engel replied: A cursory examination of our algorithm using a model system indicates that the linear prediction z-transform fails at detecting the same frequency with multiple decay rates. I am not sure if this a general failure of the algorithm, or a failure of the particular method of linear prediction that is utilized. However, that does not preclude using this as a method of examining different cuts through two dimensional spectra, either looking for locations where the signal decay is similar, or using the natural improvement in resolution to dissect the relationship between features that beat within a narrow region of frequency and decay space. This will be the topic of a future paper.

We had not thought about using harmonic analysis, though it is probably suitable for improving the resolution of the signal. We chose the linear prediction z-transform as it appeared that the signal was comprised mostly of exponentially decaying sinusoids, and this method allowed us to control the number of signals and build a directly interpretable frequency-decay rate representation of a time vector. That being said spectral analysis is a widely discussed problem, and a number of solutions may be superior to using linear prediction, including filter diagonalization and spectral estimation. We are now examining these possibilities, and we very much appreciate your suggestion.

Professor de Vivie-Riedle remarked: What type of molecular vibrations do you think are involved in the long lived electronic coherence observed in electronic energy transfer from chromophore to chromophore process? Are intramolecular vibrations of the individual chromophores involved or is the energy transfer also mediated by low frequency inter-chromophore vibrations? In any case the appearance of long lived electronic coherences requires that the accompanying vibrational motion in the coupled electronic states (that are responsible for the electronic coherence) stays synchronously. Otherwise the electronic coherence is damped.

Professor Engel replied: This is an excellent question and an open area of research. I have no experimental data at this time that would allow me to answer your question. In fact, we have run many experiment with isotopic substitution that have given negative results. I agree with your analysis that correlated noise on the excitons would give rise to long-lived coherence. This idea was first proposed by Lee, Cheng and Fleming in 2007.[1] However, microscopically, I do not understand how to create such correlated fluctuations in a protein environment.

1 H. Lee, Y.-C. Cheng and G. R. Fleming, *Science*, 2007, **316**, 1462–1465.

Dr Stolow commented: Is it possible in general to distinguish electronic from vibronic (coupled vibrational-electronic) motions?

Professor Engel replied: In FMO, the fluorescence lifetime is long, and the reorganization energy of the excited state is small implying that the derivative coupling is likely small. It would be very interesting to look for signatures of vibronic coherence in a two dimensional spectra. We have not yet attempted a sample where we expected strong nonadiabatic coupling though Kaufmann and Moran have interrogated samples with a clear vibronic progression.[1,2] I expect that non-adiabatic coupling would have very different character compared to an electronic coherence. An electronic coherence between two excited states (or a zero quantum coherence) manifests itself as a "quantum beat" in a particular region of our two dimensional spectra. This signature beats at the energy difference between the two different excited states. A vibronic signature would involve coupling between the vibrational and electronic modes induced by a dissipative or quickly altering excited state (IE one where the Born–Oppenheimer approximation breaks down). You might anticipate seeing a beating signature between a ground and excited state at the energy difference in the vicinity of a conical intersection or avoided crossing. However, this is not a pathway that we can detect given the frequency range of our laser pulse. Further, the fast dynamics make seeing "stable" beats difficult or impossible.

1 N. Christensson *et al.*, *J. Phys. Chem. B*, 2011, **115**(18), 5383.
2 J. M. Womick and A. M. Moran, *J. Phys. Chem. B*, 2009, **113**, 15771–15782.

Dr Stolow opened the discussion of the paper by Dr Shalashilin: The Ehrenfest trajectories seem to be like a 'happy family' where the trajectories enjoy each other's 'close company'. I can imagine that this will not always be the case. In which situations will the multiconfigurational Ehrenfest approach have the greatest difficulties?

Dr Shalashilin replied: MCE approach aims at the situation when the system is the most quantum. Thus, I would say that the MCE method is complementary to the existing Multiple Spawning technique, which works well in the semiclassical limit.

Professor Shapiro asked: The Ehrenfest approach you have been describing, relying on an "average" trajectory, will surely fail under bifurcating situations where the system must be either at one part of configuration space or another, but cannot be in both.

Dr Shalashilin replied: I agree that the standard Ehrenfest approach would often fail. An example of such failure is given by the line 1 in Figure 1 in the paper. However the current MCE method is multiconfigurational and employs an ensemble of Ehrenfest trajectories. The whole ensemble contains trajectories which end up predominantly on each of the surfaces. Thus, the whole ensemble bifurcates quite efficiently and reproduces accurately the wave function. The MCE technique also couples the amplitudes of Ehrenfest trajectories. It is formally exact and many tests have shown that numerical accuracy of the MCE basis can be quite good. What lacks in the proposed MCE technique is the physically motivated surface hoping based picture of Multiple Spawning method. An ideal technique would probably combine both Ehrenfest and spawning pictures with a switch between the two.

Professor Tannor remarked: The Ehrenfest prescription for determining the classical motion is a mean field approach. One of the most difficult challenges for the Ehrenfest method is the case where one potential energy surface is bound and one is dissociative. In this case, at long times the trajectory should be localized on one or the other surface, and the mean field trajectory is physically meaningless. This breakdown is known to happen when there is a single configuration. Does the multi-configuration Ehrenfest method solve this problem?

Dr Shalashilin responded: Unlike standard Ehrenfest method which uses single trajectory the MCE wave function employ an ensemble of trajectories. It is a much more flexible wave function, because the "final destiny" of those trajectories can differ. They can end up on either of the states and if say the state 1 is the dominant channel those trajectories which have higher amplitude of the state 1 will have greater weight (*i.e.* the coefficient D) in the final wave function.

Professor Kosloff asked: You described a method based on a moving Gaussian representation based on Eherfest dynamics. What is missing in this picture for large systems is the influence of the environment. The environment causes localization and I would think that in this case will lead to collapse of the wavefunction to one of the basis Gaussians.

These ideas are advocated by Zurik[1] and termed pointer states. These are the states that are most immune to the environment. We have used such ideas in a model simulation method for large Hilbert space dimensions. These ideas could be used to improve your approach.[2]

1 W. H. Zurek, *Rev. Mod. Phys.*, 2005, **75**, 715.
2 M. Khasin and R. Kosloff, *Phys. Rev. A*, 2010, **81**, 043635.

Dr Shalashilin answered: Previously very large systems have been treated with MCE. For instance spin–boson model with two electronic states and the bath discretised with thousands of vibrational modes was shown to be within the reach of the method. All those modes were treated on a fully quantum level. Thus the bath can be large enough to simulate the effects of environment but I agree that adding another "layer" of environment treated with a stochastic approach is a very good idea.

Professor de Vivie-Riedle asked: I assume that one prominent aim of the multiconfigurational Ehrenfest approach you just presented is to mimic the behaviour of a nuclear wavepacket in a classical simulation. In this sense, can the feature of broadening and contracting of wavepackets be simulated?

Dr Shalashilin answered: Although each coherent state is a frozen Gaussian with a fixed width, which is indicated by the word "frozen" the wave function is represented by the whole ensemble of CS. Their trajectories diverge and later they can get closer to each other again, which effectively mimics the broadening and contraction of the wave packet. This is very similar in spirit to the semiclassical approximation known as Heller's frozen Gaussians. In addition to that the exchange of their amplitudes helps to describe the behavior of the wave packet quantitatively on a fully quantum level.

Professor Rabitz commented: Can these computational techniques be used to design control pulses in realistic laboratory circumstances?

Dr Shalashilin answered: The advances of multidimensional quantum simulations push the limits of quantum mechanics and now perform quantum dynamics for larger and larger systems. MCE is one of the competing new quantum multidimensional techniques. Currently we are aiming at doing MCE dynamics together with "on the fly" calculations of the potentials. This should allow applications beyond simple models and the hope is that MCE could be used just like classical Molecular Dynamics. Potentially there are many applications and coherent control is one of them. Whether we will be able to design a pulse one day remains to be seen.

Professor Shapiro commented: Picking up on the challenge formulated by Prof. Rabitz, I would like to point out that we have recently shown,[1,2] that even if calculations are too complex to be carried out in an *ab initio* way, it is possible to use experimental data (such as absorption or Raman spectra) to determine the Kernel of a set of integral equations describing the response of molecular systems to the action of strong external fields. The numerical solution of these integral equations then becomes perfectly feasible, thus enabling the calculations of the response of a complex molecular systems to strong external fields under many (though not all) situations.

1 I. Thanopulos, P. Brumer and M. Shapiro, *J. Chem. Phys.*, 2010, **133**, 154111.
2 X. Li, I. Thanopulos and M. Shapiro, *Phys. Rev. A*, 2011, **83**, 033415.

Dr Stolow asked: Theorists can calculate and plot both wavefunctions and wavepackets. Unfortunately, experimentalists can measure only observables—matrix elements involving these wavefunctions and/or wavepackets. In order to make comparisons between experiment and theory, therefore, it would be most helpful if the emerging quantum dynamics methods could also calculate observables—for example the photo-electron spectrum—also 'on the fly'. Is this possible using the MCE approach?

Dr Shalashilin responded: From the experience of the related coupled coherent states method we know that quantum equations can be solved not just for the wave function but also for operators, which represent the observables. For example in an article by D. V. Shalashilin, M. S. Child, and D. C. Clary[1] the infrared absorption spectrum of water trimer has been calculated by propagating the dipole moment operator rather than wave function itself. This can be more efficient than a more traditional way when observables are extracted from wave functions. The present article propagates wave functions and then obtains the observables by proper statistical averaging. Populations of the states of the spin–boson model, Frank-Condon spectrum of pyrazine and probabilities of sticking on the surface are calculated. The examples have been chosen to show that MCE can be used in many physical situations, which are extremely diverse.

D. V. Shalashilin, M. S. Child, and D. C. Clary, *J. Chem. Phys.*, 2004, **120**, 5608.

Mr Mendive-Tapia communicated: In trajectory-based methods the initial conditions, initial shifting in position, momentum and widths of the coherent state basis functions should naturally emulate the initial state of the system. Could you elaborate on the election of initial conditions? Are the coherent state basis disposed randomly of following a certain distribution?

Dr Shalashilin responded: Yes sampling is a crucial part of the trajectory based methods. The basis should cover all parts of the phase space which are important initially or at later times. Several types of sampling have been developed in the past including Coherent States swarms, trains, and pancakes. See for example ref. 1. Typically samplings are random but as mentioned above are biased towards the most important areas. In addition to that the chapter 3.2 gives an example of a sampling which combines both the elements of randomness and regularity.

1 D. V. Shalashilin and M. S. Child, *J. Chem. Phys.*, 2008, **128**, 054102.

Mr Mendive-Tapia addressed Dr Shalashilin and Professor Sension: I would like to comment on the utility of coherent state basis set based simulations since this has been questioned. Trajectory based modelling gives mechanistic information in semi-classical terms complementary to experiment. For example in connection with the next paper presented at this meeting,[1] where the bifurcation or not during the excited state decay of provitamin D_3 is an open question, these kind of simulations could help on the interpretation of the mechanism. It is a big challenge of course, but potentially one could run trajectories "on the fly", using MCE[2] or other available methodology,[3–5] and see whether they branch or not on the excited state. This is a clear example where a pictorial and intuitive classical-like mechanistic interpretation could be obtained via theoretical modelling.

1 K.-C. Tang and R. J. Sension, *Faraday Discussions*, 2011, DOI: 10.1039/C1FD00035G.
2 D. V. Shalashilin, *Faraday Discussions*, 2011, DOI: 10.1039/C1FD00034A.
3 C. Tully, *Faraday Discussions*, 1998, **110**, 407–419.
4 M. Ben-Nun, T. J. Martinez, *Advances in Chemical Physics*, 2002, 439–512.
5 B. Lasorne, G. A. Worth, M. A. Robb, *Wiley Interdisciplinary Reviews: Computational Molecular Science*, 2011, **1**(3), 460–475.

Professor Tannor opened the discussion of the paper by Professor Sension: Do you have any indication, either from experiment or simulation, of which vibrational modes lead to rapid product formation and which lead to slower product formation?

Professor Sension answered: Not at this time, the data provides no direct information. We hope that continued work on theoretical simulations can shed some light on this question.

Professor Weinacht remarked: I am wondering if measurements of the chirp dependent branching ratio at different laser energies might help distinguish between linear and nonlinear control mechanisms. Also, it seems from figure 7 in the paper that the control over the branching ratio is most pronounced for very large chirps where the pulse duration is long enough to influence the wave packet dynamics as they evolve on the PES, perhaps by dressing the potential energy surface. Are there indications that this might be the case?

Professor Sension replied: I doubt if the laser intensity is sufficient to dress the potential energy surface at the largest chirps used in these experiments. There is no clear indication in the data that this could be the case. However, it would be a good idea to measure the influence of chirp on the excited state dynamics as a function of laser energy. This would be an important experiment to help identify the control mechanism and the role that intensity could play in control.

Professor Baumert said: Could you think also of other physically motivated parameterizations for your pulse in order to uniquely address your specific reaction pathways?

Professor Sension replied: More theoretical work is necessary to build a good understanding of the specific reaction pathway in these systems. As an initial thought, a parameterization that combined short pulse trains with linear chirp would be relatively easy to implement, lead to pulses that could be characterized, and would allow the exploration of a broader range of control mechanisms.

Professor Kosloff asked: In your paper you describe a scenario where you control by positive chirp the branching ratio between the fast and slow decay process. It seems that your control scenario falls in the category of weak field where control is induced by the interaction with the bath.

We have addressed this issue theoretically by constructing a model of a ground state and two coupled excited electronic states. When the system was isolated no control via changed in the chirp rate was observed. Once coupled to the bath the change in chirp affected the outcome.[1] It seems that your measurements fit this scenario. Our model relied on a vibrational focused wavefunction on the crossing point between two electronic surfaces. Could the control in your case be a result of dynamics on more than one electronic surfaces in the excited state.

1 G. Katz, M. Ratner and R. Kosloff, *New J. Phys.*, 2010, **12**, 015003.

Professor Sension responded: Our experiment was carried out in an intensity regime where it is difficult to determine if it is strictly weak field or not. When the solvent alone was probed there was a weak signal due to two-photon ionization of the solvent. On the other hand the experiments were carried out in a regime where only ~2–3% of the provitamin D_3 molecules were excited by the pump pulse; the one-photon transition was not saturated. On the second point, excited state crossings and conical intersections play a major role in the ring-opening reaction of the isolated cyclohexadiene chromophore. It is possible that there are two excited electronic states playing a role in provitamin D_3 as well, but the mechanism for control needs to be investigated more carefully.

Professor Miller asked: Your demonstration of coherent control in the UV spectral region is quite important as this is the spectral window where most photochemistry lies. The photon energy is comparable or exceeding most chemical bonds such that virtually every class of chemical reaction can be explored. It is really a new playground for the coherent control field. Now that you have broken ground, can you elaborate on the challenges? I would expect that multiphoton problems are compounded as now ionization of solvent or sample involves only a 2-photon process rather than an n-photon process at longer wavelengths. There are also problems in characterizing the pulses as well as dispersion management. Could you please comment on these challenges.

Professor Sension answered: You are right about the major challenges. Two photon ionization of the solvent is a constant concern. The measurements reported here were at the threshold for measurable solvent ionization, with the signals from the solvent about 2 percent the size of the signals from the excited state absorption of the DHC molecule. The biggest problem however, is characterization of the weak UV pulses. Pulses with wavelengths longer than 300 nm or so can be characterized as described by Motzkus and coworkers.[1] Self-diffraction FROG can be used to characterize relatively intense UV pulses as shown by Weinacht and coworkers for 260 nm pulses.[2] When the pulses are <µJ and shorter than 300 nm the characterization problem is much harder. This will be the relevant regime for wavelength and intensity in many experiments. Our pulses were characterized in part by cross-correlation and the experimental widths matched expectations from the chirp programmed into the Dazzler.

1 J. Möhring, T. Buckup, and M. Motzkus, *Optics Lett.*, 2010, **35**, 3916–3918.
2 M. Kotur, T. Weinacht, B. J. Pearson, and S. Matsika, *J. Chem. Phys.*, 2009, **130**, 134311.

Professor Miller commented: In this wavelength range, most solvents, except water, absorb weakly and there is a long absorption tail into the 300 nm region for proteins and other biomolecules. These absorption bands will give resonant enhanced 2-photon cross sections. Even for water, we find in the visible range, that peak powers above 30 GW/cm^2 lead to ionization.[1] The cross section for 2 photon absorption is orders of magnitude larger than 3–4 photon transitions that give rise to the same effect at longer wavelengths. Can you comment on the peak power limitations for coherent control in the UV? I would expect that most experiments will have to be executed in the weak field control regime.

1 V. I. Prokhorenko, A. Halpin, P. J. M. Johnson, L. S. Brown and R. J. D. Miller, *J. Chem. Phys.*, 2011, **134**, 085105.

Professor Sension replied: As mentioned in the answer to your previous question, two-photon ionization is a serious problem. I agree that most useful experiments in condensed phase environments will have to be executed in the weak field control regime, or at least close to the weak field regime.

Professor Ashfold remarked: This following comment could be directed at any one of the papers in this Discussion that appeal for supporting high level quantum chemical calculations of excited state potential energy surfaces (PESs), and for simulations of nuclear motions on such surfaces and/or of the interaction of optical pulses with the material system. The calculation of excited state PESs over large regions of configuration space with sufficient accuracy to aid interpretation of detailed experiments is very time consuming. There is a clear need for experimentalists and quantum chemists to liaise closely in order to identify systems meriting such concerted, detailed investigation. I now turn to the specific study of Tang and Sension. Given that 1,3-cyclohexadiene (CHD) is the UV chromophore in DHC, and that the accepted mechanism for ring opening in the former involves population transfer via two excited states, would it not be rather surprising if the related ring opening in the (more complex) DHC system involves just one excited electronic state?

Professor Sension replied: Our initial assumption was that the electronic states involved in the ring-opening reaction of DHC would be similar to those involved in the same reaction in CHD. In our recent paper on the dynamics of DHC[1] we considered a two-state model consistent with the accepted mechanism for CHD in addition to the one-state model preferred here. The data could be consistent with either model although the interpretation within the one-state model is much simpler. This is a question that will only be resolved by close collaboration between theorists and experimentalists. The DHC ring-opening reaction, because of its importance in the natural synthesis of vitamin D, has attracted the interest of theorists. Calculations are being carried out on models that include the ring-structure, eliminating only the hydrocarbon tail. These calculations should help the interpretation of our experimental results. We are also conducting experiments on smaller substituted CHD molecules to facilitate simulation and collaborating with theorists to interpret the results. A close collaboration, selecting systems for ease of experimental measurement and ease of simulation is certainly called for.

1 K.-C. Tang, A. Rury, M. B. Orozco, J. Egendorf, K. G. Spears, and R. J. Sension, *J. Chem. Phys.*, 2011, **134**, 104503.

Professor Matsika commented: I would like to comment on the difficulty to calculate the second excited state in this molecule. The excited state that is missing from the calculations has significant two electron excitation character. This state is present in 1,3-cyclohexadiene (CHD) and other conjugated molecules. Doubly excited states are very difficult to calculate with electronic structure methods. Specifically, single reference methods cannot account for these states, and multireference methods are needed. Consequently, TDDFT, which is a single reference method, cannot describe this state, and it is natural that a calculation that has employed TDDFT will not find this state. Developing methods to improve the ability of TDDFT to describe doubly excited states is an active area of research.

Professor Miller asked: The theoretical work you discussed found similar branching ratios of fast and slow processes. Was the predication based on excited state dynamics or different conformations in the ground state?

Professor Sension replied: The prediction was based on calculations of the excited state dynamics and did not rely on different ground state conformations.

Professor Tannor asked: You have shown us that there is a biexponential decay when a transform limited pulse is used, and that in the presence of positive chirp one of these decay channels in increased and one is decreased. But there is no explicit evidence that chirping changes the product ratio. Hence, your results are consistent with weak field control, even without invoking dissipation, since there is no evidence that the product ratio is being affected.

Professor Sension responded: These experiments probed only the excited state behavior. As such you are right that there is no explicit evidence that the product ratio is changed by the excitation pulse. Our next series of experiments will probe the product formation explicitly.

Professor de Vivie-Riedle remarked: In our *ab inito* calculations (on CASSCF/CASPT2 level of theory) we observed that the excited state electronic structure changes significantly going form the cyclohexadiene to larger systems like fulgides with cyclohexadiene as chromophore. The two-electron excited state which is a dark state in cyclohexadiene becomes mixed with optically bright states. The transition between single excited state and doubly excited state may be not recognizable anymore in the fluorescence signal. For example in the larger fulgid-systems we find a non-vanishing dipole transition moment far beyond the Franck–Condon area, in region approaching the low-lying conical intersection seam.

Professor Sension replied: Our experiments demonstrate that the excited state responsible for the short-lived visible absorption is optically bright and responsible for the observed fluorescence spectrum (Fig. 1).

The fluorescence spectrum in n-heptane excited at 266 nm (green) is broad and relatively structureless, but has the breadth and peak position expected for (near) mirror image emission from the initially excited state when compared with the absorption spectrum (blue). This precludes rapid relaxation to a lower energy state with red-shifted fluorescence.

In addition, the anisotropy of the excited state absorption is constant (Fig. 2).

Parallel ($I_\parallel$, red) and perpendicular ($I_\perp$, blue) transient absorption signals obtained following excitation of DHC in 2-butanol. The data shown are line-outs at two specific wavelengths, the anisotropy (green) is calculated using $r(t) = \dfrac{I_\parallel - I_\perp}{I_\parallel + 2I_\perp}$. The quantity $r(t)$ relates the polarization of the absorption to the average angle between the pumped and probed transition dipoles: $r(t) = \dfrac{2}{5}\langle P_2(\cos\theta)\rangle$ where $P2(\cos(\theta))$ is the second Legendre Polynomial, θ is the angle between the transition dipole moments, and the brackets, $\langle \rangle$, denote the ensemble average. There is no significant time-dependence to the anisotropy in 2-butanol or n-heptane for any wavelength between 415 nm and 750 nm. This observation places important constraints on the possible involvement of multiple excited states, the transition moment direction does not change during the lifetime of the excited state.

Professor Miller remarked: This question follows up on the discussion on coherent control through conical intersections. My understanding is that it is very difficult to treat the conical intersection. The breakdown in the Born–Oppenheimer approximation at the conical intersection really complicates the level required to properly capture the coupled electronic-nuclear motion. Can you comment on the best approach to treat the dynamics in this region?

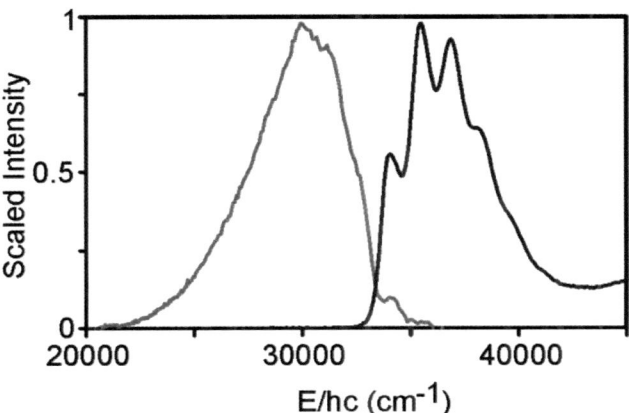

Fig. 1 Absorption (right, blue) and fluorescence (left, green) spectra of DHC in heptane. The spectra are scaled to the same peak height. The fluorescence quantum yield is $\sim 2\times 10^{-4}$.

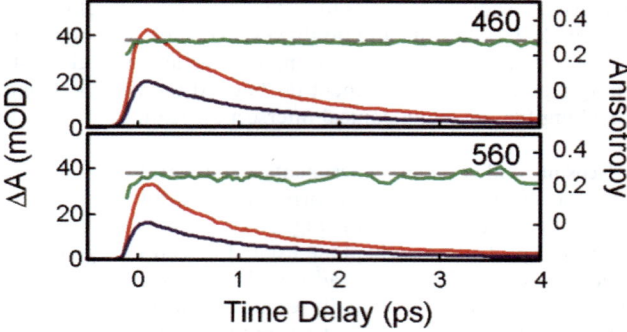

Fig. 2 Absorption anisotropy for DHC in 2-butanol at two select wavelengths given in the figure. The anisotropy (green) is constant over the lifetime of the excited state absorption. The grey dashed line is a constant value for comparison. The parallel (red) and perpendicular (blue) absorption decay traces are also shown.

Professor de Vivie-Riedle answered: The best approach is certainly to propagate (quantum dynamically) vibartional wavepackets on the coupled potential energy surfaces, including the non-adiabatic coupling elements and ensuring that the quality of the electronic structure calculation is high. This approach however is limited in dimension, and for the treatment of photochemical reactions often reactive coordinates (up to five) are selected. In the case of photophysical reactions the Heidelberg MCTDH Program allows quantum dynamical simulations of up to 30 vibrational degrees of freedom. The second best choice allowing to take into account all vibrational degrees of freedom are (at the moment) on-the-fly dynamics strategies, where the motion of the nuclei is treated classically and surface hopping is used for the nonadiabatic transitions.

Professor Motzkus opened the discussion of the paper by Professor Tannor: Could you comment on the experimental constraints of your approach? *E.g.* in your theoretical approach you assume laser pulses as delta-function? How does a realistic bandwidth limited pulse affect the accuracy of your wavepacket/excited state reconstruction?

Professor Tannor responded: If the excitation laser pulse is much shorter than an excited state vibrational period then the delta-function description should be a reasonable approximation. Similarly, if the dump-pulse is much shorter than a ground state vibrational period, the delta-function description should be reasonable. Thus, as an order of magnitude estimate, 5 fs pulses for iodine motion or 1 fs for hydrogen motion should be approximately a delta-function. In cases where the delta-function approximation is not valid, much can still be done. The first-order wavepacket that is created on the excited state surface (the wavepacket we are trying to probe) is actually the convolution of the excitation pulse with the delta-function wavepacket. The entire inversion methodology which we presented for delta-pulse excitation can be applied to finite pulse-duration excitation with this redefinition of the first-order wavepacket that this being probed. The trickier aspect is the effect of finite pulse duration in the dump pulse. Finite duration implies a narrower frequency spectrum, and hence it may be that the excited state wavepacket is dumped only within part of the frequency window spanned by the Franck–Condon overlaps of the moving wavepacket with the ground vibrational states. In this case one may still be able to reconstruct the excited state wavepacket by repeating multiple times the inversion procedure we have laid out, using different central frequencies for the dump pump in order to ultimately span the full frequency window spanned by the Franck–Condon overlaps.

Professor Motzkus asked: In polyatomic molecules you have usually very complex excited state energy surfaces and in order to describe the dynamics correctly one has to include other states which interact the originally excited state. Is the presented method for example able to reconstruct the potential when the originally excited potential is strongly coupled to a dark state?

Professor Tannor responded: In this case it may be possible to make some progress by replacing the dump pulse by an ionization pulse, e.g. to the ground state of the molecular ion. Photoionization is a well-known method for probing dark states, and the vibrational state of the ground electronic state of the ion would then serve as the basis for reconstructing the excited state wavepacket. However, if photoionization occurs with the second pulse it is not clear that a CARS signal can be generated with the third pulse, since the third pulse would somehow have to recombine the ion and the electron.

Dr de Miranda asked: A natural question about the method you have described is: could it work when the dynamics involves several excited states, and if so, how would it work? You have briefly addressed this issue in your paper, but I would like to ask you to clarify the following. Would it be the case that straightforward application of your method to what you have called the "effective" wavepacket resulting from multiple-surface dynamics would lead to a single, "effective" potential energy surface? If so, could that possibly be correct? If not, is it the case that the technical details or the mathematics of the procedure would stop it from producing what in my view would be an unphysical result, that single "effective" potential energy surface?

Professor Tannor answered: At this point, we do not really have a complete picture how to extend our inversion method to multiple excited state potentials. I tend to agree with you that a single "effective" potential energy surface would most likely be unphysical.

Professor Shapiro asked: In the past we have published[1] a formula that allows one to derive the excited potential of any molecule given the frequency resolved fluorescence lines (positions and probabilities) and the (ground state) potential to which the fluorescence occurs. On the basis of the excited state potential thus derived and the time resolved total fluorescence it is then possible to image the (magnitude and phase of a) moving wave packet that gives rise to the time resolved fluorescence. The only stumbling block was our inability at the time to derive the phase (*i.e.*, signs) of the transition-dipole matrix elements since the experiment only gives their magnitudes. Last year[2,3] we were able to overcome this problem and solve this phase problem in an iterative way by starting with a rough guess of the potential, calculating from it the signs of the dipole matrix element, using these signs and the (experimentally determined) magnitudes to obtain the transition dipole amplitudes which when substituted in our potential inversion formula yield in a point by point manner an improved potential. This potential is then used to re-calculate the transition amplitude signs, with the whole procedure repeated until convergence which occurs very rapidly and robustly. More recently we have used this procedure[4] to invert for the first time a multidimensional potential - pertaining to a linear triatomics that has as shown in the attached figures one well and double well structures.

1 M. Shapiro, *J. Chem. Phys.*, 1995, **103**, 1748.
2 X. Li, C. Menzel-Jones and M. Shapiro *J. Phys. Chem. Lett.*, 2010, **1**, 3172.
3 X. Li, C. Menzel-Jones, D. Avisar and M. Shapiro *Phys. Chem. Chem. Phys.*, 2010, **12**, 15760.
4 X. Li and M. Shapiro, *J. Chem. Phys.*, 2011, **134**, 094113.

Professor Tannor replied: Indeed, I think there are certain aspects in common between our methods and certain clear differences. First the obvious differences.

1. Our method is cast in the time domain, where the central object is the moving wavepacket, and your method is cast in the frequency domain where the central objects are the Franck–Condon factors. 2. In our method the wavepacket is recovered first, and from the wavepacket the potential is reconstructed. In your method, the potential is reconstructed first, and from the potential the wavepacket is recovered. Despite these differences, there may be some deeper relationship between the methods. Both methods require knowledge of the ground state potential, and both method belong to a very limited group of wavepacket/potential inversion methods that have the two properties of a) uniqueness and b) not requiring any *a priori* knowledge of the excited state potential. Moreover, it is striking that in both our methods, the phase problem took the same form: finding the sign of the square root of a series of squared coefficients, and that the solution of this problem was the impediment to getting the method to finally work. These similarities may hint at some formal equivalence in the informational content in the two methods, reminiscent of the interrelationship of the interfering pathways and the pump-dump perspectives for laser control. Still, I would like to emphasize certain practical differences in our schemes. 1) In our method, the potential is reconstructed locally—in regions where the wavepacket has significant amplitude. In your method, the reconstruction is global. While global reconstruction of the potential has advantages, it is plausible that the locality of reconstruction in our method gives superior accuracy in the regions where the wavepacket is large. Also, this local reconstruction provides information on precisely the part of the excited state potential that is important for controlling photochemical reactions—the places where the excited state wavepacket visits. 2). Our method can be implemented the same way whether the excited state potential is bound or dissociative. In your method, if the excited state potential is dissociative it becomes less intuitive to work with high resolution Franck–Condon factors. 3) The time-locality in our method may be useful for processes in which the dynamics of interest ends relatively quickly (dissociation, strong radiationless transitions, etc.). The frequency domain may be less convenient for these short-lived processes. 4) In your methods, if I understand correctly, formally one needs the Franck–Condon factors for state-to-state transitions between all ground and all excited vibrational levels. Even for small polyatomics, this requires the measurement of a very large matrix of amplitudes, where the density of states typically becomes enormous at even fairly low energies. In order to measure all these state-to-state transitions one must pre-excite very high overtones and combinations on the ground electronic state, before optical excitation to the electronically excited state. The multi-dimensional CARS experiment we propose for the inversion, although admittedly not easy, may have certain advantages in terms of experimental implementation.

Dr Leibscher asked: Is it possible to apply the proposed method if the excited electronic state for which the potential energy surface shall be reconstructed is coupled to the electronic ground state, for example via a conical intersection?

Professor Tannor replied: In this case, the ground electronic state plays two roles: first, the basis for expanding the time-evolving wavefunction and second, the state on which at least part of the time-evolving wavepacket evolves. I don't see any simple way of applying our method to the portion of the wavepacket that has migrated back to the ground electronic state.

Professor Ashfold asked: The proposed method for reconstructing excited state wavefunctions and potentials for polyatomic molecules from resonant coherent anti-Stokes Raman spectroscopy assumes that the transition dipole moment (TDM) for the second, 'dump', step is independent of nuclear geometry. Such will rarely be the case if the excited molecule samples large regions of configuration space; indeed the TDM will typically approach zero in regions of conical intersection or when approaching a common dissociation asymptote. How restrictive is this limitation likely to be?

Dr Stolow commented: The Condon approximation is a severe one. How will your method deal with the more general coordinate dependence of the electronic wavefunction?

Professor Tannor responded: Several comments. 1) The Condon approximation may be severe in certain cases and not so severe in other cases. The Condon approximation is made quite routinely in studies of electronically allowed transitions in medium-sized molecules. Nevertheless, we hope to be able to extract the coordinate-dependent dipole in our work as well. 2) The coordinate-dependent dipole enters formally in the CARS process both in each of the three interactions with the laser. The first and third interaction present no problems in our formalism: the initial wavepacket is simply defined as $\phi(R)=\mu(R)\psi(R)$. We then invert to $\phi(R,t)$ using our scheme. From $\phi(R,0)$ and knowledge of $\psi(R,0)$ (since the ground state potential is known, so is the form of the initial vibrational state) we can calculate $\mu(R)$ by division. As you indicate, the problem with the coordinate-dependence of $\mu(R)$ is in the second interaction with the laser, since now the projections of the excited state wavepacket are no longer on $\psi_g(R)$ but on $\mu(R)\psi_g(R)$, with unknown $\mu(R)$. 3) We have some ideas for extracting $\mu(R)$ that came up only after we submitted our manuscript to this meeting. Note that the cross-correlation functions $\langle\psi_g|\mu e^{-iHt/\hbar}\mu|\psi_0\rangle$ at $t=0$ are zero (for $g \neq 0$) *unless* there is a non-constant $\mu(R)$. The idea is therefore to use the $t=0$ correlation functions recovered from the CARS signal to find $\langle\psi_g|\mu^2|\psi_0\rangle$. These are the coefficients of the quantity $\mu^2|\psi_0\rangle = \sum |\psi_g\rangle\langle\psi_g|\mu^2|\psi_0\rangle$. Since all the ψ_g are assumed known, and since ψ_0 is assumed known, this seems to be a feasible way to extract μ^2 and ultimately μ. Several details remain to be worked out, including how to extract the +/− sign of μ and issues connected with the CARS signal at zero time delay, but this looks like a promising direction.

Professor Kosloff remarked: You are describing an inversion scheme starting from the CARS signal to obtain the excited state potential. An inversion scheme from experimental data can be extremely sensitive to noise. We found that different parts of the potential have different sensitivity.[1,2] How does your inversion scheme perform.

1 R. Baer and R. Kosloff, Obtaining the Excited State Potential by Inversion of Photodissociation Absorption Spectra, *Chem. Phys. Lett.*, 1992, **200**, 183–191.
2 R. Baer and R. Kosloff, Inversion of Ultrafast Pump Probe Spectroscopic Data, *J. Phys. Chem.*, 1995, **99**, 2534–2545.

Professor Tannor responded: We have done some tests with respect to noise, but so far these tests have not been systematic and they were done only for the 1-d potential inversion. Nevertheless, our preliminary tests indicate that our inversion procedure is quite robust with respect to noise. We would have to check, but it may be the case that in contrast with your observation we will see a uniform sensitivity of the potential in our inversion method. The reason is that our inversion procedure does not extract any spatial information—all of that is already contained in the vibrational eigenstates of the ground electronic state. Our inversion only extracts the time-dependent coefficients. It may be the case that these coefficients have a smooth sensitivity with respect to different parts of the excited state potential.

Professor Miller remarked: The use of a well defined ground state only works for relatively small systems, few atoms. Can you comment on how the accuracy required scales with the size of the molecular system? As a follow up point, if we know the ground state accurately why not do strong field Coherent Control on the ground state potential energy surface. This is where most chemistry occurs. There has been a long standing problem in doing what was previously referred to as laser selective chemistry using IR pulses. The excitation process to access far from equilibrium fluctuations needed to induced the desired chemistry leads to strong anharmonic coupling and rapid IVR. I understand that electronic excited states provide access to far from equilibrium displacements that are key to controlling chemistry but we then have the problem of inversion. I think we need to reconsider strong field control in the ground state. Previous experiments all used a single IR frequency with chirp to try to vibrationally ladder climb up a single coordinate that is generally not the strongest mode coupled to the reaction coordinate. The technology is becoming available to do multidimensional mode coupling. It is this aspect of Coherent Control and its importance in opening up new systems to study using next generation light and electron sources to atomically resolve molecular dynamics that I referred to in an earlier comment.

Professor Tannor answered: I am not sure why you say that the ground electronic state is well-defined only for relatively small systems. As far as I am aware, the concept of a lowest adiabatic electronic state is well-defined for any number of atoms. With respect to the accuracy as a function of system size, at this point we have experience only with one and two-degrees of freedom. The difficulty with increasing the number of degrees of freedom comes primarily from the increase in the vibrational density of states on the ground electronic state. This higher density of states has to be resolved using the Fourier transform of time t_2, the propagation time on the ground state; to achieve higher resolution requires longer time propagation. As you indicate, the problem with strong field coherent control on the ground electronic state is the weak coordinate dependence of the dipole which leads to a picosecond timescale for ladder climbing which is generally commensurate with IVR (intramolecular vibrational relaxation). The advantage of optical control is that if the difference of ground and excited electronic state equilibrium geometries is large, energy can be deposited quickly and in a way that is phase-space localized. This was the motivation behind the original control proposal of Tannor and Rice[1] and Tannor, Kosloff and Rice[2] to use electronic transitions. There is no reason that the vibrational ladder climbing schemes shouldn't be revisited with new technology; at the same time, the bottleneck to optical control has largely been the scant knowledge of excited state potentials, and hence the timeliness of the work I presented here on the inversion process. I would like to reemphasize that the pulse sequence we propose using in our inversion process is closely related to the pulses that might actually be used to control the outcome of the photochemical reaction.

1 D. J. Tannor and S. A. Rice, *J. Chem. Phys.*, 1985, **83**, 5013.
2 D. J. Tannor, R. Kosloff and S. A. Rice, *J. Chem. Phys.*, 1986, **85**, 5805.

Professor de Vivie-Riedle remarked: I do not agree that it is possible to extrapolate form triatomics to polyatomic systems. Three atoms define a plane, four atoms open up the third dimension, including the dihedral angle as new type of motion and so on.

Professor Tannor answered: I confess that this is the first time that I have heard that triatomics are not polyatomics. As far as I know, being married to three wives (or husbands!) constitutes polygamy, without having to take the nuptial vows a fourth time, so why shouldn't a triatomic be considered a polyatomic? Kidding aside, it is clear that triatomics are the place to begin with any new type of polyatomic inversion method. Triatomics already illustrate the basic problem that the number of coordinates is larger than the number of spectroscopic variables in a high resolution absorption spectrum, and therefore the conventional RKR types of spectroscopic inversion methods already break down. This problem is overcome in our inversion scheme by assuming knowledge of the ground state potential in all the coordinates. Although we haven't tested it yet I don't see any reason why the approach won't apply to nonplanar modes as well. Triatomics support conical intersections, and triatomics provide the basic prototype for control of chemical bond breaking since there are already two different bonds. I agree that there are new types on motion once one gets to four atoms, notably torsions and out-of-plane bends. But still I don't see any reason why one cannot in principle have a complete set of vibrational eigenstates that spans all these coordinates and therefore provides the basis for the excited state wavepacket motion. Perhaps your comment is motivated by the importance of out of plane modes in leading to conical intersections with the ground electronic state, but conical intersections with the ground electronic state can occur in triatomics as well.

Dr Marquetand asked: Your method is based on the knowledge of the ground state. How accurately do you have to know the ground state? Would force fields suffice, which would allow you to treat big molecules? Can you give any numbers for the necessary accuracy?

Professor Tannor answered: Generally speaking force fields will not be enough. Force fields are designed to work around the ground state equilibrium geometry. Our method requires the ground state basis to span the Hilbert space far from the Franck–Condon region, in order to provide a suitable basis for expanding the excited state wavepacket when it gets far from the Franck–Condon region.

Professor Motzkus commented: Ground state control using shaped pulses in the mid-IR is still a very promising approach but the reason why this approach still plays a minor role in coherent control of chemical reactions is based on two reasons. One is the available pulse energy in the mid-Ir range and the second reason is the limited bandwidth. The pulse energies of shaped mir-Ir pulses are still in the range of few Microjoules which allow ladder climbing only in those vibrational bonds which feature very strong dipole moments. The prototype system here is still the carbonyl bond. This vibration is also very harmonic so that vibrational transitions between higher quantum numbers are covered by the laser spectrum of a 100 fs pulse. However this ladder climbing considers only one vibrational mode so far. Since the reaction coordinate is usually a superposition of several vibrational modes the excitation of at least two phase correlated modes is necessary which are typically well separated in frequency, e.g. a high frequency asymmetric stretching mode and low frequency bending mode. The corresponding experiment is quite challenging particular with respect to the currently available laser technology. However there is no doubt that this will be only a question of time.

Professor Miller responded: This discussion is related to my earlier comment on the importance of coherent control along the ground state surface. In the IR spectral region, it is possible to execute strong field control of molecular systems by virtue of steering vibrational wavepackets without artifacts from multiphoton ionization as occurs using electronic transitions. As stated by Professor Motzkus, the femtosecond IR laser technology is improving. I hold out great hopes for this prospect as it will open up all molecular reactions to atomic level inspection using the new "ultrabright" electron and X-ray structural probes to directly observe the light driven atomic motions. The one concern I have is the seemingly general involvement of strongly damped low frequency modes to couple reaction coordinates (see discussion on the role of low frequency modes in the 200 cm^{-1} range with respect to paper 13).[1] This frequency range will not be accessible in the foreseeable future with sufficient bandwidth and energy to effectively couple reactive modes. My hope is that the anharmonic coupling to these low frequency modes can be still be manipulated through addressing the higher frequency vibrational modes and overtones to take advantage of this interaction. I think we are in for some interesting times.

1 T Buckup, J. Hauer, J. Voll, R. de Vivie-Riedle and M. Motzkus, *Faraday Discussions*, DOI: 10.1039/C1FD00037C.

Professor Ohmori opened the discussion of the paper by Professor Shapiro: Suppose you make a coincidence measurement of the photon and material states with a photon number detector whose resolution is not good enough to distinguish those one and two photon processes. Does the interference appear in such a measurement?

Professor Shapiro replied: If in a given setup we have the possibility of obtaining the "which way" information then no interference fringes should exist. If however our device (e.g., an electron going through two gates with variable charging), is such that we only have partial "which way" information no matter how good the detector is, then interference fringes should begin to appear, depending on the degree of inability to tell which pathway the particle took.

Professor Brumer said: It is irrelevant whether your measurement apparatus has, or has not, enough resolution to distinguish between the two pathways. Quantum theory is very clear: if it is in principle possible to distinguish the two pathways, then the interference does not appear.

Professor Tannor remarked: In the field of quantum information, we are often given the impression that entanglement is desirable. Your work drives home the point that entanglement can be undesirable, since it can kill quantum coherence and hence coherent control. In fact, it seems to follow from your work that from the point of view of controlling the matter, the best possible situation is to create a final state that is separable in the matter and the field. This suggests defining as your objective a separable product state in the matter and the field. Have you considered defining a control problem where this is the objective?

Professor Shapiro replied: I agree. We have in fact made this point several times in the past but have not tried to pose this separability condition as an objective in an optimal control scheme. Such a study would definitely be of interest.

Professor de Vivie-Riedle said: You said that entanglement is equal to correlation. Is then electron correlation a form of entanglement?

Professor Shapiro answered: Entanglement implies correlation but we usually reserve this term to quantum correlation between non-interacting particles. If the particle are close enough to interact there is no "wonder" that they are correlated. The aspects of correlation between non-interacting particles (resulting from their mode of preparation) is what is usually termed entanglement.

Professor Rabitz asked: In Section 3 of your paper is the argument equivalent to stating that pure states may not be converted to mixed states under unitary evolution?

Professor Shapiro replied: The issue is not the purity but the ability to execute interference based control. In section 3 of the paper we show how the interference terms naturally decay as we add more channels to the subspace we wish to control. This decay of the interference terms has nothing to do with purity since in this derivation all the states, even those for which the interference terms nearly vanish, are pure.

Dr Leibscher asked: What are the experimental conditions to create quantum light in a number state? Is it possible to experimentally observe the switch from a regime of quantum light (in a number state) where coherent control is not possible, to classical light (or coherent photon state) where coherent control works?

Professor Shapiro responded: Number states have been created by a variety of techniques such as in cavity QED. (see the works of H. Walther and also of S. Haroche on this topic).[1–3] I suppose that the best way of seeing the disappearance of coherent control is to go to very low number of photons where the description of the field becomes more and more quantized, making it possible to tell apart the states of the field upon absorption of one vs. two photons.

1 H. Walther, *Fortschritte der Physik*, 2003, **51**(4–5), 521–530 and references therein.
2 J. M. Raimond, M. Brune, J. Lepape and S. Haroche, "Rydberg atoms in a cavity: a new method to generate photon number states" in *Laser Spectroscopy IX*.
3 M. Brune, S. Haroche, V. Lefevre, J. M. Raimond and N. Zagury, *Phys. Rev. Lett.*, 1990, **65**, 976–979.

Professor Kosloff said: You describe very interesting results of the influence of quantum light on coherent control. How would you extend your theory to broadband light fields?

Professor Shapiro answered: The results can be immediately extended to multimode broadband pulses. If such pulses are constructed from number states and their measurement enables the determination of the "which way information" no interference-based control is possible.

Dr Amitay opened the discussion of the paper by Mr von den Hoff: The initial laser field that was chosen for the present OCT optimization of the $4^1\Sigma_g^+$ selective population is a weak 911 nm 20 fs Gaussian pulse. Could you please comment how robust you think is the obtained SPODS pulse solution with respect to the chosen initial laser field?

Mr von den Hoff responded: In my calculations I found out, that the optimization of the $4^1\Sigma_g^+$ via the SPODS mechanism is very robust with respect to the initial intensity, pulse duration and central frequency. In case of the $5^1\Sigma_g^+$ state optimization, a high intensity the initial laser field is needed to populate this state, which is necessary for the OCT algorithm to perform. Due to this high intensity initial Rabi oscillations as well as intermediate population in the $4^1\Sigma_g^+$ state are unavoidable. To filter out the pure SOPDS mechanism the most easiest way is to reverse the temporal phase of the optimal pulse found for the $4^1\Sigma_g^+$ state and select it as initial pulse for the $5^1\Sigma_g^+$ state optimization. In this case the solution is found in few cycles also for the $5^1\Sigma_g^+$ state.

Professor Tannor asked: I find the mechanism you have told us about, for selective population of the $4^1\Sigma_g^+$ vs. the $5^1\Sigma_g^+$ state fascinating. But since these two electronic states are non-degenerate, if one's goal is simply the selective population of one of these states, couldn't one reach this goal just by using CW excitation (either one- or two-photon) and changing the excitation frequency to resonance with the $4^1\Sigma_g^+$ or $5^1\Sigma_g^+$ state?

Mr von den Hoff answered: Certainly the selective population of the $4\Sigma_g$ or $5\Sigma_g$ state is also possible by using CW excitation. There will be also different solutions in the OCT search space, depending on the initial conditions and laser interaction time chosen. In our presented OCT studies we were interested in the question whether, or under which circumstances, the algorithm finds the SPODS mechanism as a robust pathway to populate the target states. In addition we think that the SPODS scheme is a strategy which nicely combines the routes of phase and strong field control and which opens a wide spectrum of applications ranging form reaction control within molecules up to discrimination between different molecules in a mixture.

Dr Stolow remarked: To what extent do the linear (dipole) *versus* quadratic (polarizability) dynamic Stark effect play a role in these dressed state (or Floquet) mechanisms? In near resonant cases, I would expect the linear Stark interaction to dominate, in which case rapid passage (crossing of Floquet states) will likely obtain. But there still may be strong non-resonant contributions which affect the evolution in the field. In other words, in terms of Floquet states, there may be both adiabatic and non-adiabatic evolution, as discussed, for example ref. 1 and following papers. Can not the molecular dynamics itself, via the second order Stark interaction for example, modify spacings between Floquet states and therefore lead from one regime of Floquet evolution to another?

1 H. P. Breuer, K. Dietz, and M. Holthaus, *Nuovo Cimento*, 1990, **105B**, 53.

Mr von den Hoff responded: The mechanism "selective population of the dressed states" only works, when the laser frequency of the second sub pulse is exactly in resonance to the previous electronic excitation and thereby with the induced oscillating dipole moment. From our simulations we learned that the selectivity is essentially lost, when this requirement is not fulfilled. Thus we expect the linear dynamic Stark effect to play the major role in the SPODS mechanism.

Professor Baumert asked: During the discussion of this paper you raised the question how to distinguish between high order spectral interference (perturbation theory) and other strong field effects like the selective population of dressed states.[1] Experimentally we usually perform intensity dependent measurements to that end. If the signal scales in a nontrivial way (especially beyond power laws), it is likely that a strong field scenario is at play. We studied this systematically on the strong field excitation of potassium atoms (see Fig. 3 and 4 in ref. 2). I showed in my lecture also, that in our molecular experiments on potassium dimers that triggered this theoretical contribution our molecular signal scales in such a non trivial way.

1 M. Wollenhaupt, D. Liese, A. Präkelt, C. Sarpe-Tudoran and T. Baumert, Quantum control by ultrafast dressed states tailoring, *Chem. Phys. Lett.*, 2006, **419**, 184–190.
2 M. Wollenhaupt, A. Assion, O. Bazhan, C. Horn, D. Liese, C. Sarpe-Tudoran, M. Winter and T. Baumert, Control of interferences in an Autler-Townes doublet: Symmetry of control parameters, *Phys. Rev. A*, 2003, **68**, 015401–015404.

Professor Weinacht addressed Professor Shapiro and Mr von den Hoff: I don't understand why coupling to the continuum is required in the calculations. I would have thought that for this problem including only the bound states is sufficient for an accurate description of the dynamics.

Professor Shapiro replied: The effect has two aspects: 1) The existence of (Autler-Townes) splitting between two levels due to the presence of a coupling laser; and 2) each split level that is being broadened by the interaction with the ionization continuum due to the ionizing laser becoming a "resonance". This causes each level to interfere with the other in a phenomenon known as "overlapping resonances". Since the broadening that brings about this interference is influenced by the (ionizing) laser, any change in the (shape, strength) of the ionizing laser will affect the way these two resonances overlap and interfere.

Mr von den Hoff replied: The shown dynamics of the potassium dimer is indeed governed by the bound state dynamics. Nevertheless, in our numerical treatment the continuum states are inherently included and used in the transition mechanism whenever nneeded

Professor Kosloff said: What is the efficiency of the process?

Mr von den Hoff replied: The optimization of the $4\Sigma_g$ target state, starting from an unshaped gaussian laser pulse, delivers an efficiency of 98.3%. The optimization of the $5\Sigma_g$ target state using the optimized pulse for the $4\Sigma_g$ target state with reversed phase as initial guess, delivers an efficiency of 96.7%. The optimization of the $5\Sigma_g$ target state, starting from scratch with an unshaped gaussian laser pulse, delivers an efficiency of 96.5%. But in the latter case the SPODS mechanism is not the dominant mechanism any more.

Professor Whitaker opened the discussion of the paper by Professor Janssen: In Fig. 9 of your paper you suggest that two photon excitation around 540 nm can excite a wavepacket on the A state of CF_3I which will rapidly evolve on the dissociative potential energy surface where it can continue to interact with the laser field to be promoted to higher lying Rydberg states and eventually the ionization continuum. The fact that you observe (Fig. 4 of your paper) an increase in the CF_3^+ signal with respect to the parent ion signal that is symmetric in linear chirp, positive or negative, suggests that lengthening the laser pulse duration enhances the fragmentation signal. This would be consistent with wavepacket motion with a time delayed resonance in one of the subsequent excitation steps. By contrast, a wavepacket following mechanism should exhibit a signed chirp dependence. One way to test this would be to apply a pulse pair or an asymmetric phase mask (leading to a double pulse in the time domain). It might also be informative to scan a pi-step phase flip across the laser excitation spectrum as this would have a differential effect on any two and three photon resonances. I wonder if you have attempted such experiments yet or have plans to do so?

Professor Janssen responded: So far, we have only been able to do a few experiments using double pulses or a pulse train. We do see an increase of the CF_3^+ signal for certain pulses as shown in Fig. 5 of the paper. We are planning to do more extensive experiments also including pi-step phase flip experiments.

Professor Fielding asked: When you determine the maximum electron kinetic energies expected from a particular ionisation process, it seems (from figure 9 of the paper) that it is assumed that the potential energy surface of the ion is flat, i.e. independent of the nuclear coordinates.

Professor Janssen answered: To calculate the maximum available energies we use the available data from the literature on the ionization energies of the CF_3I^+ parent molecule and the various fragment channels. We do not assume anything about the nuclear dependence of the potential energy surface of the ion states. The lines shown in Fig. 9 in the paper are schematic only and used to indicate the position of the ionization energy as reported in the literature.

Professor Shapiro asked: Y. Silberberg suggested a simple mechanism by which a negative (positive) chirp can change the photo-ions kinetic energy distributions: The pulse starts by exciting the high (low) energy photo-fragments causing a depletion in the ground state population. As time goes on and the chirped frequency goes down (up), the frequencies in the later part of the pulse become less effective. The net effect is that kinetic energies of products created at the early part of the pulse are being enhanced at the expense of energies created in the later part of the pulse.

Professor Janssen replied: In our experiments we do not observe much change in the kinetic energy of the CF_3^+ fragment. They all come out with very low kinetic energy. This seems to suggest a much more statistical dissociation mechanism on the ground state potential of the parent ion.

Professor Baumert remarked: Regarding the symmetric behaviour of your signal with respect to chirp hints to a pulse duration or intensity effect instead of a dynamical resonance. However the pulse sequence experiments hint to dynamics. So far both observations together do not give a conclusive answer. There is more information needed either being experimental or on the modelling side. If experimentally, what additional experiments perhaps also with other designed pulse shapes would have to be performed?

Professor Janssen responded: We plan on doing more extensive experiments with pulse pairs, and especially at various center wavelengths. In addition we do hope that theoretical calculations on potential energy surfaces and dynamics will further help us to guide more pulse shaping experiments. We are very open to any other suggestions of possible pulse shapes that may help to elucidate the control mechanism.

Dr Stolow asked: How can you make use of the fully correlated 6-dimensional photoion-photoelectron vector information that you have available in order to discern control mechanisms? As one example, in cases where axial recoil applies, will transformation to the recoil frame (as in O. Gessner *et al.*[1]) assist in understanding control mechanisms and, if so, in what way will this work?

1 O. Gessner *et al.*, 2006, *Science*, **311**, 291.

Professor Janssen responded: For the CF_3I and CH_2BrCl systems that we studied so far with pulse shaping and coincidence imaging, the fragment ion is formed with almost zero kinetic energy, very slow. The lab frame angular distribution is very isotropic. Therefore, it is very difficult to make meaningful recoil-frame photoelectron (RFPAD) angular correlations, as the labframe velocity of e.g. the CF_3^+ fragment is near zero. We think that to make meaningful RFPAD one needs a fragment with non-zero velocity, representing much more of a (fast) axial recoil. It such systems it may be much more helpful to also analyze the RFPAD and compare this with theoretical calculations.

Professor Weinacht commented: I wonder if a good test of whether it is simply the pulse duration which is driving the control (instead of the time varying frequency of the chirped pulse) would be to limit the bandwidth of the control pulse such that you get a pulse of similar duration to a chirped pulse, but with a limited bandwidth and no time varying frequency.

Professor Janssen answered: That is a good suggestion that we haven't tried yet, but will do so in the near future with our LCD pulse shaper.

Professor Ashfold asked: Janssen and co-workers describe a welcome step towards unravelling the mechanism(s) that can influence the branching between parent and fragment ions in the multiphoton excitation of polyatomic molecules like CF_3I. Analysis suggests that the achieved control (in CF_3^+/CF_3I^+ branching ratio) is imposed by excited states of the neutral molecule resonant at the energy of two and three absorbed photons, and the authors conclude with an appeal for high level ab initio calculations of the excited state potentials at the relevant energies. A-band absorption is a characteristic feature of all organic iodides, and has now been investigated in some detail for CH_3I,[1,2] CF_3I,[3] C_6H_5I,[4] and cyclo-$C_6H_{11}I$.[5] In each case, eight states correlate to ground ($^2P_{3/2}$) state I atoms upon C–I bond fission. A further four states correlate to the spin-orbit excited (I*) limit. One photon excitation of

CF_3I at $\lambda \sim 270$ nm populates predominantly the $^3Q_{0+}$ state; subsequent dissociation yields mainly I* products with a quantum yield $\Phi_{I^*} \sim 0.9$.[2] The present experiments access this same energy region by two photon absorption. Excitation to the $^3Q_2$ repulsive state would also be allowed under such circumstances, though the cross-section for this excitation (relative to that for two photon absorption to the $^3Q_{0+}$ state) remains to be established.

1 A. T. J. B. Eppink and D. H. Parker, *J. Chem. Phys.*, 1999, **110**, 832.
2 A. B. Alekseyev, H. P. Liebermann and R. J. Buenker, *J. Chem. Phys.*, 2007, **126**, 234102–234103.
3 G. Hancock, A. Hutchinson, R. Peverall, G. Richmond, G. A. D. Ritchie and S. Taylor, *Phys. Chem. Chem. Phys.*, 2007, **9**, 2234 and references therein.
4 A. G. Sage, T. A. A. Oliver, D. Murdock, M. B. Crow, G. A. D. Ritchie, J. N. Harvey and M. N. R. Ashfold, *Phys. Chem. Chem. Phys.*, 2011, **13**, 8075.
5 D. K. Zaouris, A. M. Wenge, D. Murdock, T. A. A. Oliver, G. Richmond, G. A. D. Ritchie, R. N. Dixon and M. N. R. Ashfold, *J. Chem. Phys.*, 2011, **135**, 094312.

Professor Janssen answered: I would like to add that besides the many studies that have been done on the A-band there is also a lot done on the various Rydberg states of CF_3I. We hope that this will also help in elucidating the mechanism of pulse shaping control in CF_3I.

Professor Kosloff asked: Examining the chirp effect in Figure 4 of your paper we see that both positive and negative chirp enhance the fragmentation. In our calculations on Mg_2 photoassociation we find such a chirp effect in strong field two-photon transitions. Would you support such an interpretation for your system?

Professor Janssen answered: We know from all the spectroscopic and photodynamics studies on CF_3I that at the two-photon level we are resonant with the A-band. See also the earlier comment by Professor Ashfold. I cannot really say if there are similarities between your Mg_2 photoassociation study and the control experiment on CF_3I. In our study on the ionization and fragmentation dynamics of CF_3I we learn from the energetics and our data that we are dealing with a five photon process. We see effects on the fragmentation ratio not only from chirped pulses but also from double pulses and pulse trains. In the absence of theoretical calculations we cannot say much more conclusive at this moment on the exact mechanism, so we can neither support nor deny if there is a similar interpretation as in your Mg_2 photoassociation study.

Professor Tannor asked: There are a number of dihalogenated molecules that have been the subject of studies of photoselective chemistry, e.g. CH_2IBr and C_2F_4IBr. I realize that the experiments you have done are still preliminary and already quite challenging and pose interpretational problems. Nevertheless, can you give us some prognosis of applying your imaging method to these dihalogenated molecules and therefore to the control of chemical branching?

Professor Janssen responded: In a recent pulse shaping study[1] on CH_2BrCl, we demonstrated a strong enhancement of the CH_2Cl^+/CH_2BrCl^+ ion ratio with up-chirp. We need to study more different pulse shapes to investigate the effect of specific pulse sequences on the chemical branching. Furthermore, we hope that theoretical studies on the potential energy surfaces of neutral excited states (collaboration with Professor Leticia Gonzalez and coworkers) will provide further help and guidance. We are still in the early stages of studies, so it is still somewhat challenging to elucidate the precise mechanism in the chemical branching control in these systems.

1 D. Irimia and M. H. M. Janssen, *J. Chem. Phys.*, 2010, **132**, 234302.

Professor de Vivie-Riedle said: In the early days of coherent control experiments wavelengths around 800 nm were used for the initial pulse due to experimental constraints and regardless whether they are in resonance or out of resonance with the optical transitions involved in the molecular processes under control. Few groups like the group of Marcus Motzkus specialized on pulse shaping with tunable lasers. Would your control experiment benefit, when starting with light fields that are resonant with the prominent transitions?

Professor Janssen answered: We see significant effects of the centre wavelength on the fragmentation ratio, see Fig. 3 in the paper. This seems to suggest to us that resonances do play a role.

Professor Baumert remarked: During the discussion of this paper you raised the general question about pros and cons for UV *vs.* VIS/IR control. To my opinion it is advantageous in a closed loop experiment to stay with VIS/IR excitation as the density of states is increasing with increasing energy. So once in this region, there are many potentials that can be exploited in multiple pump-dump steps or resonantly changed via the Stark effect to guide the system into the desired target state. If there are worries that in such a case the initial non resonant steps require such an intensity that the system will be immediately ionized, an initial excitation with a UV photon could be employed. As soon as ultra broadband laser pulses can be generated and properly shaped, this topic could be addressed also experimentally. Until then, dedicated OCT simulations might also give in sight.

Professor Weinacht remarked: Our experience with similar molecular systems seems to be consistent with what Maurice sees, in that at 800 nm, coupling between ionic states plays a more important role than coupling between neutral states. Ionization proceeds rapidly in a strong non-resonant laser pulse and then resonances in the ion can play an important role in the fragmentation of the molecule.

Professor Janssen answered: I fully agree that the specific dynamics is very dependent on the wavelength. For instance, if we compare the TOF-mass spectra in the pulse shaping experiment of Gerber and coworkers[1] on CH_2BrCl we observe that they are very different. The mass spectrum at 800 nm only shows both CH_2Cl^+ and CH_2Br^+ fragment ions and no parent ion, whereas the spectrum at 400 nm shows both the CH_2Cl^+, CH_2Br^+ fragments and the CH_2BrCl^+ parent ion. The CH_2Br^+ yield is very small, and the CH_2Cl^+/CH_2BrCl^+ ion ratio is about 3 : 1 at 400 nm, very similar to the CH_2Cl^+/CH_2BrCl^+ ion ratio that we observed in pulse shaping control experiments at 509 nm.[2] It seems very clear that the centre wavelength at which the control experiment is carried out plays a very important role in the mechanism.

1 N. H. Damrauer, C. Dietl, G. Krampert, S.-H. Lee, K.-H. Jung and G. Gerber, *Eur. Phys. J. D*, 2002, **20**, 71.
2 D. Irimia and M. H. M. Janssen, *J. Chem. Phys.*, 2010, **132**, 234302.

PAPER

A General control mechanism of energy flow in the excited state of polyenic biochromophores

Tiago Buckup,[a] Jürgen Hauer,[b] Judith Voll,[c] Regina Vivie-Riedle[c] and Marcus Motzkus[*a]

Received 8th March 2011, Accepted 24th March 2011
DOI: 10.1039/c1fd00037c

Quantum dynamics in photobiology is a highly controversial subject of modern research. In particular, the role of low-frequency vibrational coherence of biochromophores has been intensely discussed. Coherent control of polyenic chromophores, like carotenoids and retinoids, has been showing that the manipulation of such low frequency coherences may play a crucial role in the evolution of excited population and therefore in the efficiency of photosynthesis. However, no precise control mechanism has been derived. In order to clarify this open question, we combined quantum dynamical modelling with a sensitive experimental technique, namely Pump-Degenerate Four Wave Mixing (Pump-DFWM). In this work we investigate in detail the internal conversion channel of β-carotene, an important polyenic chromophore, under multipulse excitation and focus on the role of the non-adiabatic coupling between excited-state potentials and the internal energy loss. Our control mechanism is based on the interference between wavepackets in the excited state, which leads to a transient evolution of the vibrational population dependent on the relative phase between excitation sub-pulses. Such a transient evolution can affect the branching ratio between competing channels in the excited state. Therefore, our results are able to rationalize pulse shapes found in a whole class of coherent control experiments involving polyenic biochromophores, like in light harvesting complexes and in bacteriorhodopsin.

1 Introduction

Understanding ultrafast energy flow in the excited state of biological chromophores is an important step in the technological utilization of photosynthesis. The efficiency of the transformation of light into chemical energy is mainly determined by the competition between the ultrafast energy loss and reactive channels.[1,2] In this regard, intramolecular deactivation processes in the excited reactive states of donor molecules are not desirable and should be identified and possibly suppressed for an efficient photosynthesis.

The identification, however, of such deactivation and reactive channels in biomolecules and how they interplay to make photobiological processes highly efficient are still a challenge in time-resolved spectroscopy.[3-5] Nature through evolution has been exploiting efficient energy transfer mechanisms by selecting ultrafast processes

[a]*Physikalisch-Chemisches Institut, Ruprecht-Karls-Universität Heidelberg, Im Neuenheimer Feld 229, D-69120 Heidelberg, Germany. E-mail: marcus.motzkus@pci.uni-heidelberg.de; Fax: +49 6221 54-8730; Tel: +49 6221 54-8726*
[b]*Faculty of Physics, University of Vienna, Strudlhofgasse 4, 1090 Vienna, Austria*
[c]*Department Chemie, Ludwig-Maximilian-Universität München, Butenandt-Strasse 11, D-81377 München, Germany*

between acceptor and donor chromophores to minimize the effect of interaction with the bath. Unfortunately, such a short time scale leads to a very narrow experimental time window where the desired dynamics can be detected. An additional challenge is posed by the transition dipole moments of the involved states: often reactive excited states show strong light absorption from ground state but none or weak excited state absorption. Such weak transition dipoles can make the probing of the respective reactive states particularly demanding. Furthermore, even when these issues are not the main concern or can be avoided, the ultrafast dynamics of photobiological processes are often difficult to disentangle: Transient data may not just contain contributions from degrees of freedom decisive for the reaction itself, but may also include contributions from, *e.g.*, spectator vibrational modes or unconnected energy relaxation.

A modern method to circumvent the issues discussed above and, therefore, assist the investigation of the initial ultrafast steps of photosynthesis is based on the combination of time-resolved spectroscopy and tailored excitation.[6–12] In this field, specially shaped laser fields can select or suppress a transient spectral signature, leading potentially to the identification of all internal degrees of freedom directly involved in an energy transfer reaction.

In this context, the ultrafast energy transfer and losses in biological complexes involving a main chromophore with polyenic structural design are still intensely debated. A major point regards the role of reactive vibrational modes. The excitation of natural[9,13] as well as artificial[14] light harvesting complexes with different carotenoids showed that deactivation and, in one case, energy transfer can be controlled with tailored pulses in form of a sequence of pulses (multipulse excitation—see Table 1 for optimal and anti-optimal sub-pulse separations). More recently,[15] an independent experiment on the isomerization of retinal showed a similar dependence on the multipulse excitation. The optimal pulse shapes obtained in these control experiments suggested a general mechanism where low

Table 1 Coherent control on polyenic biochromophores: Optimal results are for optimization of energy transfer over internal conversion on the donor chromophore. Frequency values in cm^{-1} are obtained from the sub-pulse separation in the optimal or anti-optimal multipulse shape. N is the number of conjugated double bonds

System	Natural Light Harvesting Complex LH2[13]	Artificial Light Harvesting Complex[14]	Bacteriorhodopsin[15]
Chromophore	Rhodopin glucoside ($N = 11$)	β-carotene derivative ($N = 10$)	All *trans*-retinal ($N = 6$)
Chromophore structure			
Sub-pulse separation optimal	N.A.	300 fs (110 cm^{-1})	145 fs (230 cm^{-1})
Sub-pulse separation anti-optimal	210 fs (160 cm^{-1})	200 fs (166 cm^{-1})	215 fs (155 cm^{-1})

frequency modes play a crucial role in the evolution of excited population. However, the dynamical details of the excitation mechanism remain unclear. Historically, multipulse excitation has been interpreted in a "child-on-a-swing" picture, where the vibrational wavepackets temporally in phase with the sub-pulse separation will be promoted while out-of phase vibrational wavepackets will be suppressed.[16] In spite of this mechanistic explanation, this model misses an important point: Although out-of-phase vibrational wavepackets are suppressed by an appropriately spaced multipulse, it still bears the same linear spectrum as a Fourier-limited pulse. Therefore, the respective vibrational levels are still populated even if no wavepacket is formed.[17,18] This is an important aspect given that energy flow in the excited state will take place even if vibrational coherence is weak.

Another open question concerns the electronic state of origin of the observed wavepacket. Via impulsive stimulated Raman scattering, a sufficiently broadband excitation pulse may generate wavepackets either in the electronic excited- or in the ground-state. In the case of a multipulse excitation, as in control experiments for polyenic chromophores, this can lead to different interference effects between the wavepackets generated by each sub-pulse of the sequence because Franck–Condon modes may have different frequencies in each potential surface. Therefore, the "child-on-a-swing" picture does not give any mechanistic explanation for how the interference between these wavepackets may lead to a modulation of energy transfer or loss in the excited-state. Moreover, the nature of the interplay between excited state lifetime and sub-pulse separation remains elusive. Sub-pulse separations much longer than excited-state population lifetime or the electronic coherence time should not have any effect on the molecular dynamics since the system already relaxed before the next sub-pulse arrives at the sample. In situations where strong control over population loss/energy transfer is observed but the sub-pulse separation is on the order of excited population or coherence time, it is not apparent whether the "child-on-a-swing" picture is applicable as it fails to describe the quantum nature of the process.

In order to address this fundamental question about the control of energy flow in biochromophores, we combined quantum dynamical modelling with a very sensitive experimental technique, namely Pump-Degenerate Four Wave Mixing (Pump-DFWM).[19–23] Within this method, the multi-beam DFWM acts as a Probing Sequence (PS) which is free of ground state contributions since Pump-DFWM exploits excited state resonances missing in the static absorption spectrum. In this work we investigate in detail the internal loss channel of β-carotene, an important polyenic chromophore, under multipulse excitation and focus on the role of the non-adiabatic coupling between excited-state potentials and the internal energy loss.

2 Experimental details

Theoretical approach[24]

β-carotene, and carotenoids in general, present two low-lying excited states, namely S_2 and S_1 states.[3] The S_2 state is the first one-photon allowed electronic state. In nature and artificial systems, the competition between S_2–S_1 relaxation and the energy transfer between the S_2 state and acceptor molecules (e.g. (bacterio-) chlorophylls or purpurins) is a decisive step for the photosynthetic efficiency. While the energy transfer is described as a dipole–dipole interaction (Förster energy transfer), the mechanism of internal relaxation is still not fully understood. In spite of the controversial presence of additional electronic states between S_2 and S_1 states, we have chosen to describe the coupling between S_2 and S_1 states without additional excited electronic states.[25] The validity of such a choice will be discussed later.

In order to describe the excited state dynamics on S_2 surface in the context of wavepacket tailoring, two representative modes ($x = 178$ cm^{-1}—the electronic coupling process driving mode, and $y = 1157$ cm^{-1}—a fast spectator mode) out of a ground state normal mode analysis (DFT/b3lyp, 6-31G*) of β-carotene, are selected to set up the Hamiltonian in the diabatic representation:

$$H(x,y) = T(x,y) + \begin{pmatrix} S_0(x,y) & 0 & \varepsilon(t)\mu_{02} \\ 0 & S_1(x,y) & C_{21}(x,y) \\ \varepsilon(t)\mu_{02} & C_{21}(x,y) & S_2(x,y) \end{pmatrix} \quad (1)$$

Included are the three electronic potential energy states (PES) of β-carotene (S_2, S_1 and S_0), the transition dipole moment μ_{02} between S_0 and S_2 states and the non-adiabatic coupling element $C_{21}(x, y)$. The intersection between S_2 and S_1 states, defined by the choice of their displacement parameters, builds the centre of the Gaussian shaped coupling element C_{21}. Finally, the simulations are performed by solving the open system Liouville von Neumann equation:

$$\dot{\rho}(t) = L\rho(t) = -i[H, \rho(t)] + L_D(\rho(t)) \quad (2)$$

Solvent effects are described by the dissipative part of Lindblad form introducing vibrational relaxation times of $T_{1v} = 5$ ps, pure vibrational dephasing times of $T_{2v} = 8$ ps and pure electronic dephasing times of $T_{2e} = 600$ fs for the appropriate fundamental transitions. Dissipation rates for higher transitions inside one mode are assumed to be proportional to the vibrational quantum number respective the energetic distance. To simulate the fast internal vibrational redistribution (IVR) processes on S_1, we introduced an additional fundamental vibrational relaxation time of $T_{1,IVR} = 20$ fs deduced from the experimentally observed exponential rise of the DFWM signal intensity. Inclusion of $T_{1,IVR}$ leads to an effective electronic dephasing time of 200 fs between the two excited states S_2 and S_1.

Experimental setup

In order to simulate/replicate the control experiments in polyenic chromophores with multipulse excitation we combined an open-loop control setup with the

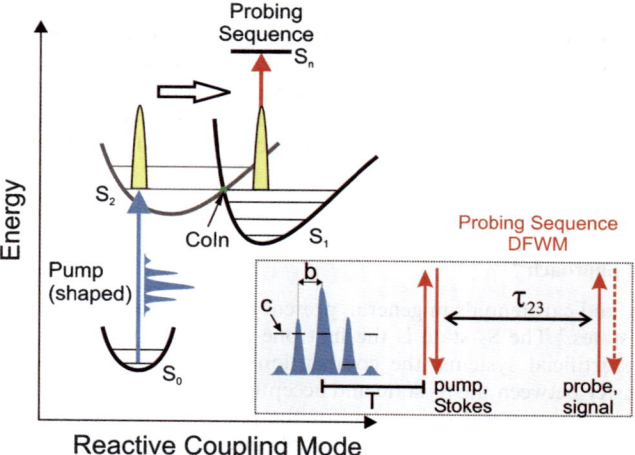

Fig. 1 Potential energy surfaces describing the early photophysics of β-carotene in solution. The dynamics are defined by a conical intersection (CoIn). The employed excitation scheme involves a pump pulse (blue), phase modulated for coherent control, and a probing sequence (red) used for detection purposes only. Inset shows schematically the sub-pulse separation b and the phase parameter c. The initial pump pulse is tailored with a sinusoidal phase. After a delay T, a degenerate four-wave-mixing is used as probe sequence. The delay between pump and Stokes was kept at 0 while the delay τ_{23} was 170 fs.

Pump-DFWM technique. In the Pump-DFWM experiment a tailored pump-pulse (Initial Pump (IP)) pre-excites β-carotene in the Franck–Condon (FC) region of the S_2 electronic state. The relaxation dynamics is probed at a delay T with a probing sequence (PS) which consists of three beams (pump/Stokes/probe) forming a time-resolved degenerate four-wave mixing (DFWM) scheme (Fig. 1). The probe pulse of this FWM sequence is independent and can be delayed (τ_{23}) with respect to the timely coincident pump and Stokes pulses ($\tau_{12} = 0$). The use of the DFWM sequence as probe instead of a single pulse, like in transient absorption (TA), allows the detection of small population variations in the excited state. DFWM, as a background-free nonlinear time-resolved technique, shows a quadratic dependence on the population variation in the excited state (Δn^2). A single probe pulse as in TA is not as sensitive to signal variations as the DFWM method because it shows a linear dependence on the population variation (Δn).

The initial Pump and DFWM sequence were generated using two separate single staged non-collinear parametric amplifiers (nc-OPA). The initial pump was centred around 510 nm with a temporal pulse width of 15 fs as determined by an autocorrelator. The DFWM sequence was obtained by splitting the output of another nc-OPA into three parts of equal intensity. The spectrum of the DFWM sequence was centred around 560 nm with duration of 13 fs. Before interacting with the sample, the beams were focused by a concave mirror with 30 cm focal length, resulting in a beam radius of 50 μm in the focus. The DFWM signal was detected by a photomultiplier after an interferometric filter centred at 610 nm. Average excitation energies were 40 nJ for the initial pump (1.3×10^{15} photons cm^{-2}) and below 10 nJ (3.5×10^{14} photons cm^{-2}) for the pulses in the DFWM-sequence, which ensured a linear regime of excitation. At these experimental conditions, the initial pump promoted less than 4% of the molecules to the S_2 state. Fig. 2 shows the absorption of β-carotene dissolved in cyclohexane and the experimental excitation spectra. There is no spectral overlap between the linear absorption spectrum and the DFWM spectrum. The DFWM spectrum is resonant with the excited state S_1–S_n

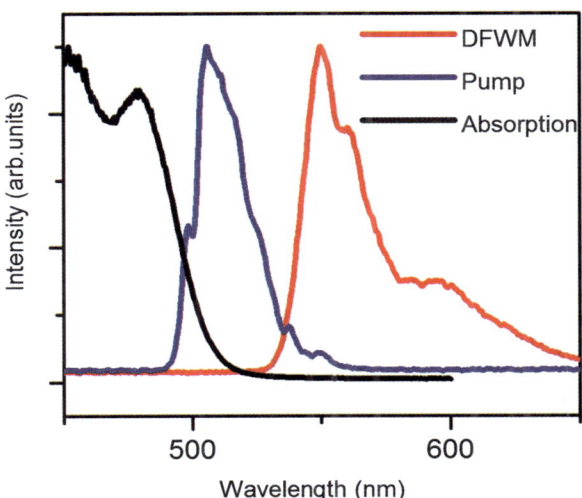

Fig. 2 β-Carotene absorption and Pump-DFWM excitation spectra. The spectrum of the initial pump is near resonant with the ground-state absorption spectrum of β-carotene. The DFWM, on the other hand, is not resonant with any absorption of the ground-state: Its spectrum is solely resonant with the excited-state absorption of the S_1 state, which has a maximum at about 550 nm in cyclohexane.

absorption (not shown), and, therefore, probes the arrival of population directly after the non-adiabatic coupling between the S_2 and S_1 states.

In our open-loop control approach, a pure phase modulation is exerted on the initial pump pulse. Its spectral phase $\Phi(\omega)$ is manipulated *via* a spatial light modulator (SLM) with 128 pixels in a 4f-arrangement. Application of $\Phi(\omega) = a \sin(b\ \omega + c)$ yields a well defined pulse train.[16,26,27] The parameter a is kept fixed at 1.23 and parameter b defines the sub-pulse separation. Note that the parameter c defines the *relative* phase between the sub-pulses as shown in Fig. 3. The *absolute* phase of each sub-pulse is not controlled in this experiment. In all measurements, a non-zero T-delay between the tailored initial pump pulse and DFWM sequence was chosen to avoid the shaping window.[28] Since the delay T is defined between the centre sub-pulse of the multipulse sequence and the pump/Stokes arrival time, a T delay over 112 fs ($T > 2b$) guarantees probing outside the shaping window in the case of $b = 56$ fs.

3 Results

Experiment

Usually, in the Pump-DFWM technique, DFWM transients are measured by scanning the probe delay τ_{23} for different delays T between initial pump and the DFWM sequence.[21–23] In what follows, the probe delay was kept fixed while the initial pump shaping was varied by tailoring the parameter c. In Fig. 4, we present these results in a 2D graph, where the Pump-DFWM signal along the T-axis was plotted for different values of the parameter c. In this way, it allows the investigation of the effect of excitation phase on the internal conversion between S_2 and S_1 states. Fig. 4 (a) depicts such a kind of Pump-DFWM signal. The excitation pulse was shaped into a pulse train with a sub-pulse spacing $b = 56$ fs and varying the parameter c between the sub-pulse. This value of parameter b guarantees three well separated sub-pulses within the life time of S_2 (180 fs). Along the y-axis of Fig. 4 (a), the parameter c between the sub-pulses is varied between 0 and 2π. The signal increases exponentially from cold colours (low intensities) to hot colours (high intensities) along the T-axis. This is due to the fact that the population on S_2 relaxes towards the S_1 entering the Franck–Condon region where we probe the dynamics with the DFWM sequence. Fig. 4 (a) shows that the DFWM signal can be modified by the parameter c. In other words, the flow of population from S_2 state into S_1 state can be controlled, *i.e.*, can be delayed or speed up by the correct choice of parameter c. The white dotted line at $T = 270$ fs in Fig. 4 (a) exemplifies it (shown in Fig. 4 (b): While the population for *e.g.* $c = 3/2\ \pi$ has almost reached its maximum, the

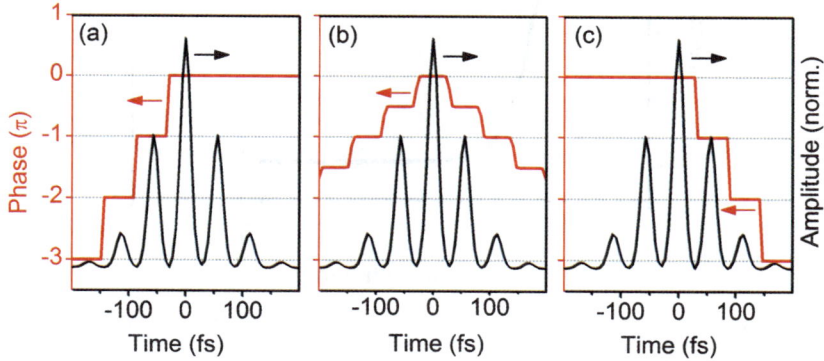

Fig. 3 Simulation of the effect of parameter c on the relative phase between the sub-pulses generated by a sinusoidal phase shaping ($b = 56$ fs). (a) $c = 0$ rad. (b) $c = \pi/2$ rad. (c) $c = \pi$ rad.

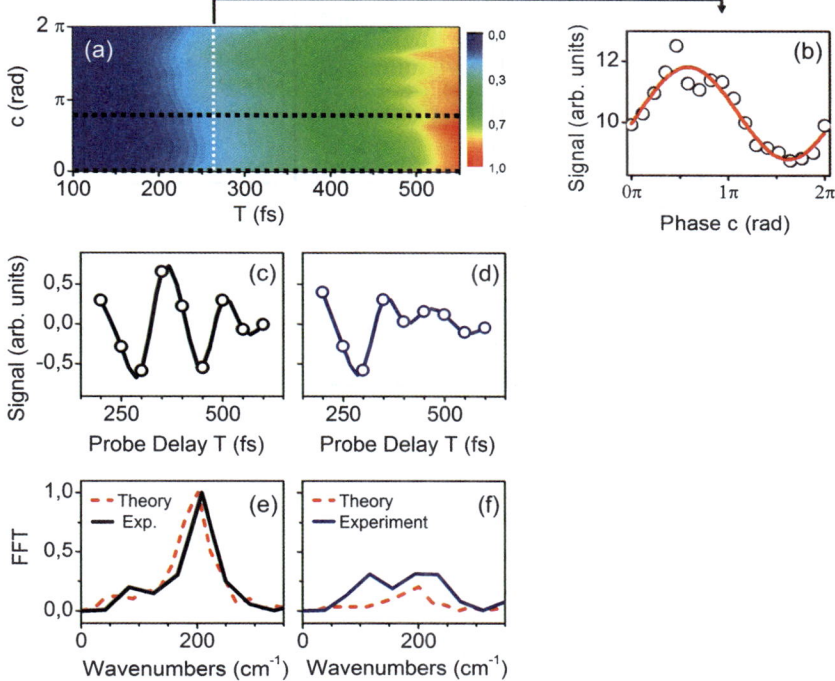

Fig. 4 (a) Pump-DFWM signal at $\tau_{23} = 170$ fs for various values of T and the excitation parameter c. (b) Signal at about T = 270 fs at different values of parameter c. (c) and (d) show the oscillatory component obtained after subtraction of the slow non-oscillatory exponential rise time for $c = 0$ rad and 2.3 rad, respectively. (e) and (f) show the respective Fourier transformations for experimental and simulation data.

population for *e.g.* $c = 1/2\ \pi$ is still rising. Eventually, independently of the choice of parameter c, the whole population excited from the S_0 into the S_2 will reach the cold S_1 state at T delays greater than 600 fs. This can be easily understood, since the excited-state population in β-carotene, as a closed system (*i.e.* without energy transfer), must relax back to ground-state *via* the S_1 deactivation channel. Parallel channels like fluorescence are present but are much less efficient than the ultrafast internal relaxation S_2–S_1.

The sensitive DFWM probing sequence is also able to unravel an oscillatory component on the rise of the signal along the T-axis. This component can be clearly observed when the slow non-oscillatory exponential rise time is subtracted. This is demonstrated for two values of the parameter c (traced lines at Fig. 4 (a)). The two oscillatory components for $c = 0$ rad and 2.3 rad are shown at Fig. 4 (c) and (d), respectively. A Fourier transformation of the signal depicted in Fig. 4 shows a clear peak around 200 cm^{-1} (Fig. 4 (e)), which is not observed for the other $c = 2.3$ rad (Fig. 4 (f)).

Simulation

The simulations show that the composition and phase of the wavepacket depends strongly on the parameters of the exciting multipulse. We observe that partial wavepackets, differing in phase and vibrational components, can be generated by each sub-pulse. The choice of b defines the phase difference and therewith the number of induced sub wavepackets. Defining T_{vib} as the oscillation period of the

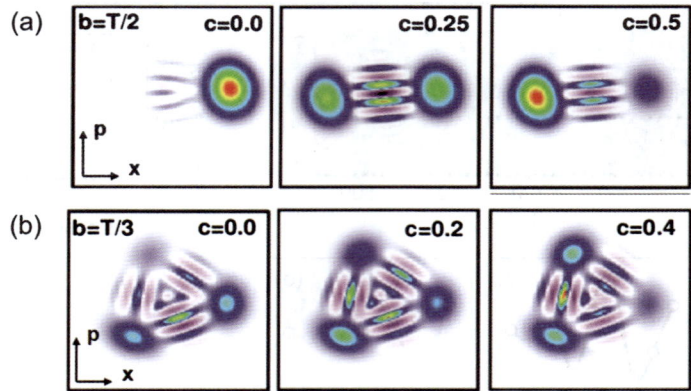

Fig. 5 Simulation of wavepackets in Wigner representation and its dependence on the sub-pulse separation (parameter b) and their relative phase (parameter c). (a) Wavepacket for a sub-pulse separation equal to half of the period of low-frequency molecular mode. (b) Wavepacket for a sub-pulse separation equals to one third of the period of low-frequency molecular mode.

wavepacket, the choice of $b = (\beta/\alpha)T_{vib}$ ($\alpha, \beta \in \mathbb{N}$) leads to the formation of α sub-wavepackets, phase shifted by $2\pi \times \beta/\alpha$.[24] The maximum value of α is limited by the number of vibrational levels of the coupling mode contributing to the excited wavepacket. Fig. 5 shows two examples of these wavepackets in the Wigner representation for $b = T_{vib}/2$ and $T_{vib}/3$.

The generated phase space structures in the dynamically active mode imprint themselves in the population transfer between S_2 and S_1. Analogously to the experimental observations, we observe tailored population decay depending on the parameter c. This is shown in Fig. 4 (e) and (f). By choosing the same experimental values, i.e. $b = 56$ fs and $c = 0$ rad and $c = 2.3$ rad, the same oscillatory component and its dependence on the relative phase between the sub-pulses can be perfectly simulated. In perfect agreement with the experimental results (Fig. 4), the electronic coupling manifests itself in a blue shift from 178 cm^{-1} for the uncoupled mode, to 200 cm^{-1}. The Fourier transformation of this oscillatory component has also a maximum for $c = 0$ rad, while for $c = 2.3$ rad it can be almost suppressed. The wavepacket generated in the case of $b = 56$ fs corresponds approximately to $T_{vib}/3$ of the electronically coupled mode and leads to the generation of three sub-wavepackets in the S_2 potential surface, i.e. $\alpha = 3$.

Contrasting to the experiment and its inherent constraints, the simulation is also able to investigate directly the transient vibrational population during the multipulse excitation. This description is very insightful since it explains the relative time dependent weight of the S_2 vibrational states constituting the wavepacket during the pulse train excitation. A generalization of the described observations to any choice of α in the valid range is possible. For simplicity, we discuss the simplest case of $b = T_{vib}/2$, hence $\alpha = 2$. For $b = T_{vib}/2$, we observe that vibrational states with odd quantum numbers are much less sensitive to the phase parameter c. The population of the vibrational levels $v = 0$ up to $v = 4$ is compared in Fig. 6 (a) and (b) for two values of c (1.4 rad and 1.75 rad, respectively). Their population follows a Gaussian distribution, initially induced by the first sub-pulse. While scanning c, we observe an oscillatory behaviour of the vibrational populations with a periodicity of nearly $2\pi/(2\alpha)$, i.e. twice the number of sub-wavepackets. The maximum asymmetry in distribution is obtained intermediately in the second half of the pulse train, as indicated by the dotted lines in Fig. 6 (c) and (d). In this case, depending on c, the even vibrational states are either enhanced or depopulated by the sub-pulse following the central

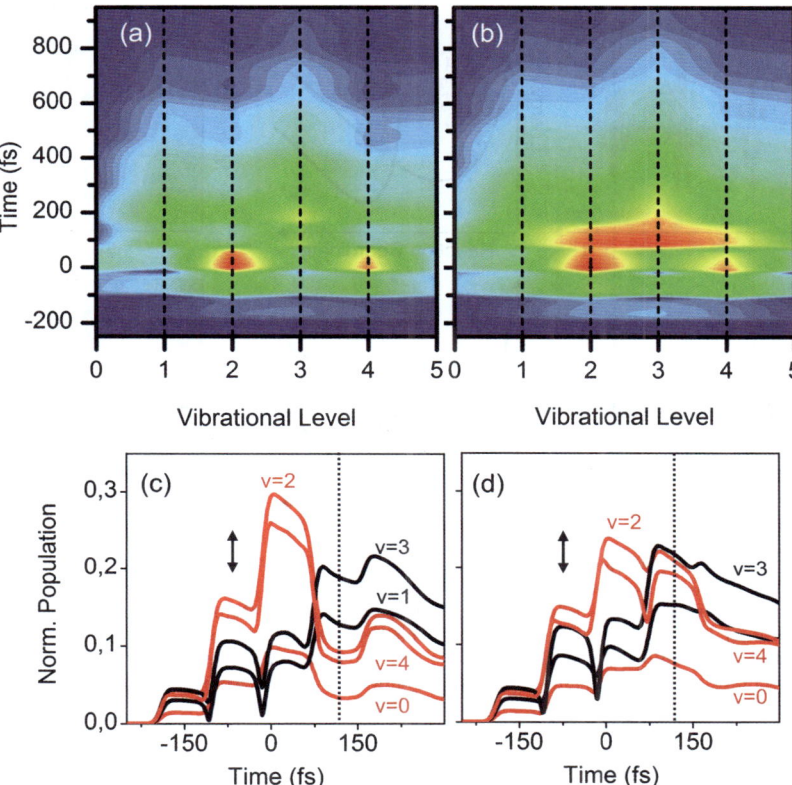

Fig. 6 Simulation of the transient evolution of the vibrational population and its dependence on parameter c for $b = T_{vib}/2$. (a) and (b) show the vibrational population for $c = 1.4$ rad and 1.75 rad, respectively. (c) and (d) show the same data but with an individual curve for each vibrational level. Time zero defines the centre of multipulse excitation. Red curves are associated with even vibrational levels ($v = 0$, 2 and 4), while black dashed curves are related to odd vibrational levels ($v = 1$ and 3). The difference in the evolution of even vibrational levels in dependence of c can be clearly observed, e.g., at about 100 fs (dotted lines).

one, imprinting holes in the overall Gaussian distribution. We observe maxima of FFT amplitudes for $c = 2n\,\pi/(2\alpha)$, $n \in \mathbb{N}$ and minima at $c = (2n + 1)\pi/(2\alpha)$.

Discussion

Understanding the control mechanism

In order to rationalize the control mechanism, two points must be discussed. The first is related to the effect of multipulse excitation on wavepackets. Our simulations show that the interference between wavepackets generated by different sub-pulses depends on their relative phase. Of course, this interference, which can be constructive or destructive, is not equal for all vibrational levels on the excitation bandwidth of the laser since each level has different time evolution factors. This can be clearly observed by comparing Fig. 6 (c) and (d): By changing the relative phase between the sub-pulses, a different transient evolution for each vibrational level can be tailored.

The second point is related to the interaction of this transient vibrational population with a non-adiabatic coupling. In general, the degree of coupling between two electronic surfaces is not necessarily homogeneous but depends on the vibrational

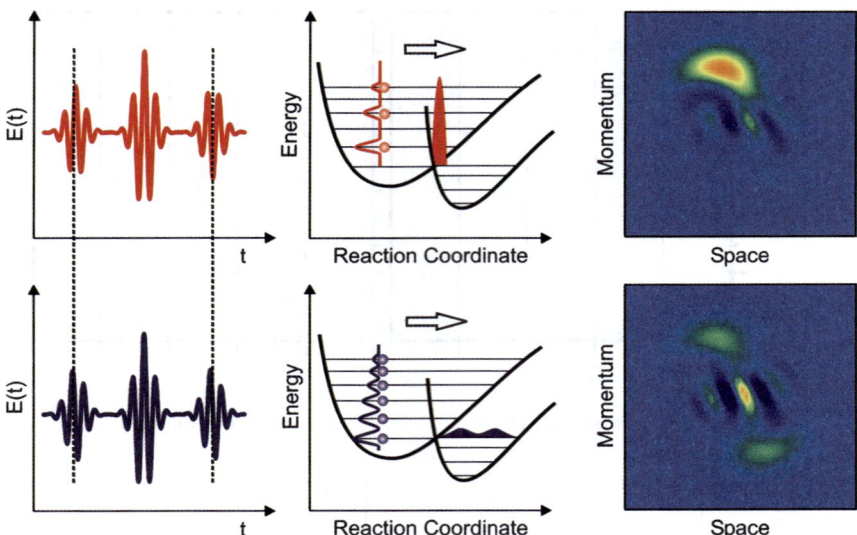

Fig. 7 Effect of the relative sub-pulse phase for the case of $b = T_{vib}/2$ ($\alpha = 2$) on the composition of the resulting wavepacket in the vicinity of the CoIn. The left column shows the excitation pulse trains with differing parameter c, the middle column shows a scheme of the vibrational populations of the excited wavepacket after the sub-pulse following the central one and the resulting wavepacket, the right column shows the wavepacket in the phase space representation of the coupling mode. In the upper row a properly spaced pulse train creates a wavepacket, composed mainly of odd wavefunctions. The pulses in lower and upper rows differ by a relative phase shift of $\pi/4$. The blue pulse train yields a delocalized wavepacket with evenly distributed eigenstate-coefficients. In the phase space representation we observe the phase shift of π between the two components consisting of odd and even eigenstates and an interference pattern in between.

level. In other words, the magnitude of the vibronic coupling matrix elements depends on the vibrational quantum number in the respective coupling mode. If in such a scenario a wavepacket is prepared which lacks contributions with strong coupling elements,[29,30] the overall non-adiabatic transition will be less effective.

These two points are summed up in Fig. 7. We depict the excitation pulse trains and their relative phase differences (left column; see the shape of the electric fields), the distribution of the S_2 vibrational states after the sub-pulse following the central one (middle column) and the Wigner representation of the corresponding wavepackets after the pulse train (right column). The top row represents the scenario for sub-pulse spacing $b = T_{vib}/2$ and relative phase $c = 2n\pi/4$. In the vibrational distribution (middle column) we observe holes at the even levels as discussed above. The annihilation of the corresponding sub-wavepackets, visible in the Wigner representation (right column), leads to the formation of a sharp FFT peak. Therefore, the experimental data in Fig. 4 (c), obtained for $c = 0$, can be assigned to the related scenario with $\alpha = 3$. The lower row in Fig. 7 shows the reversed case for $c = (2n + 1)\pi/4$, where the vibrational population follows again a Gaussian distribution. The appearance of two phase shifted sub-wavepackets leads to the depletion of the FFT-signal. This case again is in good agreement to the experimental data shown in Fig. 4 (f), where for the case of $\alpha = 3$ and $n = 2$ the formation of three wavepackets leads to the observed depletion.

Extending the control mechanism to competing channels

The transient tailoring of vibrational population should ultimately affect the electronic lifetime of the S_2 state, *i.e.*, the internal conversion between S_2–S_1 states. In this situation, the population will remain longer in the higher lying electronic state compared to Fourier-limited excitation. Not surprisingly, the tailoring of wavepackets does not affect the energy flow in β-carotene, since the excited state population experiences just one effective deactivation channel *i.e.* via S_1 state. In this regard it can be considered as a closed system, where control *via* one-photon excitation is not possible.[31,32] Phase control with one-photon transition is, however, possible if strong dissipative coupling (with *e.g.* the environment) takes place and assists the control mechanism,[33] as has been demonstrated to constrain the control of population transfer from the ground- to the excited-state in closed systems.[17,18,34]

In the case of open systems with competing channels, like those found in photosynthetic systems, the branching ratio may be directly affected *via* one-photon phase control.[33] In that sense, our findings are able to rationalize pulse shapes found in the coherent control of energy flow in polyenic biochromophores, where the excited state lifetime is at the order of the total duration of the excitation pulse train. A prerequisite for the mechanism proposed here is a low frequency mode coupling the initially excited state and its successor. This criterion is met by many large and medium sized organic molecules and is already confirmed by theory[35] for polyenes. Besides our explanation does not only describe the findings in bacteriorhodopsin,[15,29] but also provides a commonly applicable mechanism for the control of ultrafast photoisomerizations, where the decisive conical intersection lies in the vicinity of an excited state minimum.[36]

Moreover, the good agreement of our experimental data with the simulations unravels a new feature of carotenoid's ultrafast photochemistry. The early photochemistry of carotenoids is determined by strong vibronic coupling like that given by a conical intersection (CoIn). By the aid of coherent control and quantum dynamical modelling, through the observed phase dependencies, we reveal the matter wave character of the process, even preserved through the CoIn, and identify an effective low-frequency coupling mode of 200 cm^{-1} driving this ultrafast internal conversion process. Such out-of-plane backbone motion in the carotenoid as part of a Light harvesting complex (LHC) has been discussed as the source of conformational change affecting the biological function.[9,35]

Conclusion

In this paper we addressed for the first time a general mechanism to explain the control of energy flow reported for a whole class of biochromophores. In particular, we show how a specifically designed light pulse tailors the induced wavepacket dynamics. By a detailed and accurate simulation of our results, we identify the nuclear degree of freedom which couples the two relevant excited electronic states in the investigated chromophore. Our control mechanism is based on the interference between wavepackets in the excited state, which leads to a transient evolution of the vibrational population dependent on the relative phase between excitation sub-pulses. Such a transient evolution can affect the branching ratio between competing channels in the excited state. Thus, we are able to rationalize pulse shapes found in a whole class of coherent control experiments involving polyenic biochromophores, like in light harvesting complexes and bacteriorhodopsin. Furthermore, we show that the coherence imprinted by the pulse train is preserved through the conical intersection pointing out how coherence could be exploited to manipulate the energy transfer process in these systems.

We note that in the present study, the use of coherent control goes beyond the scope of product optimization demonstrated in numerous studies.[37,38] In several works, light was used as a photonic reagent,[39,40] catalyst[41,42] or as tomographic tool.[43] The present investigation provides a case where tailored light pulses are employed as a highly specialized diagnostic tool for photophysical processes impossible to describe by the use of unmodulated light pulses. In contrast to conventional coherent control, we would like to call this approach Quantum Control Spectroscopy (QCS).

References

1. R. Blankenship, *Molecular Mechanisms of Photosynthesis*, Blackwell Publishers, Malden, MA, 2002.
2. H. A. Frank, A. J. Young, G. Britton and R. J. Cogdell, ed., *The Photochemistry of Carotenoids*, Kluwer Academic Publishers, Govindjee, 1999.
3. T. Polivka and V. Sundstrom, *Chem. Rev.*, 2004, **104**, 2021–2071.
4. T. Buckup, J. Savolainen, W. Wohlleben, J. L. Herek, H. Hashimoto, R. R. B. Correia and M. Motzkus, *J. Chem. Phys.*, 2006, **125**, 194505.
5. A. Ishizaki, T. R. Calhoun, G. S. Schlau-Cohen and G. R. Fleming, *Phys. Chem. Chem. Phys.*, 2010, **12**, 7319–7337.
6. C. J. Bardeen, V. V. Yakovlev, J. A. Squier and K. R. Wilson, *J. Am. Chem. Soc.*, 1998, **120**, 13023–13027.
7. P. F. Tian, D. Keusters, Y. Suzaki and W. S. Warren, *Science*, 2003, **300**, 1553–1555.
8. D. Abramavicius and S. Mukamel, *J. Chem. Phys.*, 2004, **120**, 8373–8378.
9. W. Wohlleben, T. Buckup, J. L. Herek and M. Motzkus, *ChemPhysChem*, 2005, **6**, 850–857.
10. T. Buckup, T. Lebold, A. Weigel, W. Wohlleben and M. Motzkus, *J. Photochem. Photobiol., A*, 2006, **180**, 314–321.
11. G. Vogt, P. Nuernberger, T. Brixner and G. Gerber, *Chem. Phys. Lett.*, 2006, **433**, 211–215.
12. D. G. Kuroda, C. P. Singh, Z. H. Peng and V. D. Kleiman, *Science*, 2009, **326**, 263–267.
13. J. L. Herek, W. Wohlleben, R. J. Cogdell, D. Zeidler and M. Motzkus, *Nature*, 2002, **417**, 533–535.
14. J. Savolainen, R. Fanciulli, N. Dijkhuizen, A. L. Moore, J. Hauer, T. Buckup, M. Motzkus and J. L. Herek, *Proc. Natl. Acad. Sci. U. S. A.*, 2008, **105**, 7641–7646.
15. V. I. Prokhorenko, A. M. Nagy, S. A. Waschuk, L. S. Brown, R. R. Birge and R. J. D. Miller, *Science*, 2006, **313**, 1257–1261.
16. A. M. Weiner, D. E. Leaird, G. P. Wiederrecht and K. A. Nelson, *Science*, 1990, **247**, 1317–1319.
17. J. Hauer, T. Buckup and M. Motzkus, *J. Chem. Phys.*, 2006, **125**, 061101.
18. T. Buckup, J. Hauer, C. Serrat and M. Motzkus, *J. Phys. B: At., Mol. Opt. Phys.*, 2008, **41**, 074024.
19. J. Oberle, G. Jonusauskas, E. Abraham and C. Rulliere, *Chem. Phys. Lett.*, 1995, **241**, 281–289.
20. M. Motzkus, S. Pedersen and A. H. Zewail, *J. Phys. Chem.*, 1996, **100**, 5620–5633.
21. T. Hornung, H. Skenderovic and M. Motzkus, *Chem. Phys. Lett.*, 2005, **402**, 283–288.
22. J. Hauer, T. Buckup and M. Motzkus, *J. Phys. Chem. A*, 2007, **111**, 10517–10529.
23. T. Buckup, J. Hauer, J. Mohring and M. Motzkus, *Arch. Biochem. Biophys.*, 2009, **483**, 219–223.
24. J. Voll and R. de Vivie-Riedle, *New J. Phys.*, 2009, **11**, 105036.
25. T. Polivka and V. Sundstrom, *Chem. Phys. Lett.*, 2009, **477**, 1–11.
26. D. Zeidler, S. Frey, K.-L. Kompa and M. Motzkus, *Phys. Rev. A: At., Mol., Opt. Phys.*, 2001, **64**, 3421–3433.
27. J. Hauer, H. Skenderovic, K. L. Kompa and M. Motzkus, *Chem. Phys. Lett.*, 2006, **421**, 523–528.
28. T. Buckup, J. Hauer and M. Motzkus, *New J. Phys.*, 2009, **11**, 105049.
29. V. I. Prokhorenko, A. M. Nagy, L. S. Brown and R. J. D. Miller, *Chem. Phys.*, 2007, **341**, 296–309.
30. P. S. Christopher, M. Shapiro and P. Brumer, *J. Chem. Phys.*, 2006, **125**, 124310.
31. P. Brumer and M. Shapiro, *Chem. Phys.*, 1989, **139**, 221–228.
32. M. Shapiro and P. Brumer, *Principles of the Quantum Control of Molecular Processes*, John Wiley & Sons, Hoboken, 2003.
33. M. Spanner, C. A. Arango and P. Brumer, *J. Chem. Phys.*, 2010, **133**.
34. V. I. Prokhorenko, A. M. Nagy and R. J. D. Miller, *Journal of Chemical Physics*, 2005, 122.
35. W. Fuss, Y. Haas and S. Zilberg, *Chem. Phys.*, 2000, **259**, 273–295.
36. B. Dietzek, B. Bruggemann, T. Pascher and A. Yartsev, *Phys. Rev. Lett.*, 2006, **97**, 258301.
37. C. J. Bardeen, V. V. Yakovlev, K. R. Wilson, S. D. Carpenter, P. M. Weber and W. S. Warren, *Chem. Phys. Lett.*, 1997, **280**, 151–158.
38. A. Assion, T. Baumert, M. Bergt, T. Brixner, B. Kiefer, V. Seyfried, M. Strehle and G. Gerber, *Science*, 1998, **282**, 919–922.
39. H. Rabitz, *Science*, 2003, **299**, 525–527.

40 C. Daniel, J. Full, L. Gonzalez, C. Lupulescu, J. Manz, A. Merli, S. Vajda and L. Woste, *Science*, 2003, **299**, 536–539.
41 B. J. Sussman, D. Townsend, M. Y. Ivanov and A. Stolow, *Science*, 2006, **314**, 278–281.
42 R. J. Levis, G. M. Menkir and H. Rabitz, *Science*, 2001, **292**, 709–713.
43 M. Wollenhaupt, M. Krug, J. Kohler, T. Bayer, C. Sarpe-Tudoran and T. Baumert, *Appl. Phys. B: Lasers Opt.*, 2009, **95**, 647–651.

PAPER

Coherent control of vibrational transitions: Discriminating molecules in mixtures

A. C. W. van Rhijn, A. Jafarpour, M. Jurna, H. L. Offerhaus and J. L. Herek

Received 13th March 2011, Accepted 11th May 2011
DOI: 10.1039/c1fd00040c

Identifying complex molecules often entails detection of multiple vibrational resonances, especially in the case of mixtures. Phase shaping of broadband pump and probe pulses allows for the coherent superposition of several resonances, such that specific molecules can be detected directly and with high selectivity. Our particular implementation of coherent anti-Stokes Raman scattering (CARS) spectroscopy and imaging employs broadband pump and probe fields in combination with a narrowband Stokes field. We describe our approach for combining spectral phase shaping and closed-loop optimization strategies to perform chemically-selective microscopy. To predict the optimal excitation profile we employ evolutionary algorithms that use the vibrational phase responses of five distinct molecules with overlapping resonances and investigate the effect of phase instability on the optimization. We have recently shown that modified polynomials and orthogonal rational functions can give rise to improved contours for CARS fitness landscapes. Now, by considering the landscapes associated with different basis sets, we introduce two figures of merit to quantitatively rank basis functions in terms of their "appropriateness" for modeling nonlinear phase-shaped processes.

Introduction

Coherent anti-Stokes Raman scattering (CARS) has been used successfully in spectroscopy and microscopy since the development of (tunable) pulsed laser sources.[1] In CARS, molecular vibrations are excited coherently by a combination of a pump (ω_p) and Stokes (ω_s) pulse. Subsequently, a probe (ω_{pr}) pulse, which is often derived from the same pulse as the pump, generates the anti-Stokes (CARS) signal ($\omega_c = \omega_p - \omega_s + \omega_{pr}$). By obtaining its contrast from molecular vibrations, CARS is a label-free imaging technique.

A vibrational resonance can be considered as a damped driven harmonic oscillator with a certain resonance frequency, resonance strength, and damping factor. The response of such a harmonic oscillator yields a peak in the amplitude spectrum and a phase step from 0 to $-\pi$ in the phase response at the resonance frequency. As multiple resonances are combined, the peaks in the amplitude spectrum add up linearly, but the phase responses combine in such a way that the phase remains between 0 and $-\pi$. The vibrational phase of the resonances can be used to suppress non-resonant contributions to the CARS signal.[2]

Here we show how phase-shaped pulses that specifically address the vibrational phase of the molecules can be used to enhance desired CARS signals and simultaneously suppress non-resonant contributions. To predict optimal response functions

---

Optical Sciences group, MESA+ Institute for Nanotechnology, Faculty of Science and Technology (TNW), University of Twente, The Netherlands

we employ evolutionary strategies that use the vibrational phase responses of five distinct molecules with overlapping resonances. We first address the basic concepts and then report the following new contributions: 1) extension of phase-shaped CARS strategy to a larger library of mixtures, 2) a coherence sensitivity analysis to quantify the negative contribution of parameters such as the jitter (between pump and Stokes beams) and phase noise on the performance of these optimizations, and 3) quantitative ranking of basis sets in terms of their "appropriateness" for modeling a phase-shaped CARS process.

1. Phase-shaped CARS

Introduction

Although successful, narrowband CARS (Fig. 1(a)) has disadvantages, such as the need for a unique identifying spectral feature to identify a constituent in a mixture of resonant compounds and its need for a tunable laser source to access different vibrations. Broadband CARS, where one or more of the input pulses are broadband (femtosecond) pulses,[3,4] especially in combination with pulse shaping,[5–11] can circumvent these problems.

Our approach is based on a double broadband interaction, where the degenerate pump and probe pulses are broadband and the Stokes pulse is narrowband (Fig. 1 (b)). Due to this double broadband interaction there are many different combinations of pump and probe wavelengths which lead to the same wavelength CARS signal. Even though the CARS wavelength is the same for these combinations, the vibrational frequency that is addressed is different for each combination, leading to a difference in amplitude and phase of the CARS signal, giving rise to interference. We can influence this interference by changing the spectral phase profile of our pump and probe pulses. We previously reported on an intuitive shaping strategy, based on mimicking the π-phase step of a vibrational resonance.[12,13] Using more complex phase shapes, further control over the CARS signal is possible, but in order to accurately calculate the CARS signal that is generated by such phase shaped pulses, detailed knowledge of the vibrational response (including the phase information) is needed.

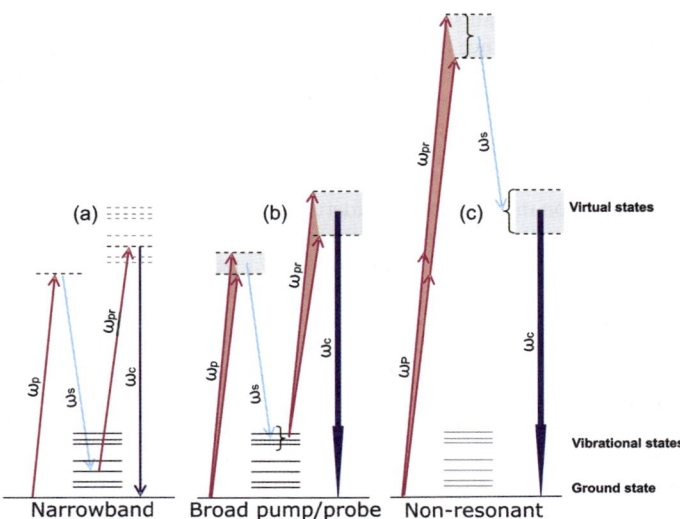

Fig. 1 Energy level diagram for (a) narrowband CARS, (b) broadband pump and probe CARS, and (c) non-resonant four-wave mixing.

The vibrational phase response

The vibrational phase response can be measured directly using heterodyne detection[14] or vibrational phase contrast CARS.[15] Several other indirect techniques exist that retrieve the vibrational phase from a recorded broadband CARS spectrum.[16–19] Here, we determine the vibrational phase response by fitting with an evolutionary algorithm[20] of the spontaneous Raman scattering spectrum.[21,22] The spontaneous Raman spectrum corresponds to the negative imaginary part of the individual vibrational lineshapes (eqn (1)). By fitting the negative imaginary part of a sum of vibrational lineshapes to the spontaneous Raman spectrum, the vibrational phase can be retrieved. We employ a covariance matrix adaptation evolution strategy (CMA-ES)[23] to solve the fitting problem. Details of the code implementing CMA-ES are reported elsewhere.[24] An example fit for polystyrene is shown in Fig. 2(a,b).

$$I_{\text{Raman}}(\omega) \propto -\Im[\chi^{(3)}(\omega)] = -\Im[\chi^{(3)}_{\text{NR}} + \sum_R \frac{A_R}{(\omega_R^2 - \omega^2 + 2i\omega\gamma_R)}] \quad (1)$$

Pulse optimization

Using the complete vibrational response to calculate the CARS signal, CMA-ES is used to numerically optimize the spectral phase of the pump (=probe) pulse. The CMA-ES optimizes the spectral phase $\phi(\omega)$ of the pump and probe pulses such that the difference in CARS signal, integrated over the full bandwidth, generated by a pump- and probe-pulse with phase $\phi(\omega)$ and a pump- and probe-pulse with phase $-\phi(\omega)$ is maximized. By subtracting the CARS signals generated by a phase profile $\phi(\omega)$ and its inverse $-\phi(\omega)$, the purely non-resonant background is removed.[12,13] Using the difference as the target for the optimization yields a phase profile that enhances the desired molecule's response, while suppressing other contributions.

The pump (and probe) pulse is assumed to have a Gaussian spectral power distribution with a center wavelength of 806.5 nm (12400 cm^{-1}) and a FWHM of 26 nm (400 cm^{-1}). The Stokes pulse is assumed to have an infinitely small bandwith with a center wavelength of 1064.3 nm (9395.85 cm^{-1}). The spectral phase profile is optimized by the CMA-ES on 40 points, divided evenly over 1021 cm^{-1}. This phase profile is extended to 4096 points by cubic spline interpolation in order to improve the resolution when calculating the CARS response. The CMA algorithm uses 20 parents and a population size of 40 per generation. The non-resonant background is assumed to be constant with a ratio between resonant and non-resonant response of 5 : 1 (peak to baseline). An example of an optimized phase profile for maximum signal from pure polystyrene is shown in Fig. 2(c). The shape of the excitation profile clearly reflects the molecular phase response 2(b), as the optimal pulse compensates the phase profile of the molecule. Optimizations with different ratios of

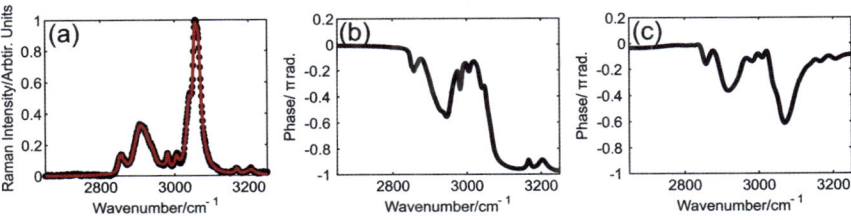

Fig. 2 (a) Spontaneous Raman spectrum (black) and fit (red) of polystyrene and (b) retrieved vibrational phase. Panel (c) shows the optimal excitation phase for maximum CARS signal from polystyrene.

non-resonant background showed similar results, as the purely non-resonant signal is discarded by looking at the difference CARS signal. The mixing between resonant and non-resonant signal influences the depth of the phase jumps in the optimized phase, but the general shape remains the same.

Besides optimizing the phase for maximum signal from a compound, it is also possible to optimize for minimum signal. Combining these two ideas, it is possible to optimize the excitation phase for selective excitation of a single constituent in a mixture of resonant compounds.[22] Here, we consider a mixture of five resonant compounds; polystyrene, PMMA, polyethene, toluene, and ethanol. These five substances have strongly overlapping resonances in the region around 3000 cm^{-1}, which is accessed by our combination of input pulses (pump-Stokes). Using CMA-ES, we calculate five different excitation phases that selectively excite one of the resonant compounds, while minimizing the contributions from the other four compounds. An example of such a selective excitation phase for optimizing the CARS signal from polyethene is shown in Fig. 3. The complicated spectral phase pattern (red curve) that results from the optimization bears little resemblance to the vibrational phase profile of polyethene (black curve). Fig. 3(b) shows the corresponding learning curve, truncated after 400 generations to better illustrate how contributions from the other molecules are actively suppressed. After 1000 generations, the CARS signal from polyethene in the mixture reaches a contrast ratio of 500 : 1. Remarkably, the signal is diminished by only 30% relative to the optimized signal for the pure substance. The obtained results and contrast ratios for all five substances can be found in Table 1.

It can be seen from Table 1 that the contrast and total signal is lowest for polystyrene and toluene. These two molecules have very similar chemical structures, and hence the vibrational response is also very similar. Therefore, it is more difficult to differentiate between the two and the obtained contrast ratio and total signal is lower. It should be noted that the obtainable contrast ratio might be lower if multiple resonant compounds are present in the focal volume simultaneously. This potential decrease in contrast ratio is due to homodyne mixing of the CARS signals in the focal volume.[22]

Effect of noise

In order to gain insight into the robustness of this optimization approach under laboratory constraints, we consider the effect of noise on the optimization process. We simulate fluctuations caused by, for example, the spatial light modulator or jitter between the different laser sources by adding a random phase noise to each point of the phase pattern that is simulated. The resulting phase profiles are described by

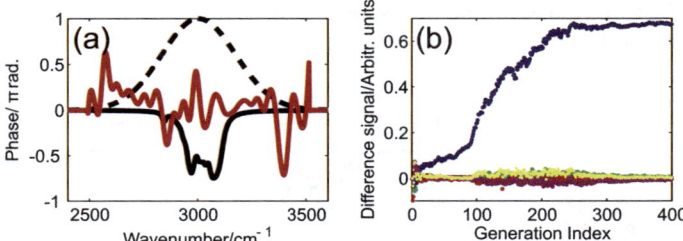

Fig. 3 (a) Optimized excitation phase (red) for maximum signal from polyethene and minimum signal from PMMA, polystyrene, toluene and ethanol. Furthermore, the pump (=probe) pulse envelope (dotted black line) and the vibrational phase of polyethene (solid black line) is shown. (b) Integrated difference CARS signals as a function of generation of the CMA algorithm for polystyrene (green), PMMA (red), polyethene (blue), toluene (yellow), and ethanol (magenta).

Table 1 CARS signal and contrast ratios for selective excitation of a single compound in a mixture of 5 resonant compounds. The first column indicates the optimized compound and the cells of that row show the amount of integrated difference CARS signal normalized to the amount of signal obtained from a pure sample of the optimized compound with a pulse optimized for its maximum signal

Compound	Polystyrene	PMMA	Polyethene	Toluene	Ethanol	Contrast
Polystyrene	0.05	$< 4 \times 10^{-4}$	$< 4 \times 10^{-4}$	$< 4 \times 10^{-4}$	$< 4 \times 10^{-4}$	132 : 1
PMMA	$< 3 \times 10^{-4}$	0.66	$< 3 \times 10^{-4}$	$< 3 \times 10^{-4}$	$< 3 \times 10^{-4}$	2200 : 1
Polyethene	$< 1.5 \times 10^{-3}$	$< 1 \times 10^{-3}$	0.71	$< 1.5 \times 10^{-3}$	$< 1.5 \times 10^{-3}$	470 : 1
Toluene	$< 6 \times 10^{-4}$	$< 2 \times 10^{-4}$	$< 2 \times 10^{-4}$	0.10	$< 4 \times 10^{-4}$	168 : 1
Ethanol	$< 1 \times 10^{-3}$	$< 1 \times 10^{-3}$	$< 1 \times 10^{-3}$	$< 1 \times 10^{-3}$	0.42	415 : 1

$\phi_{total} = \phi_{orig} + \Phi_{noise} n(\omega)$, where ϕ_{orig} denotes the original (noiseless) phase pattern, Φ_{noise} is a scaling factor for the noise level (in radians), and $n(\omega)$ is a random sequence of noise values with a uniform distribution between -1 and 1. The obtained contrast ratio and CARS signals for different levels of noise strength are presented in Table 2.

Table 2 shows the devastating role of phase instability on these coherent processes by a decrease of the contrast ratio from 470 : 1 to 5 : 1, corresponding to a noise span of $\Phi_{noise} = 0$ to $\Phi_{noise} = 0.5\pi$. This effect can be compared and correlated with the smear out of coherent control landscapes, as shown in[25] (see Fig. 11 therein). With a given level and type of hardware noise, alternative formulations, such as delay-based pulse shaping,[26] are expected to provide some (software-based) noise robustness and maintain reasonably high contrast ratios.

Comparing previous experimental results based on control with a π-phase step[13] with the noise model presented in Table 2, we attribute an effective noise level of $\Phi \approx 0.23\pi$ as a rough estimation of noise in our previous experiment.

2. Quantitative ranking of basis sets

Introduction

Identification and ranking of an "appropriate" set of basis functions to describe a (nonlinear) pulse shaping problem is one of the key questions in both open- and closed-loop coherent control. In linear operator theory, this is a well-established classical problem. By considering functions as generalized vectors in an infinite-dimensional (Hilbert) space, an appropriate basis set (of eigenfunctions) can be considered as the independent (orthogonal) axes of the space. Such eigenfunction decompositions have been successfully used in quantitative and intuitive

Table 2 CARS signal and contrast ratios for selective excitation of polyethene in the presence of various levels of noise. The normalized signal is once again relative to the signal obtained from a pure sample of polyethene with a pulse optimized for maximum signal from polyethene

Noise Level	Contrast Ratio	Norm. CARS signal
0	470 : 1	0.71
0.01π	300 : 1	0.71
0.02π	79 : 1	0.71
0.05π	47 : 1	0.71
0.1π	28 : 1	0.71
0.2π	14 : 1	0.56
0.5π	5 : 1	0.07

formulation of problems such as wavepacket dynamics in molecules and light propagation in optical fibers with both fundamental and practical significance.

Our motivation for finding alternative basis sets for pulse shaping spectroscopy is to create an intuitive approach that is easily applicable to experiments. As such, we have recently suggested the relative convex shape of contours in the energy landscape as a measure of "appropriateness" of a basis set. For a real non-degenerate linear operator with orthonormal eigenfunctions Ψ_n and a unity weight function, an arbitrary function f can be written as $f = \sum c_n \Psi_n$. According to Parsevall's theorem, the (statistical) energy of the function f can be described as $E_f = \sum c_n$, which is the equation of a hyper-sphere. This is the ideal optimization landscape, as the contours $E = E_n$ are symmetric, isotropic, convex (see Fig. 4), connected, and without relative rotation.[25] With the convex contour as a qualitative figure of merit, we have shown the superiority of orthogonal rational functions for a third-order resonant process[27] and modified polynomials for a second-order non-resonant process.[25] However, these comparisons have been done by visual inspection and in a qualitative way. Here, we report on two figures of merit that can be used for quantitative and automated comparison of different basis sets in spaces with arbitrary dimensionalities.

First figure of merit: isoperimetric quotient

A closed string will enclose the largest area when it is stretched uniformly to form a circle. This intuitive property, referred to as the isoperimetric inequality in a more formal terminology, is the basis of our approach. Also note that any concave shape can be modified towards a convex shape with the same circumference, but a larger area, by "flipping the dents" (see Fig. 4(C)). So, if P and A denote the circumference and the area of a closed shape, the ranking factor $IPQ = 4\pi A/P^2$ (referred to as the isoperimetric quotient) ranks convex shapes on top of "comparable" concave ones (see Fig. 4(C)) and a circle on top of all convex shapes; which is exactly what we are looking for. The IPQ can also be expressed in three- and higher-dimensional spaces, where visual inspection is difficult or even impossible. It is specifically useful for the automated global search of weight factors in Gram–Schmidt-like modifications of a basis set[25] in high-dimensional spaces. Note that the search for such alternative bases does not require new simulations or measurements, except possibly for improving the resolution.[25]

While the IPQ is a useful figure of merit to evaluate the convex shape and the apparent ellipticity of a contour, it can yield similar values for a concave contour and an elongated convex one. A complementary figure of merit is one that only

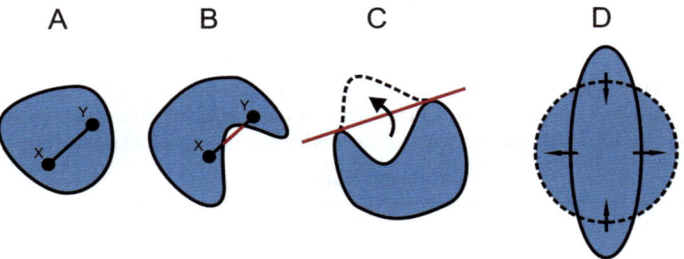

Fig. 4 (A) A convex curve; the line segment connecting any two arbitrary points within the enclosed area will lie entirely within the enclosed area. (B) A concave curve; the line segment connecting at least one pair of points within the enclosed area will lie partially outside the enclosed area. (C) By "flipping a dent", it is possible to convert a concave shape to a convex one with the same circumference, but a larger area. (D) Among convex patterns with the same circumference, a circle encloses the largest area.

differentiates between convex and concave contours with no sensitivity to the elongation of convex contours.

Second figure of merit: convex hull quotient

When a stretched elastic band encompassing an object is released, it is (partially) attached to a cross section of the object. The resulting shape of the released band is referred to as the convex hull (minimal convex set) of the cross section. If the cross section is convex, the convex hull is the cross section itself. The difference between the areas enclosed by a contour and its convex hull is a measure of deviation from a convex shape. We define a second figure of merit, referred to as Convex Hull Quotient (CHQ), to be the ratio of the areas enclosed by a contour and its convex hull. CHQ is a figure of merit complementary to the IPQ. For CHQ values (considerably) smaller than unity, the contours have (considerable) deviation from a convex shape. If the CHQ is (close to) one, then the IPQ is (mainly) a measure of elongation of a convex contour.

Quantitative comparison of basis sets

We have recently shown how the choice of an appropriate basis set can lead to landscapes with visibly smoother, more convex contour lines, for a given CARS configuration.[27] Here we compare the CARS landscapes associated with polynomial and Kautz bases[27] (quantitatively), using the IPQ and CHQ figures of merit introduced in this contribution. All simulation parameters are the same as provided in ref. 27.

The number of data points along each coordinate is chosen to be 200 in order to have a reasonably small numerical error in discretization of contours with small areas. The first analyzed contour corresponds to 88% of the maximum fitness, which has been discretized with 35 points along the periphery. The last analyzed contour corresponds to 39% of the maximum fitness, and has been discretized with 245 points. It corresponds to the lowest fitness, for which the contour in the polynomial landscape is completely accommodated in appropriately-chosen range of the landscape.[27] Further numerical considerations are beyond the scope of this contribution, and will be addressed separately.

As seen in Fig. 5, the contours in the polynomial landscape deviate more from a convex shape (lower IPQ), as the fitness drops (compared to the contours in the other landscape). A higher value of IPQ for a given contour level implies more insensitivity to the direction approaching the contour. From a practical perspective, if a laser setup starts from similar energy levels, but different initial (background) phase every day, the optimizations will remain essentially repeatable, and also take approximately the same amount of time. Also note that if the contours of a landscape are obtained from each other by isotropic scaling, then the IPQ will be a constant function of contour level (almost similar to Fig. 5, bottom right). While the convex (concave) shape of contours in the Kautz (polynomial) landscape is noticeable by visual inspection in Fig. 5(A) and 5(B), the same information is quantitatively demonstrated using the aforementioned figures of merit in Fig. 5(C) and 5(D). The value of this approach will be especially apparent when comparing multiple solution landscapes generated by a variety of basis functions.

Summary and discussion

We have shown the feasibility of using an evolutionary algorithm to extract vibrational phase information from spontaneous Raman scattering data and using this information to optimize broadband coherent anti-Stokes Raman scattering (CARS). By optimizing the spectral phase profile of our broadband pump (=probe) pulse we can obtain selective excitation of a single compound in a mixture of resonant compounds. We predict contrast ratios ranging from 170 : 1 to 2200 : 1 for each separate compound in a mixture of five compounds. The obtained contrast

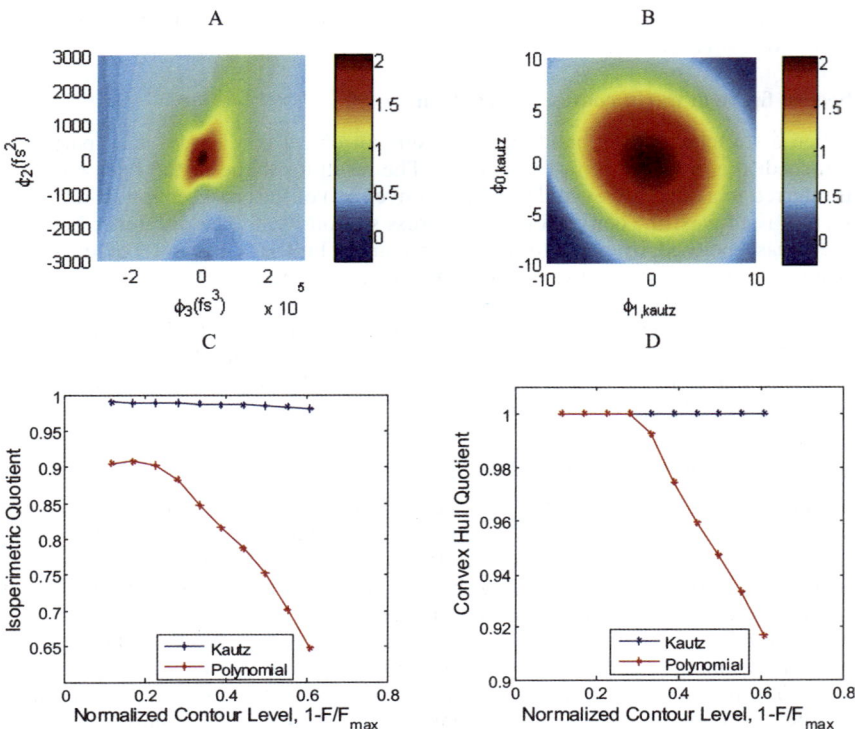

Fig. 5 Two-dimensional landscapes of the CARS process with (A) polynomials and (B) Kautz functions, and (C) isoperimetric quotient and (D) the convex hull quotient values, associated with the contours in both landscapes.

ratio depends on the uniqueness of the complex vibrational response of the molecule of interest compared to that of the surrounding molecules. In the presence of phase noise an optimal phase profile can still be found, but the obtainable contrast ratio will decrease as the noise level increases. Furthermore we have shown the advantages of considering alternative basis sets and we have introduced a method of quantitatively ranking these basis sets based on the isoperimetric quotient and the convex hull quotient.

The results of this study have both fundamental and practical implications for coherent control:

• A fundamental question in spectroscopy with shaped pulses is the nature of the control mechanism.[28] Enhancing the CARS signal generated in our experiments can be achieved by either a transform limited pulse or a pulse that compensates the phase of the vibrational resonance. In practice, an optimization seeks a compromise to these two effects (see section 2.2 of ref. 27). As such, the optimal solutions of CARS signals from a mixture are not only about a trade-off between the signals of individual compounds, but also about a compromise between competing mechanisms in each compound.

• The strategies discussed in this paper all fall in the category where the temporal dynamics of the excited vibrational states do not play a direct role. A first step toward control where the dynamics are directly addressed would be to include the effects of vibrational relaxation and create pulses that excite one vibrational level and probe another level at some later time.

• While the employed technology is relatively new, the notion of using temporal interferences of all resonances (rather than a spectral scan of individual resonances) is a familiar concept in vibrational spectroscopy. One may consider phase-shaped CARS a nonlinear generalization (or analogue) of FTIR or FT-Raman spectroscopy.

- In general, evaluation of the IPQ and the CHQ of different contour lines (geometrical changes) should be done along with an explicit verification of local optima or lack thereof (possible topological changes). Regarding the two specific cases of polynomials and Kautz functions, this issue has been addressed in Section 6.4 of ref. 27.

Acknowledgements

This research is supported by the Stichting voor Fundamenteel Onderzoek der Materie (FOM, grant number 03TF78-3), which is supported financially by Nederlandse Organisatie voor Wetenschappelijk Onderzoek (NWO). This work is also supported by the IOP (Innovatiegerichte Onderzoeksprogramma's) Photonic Devices program managed by the Technology Foundation STW (Stichting Technische Wetenschappen) and AgentschapNL.

References

1 P. D. Maker and R. W. Terhune, *Phys. Rev. Lett.*, 1965, **137**, A801.
2 E. O. Potma, C. L. Evans and X. S. Xie, *Opt. Lett.*, 2006, **31**, 241.
3 H. Kano and H. Hamaguchi, *Opt. Express*, 2005, **13**, 1322.
4 G. W. H. Wurpel, J. M. Schins and M. Müller, *J. Phys. Chem. B*, 2004, **108**, 3400.
5 D. Oron, N. Dudovich and Y. Silberberg, *Phys. Rev. Lett.*, 2002, **89**, 273001.
6 D. Oron, N. Dudovich and Y. Silberberg, *Phys. Rev. A: At., Mol., Opt. Phys.*, 2004, **70**, 023415.
7 S. O. Konorov, X. G. Xu, J. W. Hepburn and V. Milner, *Phys. Rev. A: At., Mol., Opt. Phys.*, 2009, **79**, 031801.
8 V. V. Lozovoy, B. Xu, J. C. Shane and M. Dantus, *Phys. Rev. A: At., Mol., Opt. Phys.*, 2006, **74**, 041805.
9 J. Konradi, A. K. Singh and A. Materny, *Phys. Chem. Chem. Phys.*, 2005, **7**, 3574.
10 J. Konradi, A. Scaria, V. Namboodiri and A. Materny, *J. Raman Spectrosc.*, 2007, **38**, 1006.
11 S. D. McGrane, R. J. Scharff, M. Greenfield and D. S. Moore, *New J. Phys.*, 2009, **11**, 105047.
12 S. Postma, A. C. W. van Rhijn, J. P. Korterik, P. Gross, J. L. Herek and H. L. Offerhaus, *Opt. Express*, 2008, **16**, 7985.
13 A. C. W. van Rhijn, S. Postma, J. P. Korterik, J. L. Herek and H. L. Offerhaus, *J. Opt. Soc. Am. B*, 2009, **26**, 559.
14 M. Jurna, J. P. Korterik, C. Otto, J. L. Herek and H. L. Offerhaus, *Opt. Express*, 2008, **16**, 15863.
15 M. Jurna, J. P. Korterik, C. Otto, J. L. Herek and H. L. Offerhaus, *Phys. Rev. Lett.*, 2009, **103**, 043905.
16 S. H. Lim, A. G. Caster, O. Nicolet and S. R. Leone, *J. Phys. Chem. B*, 2006, **110**, 5196.
17 S. H. Lim, A. G. Caster and S. R. Leone, *Opt. Lett.*, 2007, **32**, 1332.
18 E. M. Vartiainen, H. A. Rinia, M. Müller and M. Bonn, *Opt. Express*, 2006, **14**, 3622.
19 Y. Liu, Y. J. Lee and M. T. Cicerone, *Opt. Lett.*, 2009, **34**, 1363.
20 W. L. Meerts and M. Schmitt, *Int. Rev. Phys. Chem.*, 2006, **25**, 353.
21 M. H. Hennessy and A. M. Kelley, *Phys. Chem. Chem. Phys.*, 2004, **6**, 1085.
22 A. C. W. van Rhijn, M. Jurna, A. Jafarpour, J. L. Herek and H. L. Offerhaus, *J. Raman Spectrosc.*, 2011, DOI: 10.1002/jrs.2922.
23 N. Hansen, "The CMA evolution strategy: a tutorial" http://www.lri.fr/~ hansen/cmatutorial.pdf, 2010 description.
24 R. Fanciulli, L. Willmes, J. Savolainen, P. van der Walle, T. Bäck and J. L. Herek, *Lect. Notes Comput. Sci.*, 2008, **4926**, 219.
25 P. van der Walle, H. L. Offerhaus, J. L. Herek and A. Jafarpour, *Opt. Express*, 2010, **18**, 973.
26 D. Yang, D. P. Sprünken, A. C. W. van Rhijn, J. Savolainen, T. L. Chen, H. L. Offerhaus, J. L. Herek and A. Jafarpour, *Opt. Commun.*, 2011, **284**, 2748.
27 A. C. W. van Rhijn, H. L. Offerhaus, P. van der Walle, J. L. Herek and A. Jafarpour, *Opt. Express*, 2010, **18**, 2695.
28 T. Brixner, N. H. Damrauer, B. Kiefer and G. Gerber, *J. Chem. Phys.*, 2003, **118**, 3692.

Coherent control of the motion of complex molecules and the coupling to internal state dynamics

Paul Venn and Hendrik Ulbricht*

Received 13th April 2011, Accepted 2nd June 2011
DOI: 10.1039/c1fd00066g

We discuss coherent control of the centre of mass motion of complex molecules by de Broglie interferometry. We describe an experiment to couple the dynamics of internal state population of complex molecules to their centre of mass motion. We discuss how this can be used to probe state population and transition, especially the photo-switching of flourinated di-azobenzene molecules between their *cis*- and *trans*-configuration. We propose an experiment to photo-isomerise complex di-azobenzene molecules in the gas-phase, including the selective detection of molecules in different conformations. In addition we discuss possible ways of optimising the conformation detection through cooling, and optical techiques.

1 De Broglie interference of complex particles with Talbot–Lau Interferometer

Over the past 10 years matter wave interferometry experiments have been extended to complex molecules such as fullerenes, di-azobenzene derivatives and organic molecules in the mass range 10^4 dalton.[1,2] These experiments utilise a three grating interferometer, employing the Talbot and Lau effects to maximise the number of molecules contributing to the interference pattern.[3] A Talbot–Lau interferometer (TLI) operates in the near-field regime where the interference occurs as spatial modulation of the molecular beam, of comparable scale to the grating period g of the interferometer. We here give a short introduction to the principles behind the TLI. Each slit of the first grating collimates a fraction of the incident molecular beam and therefore becomes a beam source with higher spatial transverse coherence length on the order of 10^{-6} m. Matter waves are effectively prepared with plane wavefronts when they reach the second grating. According to the Talbot effect, spatial coherent illumination of a grating generates a self-image at the Talbot distance $L_T = g^2/\lambda_{dB}$ behind the grating. This self image contains transverse spatial intensity modulations of the molecular beam from the second grating. The third grating, with equal period to the second, is placed at L_T after the second grating and scans over the transverse direction, acting as a detection mask. A molecular detector, such as a mass spectrometer, is then used to measure the total flux of the molecular beam as the grating scans. Due to the Lau effect all slit sources of the first grating, which are incoherent to each other, contribute to the same interference self-image pattern for certain distances between the three gratings. More details about the principle of the TLI scheme can be found elsewhere.[4] Typically, the de Broglie wavelength of the molecules λ_{dB} is on the order of 1 pm and the grating period is on the order of 100 nm, quantum interference fringe visibilities of 30% to 40%

School of Physics and Astronomy, University of Southampton, Highfield, Southampton, UK. E-mail: h.ulbricht@soton.ac.uk; Fax: +44 (0)23 80593910; Tel: +44 (0)23 8059 2073

have been reported. When using material gratings, such as those made from silicon nitride or gold, the influence of dispersive van der Waals interaction between the molecule and the grating wall will be observed as dephasing, reducing the visibility.[5] An extension to the TLI is the Kapiza–Dirac–Talbot–Lau interferometer (KDTLI), which replaces the second material grating with a grating formed from a standing light wave.

The free centre of mass motion interference

Molecular beams are typically generated by a thermal source operating at temperatures of 400 K to 900 K to enable sublimation of molecules. The internal temperature and the temperature of the longitudinal external motion of the molecules are given at source temperature, while the temperature of the transverse motion is quadratically reduced by the collimation factor of the first TLI grating. Molecules have highly excited internal electronic, rotational and vibrational states. The temperature also affects the conformation of the molecules, which is rapidly changing on time scales of 10^{-9} s, as can be seen from *ab initio* computations.[6] Even so, in all molecule interference experiments there is no indication that the interference pattern is influenced by these internal state distributions.[1,7] Free molecular beam propagation means that the molecules do not interact with external fields, particles or surfaces. The development of matter wave interferometry with complex molecules was primarily triggered by the interest in the foundations of physics.[8] While the aim is to demonstrate quantum superposition of massive particle to test the standard interpretation of quantum mechanics, recent developments have given us new techniques for precise molecule metrology.

2 Interferometric molecule metrology and coupling to internal state dynamics

Matter wave interference of free beams only provides information about the coherence of the centre of mass motion, however we here are interested in extending this to enable control of internal properties in order to develop an understanding of the coupling between the external and internal degrees of freedom of molecules. In principle any coupling to the centre of mass motion can be precisely measured, which is at the heart of applications of matter wave interferometry. Such applications include absolute absorption cross sections[9] and van der Waals/Casimir–Polder interaction potentials and effects,[5,10] but may also be extended to measurements of gravitation effects beyond the scope of this discussion. We will discuss how introducing electrostatic deflectors into the TLI as shown Fig. 1, allows the investigation of the internal dielectric properties of the molecules through their deflection during interference.[11,12] The deflectors create a phase shift Δx and a fringe visibility reduction of the interference pattern, which are proportional to the electric susceptibility of the molecule χ.

$$\Delta x = \frac{\chi}{m} \frac{E\nabla E}{v^2} d\left(\frac{d}{2} + L\right), \qquad (1)$$

with E the electric field strength, v the velocity and m the mass of the molecules, d and L are geometric dimensions of the setup as illustrated in Fig. 1. In general, the deflection of a molecule in an external field E acts according to its electric susceptibility χ, which includes the static (electronic) polarizability α (with (αE) being the induced dipole moment) as well as the thermal average of the orientation of the permanent electric dipole moment μ_0 and is given by the Debye–Langevin formula:

$$\chi = \alpha + \frac{\langle \mu_0^2 \rangle}{3k_\mathrm{B} T}. \qquad (2)$$

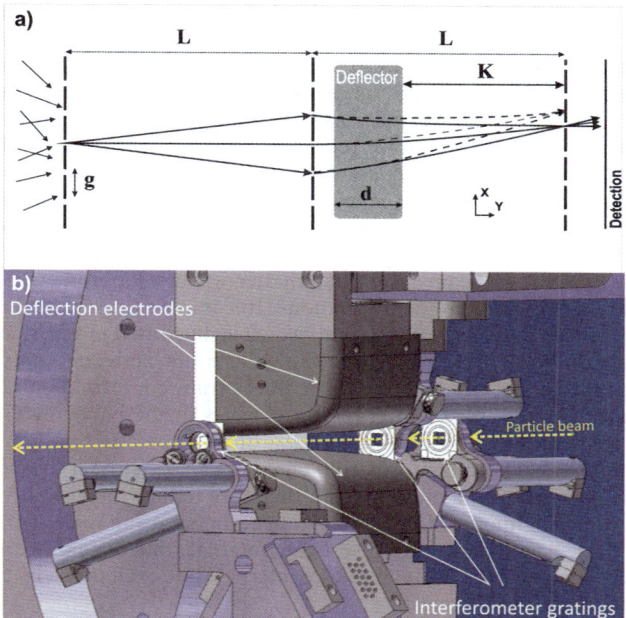

Fig. 1 Talbot–Lau interferometer setup with deflection electrodes for electric field coupling of internal state dynamics to particle motion. a) shows the schematics of the interferometric deflector giving the dimensions and b) is an illustration of the setup at Southampton.

Both, α and μ_0, depend on internal excitation of the molecule and on variation of the conformation of the molecule.[6]

Average over thermally activated states

Molecular beams for interference experiments are launched by thermal sources at high temperatures. As a result, each molecule in the beam is excited in a different internal state and a single molecule rapidly switches between different conformations. The internal state population is not controlled. Therefore, the measured values of polarizability and dipole moment with the described interferometric deflection method always include an average over many thermally activated internal states. Careful molecular dynamics simulations have to be performed to interpret experimental values.[13]

Coupling to internal state dynamics

The deflection of molecular interference pattern can be interpreted as coupling to internal molecular properties. These properties can be collective properties such as the polarizability, where many internal states contribute to the actual value. Alternatively, the coupling can be very specific to one or many internal molecular states depending on the specific preparation method. The latter requires sophisticated cooling techniques, which are unavailable for the complex molecules we discuss. The coupling of the centre of mass motion to the static polarizability has been demonstrated for a di-azobenzene derivative (DAD). Using a KDTLI, the optical polarizability at frequency equal to the grating laser can be extracted. As the laser power is increased a variance in the interference pattern, which due to the high frequency of the grating (5×10^{14} Hz) is only dependent on optical polarizability and not the dipole moment. For large molecules with all molecular absorption lines in the ultra-violet region, the static polarizability is approximated well by the optical polarizability. From the experiment it was shown that the electric dipole moment was dramatically affected by the thermally triggered change of conformation of the DAD molecule, while the polarizability was almost unaffected. Computed average values for χ, α and μ_0 were reproduced by the experiments. More details on this investigation can be found elsewhere.[6]

It is well known that DAD can change conformation after photo-activation, this has been demonstrated for molecules in solution and for those attached to solid surfaces.[14,15] We are interested in showing that this change can be observed with molecules in gas phase beams and that matter-wave interferometric techniques can be used to detect different conformations. It is important to note that the transition between *trans* and *cis* conformation, which is illustrated in Fig. 2 and changes the relative position of the two benzene rings, are not activated at typical oven temperatures. Molecules are expected to be in the *trans* conformation in all experiments to date. The observed conformation changes in the paper by Gring *et al.*[6] include the reorientation and re-configuration of the atoms in the fluorinated side chains of the DAD. Those conformation changes can generate a dramatic change in the dipole moment of the molecule and therefore reduce the visibility of the *cis-trans* transition. The internal to external state coupling using interferometric techniques may draw new pathways to investigate internal state dynamics and coherent state transfer in molecules, if we can reduce internal state excitations by cooling and optical manipulation techniques.

Example of internal to external state coupling—photoisomerisation of di-azobenzene derivative

Here we discuss details for the next step to experimentally study the coupling between external and internal molecular states in a molecule interferometer, when the conformation of the molecule changes on demand by photo-isomerisation. Using a TLI setup, we consider two different possibilities with respect to the choice of the molecular system used to demonstrate the effect. First, a cooled small molecule could be used by preparing or selecting[16–19] the molecule to be in a well defined quantum state. Here, all degrees of freedom of the molecule can be controlled and molecule interferometric deflection can be performed to couple internal to external states. The Stark shift of a single state transition could be resolved by a phase shift of the interference pattern. In this case the demonstration of internal–external coupling is clear. Secondly, we could demonstrate a photo-triggered *cis-trans* conformation change for larger, more complex molecules. DAD is an ideal candidate for this study as the dielectric susceptibility χ is unique for its *cis* and *trans* conformation states. The distinction between different structural isomers has recently been demonstrated with molecule interferometric deflection for particles of 1600 dalton.[20] Molecules of this size consist of tens to hundreds of atoms oscillating around their ground state position resulting in vibrations and micro-conformational changes as flipping of small functional groups at a given temperature. The dynamics along the internal

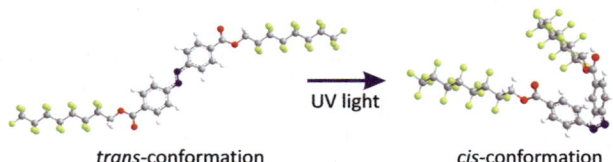

trans-conformation   UV light   *cis*-conformation

Fig. 2 The *cis* and *trans* conformation of the DAD molecule. This molecule can be photo-isomerised as has been demonstrated in solution and on solid surfaces. The molecule was synthesized in the group of Marcel Mayor at Basel and showed interference in Vienna.[1]

state manifold changes α and μ_0. It is vital to ensure that macroscopic effects still dominate, specifically the *cis-trans* conformation change effect on the electric susceptibility. Therefore it is necessary to predict the micro-conformational changes that contribute to the average values of polarizability and electric dipole moment from theory as has been done in Gring *et al.*[6] Cooling and alignment of the molecular beam would reduce the number of available micro-conformations and therefore control the effect of the micro-conformation changes. An increased molecular beam brightness by beam guiding will allow for shorter integration times during the interferometric measurements and therefore make the experiment less susceptible to noise. All effects of thermal activated internal state changes such as vibrations, rotations and the change of external micro-conformation on the susceptibility have to be smaller than the *cis-trans* conformation effect, which is on the order of 20%. We will discuss in section 3 if and how this can be achieved.

Photo-isomerisation of DAD

Photo-isomerisation of the DAD molecules has to be performed before the molecule beam enters the interferometer and the deflection electrode. The optical switching process is considered to be very similar to the observed for di-azobenzene in solution or attached to surfaces: A single photon, typically with wavelength between 350 nm and 500 nm, is sufficient to trigger the conformational change from the equilibrium *trans*- to the *cis*- conformation (see Fig. 2). The back-folding to the *trans*-state can be triggered by a different photon or will happen thermally after some minutes. The energetics for the transition from the thermal equilibrium *trans*-state to higher excited *cis*-state is described elsewhere in more detail.[21] The *trans*-isomer has a strong $\pi\pi^*$ band at 320nm, which is interpreted for the electron to go from a bonding π-state to an unoccupied anti-bonding π^*-state, and weak $n\pi^*$ band at 440 nm, where the electron goes from a *non*-bonding *n*-state to an unoccupied anti-bonding π^*-state. Typical time scales for the transitions are at some ps and the quantum efficiency of the transitions is: $\pi\pi^*$is 12% and $n\pi^*$ is at 25%. The *cis*-isomer has strong $\pi\pi^*$ band at 290nm and again a weak $n\pi^*$ band at 440nm with quantum efficiencies of $\pi\pi^*$ of 28% and for $n\pi^*$ of 51% at illumination. Therefore light of the wavelength of 320nm can be used to trigger the *trans-cis* conversion, while 290 nm for the *cis-trans* transition. The details of this photo-triggered state transfer processes are still under intense investigation and discussion[14,21,22] and our proposed experiment will contribute to this discussion.

3 Controlling internal and external states of large molecules with non-resonant light

A number of internal ro-vibrational states remain excited in a large polyatomic molecule even at temperatures of 5 K.[23] All these states and state changes result in variations of polarisability and electric dipole moment of the molecule. In order to couple the internal and external states within Talbot–Lau interferometric enhanced deflection a more complete method for preparing internal states is required, which means cooling. Here, we discuss the use of off-resonant coherent light to control the motion of molecules. So far, molecular beams are prepared by selection or collimation of particles from the broad distribution of longitudinal and transverse velocities of beams generated by a thermal source. High collimation as needed for matter wave interferometry leads unavoidably to a dramatic reduction of the number of molecules, close to the detection threshold. Generally, we need to increase control on the external motion itself in order to study effects such as internal–external state coupling by beam interferometric techniques. The main goal is to find efficient techniques to cool molecular degrees of freedom. Obviously, the complex structure of molecules, consisting of many atoms, hinders a straight forward application of cooling techniques as known from atomic physics and as well as detailed computation of its internal state structure.

Buffer gas cooling has already been used to produce intense beams (10^{13} cm^{-3}) of small molecules such as ammonia at temperatures of 20 K. This cooling scheme is simple and acts for all, external and internal, molecular degrees of freedom.[24] From our simulations we estimate that after 90 collisions with a 20 K neon buffer gas a thermal source of DAD would sufficiently thermalise, C_{60} under the same conditions requires 75 collisions. At this point we are not able to predict how many internal vibrations and rotations stay activated at 20 K, but we simulate a clear enhancement of the molecular beam brightness in optical guiding as discussed in the following. We assume, in agreement with experiments on smaller molecules such as tryptophan,[23] that the number of occupied internal states is significantly reduced a below 50 K temperatures, but that the respective ground states are not yet reached.

Optical guiding to increase molecule beam brightness

Starting with buffer gas cooling, we investigate the use of subsequent non-resonant light matter interactions in the context of external control, possibly leading to internal state cooling.[25–27] Several laser light manipulation and interaction schemes and experimental geometries have been theoretically proposed for complex molecules,[28–31] but these are yet to be realised. Small molecules have been successfully decelerated, oriented and deflected—and recently even cooled—by resonant and off-resonant optical fields.[32–36] We aim to explore if it is possible to optically manipulate large and therefore more complicated molecules. Interestingly, mechanical effects of light are common for the manipulation of much larger structures such

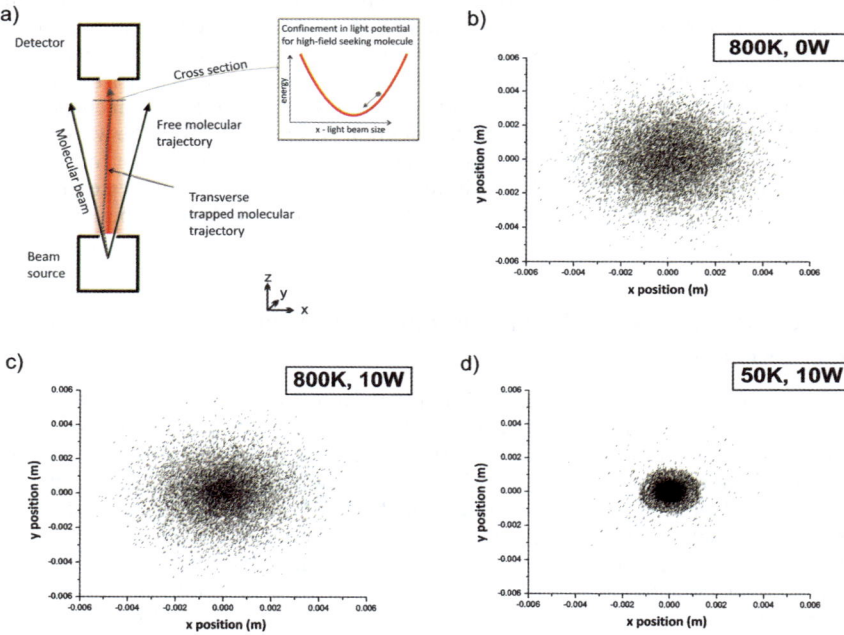

Fig. 3 Simulations show the guiding of C_{60} molecule by CW light, wavelength of laser is 800 nm, light beam diameter is 1 mm, propagation length of molecules is 0.6 m. a) shows the spatial arrangement of molecular beam and laser beam. b) shows the spatial distribution of detected molecules without the beam guide and therefore represents the natural thermal beam spread. The spatial density of molecules at detection is dramatically changed, if the laser is used to guide molecules. c) and d) show this collimation effect for hot and cold molecules.

as polystyrene and glass micro-spheres or biological cells.[37–40] We will discuss the guiding of neutral (C_{60}) molecules by a longitudinal continuous laser of reasonable power as shown in Fig. 3. Such experiments will give insight into light matter interaction in this regime, which can then be used to prepare more sophisticated laser cooling experiments of complex molecules.

In the case of molecular guiding with a continuous wave (CW) laser the optical potential is $U = \frac{1}{2}\alpha E^2$ and the dipole force is the gradient of the laser, typically a Gaussian beam. Simulations of molecule trajectories were performed for the guiding of C_{60}. The starting position of a particle was randomly chosen form the x,y plane at the position of the oven for the given oven size of 1 mm². The longitudinal velocity of the particles is fixed at the most probable velocity of the thermal source at a set temperature $v_{mp} = \sqrt{2k_B T/m}$. The transverse velocity is then given by a normally distribution generated by a random number and the transverse beam collimation multiplied to the longitudinal thermal velocity. For example, for a longitudinal velocity of v_{mp} = 180 m s^{-1} and a beam collimation of 1mrad we achieve a transverse temperature of about 1mK. The optical potential U_0 needs to be deeper than this transverse temperature to affect the transverse motion of a molecule. Particle trajectories are then calculated and plotted for the position of the detector of the given size (see Fig. 3 a). We simulated guiding for C_{60} molecules with a mass of 720 dalton and a polarizability of 79 Å^3. Molecules are propagating at a longitudinal velocity of 180 m s^{-1}, and the beam spread is assumed to be collimated to 1 mrad. The diameter of the molecular beam is 1mm.

The aim is to increase neutral molecule beam brightness in similarity to earlier experiments with atoms, as nicely illustrated for instance in the review paper by Balykin and Letokhov[41] or Chu.[42] This experiments are closely related to optical cooling of atoms[43] towards atomic Bose–Einstein Condensation (BEC).[44] We here theoretically investigate the increase in brightness by guiding beams of large molecules by light in a similar geometry as used in[45] to focus neutral atoms by the dipole force. Molecular beam guiding has been demonstrated before by hydrodynamic effects, see for instance the paper by Patterson and Doyle.[46]

A molecular beam of C_{60} can be guided over long distances by a CW off-resonant laser field of relatively low power of 10W. The arrangement for the longitudinal laser guide and the simulated detection particle density is illustrated in Fig. 3 for beam trajectories of a thermally distributed molecular beam. The guiding effect is enhanced in combination with a 50 K buffer gas cooled molecular beam, as seen in Fig. 3 d). The beam guiding effect is not a cooling effect, which means that after switching off the laser, molecules would continue to spread. The transverse kinetic energy of the particles is not dissipated, it is only compensated by the light potential. This is an efficient technique to transport a considerable number of the initial molecules over large distances. It will significantly enhance the molecular beam brightness.

Alignment of molecules to control average over internal state excitations

In general susceptibility is a tensor, which means molecular beam deflection and therefore the proposed selective detection of different conformers of the molecule depends on the orientation of the molecule in the electric deflection field. In lieu of detailed knowledge of the internal properties of DAD, we assume that this anisotropy of the molecular susceptibility will be averaged out by the rotation of our molecules around possible rotation axes in the thermal beam preventing the distinction of the *trans* and *cis* conformations. Therefore we aim to include a molecular alignment procedure within our experiment in order to efficiently resolve the different conformations. Molecular alignment has been studied in depth both theoretically and experimentally, a review of the topic and related literature is available.[35,47–49] Alignment is required for the total time DAD passes the electrostatic deflectors, for a 20 K molecular beam this is

9 ms. The electric deflection field itself is not strong enough to orient the molecules by itself, therefore we propose a cavity amplified CW laser operating longitudinally through the molecular beam. Early calculations suggest that the 3nm long DAD molecule with mass 1034 dalton, rotating at 20 K can be aligned along the polarization direction of an off-resonant Gaussian optical field of power of 2×10^4 W at a beam waist of 100 μm. This estimate takes only the alignment of the rotation axis into account. At this point we are not able to calculate the excitation of the exact rotational state as this would require much more sophisticated *ab initio* computation techniques. The same stays true for our knowledge on the occupation of vibrational states.

Absorption effects

We estimate absorption effects by using the experimental absorption cross section for C_{60} fullerenes at light wavelength of 800 nm in order to understand the importance of effects such as photo-ionisation, photo-dissociation and photo-fragmentation, which would have a negative influence on the guiding and alignment. The absorption cross section of C_{60} at 800 nm is on the order of $\sigma_{sol} = 6 \times 10^{-20}$ cm^{-2} as evaluated by absorption spectroscopy of fullerene solutions.[50] Absorption cross sections for gas-phase molecules may be slightly different with respect to solvent effects, but here σ_{sol} is expected to be a good first approximation. The resulting number of photons absorbed by a single molecule per ns, which is the relevant timescale τ for the relaxation of internal excitation, for the given laser power and laser beam waist ω_0 is: $N_{abs} = (2P\sigma\tau)/(\pi\omega_0^2 h\nu) = 6 \times 10^{-4}$. Therefore, we do not expect effects depending on photon absorption to be dominant in the regime of our experiments. Instead, the dipole force is dominating the light–molecule interaction with respect to centre of mass motion and alignment, while more careful considerations are needed to understand the effects on vibrations. All absorbed photons will result in internal heating of the molecule. While the simulations were done for C_{60} we expect a very similar behaviour for DAD molecules, where absorption cross sections have not been measured yet.

In summary, the combination of buffer gas cooling to reduce the ro-vibrational energy with optical alignment to amplify the difference in conformations and with the increase in molecular beam brightness by optical guiding techniques will allow to distinguish between the photoisomerisation of *cis* and *trans* of the DAD molecule in near future interferometric deflection experiments.

4 Conclusion

Molecule interference gives a new approach to precise coherent manipulation of the centre of mass motion. Internal state dynamics can be coupled to this transverse motion and molecule internal properties can be investigated with this same precision. We propose to combine this matter-wave interferometry technique with ultrashort-pulse coherent and optimal control experiments to guide internal state excitations coherently. As an example we discuss the effects of optical switching the conformation of di-azobenzene derivative molecules and the observation by interferometry in conjunction with a collisionally cooled molecular beam source of 20 K. In addition, we have shown simulations of the light–matter interaction for large complex molecules. Both, optical beam guiding and alignment of a buffer gas pre-cooled molecular beam of flourinated di-azobenzene will significantly increase the visibility of its *cis-trans* conformation change. This will be an important step to understand and investigate external-internal state coupling in complex molecules. Our investigations here can only be a small step into this exciting field of research.

Acknowledgements

This work has been financially supported by the University of Southampton, the South-English Physics Network (SEPNet) and by the Foundational Questions

Institute (FQXi). We acknowledge Peter Horak for help with the simulations of the optical manipulation.

References

1 S. Gerlich, L. Hackermüller, K. Hornberger, A. Stibor, H. Ulbricht, M. Gring, F. Goldfarb, T. Savas, M. Müri, M. Mayor and M. Arndt, *Nat. Phys.*, 2007, **3**, 711–715.
2 S. Gerlich, S. Eibenberger, M. Tomandl, S. Nimmrichter, K. Hornberger, P. J. Fagan, J. Tüxen, M. Mayor and M. Arndt, *Nat. Commun.*, 2011, **2**, 263.
3 J. Clauser and M. Reinsch, *Appl. Phys. B: Photophys. Laser Chem.*, 1992, **54**, 380–395.
4 S. Nimmrichter and K. Hornberger, *Phys. Rev. A: At., Mol., Opt. Phys.*, 2008, **78**, 023612.
5 B. Brezger, L. Hackermüller, S. Uttenthaler, J. Petschinka, M. Arndt and A. Zeilinger, *Phys. Rev. Lett.*, 2002, **88**, 100404.
6 M. Gring, S. Gerlich, S. Eibenberger, S. Nimmrichter, T. Berrada, M. Arndt, H. Ulbricht, K. Hornberger, M. Müri and M. Mayor, *et al.*, *Phys. Rev. A: At., Mol., Opt. Phys.*, 2010, **81**, 031604.
7 L. Hackermüller, S. Uttenthaler, K. Hornberger, E. Reiger, B. Brezger, A. Zeilinger and M. Arndt, *Phys. Rev. Lett.*, 2003, **91**, 90408.
8 S. Adler and A. Bassi, *Science*, 2009, **325**, 275.
9 S. Nimmrichter, K. Hornberger, H. Ulbricht and M. Arndt, *Phys. Rev. A: At., Mol., Opt. Phys.*, 2008, **78**, 63607.
10 K. Hornberger, S. Uttenthaler, B. Brezger, L. Hackermüller, M. Arndt and A. Zeilinger, *Phys. Rev. Lett.*, 2003, **90**, 160401.
11 M. Berninger, A. Stefanov, S. Deachapunya and M. Arndt, *Phys. Rev. A: At., Mol., Opt. Phys.*, 2007, **76**, 13607.
12 S. Gerlich, M. Gring, H. Ulbricht, K. Hornberger, J. Tüxen, M. Mayor and M. Arndt, *Angew. Chem., Int. Ed.*, 2008, **47**, 6195–6198.
13 S. Deachapunya, A. Stefanov, M. Berninger, H. Ulbricht, E. Reiger, N. Doltsinis and M. Arndt, *J. Chem. Phys.*, 2007, **126**, 164304.
14 J. Buback, M. Kullmann, F. Langhojer, P. Nuernberger, R. Schmidt, F. Würthner and T. Brixner, *J. Am. Chem. Soc.*, 2010, **132**, 16510–16519.
15 G. Mercurio, E. McNellis, I. Martin, S. Hagen, F. Leyssner, S. Soubatch, J. Meyer, M. Wolf, P. Tegeder and F. Tautz, *et al.*, *Phys. Rev. Lett.*, 2010, **104**, 36102.
16 A. Nikolov, E. Eyler, X. Wang, J. Li, H. Wang, W. Stwalley and P. Gould, *Phys. Rev. Lett.*, 1999, **82**, 703–706.
17 L. Holmegaard, J. Nielsen, I. Nevo, H. Stapelfeldt, F. Filsinger, J. Küpper and G. Meijer, *Phys. Rev. Lett.*, 2009, **102**, 23001.
18 F. Filsinger, U. Erlekam, G. Von Helden, J. Küpper and G. Meijer, *Phys. Rev. Lett.*, 2008, **100**, 133003.
19 F. Filsinger, J. Küpper, G. Meijer, J. Hansen, J. Maurer, J. Nielsen, L. Holmegaard and H. Stapelfeldt, *Angew. Chem., Int. Ed.*, 2009, **48**, 6900–6902.
20 J. Tüxen, S. Gerlich, S. Eibenberger, M. Arndt and M. Mayor, *Chem. Commun.*, 2010, **46**, 4145–4147.
21 H. Satzger, C. Root and M. Braun, *J. Phys. Chem. A*, 2004, **108**, 6265–6271.
22 P. Hamm, S. Ohline and W. Zinth, *J. Chem. Phys.*, 1997, **106**, 519.
23 P. Carcabal, R. T. Kroemer, L. C. Snoek, J. P. Simons, J. M. Bakker, I. Compagnon, G. Meijer and G. v. Helden, *Phys. Chem. Chem. Phys.*, 2004, **6**, 4546–4552.
24 D. Patterson, J. Rasmussen and J. Doyle, *New J. Phys.*, 2009, **11**, 055018.
25 G. Morigi, P. W. H. Pinkse, M. Kowalewski and R. de Vivie-Riedle, *Phys. Rev. Lett.*, 2007, **99**, 073001.
26 P. Staanum, K. Højbjerre, P. Skyt, A. Hansen and M. Drewsen, *Nat. Phys.*, 2010, **6**, 271–274.
27 T. Schneider, B. Roth, H. Duncker, I. Ernsting and S. Schiller, *Nat. Phys.*, 2010, **6**, 275–278.
28 S. Deachapunya, P. Fagan, A. Major, E. Reiger, H. Ritsch, A. Stefanov, H. Ulbricht and M. Arndt, *The European Physical Journal D-Atomic, Molecular, Optical and Plasma Physics*, 2008, **46**, 307–313.
29 T. Salzburger and H. Ritsch, *New J. Phys.*, 2009, **11**, 055025.
30 S. Nimmrichter, K. Hammerer, P. Asenbaum, H. Ritsch and M. Arndt, *New J. Phys.*, 2010, **12**, 083003.
31 P. Barker and M. Shneider, *Phys. Rev. A: At., Mol., Opt. Phys.*, 2010, **81**, 023826.
32 R. Fulton, A. Bishop and P. Barker, *Phys. Rev. Lett.*, 2004, **93**, 243004.
33 R. Fulton, A. Bishop, M. Shneider and P. Barker, *Nat. Phys.*, 2006, **2**, 465–468.
34 H. Stapelfeldt, H. Sakai, E. Constant and P. Corkum, *Phys. Rev. Lett.*, 1997, **79**, 2787–2790.
35 H. Stapelfeldt and T. Seideman, *Rev. Mod. Phys.*, 2003, **75**, 543.

36 E. Shuman, J. Barry and D. DeMille, *Nature*, 2010, **467**, 820–823.
37 A. Ashkin, *Phys. Rev. Lett.*, 1970, **24**, 156–159.
38 A. Ashkin, J. Dziedzic and T. Yamane, *Nature*, 1987, **330**, 769–771.
39 A. Ashkin and J. Dziedzic, *Science*, 1987, **235**, 1517–1517.
40 T. Li, S. Kheifets, D. Medellin and M. Raizen, *Science*, 2010, **328**, 1673.
41 V. Balykin and V. Letokhov, *Phys. Today*, 1989, **42**, 23–28.
42 S. Chu, *Science*, 1991, **253**, 861.
43 W. Phillips, *Rev. Mod. Phys.*, 1998, **70**, 721–742.
44 M. Anderson, J. Ensher, M. Matthews, C. Wieman and E. Cornell, *Science*, 1995, **269**, 198.
45 J. Bjorkholm, R. Freeman, A. Ashkin and D. Pearson, *Phys. Rev. Lett.*, 1978, **41**, 1361–1364.
46 D. Patterson and J. Doyle, *J. Chem. Phys.*, 2007, **126**, 154307.
47 M. Vrakking and S. Stolte, *Chem. Phys. Lett.*, 1997, **271**, 209–215.
48 J. Underwood, M. Spanner, M. Ivanov, J. Mottershead, B. Sussman and A. Stolow, *Phys. Rev. Lett.*, 2003, **90**, 223001.
49 H. Sakai, S. Minemoto, H. Nanjo, H. Tanji and T. Suzuki, *Phys. Rev. Lett.*, 2003, **90**, 83001.
50 N. Gotsche, H. Ulbricht and M. Arndt, *Laser Phys.*, 2007, **17**, 583–589.

PAPER

Combining dissociative ionization pump–probe spectroscopy and *ab initio* calculations to interpret dynamics and control through conical intersections

Spiridoula Matsika,[*a] Congyi Zhou,[a] Marija Kotur[b] and Thomas C. Weinacht[b]

Received 21st March 2011, Accepted 20th April 2011
DOI: 10.1039/c1fd00044f

Nonadiabatic processes play an important role in molecular dynamics, and understanding these processes better can help interpret and guide control over molecules. We are using high level electronic structure calculations in combination with intense, shaped, ultrafast laser pulses to study excited state dynamics in the nucleic acid bases, cytosine and uracil. These molecules have very short excited state lifetimes as they relax radiationless through conical intersections after absorption of UV radiation. The presence of more than one relaxation pathway provides the possibility to control which pathway can be involved in the dynamics. In our approach the molecules were excited using ultrafast laser pulses in the deep UV and then probed with strong field near infrared pulses which ionize and dissociate the molecules. Key to this approach is the fact that different fragments exhibit different dynamics and we can correlate these fragments, and their associated dynamics, to the various pathways involved in the neutral dynamics. Multiconfigurational electronic structure methods were used to calculate potential energy surfaces of the neutral and ionic states involved in the dynamics. Calculating mechanisms for fragmentation in the ion enables us to relate specific fragments to different neutral pathways, and use them as signatures to follow the dynamics. Possibilities for control are also discussed.

1 Introduction

When molecules absorb radiation, radiative or radiationless competing processes can take place on the excited state driven by the shape of the potential energy surface (PES) and the kinetic energy deposited on the wavepacket. The excited state dynamics can lead to different final products enabling photochemical reactions, or they can lead back to the initial ground state if photophysical events prevail. Conical intersections (CIs) are ubiquitous in the PESs of molecules and they play an important role in driving the dynamics through the various competing pathways.[1,2]

Designing schemes to control the outcome of dynamics through conical intersections is very advantageous, since it can lead to control over the outcome of the various competing photophysical and photochemical events. For example, going though one or more conical intersections may lead to different photoproducts and one may achieve control over their branching ratio.

[a] *Department of Chemistry, Temple University, Phialdelphia, PA, 19122, USA. E-mail: smatsika@temple.edu; Fax: +215 204 1532; Tel: +215 204 7703*
[b] *Department of Physics, Stony Brook University, Stony Brook, NY, 11794, USA*

Controlling the dynamics of excited states can be guided by learning more about the pathways involved. Theory can be very useful in this task, since nowadays methodology has advanced enough so that PESs can be calculated at a reliable level to give insight into the available pathways. Experimental verification and signatures of the theoretically proposed pathways are necessary in order to establish with certainty how the dynamics occurs. Several experimental approaches have been developed in order to follow excited state dynamics. In most cases a pump–probe scheme is used, where the pump excites the molecules to the excited state of interest and a probe after some time delay generates a signal characteristic of the excited state PES. In our approach we use a pump–probe scheme, where an ultrafast UV pulse excites the molecules to the first excited bright state and after a time delay an intense IR probe ionizes and fragments them. Key to this approach is the fact that different fragments exhibit different dynamics and we can correlate these fragments, and their associated dynamics, to the various pathways involved in the neutral dynamics. So, the ion yield of each fragment as a function of pump–probe delay contains information about the dynamics along the neutral excited state surface.

Obtaining information about how the fragments are related to the various points along the neutral PES of the molecules enables us to choose appropriate feedback signal for control in a closed loop experiment. We have demonstrated this idea previously using shaped ultrafast laser pulses in the deep ultraviolet to control the ring opening isomerization of 1,3-cyclohexadiene (CHD) to form 1,3,5-hexatriene (HT).[3] In that reaction we observed that the fragments produced by the reactant (CHD) are very different from the fragments produced by the product (HT). Specifically, when CHD ionizes it produces mainly the parent ion, while HT breaks into many smaller fragments. This information enabled us to use the ratio of parent ion to the smaller fragments as a feedback signal in the control experiment. The shaped pulse yielded a 37% increase in the isomerization over an unshaped laser pulse. Theory helped to recognize that CHD cation will not fragment significantly while HT is expected to fragment much more. We want to test this approach in other systems.

Theory is necessary to be able to interpret the observed signals and relate them to the neutral excited state dynamics. There are several steps involved in the experiment that need to be modeled theoretically in order to interpret what is being observed. The information needed for the various steps can be categorized as:
• PESs of the excited states of the neutral molecules describing the neutral dynamics
• Excited ionic states that can be reached upon ionization from the different points on the neutral PESs
• Fragmentation mechanisms in the ion.

All of these steps are complicated to be studied theoretically. We will discuss below how we approach each step in order to get an *ab initio* description of the processes involved.

We have used this approach for the pyrimidine bases uracil and cytosine. DNA bases have been found to have multiple conical intersection seams in their PESs which facilitate ultrafast radiationless decay to the ground state.[4] The presence of more than one relaxation pathway provides the possibility to control which pathway can be involved in the dynamics. Previous electronic structure theory calculations have shown multiple pathways for both uracil and cytosine.[5–16] Time dependent studies have also been reported which try to address the question of which of these pathways are dominating the dynamics.[9,16–18] The results are very sensitive to the theory used since these systems are very complicated and compromises have to be made in the accuracy of the methods. A combination of theory and experiment that can provide and detect signatures of the available pathways will be valuable in elucidating the details of the dynamics.

Besides using pump–probe data to explain the dynamics we are also interested in controlling the outcome of the dynamics in these systems. Control experiments, not only demonstrate control, buy can also provide further information to interpret the dynamics. Attempts for control in uracil will be discussed.

2 Theoretical methodology

Important stationary points and conical intersections along the neutral excited state PESs have been reported before,[5,12] and the optimized geometries are used here. The multireference configuration interaction method was used for these calculations. Details can be found in the original publications.[5,12]

Ionic states of cytosine and uracil cations have also been reported before[19] but we had not discussed them in length. These states are shown and discussed further here since they are necessary to describe the overall approach. They were calculated using the multi-reference perturbation theory (MRPT2) method,[20,21] and the 6-311+G(d, p) basis set using molecular orbitals from a state-averaged complete active space self-consistent field (CASSCF) calculation. A complete active space of 13 electrons in 10 orbitals was used, denoted as (13, 10). The 10 orbitals included the eight π orbitals and two lone pairs. For uracil the two lone pairs were on the two oxygen atoms while for cytosine they were one on oxygen and one on nitrogen. The energies of the ionic states were calculated at important geometries along the neutral excited state PESs.[5,12] Natural orbitals for each state were obtained at the CASSCF level.

Fragmentation was examined using two approaches. In the first approach we optimized the structures and energy of the separate fragments that could be produced from the parent ion and compared their combined energy to that of the parent ion. At least two possibilities were calculated where the charge could remain on either fragment. Sometimes we also tried different multiplicities to find the most stable species. These calculations give us an estimate of the energy needed to produce the fragments. They also show which fragment is more likely to carry the charge based on the relative energies of each pair. These energies are related to the ionization potentials of each fragment. The MP2 or UMP2 methods with the 6-31G(d) basis set were used to optimize each fragment in its neutral and cationic state. Ionization energies for the parent ions, uracil and cytosine, were calculated using MP2 (UMP2)/6-311+G(d,p) as well in order to obtain higher level results and compare them to the 6-31G(d) basis.

In some cases the energy needed to produce the fragments may be higher than the energy of the separated fragments. This will be the case when additional barriers exist along the dissociation pathways. In this case we need to explore in more detail the mechanisms for dissociation. Constrained optimizations where one bond at a time is stretched are used to produce dissociation curves. UMP2/6-31G(d) is used for these calculations.

The MRPT2 calculations were conducted using GAMESS.[22] MP2 calculations were done using NWChem[23] and GAMESS. Constrained optimizations were done using Gaussian.[24] Most calculations were performed on the Chinook supercomputer cluster at the Environmental Molecular Sciences Laboratory, a Department of Energy national scientific user facility located at Pacific Northwest National Laboratory in Richland, Washington.

3 Experimental techniques

Control experiments were carried out using a shaped ultrafast laser pulse in the deep UV followed by an intense infrared laser pulse. The UV laser pulses were generated using the output of our amplified titanium sapphire laser system (1 KHz, 1 mJ, 30 fs, 780 nm). The IR pulses were used to first generate UV pulses at 390 nm using second harmonic generation. These pulses were then mixed with the fundamental to produce light at 260 nm using sum frequency generation. Between the second harmonic generation and sum frequency generation stages (both using appropriately cut BBO crystals), we made use of a calcite crystal for group velocity dispersion compensation. We typically generated UV pulses at 260 nm with about 20 μJ of energy having a pulse duration just under 50 fs. These pulses are directed into an acousto-optic modulator based pulse shaper, where they are shaped under computer control. The shaped pulses had a pulse energy of about 2 μJ and a minimum pulse duration of about 50 fs.

The shaped UV ('pump') pulses were collinearly combined with the IR ('probe') pulses (40 fs, up to 100 μJ) and focused into an effusive molecular beam of uracil inside a time of flight mass spectrometer (TOFMS). The UV pulses excited the molecules from the ground state to the first bright state and the IR laser pulses ionized and dissociated the excited molecules. The energy of the UV pulses was limited in the experiments (by irising the beam) in order to achieve minimal ionization with the pump pulse alone while still exciting a large fraction of the molecules. For the closed loop control experiments, we set the pump–probe delay to a fixed value and used different peaks in the TOFMS to evaluate the effectiveness or fitness of each shaped UV pulse.

4 Results and discussion

4.1 Neutral excited state pathways

Fig. 1 demonstrates the main features of the pumb–probe approach. The plots correspond to the pyrimidine bases cytosine and uracil. A UV photon of 260 nm excites the molecules to the first bright excited state. After delay τ a probe ionizes the molecules producing the cation in some ionic state. Fragmentation can follow ionization, especially when ionizing to higher ionic states.

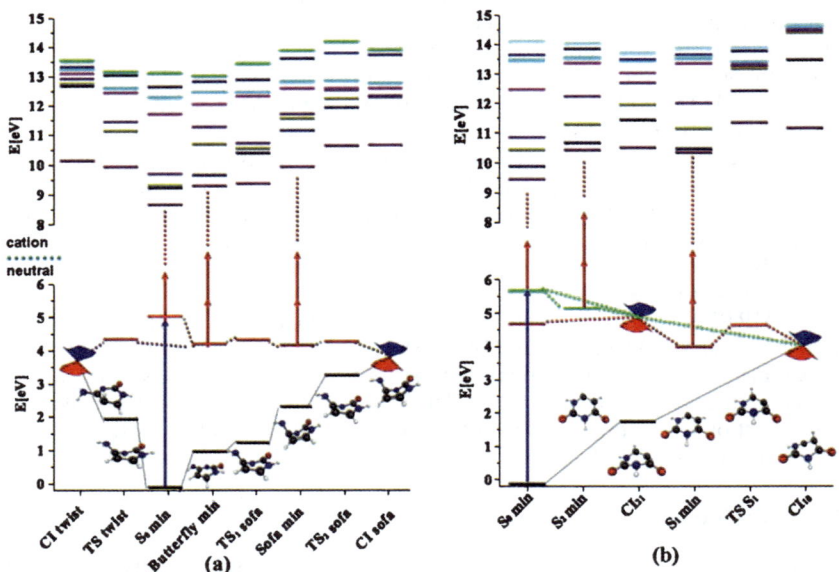

Fig. 1 Neutral and ionic energies along some of the important features of the relaxation pathway on the S_1 (and S_2 for uracil) PES of the neutral molecule for (a) cytosine and (b) uracil. Lower panel: Energies for neutral states taken from[5,12] (black: ground state, red: S_1 state, green: S_2 state). Upper panel: Energies for the eight lowest-laying ionic states.[19] Dotted lines are used to mark predicted connections between the states.

In cytosine absorption leads to the S_1 bright state which then relaxes to a minimum (butterfly min). From that minimum it can branch to two different directions which lead to conical intersections with the ground state. In one direction the molecule distorts along a 'sofa' coordinate and reaches the 'sofa' CI while in the other direction the C^5–C^6 bond twists (similarly to the CI in ethylene) until the CI is reached (called the 'twist' CI). Low barriers exist in both directions which may slow down somewhat the relaxation, but the decay is expected to be fast overall. Whether one of these pathways is preferred is not clear based on the electronic structure calculations. Based on previous dynamical studies the wavepacket will delocalize and go through multiple pathways.[16] Our previous results agree with this idea.[25]

In uracil absorption leads to the S_2 state, which is a $\pi\pi^*$ bright state, while the S_1 is an $n\pi^*$ dark state. Relaxation from the Franck–Condon (FC) region can lead either to a S_2/S_1 CI (CI_{21}) or to an S_2 minimum. If the wavepacket goes through the S_2/S_1 CI it can branch and either go directly to a CI with the ground state (CI_{10}) or relax to the S_1 minimum where it will stay for a while. Dynamical studies disagree on whether trapping on the S_2 minimum occurs.[9,18] Trapping on the S_1 minimum has also been seen both in the dynamical studies and in experimental results. This state is a dark state, so the radiative lifetime is expected to be long. It is also separated from the CI_{10} by a high barrier, so its radiationless lifetime should also be long. Long lifetimes are then the signature of trapping to this state.

Both of these molecules show fast radiationless decay to the ground state. But, as the calculations show the details are very different and specific to the system.

The x-axis in the plots in Fig. 1 represent reaction coordinates leading to the various pathways along the S_1 or S_2 PESs. In reality the plots are far from one dimensional since the reaction coordinate changes along the pathways. Furthermore the reaction coordinates are a mixture of normal modes of the molecule. Calculations of barriers give us the main reaction coordinate to surpass the barrier. Conical intersections involve at least two coordinates and thus branching can easily occur at these points. The wavepacket will need to redistribute its energy in these different modes as it evolves along the PESs.

4.2 Ionic states

The probe pulse ionizes the molecules moving them from the excited neutral PES to one or more ionic states. In order to be able to understand what signals can be generated from the probe we first need to know where the ionic states are located. We use MRPT2 to calculate the energies of the ionic states along the neutral pathways. These calculations show how the ionic PESs change as the molecule moves along the neutral PESs. Assuming a sudden ionization the pulse will create a wavepacket vertically above the place on the S_1 (or S_2) surface where the molecule is at time delay τ.

The ionic states are shown in Fig. 1. The energy of the lowest ionic state D_0 at the S_0 minimum has been set equal to the vertical ionization potential of each molecule taken from photoelectron spectra.[26–28] These energies are 8.80 eV and 9.60 eV, for cytosine, and uracil, respectively. Several features of the ionic states may be important. The ionization depends on the IP so the energy difference between the S_1 (or S_2) state and the D_0 state is crucial. Looking at Fig. 1 it is obvious that the IP increases always when moving away from the FC region. For uracil and cytosine the increase happens very fast and is more than 2 eV at some points along the PES. This increase will cause a sudden drop in the ionization yield after time 0 for all fragments including the parent ion. The gap between ionic states may be important as well. For example if the gap approaches the frequency of the probe, resonance effects may be observed. In this case there will be an increase in the population of the excited ionic states which will lead to increased fragmentation.

Besides the energies of the ionic states their character is also important when ionization occurs. In one photon ionization this can be determined qualitatively by

Koopmans picture and Dyson orbitals. The Dyson orbitals are used to describe one photon ionization from an excited neutral state to an ionic state and they are obtained from the overlap between the neutral and ionic wavefunctions. For example, when ionization occurs from an excited $\pi\pi^*$ state the easiest process is that an electron is removed from the π^* orbital and the remaining ionic state has a lone pair on the π orbital. In multiphoton ionization, as is the case here, the picture may be more complicated than what is provided by the Dyson orbitals, but the character of the ionic state should still be important.

Table 1 shows the energies and main character of the first eight ionic states for cytosine and uracil calculated at the S_0 equilibrium geometry. For each state the natural orbitals describing the wavefunction are also obtained. If there is one orbital with occupation number close to one this orbital is given in parenthesis. If there are three orbitals with occupation numbers close to one then these three orbital characters are shown in parenthesis. There are many similarities between the two molecules. D_0 in both molecules is a π state, where the hole is in the π orbital and so the π bonds are considerably weaker. The first excited state is about 0.5 eV above the ground state and it is an n state, thus it has the opposite symmetry with respect to the plane of the molecule. There are four states within 1.5 eV and then there is a big gap to the next state, D_4. The order of the first four states in both molecules is π, n, π, n. It is interesting that the order of states in the ion does not mirror the excited states in the neutral. So, although the neutral excited states in cytosine are $\pi\pi^*$ and $n\pi^*$ for S_1 and S_2 respectively, for uracil they are switched with S_1 being $n\pi^*$ and S_2 $\pi\pi^*$. Yet, for the ion the π state is lower in both cases. This points to the fact that Koopmans theorem and the simple idea that if the lowest excited state is a HOMO–LUMO excitation then the first ionic state D_0 will have an unpaired electron in the HOMO is not always right. When the electron is removed the electron density is redistributed and the multi-electron wavefunction has to be considered. When the energies are above 4 eV the states have three unpaired electrons. This will have an important effect on the probability for ionization to these states, since the overlap with the neutral excited states, represented by Dyson orbitals, will be very different. Qualitatively, there is overlap between a $(HOMO)^1(LUMO)^1$ excited state and an ionic $(HOMO)^1$ state, but the overlap of $(HOMO)^1(LUMO)^1$ with the other states with one unpaired electron is probably small, since these other states roughly correspond to removing the electron from orbitals below the HOMO. If the ionic state however is $(HOMO-1)^1(HOMO)^1(LUMO)^1$ then this will have a good overlap with the neutral $(HOMO)^1(LUMO)^1$ excited state.

The discussion so far focused on the ionic states at the S_0 equilibrium, corresponding to vertical ionization. Beyond that point ionic states have their own complicated

Table 1 Energies of the ionic states for cytosine and uracil at the S_0 equilibrium geometry calculated using MRPT2(13,10)/6-311+G(d,p). The energy of D_0 is set equal to zero. Experimental values from photoelectron spectra are shown for comparison.[27,28]

	Cytosine		Uracil	
State	MRPT2	Exp.[27]	MRPT2	Exp.[28]
D_0	0 (π)	0 (π)	0 (π)	0 (π)
D_1	0.58 (n)	0.63 (n)	0.44 (n)	0.52 (n)
D_2	0.65 (π)		0.98 (π)	0.97 (π)
D_3	1.05 (n)	1.08	1.40 (n)	1.57 (n)
D_4	3.03 (π)	2.98	3.01 (π)	3.04 (π)
D_5	3.62 (π)		4.01 ($\pi\pi\pi$)	
D_6	3.98 ($\pi\pi\pi$)		4.18 ($n\pi\pi$)	
D_7	4.43 ($n\pi\pi$)		4.64 ($\pi\pi\pi$)	

PESs, which we should consider. They often have their own CIs, which will complicate the interpretation of the signal. Conical intersections between ionic states have been found in both cytosine and uracil.[29,30] In cytosine we have located a CI between D_0/D_1 which leads to two minima on D_0, as will be discussed below. CIs in the ion can lead to a change in the character of the ionic states along the neutral pathways. So the ionization signal may change in our pump–probe data, not because of a change in the neutral (which is what we are after) but because of a change in the ion. So we always have to be aware of this possibility and check the character of the states along the neutral PES. We checked the natural orbitals of the ionic states in uracil and cytosine along the neutral PES and found that the first ionic states D_0 and D_1 do not switch along these pathways. The orbitals change shape somewhat since the molecules distort, but they do not change character nonadiabatically. This makes us more confident in interpreting the fragment dynamics observed as dynamics of the neutral molecules.

The electronic structure and energies of the equilibrium ground state in the neutral and the ion were calculated. Vertical ionization potential (VIP) and adiabatic ionization potential (AIP) of uracil calculated at the UMP2/6-311+G(d,p) level are 10.03 eV and 9.52 eV, respectively. The relaxation energy on D_0 from vertical ionization to adiabatic ionization in uracil is 0.51 eV. Experimentally this is about 0.3 eV.[26,31] The equilibrium geometry of the cation is planar with some bonds having changed compared to the neutral equilibrium structure. The changes point to the change in the electron density as the molecule ionizes, and they can help us have an initial qualitative idea of how subsequent fragmentation occurs. The C^5C^6 increases by 0.06 Å, C^2N^1 by 0.08 Å and C^6N^1 decreases by 0.07 Å. VIP and AIP of cytosine calculated at the UMP2/6-311+G(d,p) level are 9.65 eV and 9.18 eV, respectively. In cytosine the difference between VIP and AIP, corresponding to the relaxation on the D_0 surface, is 0.47 eV. More bonds are affected by ionization in cytosine. Even more interestingly, two minima (saddle points) were found. One having an A′ and one having an A″ symmetry. This is because a CI exists at low energies which can cause the existence of more than one minima, as in a Jahn–Teller system. We have optimized a CI between the two states at the MRCI level 0.37 eV above the global D_0 minimum. As discussed above, CIs in the cations are important, and we should be aware of this in our analysis.

4.3 Fragmentation

The signal that we eventually detect is various fragments as a function of time delay, so information about the ionic states is not enough. Any information on neutral dynamics that we can extract has to be contained in the fragments. This makes the link between fragments and positions on the neutral PES very important. Such a link is difficult to establish. We first need to find a fragmentation mechanism of how the fragments are produced, and then we have to establish how they can be connected to the neutral PESs.

We focus on fragments that come predominantly from one ionic state, most likely the ground ionic state. It is assumed that any observed parent ion comes from the ground ionic state since fragmentation can happen much easier in the higher ionic states, compared to the ground state. We tested this hypothesis in uracil. Since D_0 lies just within 2 UV photons from the S_0 min, but no other states do (i.e. D_n > 9.5 eV above S_0 min for all $n > 0$), the UV intensity scans can give us useful information. They show that the parent ion is linear in UV energy and all other fragments are quadratic. This indicates that at the FC point the parent is dominated by D_0 as there is single photon enhancement from S_1 which makes the process first order. An excited state of the ion would show quadratic scaling as other fragments illustrate. If the parent ion signal is dominated by D_0 at FC, then we can argue that it is dominated at all geometries since as you move away from the FC region, you just end up putting more vibrational energy into the molecule—both because of motion on S_1/S_2

and because of exciting above the minimum on the ionic state. This suggests that if you make parent through D_0 at FC (S_0 min), then any parent you make anywhere is most likely from D_0 since away from S_0 min, you will just have more energy with which to dissociate.

In order to find other fragments that come from the same ground ionic state their dependence on the intensity of the probe is examined. Fragments that show the same dependence as the parent fragment are associated with the ground ionic state. This is because if there is a different dependence that means that a different number of photons is needed for the different fragments. Same dependence shows that the same number of photons is needed for the parent ion and the given fragment. In cases where the second ionic state is very close to the ground ionic state this association becomes weaker since the same number of photons may be needed to reach D_0 and D_1, for example.

In order to examine fragmentation there are two important questions to address: (i) What is the energy needed to break the bonds and produce fragments (ii) If two fragments are generated which of the two parts is most likely to carry the charge. The IP of each fragment determines this choice. We calculate the IPs for the fragments to determine the energetically favorable channel.

The overall energy needed to fragment can be found by comparing the energy of the product fragments compared to the minimum energy of D_0. This gives the overall dissociation energy. This energy can be compared with experimental appearance energies (AEs) for the various fragments. In order to further understand the fragmentation in more detail, mechanistic calculations are needed to establish pathways for fragmentation in the ground ionic state D_0. But since fragmentation can be complicated, we also need to look for barriers along the dissociation coordinates.

4.3.1 Cytosine and uracil. The TOFMS of both of the pyrimidine bases uracil and cytosine includes the parent ion, which means it does not always fragment. Being able to identify under what conditions the parent ion does not fragment and relate these conditions to the neutral dynamics is very useful. The parent ion is one of the simplest ones to use in our analysis since it is most likely produced from D_0 rather than higher ionic states, so we don't need to worry about higher ionic states.

Experimentally it has been measured that the lowest appearance energy (AE) is 1.3–1.8 eV for uracil (the numbers vary between experiments)[32,33] and 2.2 eV for cytosine[34] above the corresponding IPs. So the parent ion will survive when the available energy is between the IP and the AE. Fig. 1 shows how the energy of the D_0 state changes along the neutral pathways. In uracil it seems that the energy to fragment will be available shortly after the FC region. In cytosine the D_0 state is not as steep along the neutral pathway so we would expect the parent ion to decay slower than in uracil. Indeed plots of the ion yields as a function of pump–probe show a much faster decay for uracil compared to cytosine, as we have shown in previous publication.[19]

Besides the parent ion other fragments that are dominant are the 68 amu (for cytosine) or 69 amu (for uracil), which in both molecules are associated with loss of HNCO. This fragmentation is the energetically most favorable to occur. For cytosine there is a unique way to lose HNCO. Previously we had identified a transition state that facilitates breaking of the first bond (N^1–C^2) only 0.37 eV above the D_0 minimum.[25] After passing through the transition state a mechanism has been proposed theoretically which leads to fragment 68 or fragment 83.[35] These are the dominant fragments in cytosine. We were able to show that there will not be enough energy to break the parent ion when ionizing from one of the minima along the S_1 PES (butterfly minimum) but there will be enough energy if ionizing from the other minimum. This enabled us to use the parent ion in cytosine as a signature of one of the pathways along S_1 where the butterfly minimum exists.[25] This is the twist pathway.

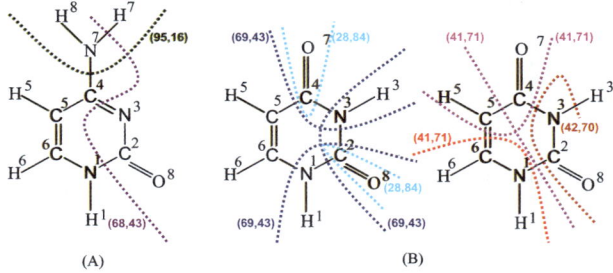

Fig. 2 Possible ways of bond cleavage that lead to the experimentally observed fragments for (A) cytosine and (B) uracil.

Focusing on cytosine for now, the second signature fragment that was used was fragment 95. This is created when the amino group is removed from the ring, as shown in Fig. 2. We were able to conclude that fragment 95 is produced mainly along the sofa minimum in cytosine.[25] So, in cytosine we have two different fragments, parent and 95, that label different pathways, twist and sofa, respectively. This brings us to the point we want to be. We can use the ratio of those two fragments as a possible feedback signal in a control experiment. Optimizing the shaped laser pulses to maximize one fragment over the other implies that the wavepacket is forced to choose one pathway over the other. Furthermore, we have gained some information about the excited state dynamics in cytosine. The above analysis shows that the wavepacket will be delocalized in the excited state and can follow both pathways, rather than preferably localize on one. This conclusion agrees with theoretical studies.[16–18]

In order to be able to do the same analysis in uracil we need to first identify mechanisms for fragmentation of the cation. Fig. 3 shows ion yields as a function of pump–probe delay for the dominant fragments in uracil. It is obvious that the various fragments show different decay signals, which could be used to interpret the neutral dynamics. As noted before, it is also quite obvious that the parent ion decays much faster than all the other fragments. So, the parent ion will be of interest in our analysis.

Another dominant fragment is the 69 amu as discussed before, and it turns out this is the one that needs the least energy to be formed. In uracil there are three different ways that fragmentation can lead to loss of HNCO and production of mass 69. Fig. 2 shows these ways. In order to identify which of these possibilities will be favored we need to know the energy needed to break the corresponding bonds. We need to know which bonds in the ring are the weaker that can break easier. The electronic structure of the cation gives some ideas on which bonds are weakened by ionization, although this is not sufficient. Dissociation curves along each bond provide a better estimate for the energy needed to break each bond. Fig. 4 shows dissociation curves obtained by constrained optimizations for uracil where one of the bonds each time is fixed and the remaining coordinates are optimized. These curves show that N^3–C^4 and N^1–C^2 require the least energy to break. N^1–C^2 is also the bond that has weakened the most upon ionization as seen by the elongated bond in the cation compared to the neutral. Further calculations of the energy of the separated fragments agrees that HNCO is produced by breaking of N^3–C^4 and N^1–C^2. The energy needed to produce these fragments is calculated to be 2.2 eV in reasonable agreement with experimental AEs. Experiments on thymine (which is very similar to uracil except of a methyl group substituent) agree with these bonds being broken first by using isotope labeling of C^2 on the ion to observe which HNCO is left.[33] We explored the dissociation to fragment 69 in more detail and found that there are no barriers higher than 2.2 eV along the dissociation pathway. A transition state exists along the pathway, and points to a concerted fragmentation.

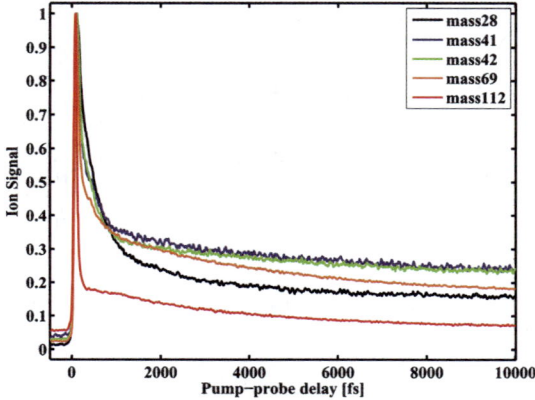

Fig. 3 Pump–probe data for several uracil fragments, taken with 0.8 mW of UV and 21 mW of IR, taken at parallel relative polarization of the two pulses.

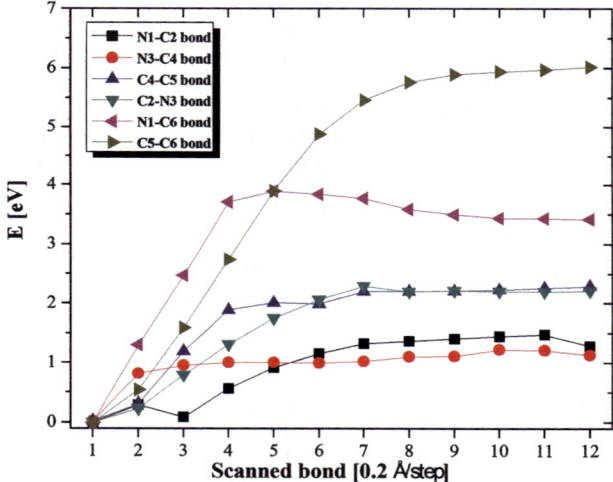

Fig. 4 Constrained optimizations of the D_0 state stretching the ring bonds in uracil. Point 1 is the equilibrium geometry of D_0. Each curve is obtained by stretching the corresponding bond by 0.2 Å per step and optimizing the remaining degrees of freedom.

Its energy is 1.85 eV above D_0, so it does not create any barriers higher than the dissociation limit. Interestingly, in cytosine the transition state leading to similar fragmentation in the ring indicates sequential, and not concerted, fragmentation.

Besides fragment 69 amu there are other fragments that appear on the TOFMS and have different dynamics. Fig. 3 shows the most important ones, 28, 41 and 42 amu. These fragments are smaller than 69. There are two possibilities of how they can be generated. The most straightforward mechanism is to be generated directly from fragmentation of the parent ion, as shown in Fig. 2. However, when we calculated the separated fragments produced by these dissociations, we found that the energy needed is too high. We calculated both cases, placing the positive charge on either fragment, and found that always the charge prefers to stay on the larger fragment. So these smaller fragments will be neutral, and undetectable in the TOFMS. This is reasonable since the IP for the larger fragments is usually larger.

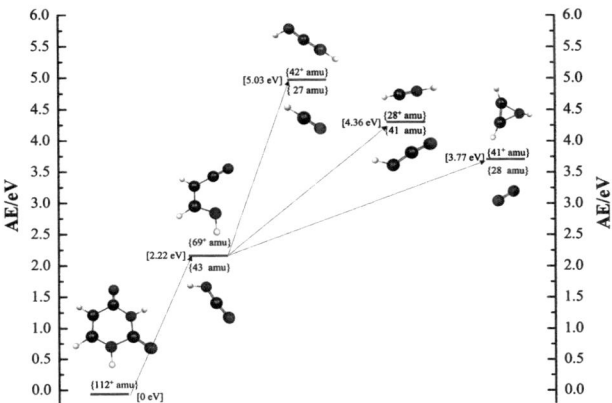

Fig. 5 Diagram showing the fragmentation mechanism in uracil cation. The energy of the pair of fragments is shown along with the optimized structures of the fragments. The zero has been set to the minimum of D_0.

Since direct fragmentation is not energetically favorable an alternative mechanism has to exist. It has been previously proposed that the smaller fragments in the mass spectrum come from further fragmentation of fragment 69. We calculated the energy needed for this further fragmentation and found that the energy needed is less than direct fragmentation from the parent ion. The energies we found are shown in Fig. 5 and agree with the experimental AEs. It should be noted however, that these energies are related to the final products only and they do not take into account the barriers that may exist. The existence of barriers may change the energetic requirements.

Interestingly our fragmentation calculations indicate that most fragments observed in uracil are generated from fragment 69, *i.e.* through a sequential fragmentation. This is unexpected since these fragments show different decay in the pump–probe plots. One would expect that once 69 is generated after ionization from a particular point on the $S_1(S_2)$ PES then how the other fragments are generated should not depend on the initial neutral geometry. This means that the other fragments should have the same dynamics as 69, since they should not depend on the neutral PESs. The data however, as seen in Fig. 3, contradict this argument. The smaller fragments have different decay signals from 69. This indicates that how 69 further fragments depends on where it came from. Fragment 69 will be generated from D_0 when *ca.* 2 eV of vibrational energy exist in the cation. But if ionization occurs on a higher ionic state and then nonadiabatic transitions move the population to D_0 then there will be more energy available to further break fragment 69 to the smaller fragments. Access to the higher ionic states will likely depend on the position on the neutral PES. Unfortunately this makes our effort to correlate fragments directly to locations on the neutral PES a lot harder.

It seems that the best choices for feedback for control in uracil are the parent ion and fragment 69. The parent ion can be used to identify very fast relaxation away from the FC region. Although fragment 69 has not been correlated to a specific location on the neutral PES, there is a special characteristic of the signal that is interesting and potentially useful. The ion yield of fragment 69, along with those of the other smaller fragments, has a long ledge that stays for much longer than the 10 ps time delay shown in Fig. 3. This delay is not present in the parent ion and it can be correlated with a previously discussed decay in neutral uracil. As discussed in Section 4.1 in uracil there is an S_1 minimum which is dark and has a high barrier towards the CI_{10}. This means if the wavepacket is trapped there a long lifetime will be observed. The long lifetime has been experimentally observed.[36,37] Then by controlling the ratio of the parent ion to fragment 69 we may be able to control the population that gets trapped in the S_1 minimum. More work is needed to confirm this interpretation.

Although we have a much better understanding of the dynamics in cytosine compared to uracil, we chose to do the initial control experiments on uracil because the signal to noise for cytosine is worse—*i.e.* fewer ions per laser shot. These experiments are discussed in the following section.

4.4 Control

We performed closed loop control experiments at several different pump–probe delays between 300 and 1000 fs using several peaks in the TOFMS to construct a fitness function for feedback. Some experiments fed back on the yield of a specific peak in the TOFMS while others fed back on a ratio of two peaks. The details of our genetic algorithm (GA) are given in an earlier paper,[38] but we note the parameters used for the GA runs in these experiments. We made use of 40 genes in a differential phase basis (where the genes encoded the phase difference between frequency components) and had 40 pulses per generation. We typically ran each GA about 50 generations while making about 250–500 measurements of the TOFMS for each pulse shape in order to acquire sufficient signal to noise for feedback. Fig. 6 shows the TOFMS for an unshaped laser pulse along with the TOFMS for a pulse shaped to optimize the ratio of the parent ion to the fragment at 69 amu. The inset shows the intensity as a function of time for the optimal pulse. Note the structure in the optimal pulse at the time delay of the probe. The TOFMS shown in the figure corresponds to an increase in the parent to 69 amu ratio of 39%. Running the GA several times at the same delay and with the same parameters, we found improvements in the parent/69 amu ratio of between 0 and 40% over an unshaped pulse. However, we also found that most of the optimal pulses had features which overlapped with the probe pulse as shown in the inset to Fig. 6. This corresponds to a somewhat trivial control as the ratio of the parent ion to the 69 amu fragment increases even for an unshaped pulse at time delays near zero. When we attempted to limit the ability of the GA to find solutions that overlapped with the probe pulse (either by limiting the phase per unit frequency or setting the time delay to larger values) then we found that the GA was not able to substantially improve the parent/69 amu ratio over an unshaped pulse. In addition to performing closed loop learning control experiments, we also measured the TOFMS as a function of quadratic spectral phase (chirp) on the pump pulse. These results were consistent with the GA runs in that we found very little variation in the fragment ion yields as a function of pump pulse shape at time delays where the pump and probe pulses were not overlapped.

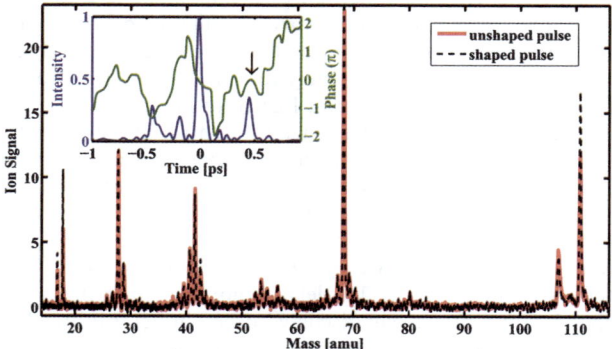

Fig. 6 TOFMS for an unshaped (pink) and a shaped (black) pump pulse followed by a near IR ionizing pulse. Inset: I(*t*) for an optimized pulse. The arrow in the inset indicates that at the time delay when the probe pulse is on the pump pulse still has significant intensity.

Our measurements suggest that either the fragment ratios are not sensitive to the neutral dynamics, or that given the limited bandwidth of our pump pulses, we were not able to influence the dynamics of the wavepacket on the excited states surface. In both uracil and cytosine, different fragments show different decay dynamics, with the fitted decay times showing substantial variation for the different fragments. The decay times are insensitive to the intensity of the probe pulse (and therefore not sensitive to the intensity dependent coupling to different ionic states), consistent with our interpretation of the different fragment decay dynamics being due to the break up and non-local evolution of the wavepacket on the neutral surface. Thus the ratio of different fragments serves as a measure of the wavepacket branching on the neutral state, and feeding back on specific peaks or ratios should in principle allow us to control the excited state dynamics by shaping the pump pulse. Given this reasoning and our measurements, we suspect that for our limited pump pulse bandwidth, the shape of the potential near the FC region dominates the wavepacket dynamics on the excited state rather than the shape of the pump pulse which prepares the wavepacket on the excited state PES. These observations are consistent with earlier calculations for laser control of CHD, in which it was found that very short pulses in the deep UV, with durations as short as 5–10 fs, were required to control the wavepacket dynamics around a conical intersection.[39,40]

5 Conclusions

We have presented an approach that can be used to explore excited state dynamics in molecules going through conical intersections and provide information for the design of control experiments over the dynamics. The approach combines a pump–probe scheme using strong-field dissociative ionization and *ab initio* electronic structure calculations. A pump pulse excites the neutral molecules while a probe intense IR pulse ionizes and fragments the excited molecules. The fragments show different dynamics indicating that they carry information about the neutral dynamics. Electronic structure calculations are used to extract this information. We have used this approach in the study of two pyrimidine DNA bases, cytosine and uracil. *Ab initio* calculations of the electronic states involved and of the fragmentation mechanisms enabled us to relate the observed ion yield pump–probe data to the neutral dynamics of these molecules. Based on the information that we got from the pump–probe study we performed closed loop control experiments in uracil where the ratio of two fragments was used as a feedback, but we were not able to change the ratio of fragments. This indicates that in this case we cannot influence the dynamics of the wavepacket on the neutral PESs.

Acknowledgements

We gratefully acknowledge support from the Department of Energy under award numbers DE-FG 02 – 08ER15983 and DE-PS 02 – 08ER08 – 01. A portion of the research was performed using EMSL, a national scientific user facility sponsored by the Department of Energy's Office of Biological and Environmental Research and located at Pacific Northwest National Laboratory.

References

1 W. Domcke, D. R. Yarkony and H. Köppel, *Conical Intersections*, World Scientific, Singapore, 2004.
2 S. Matsika and P. Krause, *Annu. Rev. Phys. Chem.*, 2011, **62**, 621–643.
3 M. Kotur, T. C. Weinacht, B. J. Pearson and S. Matsika, *J. Chem. Phys.*, 2009, **130**, 134311.
4 C. E. Crespo-Hernandez, B. Cohen, P. M. Hare and B. Kohler, *Chem. Rev.*, 2004, **104**, 1977.
5 S. Matsika, *J. Phys. Chem. A*, 2004, **108**, 7584.
6 M. Merchán, R. Gonzalez-Luque, T. Climent, L. Serrano-Andrés, E. Rodriuguez, M. Reguero and D. Pelaez, *J. Phys. Chem. B*, 2006, **110**, 26471–26476.

7 M. Z. Zgierski, S. Patchkovskii, T. Fujiwara and E. C. Lim, *J. Phys. Chem. A*, 2005, **109**, 9384–9387.
8 T. Gustavsson, A. Banyasz, E. Lazzarotto, D. Markovitsi, G. Scalmani, M. J. Frisch, V. Barone and R. Improta, *J. Am. Chem. Soc.*, 2006, **128**, 607–619.
9 H. R. Hudock, B. G. Levine, A. L. Thompson, H. Satzger, D. Townsend, N. Gador, S. Ullrich, A. Stolow and T. J. Martinez, *J. Phys. Chem. A*, 2007, **111**, 8500–8508.
10 N. Ismail, L. Blancafort, M. Olivucci, B. Kohler and M. A. Robb, *J. Am. Chem. Soc.*, 2002, **124**, 6818.
11 M. Merchán and L. Serrano-Andrés, *J. Am. Chem. Soc.*, 2003, **125**, 8108.
12 K. A. Kistler and S. Matsika, *J. Phys. Chem. A*, 2007, **111**, 2650–2661.
13 K. Tomić, T. Jörg and C. M. Marian, *J. Phys. Chem. A*, 2005, **109**, 8410–8418.
14 M. Z. Zgierski, S. Patchkovskii and E. C. Lim, *J. Chem. Phys.*, 2005, **123**, 081101.
15 L. Blancafort, *Photochem. Photobiol.*, 2007, **83**, 603–610.
16 H. R. Hudock and T. J. Martinez, *Chem.Phys.Chem.*, 2008, **9**, 2486–2490.
17 M. Barbatti, A. J. A. Aquino, J. J. Szymczak, D. Nachtigallova, P. Hobza and H. Lischka, *Proc. Natl. Acad. Sci. U. S. A.*, 2010, **107**, 21453–21458.
18 Z. Lan, E. Fabiano and W. Thiel, *J. Phys. Chem. B*, 2009, **113**, 3548–3555.
19 M. Kotur, T. C. Weinacht, C. Zhou and S. Matsika, *IEEE J. Sel. Topics Quantum Electron.*, 2011, **PP99**, 1–8.
20 H. Nakano, *J. Chem. Phys.*, 1993, **99**, 7983–7992.
21 H. Nakano, *Chem. Phys. Lett.*, 1993, **207**, 372–378.
22 M. W. Schmidt, K. K. Baldridge, J. A. Boatz, S. T. Elbert, M. S. Gordon, J. H. Jensen, S. Koseki, N. Matsunaga, K. A. Nguyen, S. Su, T. L. Windus, M. Dupuis and J. A. Montgomery, *J. Comput. Chem.*, 1993, **14**, 1347.
23 M. Valiev, E. Bylaska, N. Govind, K. Kowalski, T. Straatsma, H. van Dam, D. Wang, J. Nieplocha, E. Apra, T. Windus and W. de Jong, *Comput. Phys. Commun.*, 2010, **181**, 1477.
24 M. J. Frisch, G. W. Trucks, H. B. Schlegel, G. E. Scuseria, M. A. Robb, J. R. Cheeseman, J. A. Montgomery, Jr., T. Vreven, K. N. Kudin, K. C. Burant, J. M. Millam, S. S. Iyengar, J. Tomasi, V. Barone, B. Mennucci, M. Cossi, G. Scalmani, N. Rega, G. A. Petersson, H. Nakatsuji, M. Hada, M. Ehara, K. Toyota, R. Fukuda, J. Hasegawa, M. Ishida, T. Nakajima, Y. Honda, O. Kitao, H. Nakai, M. Klene, X. Li, J. E. Knox, H. P. Hratchian, J. B. Cross, V. Bakken, C. Adamo, J. Jaramillo, R. Gomperts, R. E. Stratmann, O. Yazyev, A. J. Austin, R. Cammi, C. Pomelli, J. Ochterski, P. Y. Ayala, K. Morokuma, G. A. Voth, P. Salvador, J. J. Dannenberg, V. G. Zakrzewski, S. Dapprich, A. D. Daniels, M. C. Strain, O. Farkas, D. K. Malick, A. D. Rabuck, K. Raghavachari, J. B. Foresman, J. V. Ortiz, Q. Cui, A. G. Baboul, S. Clifford, J. Cioslowski, B. B. Stefanov, G. Liu, A. Liashenko, P. Piskorz, I. Komaromi, R. L. Martin, D. J. Fox, T. Keith, M. A. Al-Laham, C. Y. Peng, A. Nanayakkara, M. Challacombe, P. M. W. Gill, B. G. Johnson, W. Chen, M. W. Wong, C. Gonzalez and J. A. Pople, *GAUSSIAN 03 (Revision C.02)*, Gaussian, Inc., Wallingford, CT, 2004.
25 M. Kotur, T. C. Weinacht, C. Zhou, K. Kistler and S. Matsika, *J. Chem. Phys*, 2011, **134**, 184309.
26 D. Dougherty, K. Wittel, J. Meeks and S. P. McGlynn, *J. Am. Chem. Soc.*, 1976, **98**, 3815–3820.
27 D. Dougherty, E. S. Younathan, R. Voll, S. Abdulnur and S. P. McGlynn, *J. Electron Spectrosc. Relat. Phenom.*, 1978, **13**, 379–393.
28 S. Urano, X. Yang and P. R. LeBreton, *J. Mol. Struct.*, 1989, **214**, 315–328.
29 E. Cauët, D. Dehareng and J. Liévin, *J. Phys. Chem. A*, 2006, **110**, 9200.
30 S. Matsika, *Chem. Phys.*, 2008, **349**, 356.
31 V. Orlov, A. Smirnov and Y. Varshavsky, *Tetrahedron Lett.*, 1978, **17**, 4377–4378.
32 S. Denifl, B. Sonnweber, G. Hanel, P. Scheier and T. Märk, *Int. J. Mass Spectrom.*, 2004, **238**, 54753.
33 H.-W. Jochims, M. Schwell, H. Baumgärtel and S. Leach, *Chem. Phys.*, 2005, **314**, 263–282.
34 J. Tabeta, S. Edena, S. Feil, H. Abdoul-Carime, B. Farizon, M. Farizon, S. Ouaskitd and T. Märke, *Int. J. Mass Spectrom.*, 2010, **292**, 53–63.
35 J. K. Wolken, C. Yao, F. Turecek, M. J. Polce and C. Wesdemiotis, *Int. J. Mass Spectrom.*, 2007, **267**, 30–42.
36 Y. He, C. Wu and W. Kong, *J. Phys. Chem. A*, 2003, **107**, 5145.
37 P. M. Hare, C. E. Crespo-Hernandez and B. Kohler, *Proc. Natl. Acad. Sci. U. S. A.*, 2007, **104**, 435–440.
38 F. Langhojer, D. Cardoza, M. Baertschy and T. Weinacht, *J. Chem. Phys.*, 2005, **122**, 014102.
39 D. Geppert, L. Seyfarth and R. de Vivie-Riedle, *Appl. Phys. B: Lasers Opt.*, 2004, **79**, 987–992.
40 D. Geppert and R. de Vivie-Riedle, *Chem. Phys. Lett.*, 2005, **404**, 289–295.

Nonadiabatic *ab initio* molecular dynamics including spin–orbit coupling and laser fields

Philipp Marquetand,[*a] Martin Richter,[a] Jesús González-Vázquez,[†*a] Ignacio Sola[b] and Leticia González[a]

Received 1st April 2011, Accepted 18th May 2011
DOI: 10.1039/c1fd00055a

Nonadiabatic *ab initio* molecular dynamics (MD) including spin–orbit coupling (SOC) and laser fields is investigated as a general tool for studies of excited-state processes. Up to now, SOCs are not included in standard *ab initio* MD packages. Therefore, transitions to triplet states cannot be treated in a straightforward way. Nevertheless, triplet states play an important role in a large variety of systems and can now be treated within the given framework. The laser interaction is treated on a non-perturbative level that allows nonlinear effects like strong Stark shifts to be considered. As MD allows for the handling of many atoms, the interplay between triplet and singlet states of large molecular systems will be accessible. In order to test the method, IBr is taken as a model system, where SOC plays a crucial role for the shape of the potential curves and thus the dynamics. Moreover, the influence of the nonresonant dynamic Stark effect is considered. The latter is capable of controlling reaction barriers by electric fields in time-reversible conditions, and thus a control laser using this effect acts like a photonic catalyst. In the IBr molecule, the branching ratio at an avoided crossing, which arises from SOC, can be influenced.

1 Introduction

Laser control of chemical reactions has been a target for researchers for decades. The will to exert control goes hand in hand with the desire to witness the corresponding molecular processes in real time. Since the seminal works by Zewail and coworkers,[1,2] observation of atomic motion in the femtosecond regime is possible. Nonetheless, it is still challenging to unravel complex processes in small and big systems including several electronic states. A joint effort of experiment and theory is necessary to achieve this goal. On the one hand, the theoretical description with atomistic models particularly helps to understand the complex processes as the calculations are directly carried out and visualized in the descriptive picture that we hold in our heads. On the other hand, the use of exact formulas is only possible for the simplest systems.[3] The solution of the time-dependent Schrödinger equation to represent the dynamics of molecules in full dimensionality is not feasible for most systems with today's computers. Therefore, different approximations have been established and methods like *e.g.* Multiconfigurational time-dependent Hartree method (MCTDH),[4–6] multiple spawning[7,8] or similar techniques[9–17] emerged.

[a] *Institut für Physikalische Chemie, Friedrich-Schiller-Universität Jena, Helmholtzweg 4, 07743 Jena, Germany. E-mail: p.marquetand@uni-jena.de; jgv@tchiko.quim.ucm.es; Fax: +49 3641 948302*
[b] *Departamento de Química Física I, Universidad Complutense, 28040 Madrid, Spain*

† Present address: Departamento de Química Física I, Universidad Complutense, 28040 Madrid, Spain

Here, we want to focus on another ansatz, *ab initio* molecular dynamics (MD),[18,19] where Newton's classical equations of motion are used to approximate the change of the nuclear positions while the electronic structure of the considered system is treated quantum mechanically. The gap between these two descriptions for different parts of the system is bridged by the surface hopping (SH) method.[20,21] However, the latter was originally developed to account for nonadiabatic couplings in the photo-dynamics of molecules. Recent efforts try to describe other types of coupling which may occur during or after photoexcitations.[22–26,58,59] In a recent paper, we introduced the surface-hopping-in-adiabatic-representation-including-arbitrary-couplings (SHARC) method.[27] The advantage of the latter method is that all kinds of couplings can be treated at once and on equal footing. Hence, effects like spin–orbit coupling (SOC), laser excitations and nonadiabatic coupling can be modeled for complex systems with an arbitrary number of atoms and a large number of degrees of freedom.

Especially the interaction between molecules and laser fields is important in quantum control.[28–38] In this contribution, we shall focus on a special type of control *via* the nonresonant dynamic Stark effect (NRDSE).[39–42] The Stark effect, which is ubiquitious in strong field control, see *e.g.* Ref. 43, is used to create dressed states, also called light-induced potentials (LIPs). Remarkable about NRDSE is that the laser field only shifts the potential *via* interactions with the permanent dipole moment and polarizability without inducing any resonant transitions. Therefore, reaction barriers can be altered with the NRSDE acting like a photonic catalyst.[39–42]

As a model system, we have chosen IBr since it exhibits strong SOC leading to an avoided crossing between two excited states, the $1^3\Pi_{0+}$ and the $1^3\Sigma_{0+}^-$ state, see Ref. 44 and references therein. Excited state dissociation in these two states leads to different product channels resulting in I + Br and I + Br*, respectively. The asterisk indicates the $^2P_{1/2}$ excited spin-state of the dissociating Br atom, while the ground state has the configuration $^2P_{3/2}$. After electronic excitation of the IBr molecule with a resonant laser, the branching ratio in the two channels is influenced with a second, nonresonant laser. We show that this NRDSE process can be described within the SHARC formalism. Results shall be compared to exact quantum dynamics (QD) calculations.

The methodology and the theoretical description are presented in Sec. 2; the numerical results are contained in Sec. 3, and a summary is given in Sec. 4.

2 Methodology

We use *ab initio* MD, where the electrons of a molecular system are treated quantum mechanically and the nuclei classically. On the one hand, the nuclei follow classical trajectories defined by the nuclear position $\vec{R}(t)$ and velocity $\vec{v}(t)$ at every time following the quantum potential created by the electrons. On the other hand, the quantum potential is evaluated as the expectation value of an effective Hamiltonian $V(t) = \langle \Psi(\vec{R}(t);\vec{r},t)|\hat{H}_{\text{eff}}[\vec{R}(t);\vec{r}]|\Psi(\vec{R}(t);\vec{r},t)\rangle$ and thus, depends parametrically on the nuclear coordinates.

To account for the quantum position-momentum uncertainty in the classical part of our simulations, we create a swarm of trajectories with different initial conditions. The latter are prepared in a way that resembles the probability distribution of the corresponding quantum wavepacket. Every single trajectory is propagated using the Velocity Verlet algorithm[45,46] for the solution of Newton's equations. In this algorithm, the dynamics of the nuclear coordinates $\vec{R}(t)$ is governed by the gradient of the potential at time t:

$$\vec{R}(t + \Delta t) = \vec{R}(t) + \vec{v}(t)\,\Delta t + \frac{1}{2M}\nabla_{\vec{R}}\,V(t)\Delta t^2. \quad (1)$$

Here, M represents the mass of the nuclei and $\vec{v}(t)$ is the velocity of the trajectory. The time evolution of this velocity is calculated using the gradient of the potential at times t and $t + \Delta t$:

$$\vec{v}(t + \Delta t) = \vec{v}(t) + \frac{1}{2M}\nabla_{\vec{R}}V(t)\Delta t + \frac{1}{2M}\nabla_{\vec{R}}V(t + \Delta t)\Delta t. \quad (2)$$

The definition of the potential is given by the time evolution of the electronic wavepacket following the time-dependent Schrödinger equation:

$$i\hbar \frac{\partial |\Psi[\vec{R}(t); \vec{r}, t]\rangle}{\partial t} = \hat{H}_{\text{eff}}[\vec{R}(t); \vec{r}]|\Psi[\vec{R}(t); \vec{r}, t]\rangle. \quad (3)$$

We employ a linear basis set expansion of the wavepacket to solve this equation with implicitly time-dependent basis functions at different $\vec{R}(t)$:

$$|\Psi[\vec{R}(t); \vec{r}, t]\rangle = \sum_m c_m(t)|\phi_m[\vec{R}(t); \vec{r}]\rangle, \quad (4)$$

where c_m are the amplitudes of the eigenfunctions ϕ_m.

Consequently, the time evolution of these amplitudes are given by:

$$\frac{\partial c_n(t)}{\partial t} = -\sum_m \left\{ \frac{i}{\hbar}\mathsf{H}_{nm}[\vec{R}(t)] + \mathsf{K}_{nm}[\vec{R}(t)] \right\} c_m(t), \quad (5)$$

The first term $\mathsf{H}_{nm}[\vec{R}(t)] = \langle \phi_n[\vec{R}(t); \vec{r}]|\hat{H}_{\text{eff}}[\vec{R}(t); \vec{r}]|\phi_m[\vec{R}(t); \vec{r}]\rangle$ of this equation—solved by a simple Runge–Kutta algorithm of fourth order—describes the diabatic Hamiltonian with the different potentials as diagonal elements and the diabatic couplings as the off-diagonal elements. The second term, $\mathsf{K}_{nm}[\vec{R}(t)]$, represents the change of the electronic basis functions with time, which is equivalent to the variation of the basis with the nuclear coordinates multiplied by the velocity:

$$\begin{aligned}\mathsf{K}_{nm}[\vec{R}(t)] &= \langle \phi_n[\vec{R}(t); \vec{r}]|\partial/\partial t|\phi_m[\vec{R}(t); \vec{r}]\rangle \\ &= \langle \phi_n[\vec{R}(t); \vec{r}]|d/d\vec{R}(t)|\phi_m[\vec{R}(t); \vec{r}]\rangle \vec{v}(t).\end{aligned} \quad (6)$$

If the basis functions are chosen to be the eigenfunctions of the time-independent Schrödinger equation for every $\vec{R}(t)$, then the amplitudes are directly correlated with the populations of the different electronic states. This is important since a trajectory can only be influenced by a single potential at a time and the corresponding state has to be assigned. Here, we use Tully's SH method,[20] which was originally developed to account for nonadiabatic couplings by giving the trajectory the possibility to jump from one state to another. The probability for such a hop is calculated using the time-dependent amplitudes introduced above:

$$P_{nm} = \frac{2\mathscr{R}\left\{ c_n^*(t)c_m(t)\left[\frac{i}{\hbar}\mathsf{H}_{nm}[\vec{R}(t)] + \mathsf{K}_{nm}[\vec{R}(t)] \right] \right\}}{c_n^*(t)c_n(t)} \Delta t, \quad (7)$$

where $\mathscr{R}$ denotes the real part.

The original SH methodology is widely used to simulate problems with localized couplings originating from conical intersections.[47,48] In this work, we extend SH to the situation where different types of couplings, e.g. SOCs and/or the interaction with an electric field must be taken into account. These coupling terms are typically evaluated in the diabatic representation and hence included in the potential part of the Hamiltonian. In this way, a new $H^d[\vec{R}(t)]$ matrix is introduced with elements (where the index d indicates that additional nondiagonal terms are included):

$$H_{nm}^d[\vec{R}(t),t] = H_{nm}[\vec{R}(t)] + H_{nm}^{AC}[\vec{R}(t)]. \tag{8}$$

In this equation, $H_{nm}^{AC}[\vec{R}(t),t]$ indicates the arbitrary-coupling matrix which can consist of terms like $-\vec{\mu}_{nm}\left[\vec{R}(t)\right]\vec{E}(t) + H_{nm}^{SO}\left[\vec{R}(t)\right]$, where $\vec{\mu}_{nm}[\vec{R}(t)]$ and $H_{nm}^{SO}[\vec{R}(t)]$ are the dipole moment and the relativistic SOC between the states n and m, respectively.

Such additional terms may complicate the solution of the corresponding equations since the interactions are mostly delocalized in contrast to the rather localized nonadiabatic couplings. In SHARC, we solve this problem by translating the additional elements to the $K[\vec{R}(t)]$ matrix. We choose the adiabatic (index a) representation, where the $H^d[\vec{R}(t),t]$ matrix is diagonalized and afterwards, the $K[\vec{R}(t)]$ matrix, localizing the couplings in geometries where the electronic states are degenerated, is recalculated.

Along these lines, the basis set of electronic wavefunctions $|\phi^d[\vec{R}(t);r]\rangle$ is substituted for a linear combination,

$$\left|\phi_n^a\left[\vec{R}(t);\vec{r},t\right]\right\rangle = \sum_m U_{nm}\left[\vec{R}(t),t\right]\left|\phi_m^d\left[\vec{R}(t);\vec{r}\right]\right\rangle, \tag{9}$$

where $U[\vec{R}(t),t]$ is the unitary matrix that diagonalizes the Hamiltonian $H^d[\vec{R}(t),t]$ matrix at every time t. In this new basis, the elements of the $H^a[\vec{R}(t),t]$ matrix are defined as:

$$H_{nm}^a[\vec{R}(t),t] = V_m^a[\vec{R}(t),t]\delta_{nm}, \tag{10}$$

where $V_m^a[\vec{R}(t),t]$ are the diagonal elements of $H^a[\vec{R}(t),t]$. The couplings are treated *via* the derivative of the $|\phi^a[\vec{R}(t);\vec{r},t]\rangle$:

$$K_{nm}^a\left[\vec{R}(t),t\right] = \left\langle\phi_n^{a*}\left[\vec{R}(t);\vec{r},t\right]\left|\frac{\partial}{\partial t}\right|\phi_m^a\left[\vec{R}(t);\vec{r},t\right]\right\rangle$$

$$= K_{nm}^\phi\left[\vec{R}(t),t\right] + K_{nm}^U\left[\vec{R}(t),t\right]. \tag{11}$$

Application of the chain rule easily yields the definitions for these latter terms:

$$K_{nm}^\phi\left[\vec{R}(t),t\right] = \sum_{lk} U_{ln}^*\left[\vec{R}(t),t\right]K_{lk}\left[\vec{R}(t)\right]U_{km}\left[\vec{R}(t),t\right], \tag{12}$$

which is just the rotation of the original nonadiabatic term to the new basis and

$$K_{nm}^U\left[\vec{R}(t),t\right] = \sum_l U_{ln}^*\left[\vec{R}(t),t\right]\frac{\partial}{\partial t}U_{lm}\left[\vec{R}(t),t\right] \tag{13}$$

comes from the variation of the rotation matrix.

We diagonalize the matrix $H^d[\vec{R}(t),t]$ at distance $\vec{R}(t)$, $\vec{R}(t) + \Delta\vec{R}$ and $\vec{R}(t) - \Delta\vec{R}$ to yield the gradient of the new potentials $V^a[\vec{R}(t),t]$ and the nonadiabatic coupling elements $K^a[\vec{R}(t),t]$. Hence, we obtain the gradient of the potential and the gradient of the $U[\vec{R}(t),t]$ matrix. These new matrices are used in eqn (5) and eqn (7) to calculate the nonadiabatic dynamics.

3 Numerical results

We investigate the ability of the SHARC scheme to describe the effects of excited-state dynamic Stark control. As a test system, we use a simplified model the IBr molecule. Here, a first excitation pulse $E_e(t)$ transfers some population from the ground to an excited state. The subsequent dynamics via an avoided crossing, which is present due to SOC, is then influenced by a control pulse $E_c(t)$, see Fig. 1 (Transitions (a)–(c) will be explained below). The reason for the choice of this system is that for the model used, exact QD calculations are possible and we are able to compare the outcome from the different methods.

In the quantum propagations, we use the split-operator method.[49] Stationary states are computed via imaginary time propagation.[50] Wigner distributions from the corresponding wavefunctions are employed to establish the initial conditions in the MD simulations.

For the sake of simplicity in the QD simulations, we restrict ourselves to the model potentials from Ref. 51 for IBr. They are shown in Fig. 1. As we want to model the Stark effect, we need polarizabilities and dipole moments in the direction of the laser polarization, which we assume to be the z-axis and therefore will omit vector arrows in the following. The static polarizabilites α_{nn}^{zz}, permanent dipole moments μ_{nn}^{z}, as well as the transition dipole moments μ_{nm}^{z} are fitted to the results published in Ref. 44 analytically. In deriving the formulas, our aim was to find a reasonable agreement with the published curves regardless of the number of parameters. The curves are given without SOC included, like they would be obtained from a quantum-chemical program, as:

$$\alpha_{11}^{zz} = a_1(R - a_2)e^{-a_3(R - a_2)^3} + a_4 \quad (14)$$

$$\alpha_{22}^{zz} = a_5 e^{-a_6(R - a_7)^2} + a_8 e^{-a_9(R - a_{10})^2} + a_4 \quad (15)$$

$$\alpha_{33}^{zz} = a_{11} e^{-a_{12}(R - a_{13})^2} + a_{14} e^{-a_{15}(R - a_{16})^2} + a_4 \quad (16)$$

$$\mu_{11}^{z} = a_{17}(R - a_{18})e^{-a_{19}(R - a_{18})^2} \quad (17)$$

$$\mu_{22}^{z} = a_{20}(1 - e^{-a_{21}(R-a_{22})^2}) - a_{20} + a_{23}e^{-a_{24}(R - a_{25})^2} \quad (18)$$

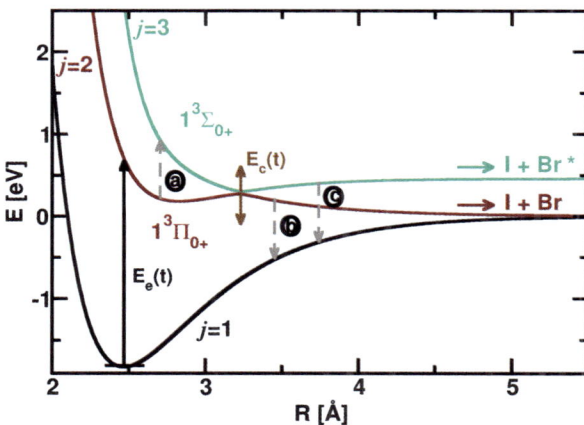

Fig. 1 Potential energy curves of the IBr molecule and excitation scheme. The IBr molecule initially in the electronic ground state (black) is excited to the $1^3\Pi_{0+}$ excited electronic state (red) and can undergo dissociation into two different channels due to an avoided crossing introduced by SOC with the $1^3\Sigma_{0+}^-$ excited state (turquoise). The control laser may shift the potential curves via NRDSE but also induce transitions at points (a)–(c), see text for explanation.

$$\mu_{33}^z = a_{26}e^{-a_{27}(R - a_{28})^2} + a_{29}e^{-a_{30}(R - a_{31})^2} \quad (19)$$

$$\mu_{12}^z = \mu_{21}^z = a_{32}(R - a_{33})e^{-a_{34}(R - a_{33})} \quad (20)$$

$$\mu_{13}^z = \mu_{31}^z = a_{35}(R - a_{36})e^{-a_{37}(R - a_{36})^2} \quad (21)$$

$$\mu_{23}^z = \mu_{32}^z = 0 \quad (22)$$

The parameters a_i are listed in Table 1. The corresponding diabatic curves are depicted in Fig. 2, left panels. They are easily adiabatized using the $U[\vec{R}(t),t]$ matrix mentioned above. If the effect of SOC is evaluated, the resulting curves nicely reproduce the ones given in Ref. 44 (Fig. 2, right panels).

The interactions between the initial diabatic states in our model system comprise SOC, dipole coupling and couplings induced by the static polarizability. To describe the interaction of the control laser with the static polarizability, the rotating wave approximation has to be employed, see, e.g., Ref. 52. Hence, the arbitrary-couplings matrix mentioned above reads:

$$H_{nm}^{AC}[R(t)] = -\mu_{nm}[R(t)](E_e(t) + E_c(t)) - \frac{1}{2}\alpha_n[R(t)]\left|E_c^0(t)\right|^2 + H_{nm}^{SO}[R(t)], \quad (23)$$

where $E_c^0(t)$ denotes the envelope function of the control laser pulse. Depending on the field strength of the control laser, the potentials of the considered system will be shifted. Here, we investigate fields of intermediate strength (compared to weak fields, which do not alter the potentials, or to strong fields that are able to induce ionization processes and Coulomb explosion). Due to the interaction with the control laser, we create light-induced potentials (LIPs) and thus, the photodynamics is altered. In order to explore how these LIPs look like, we calculate the $V_n^a[R]$ for different field strengths. The resulting curves are plotted in Fig. 3 for field intensities of 1×10^{13} W cm^{-2}, 5×10^{13} W cm^{-2} and the field free case for comparison.

As can be seen from the figure, the region of the avoided crossing shifts to larger interatomic distances with increasing field strength. For the present cases, the point where the two excited-state curves come closest is $R = 3.23$ Å for $I = 0$ W cm^{-2}, $R = 3.40$ Å for $I = 1 \times 10^{13}$ W cm^{-2}, and $R = 3.68$ Å for $I = 5 \times 10^{13}$ W cm^{-2}. The energy separation at these points does not change and is 0.037 eV. However, the gradient in the excited states is changed dramatically with increasing field strength. In this way, the dynamics of the system is expected to be influenced.

We now investigate how the branching ratio of the products from the different dissociation channels depicted in Fig. 1 is altered by the NRSDE. We use an

Table 1 Parameters a_i for analytical fit of α_n and μ_{nm}

Param.	Value	Unit	Param.	Value	Unit	Param.	Value	Unit
a_1	9.28	Å^2	a_{14}	3.15	Å^3	a_{27}	5.00	Å^{-2}
a_2	2.10	Å	a_{15}	3.50	Å^{-2}	a_{28}	2.65	Å
a_3	0.35	Å^{-3}	a_{16}	3.13	Å	a_{29}	0.45	D
a_4	8.20	Å^3	a_{17}	2.25	DÅ^{-1}	a_{30}	5.00	Å^{-2}
a_5	105.83	Å^3	a_{18}	2.07	Å	a_{31}	3.35	Å
a_6	0.11	Å^{-2}	a_{19}	0.83	Å^{-2}	a_{32}	0.59	DÅ^{-1}
a_7	−2.58	Å	a_{20}	0.80	D	a_{33}	2.07	Å
a_8	9.10	Å^3	a_{21}	4.00	Å^{-1}	a_{34}	0.67	Å^{-2}
a_9	6.43	Å^{-2}	a_{22}	2.25	Å	a_{35}	0.38	DÅ^{-1}
a_{10}	2.95	Å	a_{23}	0.30	D	a_{36}	2.14	Å
a_{11}	50.02	Å^3	a_{24}	20.00	Å^{-2}	a_{37}	3.58	Å^{-2}
a_{12}	5.62	Å^{-2}	a_{25}	2.85	Å			
a_{13}	1.68	Å	a_{26}	0.90	D			

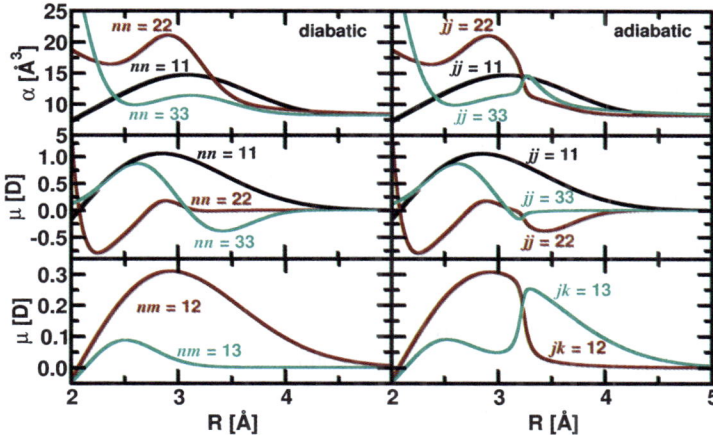

Fig. 2 Polarizabilities and dipole moments of the IBr molecule in the diabatic and adiabatic representations. The curves are analytically fitted to the results from Ref. 44

excitation pulse of 493.4 nm, a full width at half maximum of the Gaussian-shaped envelope of 50 fs and an intensity of 1×10^{13} W cm^{-2}. The control laser has a wavelength of 1.73 μm, similar to the experiment,[39] a full width at half maximum of the Gaussian-shaped envelope of 150 fs, an intensity of 1×10^{13} W cm^{-2}, and the delay $\Delta\tau$ between the pulses is scanned as indicated in Fig. 4. A negative time delay indicates the control pulse preceding the excitation pulse. In the SHARC simulations, a timestep of 0.001 fs was used to propagate 500 trajectories unless indicated otherwise. Note that the following argumentation is also valid for lower field strengths. However, the effects will not be as pronounced as presented in the given example.

The branching ratio for the dissociation products from the different channels is defined as $Q = \dfrac{[I + Br^*]}{[I + Br] + [I + Br^*]}$. The ion yield to obtain Q is represented in the QD model as the time-integrated flux[53] F through the different reaction channels:

$$F_j(R,t) = \int_0^t -i \frac{\psi_j^*(R,t')\hat{p}_R \psi_j(R,t') - \psi_j(R,t')\hat{p}_R \psi_j^*(R,t')}{2M} dt' \quad (24)$$

In the SHARC calulations, the time-integrated flux is simply evaluated as the time integration of the number of trajectories at time t' at distance R multiplied with the sign of the projection of the velocity vector on the direction of the flux. The time-integrated flux is calculated at a distance $R_F = 8$ Å. In the QD calculations, fractions

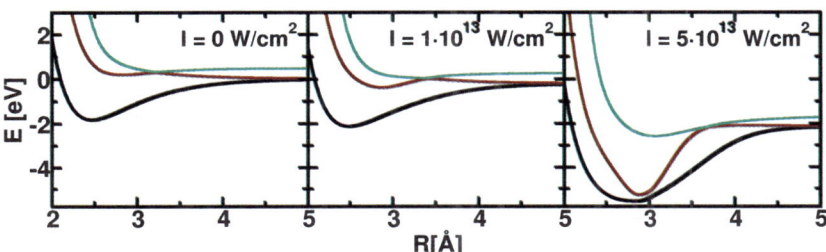

Fig. 3 Changes induced in the potential energy curves of the IBr molecule by laser fields of intermediate strength. The electronic ground state (black), the $1^3\Pi_{0+}$ excited electronic state (red), and the $1^3\Sigma_{0+}^-$ excited state (turquoise) are shifted in different manners due to their different dipole moments and polarizability.

of the wavepacket at larger distances are removed employing absorbing boundary conditions as described, *e.g.*, in Ref. 54 to avoid unphysical behavior due to limited grid size.

As can be seen from Fig. 4, results from SHARC and exact quantum dynamical simulations show the same behavior. To simulate such a curve within a surface hopping scheme is highly challenging because only a small fraction of the computed trajectories is excited from the ground state. From this small precentage, the ratio Q has to be derived, *i.e.* the small fraction of trajectories is split once more. Therefore, a large number of trajectories has to be calculated in order to minimize statistical errors. Here, 1930 trajectories are used for each delay time. The statistical noise is small in the SHARC curve and agrees well with the results from QD. There is however an issue that we would like to point out: The calculated curves do not show the same behavior as the experimental results.[39] Interestingly, curves closer to the experiment are obtained if the polarizability of the diabatic state $n = 2$ is taken to be negative as done in Ref. 55. However, also in this case, the experimental curves are not reproduced correctly. Some calculations on our side indicate that the computations of Patchkovskii[44] are correct and all static polarizabilities should be positive. Moreover, this is in agreement with the general derivation of the analytical formula for the static polarizability that can be found in textbooks.[56] A full understanding of the processes involved, which are much more complicated than they seem to be at first sight, lies beyond the scope of this work. Instead, we only focus on the ability of the SHARC method to provide the same results as exact quantum calculations given the same conditions and for this purpose, we used the simplified model described above. Nonetheless, in passing, we will point out some aspects that are relevant for further investigations of the NRDSE in IBr and the NRDSE in general.

In this model, the control acts in a way that by changing the potentials, the population of the excited states is either accelerated at delay times smaller than 50 fs or slowed down at delay times larger than 50 fs. This statement is exemplarily proven in Fig. 5, where the momentum is compared for delay times $\Delta\tau = 0$ fs and $\Delta\tau = 90$ fs to the case where no control field is present. The explanation for this acceleration and deceleration can be deduced from Fig. 6. There, the shift in the potentials curves due to the control laser interaction is visualized for the aforementioned cases. The unshifted potential curves for $j = 2$ (lower panels) and $j = 3$ are represented as dot-dashed lines and the respective potentials shifted by the maximal value of the control field strength are shown as dashed lines. These two limits mark the range between which the actual LIPs are shifting during the time-dependent

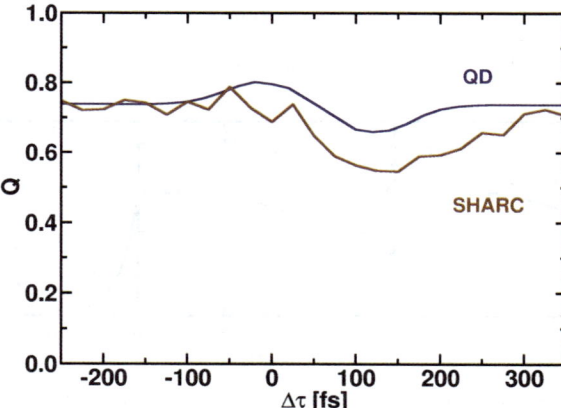

Fig. 4 Branching ratio Q of the different product channels as a function of delay $\Delta\tau$ between excitation and control pulse.

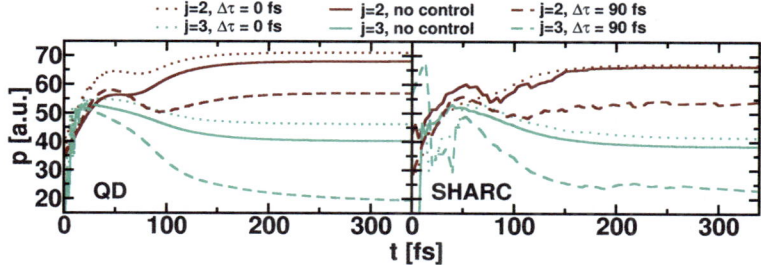

Fig. 5 Mean time-dependent momentum in the excited states. Red curves correspond to $j = 2$ and blue curves to $j = 3$. We consider three cases: dynamics without control field (solid lines), with overlapping pulses (dotted lines) and with a time delay of $\Delta\tau = 90$ fs (dashed lines). Calculations from QD and SHARC are in good agreement. A higher momentum than without control pulse is obtained for a delay of 0 fs while the momentum is lower for a delay 90 fs.

control pulse. An impression of how this shift occurs is given when the expectation value of the shifted potential $\langle V_j(t) \rangle$ at time t is plotted against the expectation value of the interatomic distance $\langle R_j(t) \rangle$. In such a curve, a larger value of $\langle R_j(t) \rangle$ also signifies a larger time t as we look at a dissociative process, where the bond distance is monotonically increasing. Additionally, the expectation value of the total energy $\langle H_j(t) \rangle$ is shown (dotted lines). The latter is not conserved during the laser interaction. The expectation values are only plotted if the population in the respective state is non-negligible.

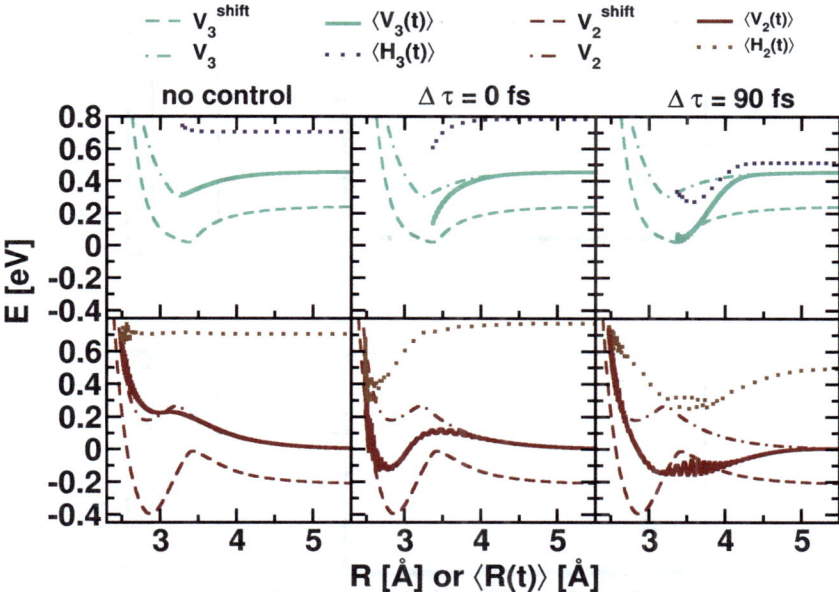

Fig. 6 Effective LIPs (solid lines), unshifted potentials (dot-dashed lines), and maximally shifted LIPs (dashed lines). The latter two types serve as extrema between which an effective LIP is changing during the interaction time of the control laser. Additionally, the total energy in the respective state is plotted (dotted lines). The curves are presented for the three cases: No control field acting, delay time $\Delta\tau = 0$ fs, and $\Delta\tau = 90$ fs. The effective LIPs and the total energy are evaluated as the expectation values $\langle V_j(t) \rangle$ and $\langle H_j(t) \rangle$ at time t plotted against the corresponding expectation value of the distance $\langle R_j(t) \rangle$ at the same time t. See text for further details.

Now, we regard different scenarios: When no control field is present (left panels), the effective potential mostly follows the unshifted curves. A slight deviation is visible due to the interaction with the pump laser at small R, i.e. at early times. The total energy is conserved. At a delay $\Delta\tau = 0$ fs (middle panels), the effective potential tracks the shifted curve at small R (early times) while it returns to the unshifted one at large R (later times, when the control pulse is over). During the laser interaction, the kinetic energy increases faster than in the field-free case, which can be deduced from the value of the total energy. For a delay of $\Delta\tau = 90$ fs (right panels), the effective potential follows the unshifted curve at small R (early times), approaches the shifted curve at intermediate R (intermediate times), and returns to the unshifted curve at large R (late times). Note that the expectation value exhibits values below the shifted curve of $j = 2$ due to population transfer to the state $j = 3$. Due to the laser interaction, the kinetic energy is diminished in this case.

Consequently, the velocity is decreased when reaching the avoided crossing for $\Delta\tau = 90$ fs. According to the Landau–Zener probability,[57] this case will result in diminished transitions to adiabatic state $j = 3$ while for $\Delta\tau = 0$ fs, a higher velocity is obtained, increasing the transition probability. At higher field strengths, the kinetic energy can be decreased so much that the dissociation limit in state $j = 3$ is not reached anymore (not shown).

So far, we have only examined excited states; however, interactions with the ground state may also be important. To explore the role of such interactions, we plot the probability density in the different states in Fig. 7. At around 90 fs and at a distance of $R \approx 3.8$ Å, population is transferred to the ground state and starts oscillating. This can be explained as a dump process from the adiabatic state $j = 3$ with the control laser corresponding to transition (c) in Fig. 1. The energy difference

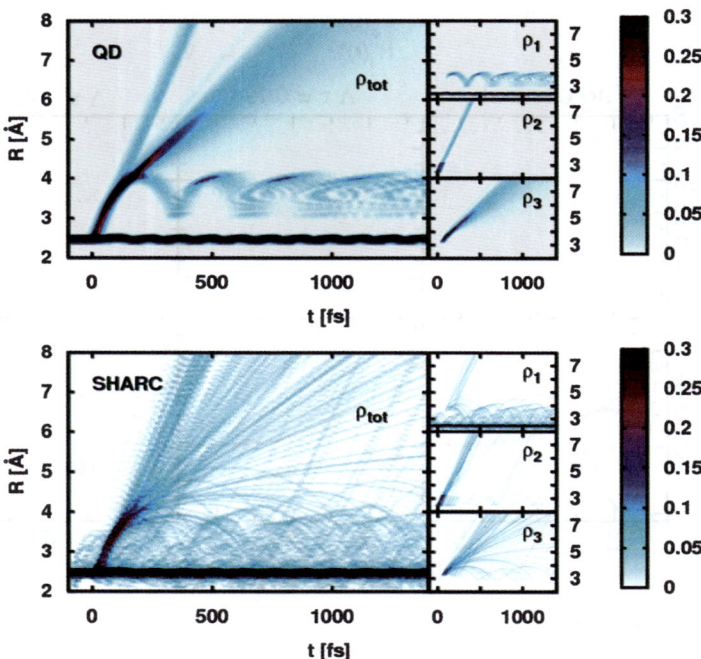

Fig. 7 Probability density for a time delay of 90 fs between excitation and control pulse. Results from SHARC are shown on the bottom, output from QD on the top. The respective panel on the left shows the total probability density and the panels on the right show the fractions of the same density belonging to the adiabatic states j as indicated.

between states $j = 2$ and $j = 3$ matches the laser frequency at (a); however, the transition dipole moment is zero between these states and a transition at this geometry is impeded. Also the transition at (b) is not possible due to a zero transition dipole moment at the corresponding bond length. Nonetheless, the control laser—although it is thought to be non-resonant in the NRDSE—may still induce resonant transitions and thus, complicate the simple picture of shifted potential curves only. Note that resonant transitions are still possible if a negative polarizability is assumed for diabatic state $n = 2$. It is gratifying to see that this effect is reproduced in the SHARC simulations, as seen in Fig. 7. Also the rest of the probability distribution from the QD calculation is nicely reproduced by SHARC.

The general conclusion is that SHARC is able to describe molecular dynamics under the given, relatively strong fields. The treatment of bigger systems, which may even include molecules in solution, is straightforward by computing the potential energies "on the fly". From this perspective, SHARC is ready to be used for large systems under complex control scenarios in the presence of SOC or nonadiabatic couplings.

4 Conclusion

To conclude, we have shown that the new surface-hopping-in-adiabatic-representation-including-arbitrary-couplings (SHARC) algorithm is able to describe photophysical processes even at intermediate field strengths. A complex control scenario including the non-resonant dynamic Stark effect (NRDSE) was modelled by our semiclassical method, where the surface hopping probabilities are calculated in terms of a unitary transformation matrix. Thus, SHARC is able to treat all kinds of couplings in molecular systems including all degrees of freedom on the same footing.

As a control target, we have chosen the influence of the NRDSE on the branching ratio of dissociation products after photoexcitation in a model of IBr. The branching at an avoided crossing, which arises from spin–orbit coupling, gives rise to the products $I + Br^*$ or $I + Br$. The mechanism of how the branching ratio is altered was rationalized in terms of the light-induced potentials (LIPs). It was also found that resonant transitions induced by the control laser which creates the NRDSE may complicate the dynamics and are important to be considered. A comparison between SHARC and exact quantum dynamics simulations show that the respective results are in good agreement. Consequently, SHARC is able to describe laser control in complex scenarios. Therefore, it is now possible to treat photoinduced dynamics in large systems including all degrees of freedom.

Acknowledgements

This work has been supported by the Deutsche Forschungsgemeinschaft (DFG) within the project GO 1059/6-1, by the German Federal Ministry of Education and Research within the research initiative PhoNa, the Dirección General de Investigación of Spain under Project No. CTQ2008-06760, the Friedrich-Schiller-Universität Jena, and a Juan de la Cierva contract, the European COST Action CM0702, and the German Academic Exchange Service (DAAD). Generous allocation of computer time at the Computer center of the Friedrich-Schiller-Universität is gratefully acknowledged.

References

1 T. S. Rose, M. J. Rosker and A. H. Zewail, *J. Chem. Phys.*, 1988, **88**, 6672–6673.
2 T. S. Rose, M. J. Rosker and A. H. Zewail, *J. Chem. Phys.*, 1989, **91**, 7415–7436.
3 D. Tannor, *Introduction to Quantum Mechanics: A Time-Dependent Perspective*, University Science Books, Sausalito, 2006.
4 M. H. Beck, A. Jäckle, G. A. Worth and H. D. Meyer, *Phys. Rep.*, 2000, **324**, 1–105.

5 J. M. Bowman, T. Carrington and H. Meyer, *Mol. Phys.*, 2008, **106**, 2145–2182.
6 G. A. Worth, H. D. Meyer, H. Köppel, L. S. Cederbaum and I. Burghardt, *Int. Rev. Phys. Chem.*, 2008, **27**, 569–606.
7 A. M. Virshup, C. Punwong, T. V. Pogorelov, B. A. Lindquist, C. Ko and T. J. Martínez, *J. Phys. Chem. B*, 2009, **113**, 3280–3291.
8 G. A. Levine, J. D. Coe, A. M. Virshup and T. J. Martínez, *Chem. Phys.*, 2008, **347**, 3–16.
9 G. A. Worth, M. A. Robb and I. Burghardt, *Faraday Discuss.*, 2004, **127**, 307–323.
10 V. A. Rassolov and S. Garashchuk, *Phys. Rev. A: At., Mol., Opt. Phys.*, 2005, **71**, 032511.
11 J. Li, C. Woywod, V. Vallet and C. Meier, *J. Chem. Phys.*, 2006, **124**, 184105.
12 R. Spezia, I. Burghardt and J. T. Hynes, *Mol. Phys.*, 2006, **104**, 903–914.
13 B. Lasorne, M. A. Robb and G. A. Worth, *Phys. Chem. Chem. Phys.*, 2007, **9**, 3210–3227.
14 D. V. Shalashilin, M. S. Child and A. Kirrander, *Chem. Phys.*, 2008, **347**, 257–262.
15 T. Yonehara, S. Takahashi and K. Takatsuka, *J. Chem. Phys.*, 2009, **130**, 214113.
16 T. Yonehara and K. Takatsuka, *J. Chem. Phys.*, 2010, **132**, 244102.
17 G. Granucci, M. Persico and A. Zoccante, *J. Chem. Phys.*, 2010, **133**, 134111.
18 *Ab Initio Molecular Dynamics: Basic Theory and Advanced Methods*, ed. D. Marx and J. Hutter, Cambridge University Press, Cambridge, 2009.
19 N. Doltsinis and D. Marx, *J. Theor. Comput. Chem.*, 2002, **1**, 319–349.
20 J. C. Tully, *J. Chem. Phys.*, 1990, **93**, 1061–1071.
21 N. Doltsinis, in: *Computational Nanoscience: Do It Yourself!*, ed. J. Grotendorst , S. Blügel and D. Marx , John von Neumann Institute for Computing, NIC Series, vol. 31, Jülich, 2006, pp. 389–409.
22 B. Maiti, G. C. Schatz and G. Lendvay, *J. Phys. Chem. A*, 2004, **108**, 8772–8781.
23 K. Yagi and K. Takatsuka, *J. Chem. Phys.*, 2005, **123**, 224103.
24 G. A. Jones, A. Acocella and F. Zerbetto, *J. Phys. Chem. A*, 2008, **112**, 9650–9656.
25 R. Mitrić, J. Petersen and V. Bonačić-Koutecký, *Phys. Rev. A: At., Mol., Opt. Phys.*, 2009, **79**, 053416.
26 I. Tavernelli, B. F. E. Curchod and U. Rothlisberger, *Phys. Rev. A: At., Mol., Opt. Phys.*, 2010, **81**, 052508.
27 M. Richter, P. Marquetand, J. González-Vázquez, I. Sola and L. González, *J. Chem. Theory Comput.*, 2011, **7**, 1253–1258.
28 P. Brumer and M. Shapiro, *Annu. Rev. Phys. Chem.*, 1992, **43**, 257–282.
29 R. J. Gordon and S. A. Rice, *Annu. Rev. Phys. Chem.*, 1997, **48**, 601–641.
30 S. A. Rice, *Adv. Chem. Phys.*, 1997, **101**, 213–283.
31 D. J. Tannor, R. Kosloff and A. Bartana, *Faraday Discuss.*, 1999, **113**, 365–383.
32 S. A. Rice and M. Zhao, *Optical Control of Molecular Dynamics*, Wiley, New York, 2000.
33 T. Brixner, N. H. Damrauer and G. Gerber, *Adv. At., Mol., Opt. Phys.*, 2001, **46**, 1–54.
34 M. Shapiro and P. Brumer, *Rep. Prog. Phys.*, 2003, **66**, 859–942.
35 M. Shapiro and P. Brumer, *Principles of Quantum Control of Molecular Processes*, Wiley, New York, 2003.
36 C. Daniel, J. Full, L. González, C. Lupulescu, J. Manz, A. Merli, Štefan Vajda and L. Wöste, *Science*, 2003, **299**, 536–539.
37 P. Nuernberger, G. Vogt, T. Brixner and G. Gerber, *Phys. Chem. Chem. Phys.*, 2007, **9**, 2470–2497.
38 V. Engel, C. Meier and D. J. Tannor, *Adv. Chem. Phys.*, 2009, **141**, 29–101.
39 B. J. Sussman, D. Townsend, M. Y. Ivanov and A. Stolow, *Science*, 2006, **314**, 278–281.
40 J. González-Vázquez, I. R. Sola, J. Santamaria and V. S. Malinovsky, *Chem. Phys. Lett.*, 2006, **431**, 231–235.
41 B. Y. Chang, S. Shin, J. Santamaria and I. R. Sola, *J. Chem. Phys.*, 2009, **130**, 124320.
42 B. Y. Chang, S. Shin and I. R. Sola, *J. Chem. Phys.*, 2009, **131**, 204314.
43 M. Wollenhaupt, A. Präkelt, C. Sarpe-Tudoran, D. Liese and T. Baumert, *J. Opt. B: Quantum Semiclassical Opt.*, 2005, **7**, S270–S276.
44 S. Patchkovskii, *Phys. Chem. Chem. Phys.*, 2006, **8**, 926–940.
45 L. Verlet, *Phys. Rev.*, 1967, **159**, 98–103.
46 L. Verlet, *Phys. Rev.*, 1968, **165**, 201–214.
47 M. Barbatti, A. J. A. Aquino, J. J. Szymczak, D. Nachtigallová, P. Hobza and H. Lischka, *Proc. Natl. Acad. Sci. U. S. A.*, 2010, **107**, 21453–21458.
48 J. González-Vázquez and L. González, *ChemPhysChem*, 2010, **11**, 3617–3624.
49 M. D. Feit, J. A. FleckJr. and A. Steiger, *J. Comput. Phys.*, 1982, **47**, 412–433.
50 R. Kosloff and H. Tal-Ezer, *Chem. Phys. Lett.*, 1986, **127**, 223–230.
51 H. Guo, *J. Chem. Phys.*, 1993, **99**, 1685–1692.
52 T. Seideman, *J. Chem. Phys.*, 1997, **106**, 2881.
53 P. Marquetand, S. Gräfe, D. Scheidel and V. Engel, *J. Chem. Phys.*, 2006, **124**, 054325/1–7.
54 P. Marquetand and V. Engel, *Chem. Phys. Lett.*, 2006, **426**, 263–267.

55 B. J. Sussman, M. Y. Ivanov and A. Stolow, *Phys. Rev. A: At., Mol., Opt. Phys.*, 2005, **71**, 051401.
56 C. Cohen-Tannoudji, B. Diu and R. Laloë, *Quantum Mechanics*, Vol. 2, Wiley, New York, 1977.
57 P. Marquetand and V. Engel, *Chem. Phys. Lett.*, 2005, **407**, 471–476.
58 M. Thachuk, M. Y. Ivanov and D. M. Wardlaw, *J. Chem. Phys.*, 1996, **105**, 4094–4104.
59 N. Shenvi, *J. Chem. Phys.*, 2009, **130**, 124117.

PAPER

Dynamic stark control: model studies based on the photodissociation of IBr

Cristina Sanz-Sanz,† Gareth W. Richings and Graham A. Worth

Received 11th March 2011, Accepted 26th April 2011
DOI: 10.1039/c1fd00039j

The Stark effect is produced when a static field alters molecular states. When the field applied is time dependent, the process is known as the dynamic Stark effect. Of particular interest for the control of molecular dynamics is the Non-Resonant Dynamic Stark Effect (NRDSE), in which the time dependent field is unable to effect a one-photon excitation. The intermediate strength laser pulse instead shapes the potential energy surfaces (PES) and so guides the evolution of the system. A prototype control scheme uses the NRDSE to change the topography of PES in regions where they intersect, thus providing control over photochemistry. Following earlier experimental work, in this paper we study the NRDSE on a new 3 state model of the IBr molecule to gain insight into the mechanism of control at the avoided crossing that governs the branching ratio of the photodissociation.

1 Introduction

In the last decade one of the main goals in chemistry, both experimental and theoretical, has been the control of chemical reactions. Modern ultrafast laser pulses are on the time scale of chemistry itself, and by shaping and timing these pulses we obtain precision tools to interact with and control the time-evolution of a system at a molecular level.[1–3]

The majority of laser control lies in the weak-field limit, in which the light-matter interaction can be taken as a perturbation on the Hamiltonian and only provides coupling to change the population of states. In the intermediate regime, the one we are interested in, the field is strong enough that perturbation theory no longer applies, but not strong enough to produce ionisation. In this limit the electric field is able to modify the PES over which the system evolves.

The change in PES is provided by the Stark effect. Traditionally, a static field is used to shift molecular states. When the field applied is time dependent, the process is known as the dynamic Stark effect (DSE) or Autler-Townes effect.[4] The applied fields may be resonant or non-resonant with the electronic transitions in question. The effect was originally observed in molecules as the splitting of spectral lines with the work of Quesada on hydrogen[5] and has subsequently been studied in Na_2,[6–9] Li_2,[10–12] N_2,[13] Cs_2,[14] NaK^{15} and CH_2.[16] Notably the DSE has been used to calculate the absolute transition dipole moments of Na_2,[8,9] Li_2^{12} and NaK.[15]

Recently, the non-resonant dynamic Stark effect (NRDSE), in which the time dependent field is unable to effect a one-photon excitation, was used by Stolow and coworkers[17,18] to control the photodissociation of IBr. The photodissociation of IBr is initiated by absorption of a visible photon, exciting the system from the

School of Chemistry, University of Birmingham, Edgbaston, B15 2TT, UK

† Present address: Institute for Fundamental Physics, CSIC, C/Serrano 113Bis, 28006, Madrid, Spain. Tel.: +34 91 5616800, Ext. 943230. E-mail: cristina.sanz@csic.es

ground electronic state, $1^1\Sigma_{0^+}^+$ to the excited state $1^3\Pi_{0^+}$. The non-adiabatic coupling between the $1^3\Pi_{0^+}$ and $1^3\Sigma_{0^+}^-$ states leads to two dissociation channels

$$\text{IBr} \rightarrow \begin{cases} \text{I} & + \quad \text{Br}(^2P_{3/2}) \\ \text{I} & + \quad \text{Br}^*(^2P_{1/2}) \end{cases}$$

The photodissociation of IBr has long been studied, with early work including that on selective magnetic alignment of the molecule by differential dissociation rates[19] and on the velocity distribution of the recoiling fragments.[20,21] Latterly work has been carried out on the alteration of the proportion of the molecules dissociating *via* the two channels noted above. This has mainly focused on the use of predissociation, where the molecule is vibrationally excited before dissociation.[22–29] As a result of this work, it has been shown that dissociation *via* the excited-state channel is favoured by vibrational excitation.[23,26] The threshold energies for dissociation *via* the competing channels have also been determined.[30] Much of the interest in IBr is down to the nature of the dissociation, which is neither described well by the adiabatic picture nor by the diabatic model,[22,24,25] making it a classic case where the Landau–Zener description of curve crossing breaks down. As such, the application of an infrared NRDSE field to modify the curve-crossing barrier at a specific time, promoting the yield of one dissociation channel over another, is of significant interest.

An avoided crossing in a one-dimensional system is the simplest version of a non-adiabatic coupled system. Non-adiabatic processes, such as internal conversion and intersystem crossing, entail charge rearrangements that occur along a reaction path at the intersections of PES. Of particular importance in photochemistry are conical intersections, also known as photochemical funnels. These provide very efficient pathways for crossing between states, and often dominate the evolution of the system.[31–33] Chemical branching ratios in non-adiabatic processes are very sensitive to the intersection geometry, and therefore the dynamic modification of these processes is an important application of the NRDSE. IBr seems to provide an ideal example of controlling a system passing through an avoided crossing by moving the crossing point.

Stolow and co-workers used a simple 3-state model of IBr to rationalise their results. The system of IBr is however not as simple as a three state model. In later studies we will treat the full manifold of states completely as a function of the electric field. In this paper, we show initially how a 3-state model can be extracted from the full manifold of states. This will provide a starting point to help understand what mechanisms are present in the NRDSE control. After showing that our new 3-state model agrees with the Stolow experiments, we will go on to investigate the dependence of the control on the system parameters. Here we look at the balance between polarisability and dipole moments, and the importance of state-coupling terms.

2 IBr electronic states

Ignoring spin–orbit coupling, there are 12 electronic states of IBr dissociating to I (^{2}P) + Br(^{2}P) in their ground states, and there are 23 states dissociating to the four channels generated when the spin–orbit couplings are considered, $I\left(^2P_{\frac{3}{2},\frac{1}{2}}\right)$, $Br\left(^2P_{\frac{3}{2},\frac{1}{2}}\right)$. The potential curves have been calculated using multiconfiguration reference internally contracted configuration interaction (MRCI), included in MOLPRO package[34] and a basis set optimised for this system by Patchkovskii.[35] The 12 non-spin–orbit-split states are shown in Fig. 1 on the left plot, while the right panel shows the 23 states dissociating to the four spin–orbit channels. Looking carefully at the plot on the right, at approximately $R \approx 6.3$ a.u, $E \approx 17000$ cm^{-1}, we can see the

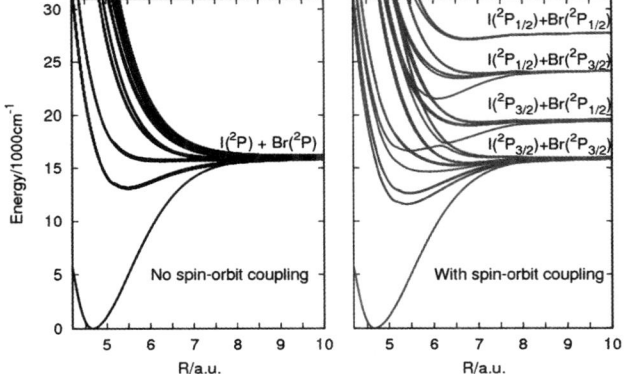

Fig. 1 Electronic states dissociating to I + Br. Left hand plot shows the 12 states dissociating to the ground states of both atoms and the one on the right the 23 states which dissociate to the four channels when spin-coupling is included.

avoided crossing which is modified by the laser pulse to control the branching ratio of the photodissociation of IBr.

In order to provide a detailed simulation of the IBr NRDSE control experiment, we have calculated the full manifold of states and couplings as a function of the electric field. Full details on the calculations and states will be presented in a forthcoming paper. In the following, a 3-state model is extracted from the full manifold of states and the dynamics presented in the next sections thus neglects the effect of the rest of the states. This is done to provide an easy to visualise set of simulations. It also provides a bridge between the earlier 3-state model of Guo and Stolow[17,36] and the full problem by adding realistic coupling in the relevant parts of the Hamiltonian.

3 3-State model hamiltonian

The 3 states considered in the model are presented in Fig. 2. The model includes the ground state, $X(^1\Sigma_{0+}^+)$, and two of the excited states dissociating to the two different spin–orbit states of Br. These are labelled $\tilde{A}(^3\Pi_{0+})$ and $\tilde{B}(^3\Sigma_{0+}^-)$ respectively and

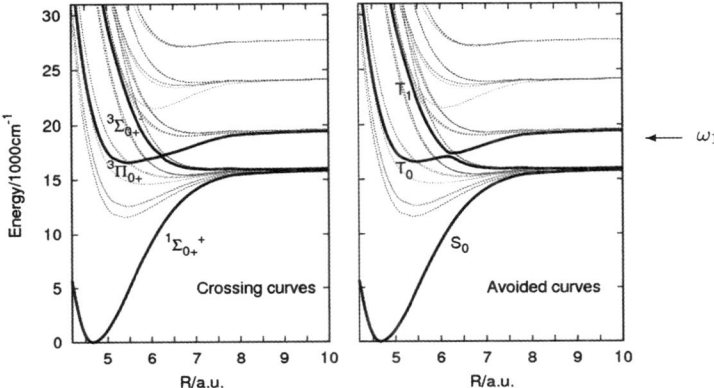

Fig. 2 3 states considered for the model study of the Stark control photodissociation of IBr. The excitation energy, $\omega_1 = 19{,}356$ cm^{-1} is marked by an arrow.

highlighted in the manifold of IBr states in the left panel of Fig. 2. These 3 states have been extracted from the set of states obtained after the diagonalisation of the whole (23 states) matrix. We refer to these 3 crossing states, in a simplified way, as *diabatic* states. A constant off-diagonal coupling of 150 cm^{-1} between the two excited states was found to provide the calculated splitting of the *adiabatic* states, highlighted in the right panel of Fig. 2. The name *adiabatic* is given to simplify the nomenclature since those states are obtained after the diagonalisation of the 3 × 3 matrix. In the adiabatic picture, the states are labelled in order of increasing energy as S_0, T_0 and T_1 so, at the Franck–Condon point $\tilde{A}$ corresponds to T_0 and $\tilde{B}$ to T_1, while at dissociation the correspondence swaps over. We use the polyatomic labelling for the *adiabatic* states as there is no rigorous correspondence to symmetry for these states and it is a simple way to distinguish the singlets and triplets and the energetic ordering.

The model Hamiltonian used in the simulations is written in the diabatic picture as

$$\hat{H} = \hat{T}\mathbf{1} + \mathbf{V} + \mathbf{D}^m + \mathbf{P}^{mn} \qquad (1)$$

with **1** a 3 × 3 unit matrix, $\hat{T}$ being the kinetic energy operator

$$\hat{T} = -\frac{\hbar}{2\mu}\frac{\partial^2}{\partial R^2},$$

and **V** the potential energy matrix

$$V = \begin{pmatrix} V_{11} & 0 & 0 \\ 0 & V_{22} & V_{23} \\ 0 & V_{32} & V_{33} \end{pmatrix}. \qquad (2)$$

The diabatic potential functions V_{11}, V_{22} and V_{33} are given by spline fits of the data from the MRCI calculations, with the constant coupling V_{23} mentioned above. In addition, $\mathbf{D}^m$ is the dipole-field interaction term

$$\mathbf{D}^m = -\begin{pmatrix} d_{11} & d_{12} & d_{13} \\ d_{21} & d_{22} & d_{23} \\ d_{31} & d_{32} & d_{33} \end{pmatrix}(\varepsilon_e(t) + \varepsilon_c(t)), \qquad (3)$$

and $\mathbf{P}^{mn}$ the polarisability-field interaction term as

$$\mathbf{P}^{mn} = -\frac{1}{2}\begin{pmatrix} \alpha_{11} & \alpha_{12} & \alpha_{13} \\ \alpha_{21} & \alpha_{22} & \alpha_{23} \\ \alpha_{31} & \alpha_{32} & \alpha_{33} \end{pmatrix}\varepsilon_c(t)^2, \qquad (4)$$

where $\varepsilon_e(t)$ and $\varepsilon_c(t)$ are the excitation and control pulses, respectively. The coupling between the excitation field and the polarisability is not included as the relative weakness of the real excitation field means this term should negligible. This allows us to increase the strength of the excitation pulse so that the simulations are reliable without drowning out the effect of the control pulse. The superscripts *mn* in the dipole and polarisability terms specify the component considered, *x*, *y*, *z* for the dipole and *xx*, *yy*, *zz*, *xy*, *xz*, *yz* for the polarisabilities. The molecule of IBr has been oriented along the *z* axis and the field is considered being applied following the same direction so only $\mathbf{D}^z$ and $\mathbf{P}^{zz}$ have been included in this Hamiltonian.

The values for the dipole moments and polarisabilities, both on- and off-diagonal, are in Table 1. The polarisabilities were calculated as the derivative of the dipole moments with respect to a field. It was seen that the calculated values were fairly constant over a range of relevant geometries and, thus, average values were selected for the model. It can be seen from the data that the $\tilde{A}$ diabat has a much larger dipole moment and polarisability than $\tilde{B}$. The $\tilde{B}(^3\Sigma_{0+}^-)$ diabat also has a much stronger

Table 1 Parameters for the 3-state model Hamiltonian. Values in au

Dipole moments (1 au = 2.5417 Debye)	Polarisabilities (1 au = 0.2798 Å²)
$d_{11} = -0.25$	$\alpha_{11} = -80$
$d_{12} = 0.02$	$\alpha_{12} = 2$
$d_{13} = 0.1$	$\alpha_{13} = 0.5$
$d_{22} = 0.35$	$\alpha_{22} = -100$
$d_{23} = 0.01$	$\alpha_{23} = 1.5$
$d_{33} = 0.07$	$\alpha_{33} = -50$
$V_{23} = 0.0006834$	

transition dipole from the ground-state as d_{12} is a factor of 5 times smaller than d_{13}. Excitation in the simulations is to $\tilde{A}$.

The excitation and control pulses are defined using an oscillating cosine and a gaussian envelope as

$$\varepsilon_e(t) = N_1 \exp\left[-\frac{1}{2\sigma_1^2}(t-t_1)^2\right]\cos(\omega_1(t-t_1)) \quad (5)$$

$$\varepsilon_c(t) = N_2 \exp\left[-\frac{1}{2\sigma_2^2}(t-t_2)^2\right]\cos(\omega_2(t-t_2)) \quad (6)$$

where N_1 and N_2 are normalization factors, σ_1 and σ_2 are the widths of the pulses, ω_1 and ω_2 are the excitation frequencies and t_1 and t_2 the position of the peaks. The difference between t_1 and t_2 is then the delay between the excitation and the control pulse.

There is in the literature another three-state model[17] based on the Morse potentials published by Guo.[36] This, which we shall call the Guo–Stolow model, is very simple. The Hamiltonian includes the dipole-field interaction only to produce the excitation between the ground and the first excited state, *i.e.* only d_{12} is non-zero. The polarisability is then included only in the diagonal terms of the two excited states, α_{22} and α_{33} so that the Stark effect operates only by shifting the two excited states relative to each other. As in our model the spin–orbit coupling between the excited state diabats provides an off-diagonal coupling term. The model we present here, being as well very simple, includes diagonal and off-diagonal terms for the dipole and polarisability as obtained in the *ab initio* calculation.

The control pulse was applied at various times relative to the excitation pulse (from 250 fs before excitation to 410 fs afterwards). The excitation pulse is timed at $t = 0$ fs, has width 50 fs and is tuned to a frequency corresponding to 2.4 eV (19,356 cm^{-1}), which excites into diabat $\tilde{A}$. This is close to the excitation laser frequency of 520 nm used in the Stolow experiment. The strength of this pulse was taken as 0.028 au, in line with earlier work.[26,28] This is much stronger than the experimental excitation pulse, but it ensures enough promotion of the wavepacket to the excited states to be quantifiable. The control pulse frequency was 0.73eV, again close to the experimental control laser which had a wavelength of 1.7 μm. The control pulse is normalised by application of a pre-factor of $s = 1.4$ to give it an intensity of 0.02 a.u., within the intermediate range mentioned above. The relationship between the strength parameter s and the normalisation factor N in Eqs. (5) and (6) is defined in Sec. 4.3.2. When the width of the control pulse was varied, the strength prefactor was altered to retain normalisation of the pulse.

The usual way to study the ratio for dissociation in different channels is by means of the calculation of the flux in every channel. The flux is the flow through a surface placed in a position where dissociation is certainly going to occur. The flux in each channel γ is computed as the expectation value of the flux operator[37]

$$\hat{F}_\gamma = -\frac{i}{2\mu R}\left(\frac{\partial}{\partial R}\delta(R - R_{\gamma c}) + \delta(R - R_{\gamma c})\frac{\partial}{\partial R}\right)$$

where $R_{\gamma c}$ is the value in the asymptotic region where the flux is computed. In the IBr photodissociation, the different channels are the diabatic states.

In order to obtain the dissociation ratio, $\frac{[\text{Br}^*]}{[\text{Br}]}$, we make time dependent wavepacket calculations using the MCTDH package,[38] and compute the flux in both channels when the wavepacket reaches the dissociation region, in our case when the distance between I and Br is 7.5 au. The Br channel contains contributions from both $\tilde{B}$ and $\tilde{X}$, whereas the Br* channel is entirely due to dissociation on $\tilde{A}$. The grid used consists of 400 fast Fourier transform points spread between bond lengths of 4 au and 10 au and the propagation used a short iterative Lanczos integration scheme. Right after the evaluation of the flux the wavepacket is absorbed by a complex absorbing potential, placed at $R = 8$ au, in order to avoid reflections.

4 Results and discussion

4.1 Control of 3-state model

In Stolow's NRDSE IBr experiment,[18] the control of the photodissociation was seen as decrease in the branching ratio when the non-resonant control pulse came concurrently with or shortly after the excitation pulse. This corresponds to a greater proportion of the ground state, $\text{Br}(^2P_{3/2})$ (hereafter referred to simply as Br), being formed than when there is no control pulse. If the delay, after excitation, is greater than about 200 fs, then an increase in the branching ratio is seen. This indicates that a larger amount of excited state $\text{Br}(^2P_{1/2})$ (in all that follows, Br*) is being formed than when no control pulse is applied. At positive time delays, longer than 500fs, the control capability of the pulse is much reduced. There is simply a second, small increase in the proportion of Br* produced between 500 fs and 1000 fs. These findings are summarised in the inset to Fig. 3(a).

Simulations using the Guo–Stolow model[18] supported the experimental data, with the same sequence of the major peaks being observed. The time scales of the experiment and theory do not, however, match. Theory predicted a change from increased Br production to increased Br* channel dissociation at control pulse delays of approximately 100 fs, as opposed to the 200 fs seen in the experiment. The model also predicts that control pulses later than 200 fs after excitation have no effect on the branching ratios, a much later cut off is seen in the experiment.

To see how the new model compares to the simpler model and experiment, Fig. 3 (a) shows the change in branching ratio for the IBr system with a control pulse (of width 150 fs and intensity 0.02 au) applied at a range of times from 241 fs prior to the excitation to 406 fs afterwards. The y axis represents the ratio of the total flux through the upper dissociation channel (Br*) to the total flux through the lower channel (Br), from which the ratio obtained without a control pulse is subtracted. The x axis represents the time of application of the control pulse relative to the excitation (at $t = 0$ fs). Comparing this plot, it is apparent that the main features of the experiment are indeed present: concurrent with the excitation and at short times afterwards, a decrease in the branching ratio is seen. At longer times after excitation, there is an increase in the proportion of molecules dissociating to Br*. As with the simpler Guo–Stolow model, the features on the plots are seen at shorter times, after the excitation, than is observed in the experiment.

Examining this plot more closely, we note negative peaks centred at $t \approx 0$ fs and $t \approx 60$ fs and a positive peak at $t \approx 90$ fs. There is also a small negative peak at $t \approx 160$ fs. The negative peaks indicate that in the region 0 fs $< t <$ 70 fs, there is an

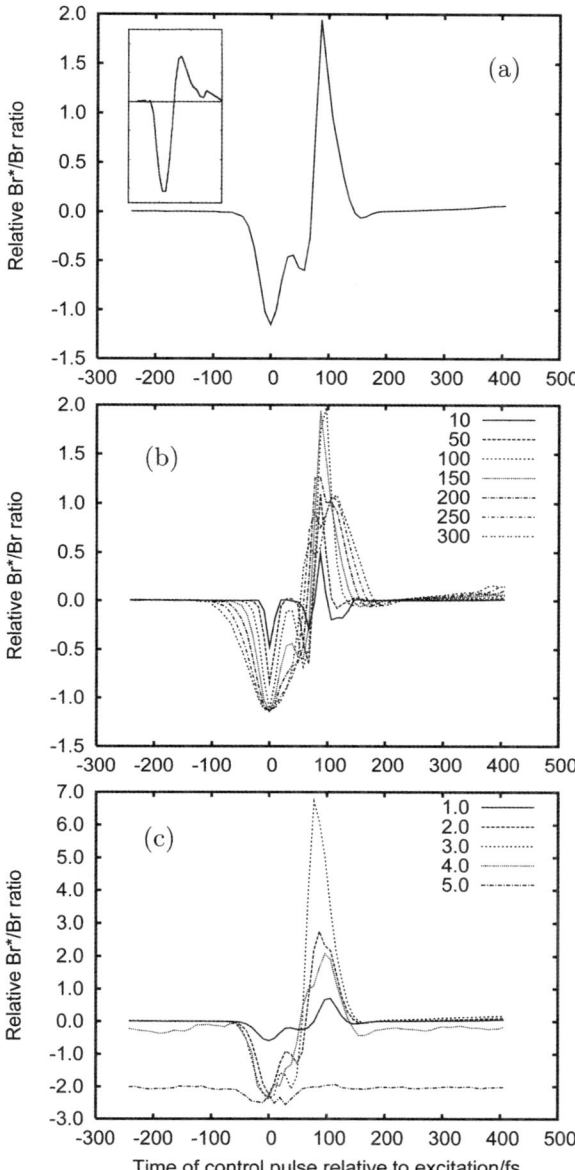

Fig. 3 Plots showing the ratio of the total flux through the Br*($^2P_{1/2}$) dissociation channel to that through the lower energy, Br($^2P_{3/2}$) channels as a function of time of application of the control pulse (with Br* the ratio calculated without a control pulse subtracted). (a) Control pulse width of 150 fs and prefactor (explained in the main text) $s_2 = 1.4$. The experimental plot is included as an inset for comparison.[18] (b) Different control pulse widths (values in fs noted in the key) with normalised intensity. (c) Control pulses with a width of 150 fs but with varying intensities. Values of the s_2 prefactor are noted in the key.

increase of the proportion of the wavepacket passing through the lower dissociation channel than when no control pulse is applied. Conversely, with control pulses applied between 70 fs–150 fs after excitation, a greater proportion of the wavepacket goes along the upper dissociation channel. In effect the wavepacket has a tendency, in the latter case, to stay on the diabat to which it was excited and, in the former, to cross to the diabats corresponding to the lower energy channels.

In the following sections we study the effect of changes in the control pulse on the branching ratio plots.

4.1.1 Control pulse width. A set of simulations were carried out where the width of the control pulse was varied. Fig. 3(b) shows the effect of this on the branching ratio plots. The shortest pulse applied was 10 fs wide and this gives relatively small magnitude, narrow extrema indicating that the pulse has a very precise, but fairly minor effect on the branching ratio. This suggests that a very accurate knowledge of the dynamics of the system would be needed before effective control could be achieved with a pulse of this width. Of added note in this case is the trough at $t \approx 120$ fs, at the tail end of the major peak. This indicates that the narrow pulse is able to affect the remaining wavepacket as it exits the crossing region, something which appears not to be possible with the wider pulses used.

As the pulse width increases to 150 fs, the peaks get stronger and gradually broaden, and the trough at 120 fs disappear. This can be rationalised as follows: as the pulse widens, its effect stretches before and after its nominal timing, meaning that the changes of the branching ratio seen at these other times also get counted within the total change at the given time. Thus, at the times of the smaller peaks, the larger, opposite, effects before and after get added in and they are subsumed.

From a width of 150 fs to 300 fs, the peak at $t \approx 90$ fs becomes shorter and with a width of 200 fs appears to split into two peaks, one at 90 fs the other at 120 fs. The latter becomes so significant at a width of 300 fs, that it appears that the original peak has shifted to later times.

Thus the branching ratio from the model matches the experiment better as the control pulse width increases. The broad peaks of the experiment could therefore be hiding the other features seen here. In a later section, we will explain what parts of the model give rise to the different peaks to give insight into the potential of NRDSE control mechanism.

4.1.2 Control pulse strength. In a second set of simulations, the intensity of the control pulse was varied, keeping the width at 150 fs. The prefactor, s_2, of the pulse was varied from 1.0 to 5.0 (the standard pulse had prefactor 1.4, corresponding to a pulse intensity of 2.8×10^{13} Wcm^{-2}). The relationship between these prefactors and the normalisation coefficient N_i in eqn (5) and (6) is that $N_i = s_i(2^{1/4}\sqrt{\pi\sigma_i})^{-1}$. The effects of such alterations are seen in Fig. 3(c). As we keep s_1 constant, we will simply refer to the strength of the control pulse, s_2, as s in the following paragraphs.

The plot with intensity $s = 1.0$ is of same basic shape as that in Fig. 3(a), but with low magnitude extrema. This is reasonable, as with a lesser magnitude there are fewer photons in the control pulse to cause the effects represented by the features in the plots. At intensity $s = 2.0$, again a similar shape is observed but with a significantly greater magnitude peak at $t \approx 90$ fs than the standard plot and more significantly enlarged negative peaks at lower times. Over this range of strengths, it appears that increasing the control pulse intensity changes the magnitude of the control.

At strength, $s = 3.0$, the positive peak becomes much larger and the second negative peak moves from $t \approx 60$ fs to $t \approx 40$ fs. At $s = 4.0$, the pattern breaks down somewhat and the intensity of the second negative and of the positive peak becomes reduced. The peak at $t = 0$ fs stays much the same, however, there is an overall shift of the baseline downwards at times earlier than -50 fs and later than 150 fs. This suggests that the effect that the control pulse has on the energy levels echoes through even when applied before excitation. It thus appears that at this intensity of pulse, the form of the plots has begun to distort, suggesting we are leaving the intermediate field region and moving into the strong field regime. The validity of our model thus begins to break down as previously unseen effects emerge. The distortion becomes complete at $s = 5.0$, where the whole plot shifts down to negative relative ratios, all peaks being lost. This suggests that crossing to the lower channel is occurring whenever the pulse is applied. At this point, where multiphoton effects are entering, the model no longer holds.

4.2 System dynamics

In order to understand the mechanism of the dissociation of IBr, the initial observations taken from the plot in Fig. 3(a) were compared to a dynamics simulation where

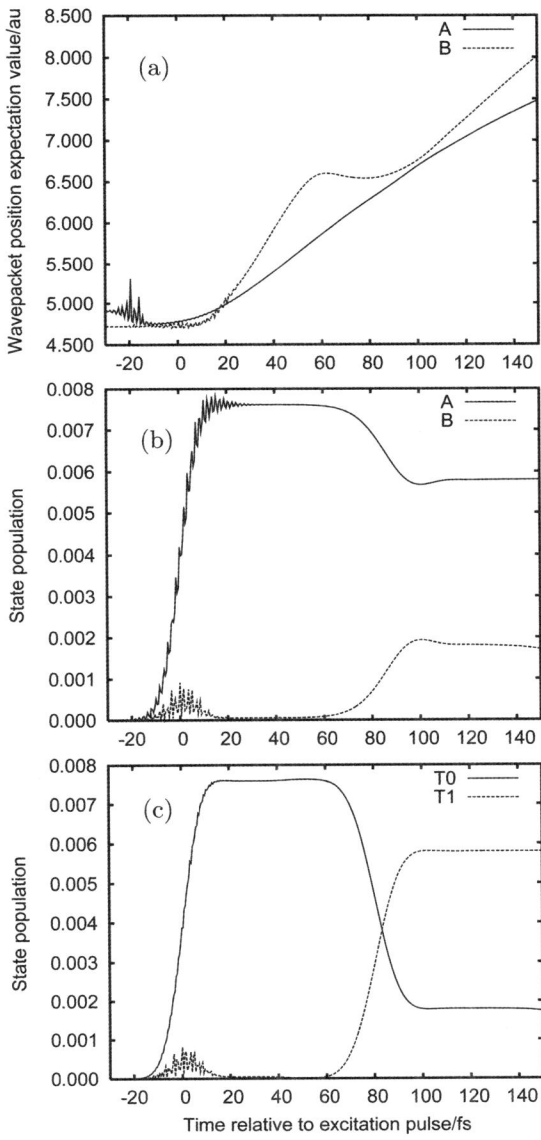

Fig. 4 As a function of time relative to the excitation pulse. (a) Centre coordinate (expectation value of R) of the wavepackets on diabats $\tilde{A}$ and $\tilde{B}$. (b) Diabatic populations of $\tilde{A}$ and $\tilde{B}$ (c) Adiabatic populations of T_0 and T_1 states.

no control pulse was applied. Fig. 4 shows the expectation values of the position of the wavepacket as well as both diabatic and adiabatic state populations as a function of time (note the similarity to Fig. 2 in an earlier paper by Vandana and Mishra[26]).

Fig. 4(a) shows the expectation values of the position of the wavepackets on the upper and lower excited surfaces with excitation at $t = 0$ fs. The molecule is excited to $\tilde{A}$ (the pulse lasts for 50 fs) and motion towards dissociation begins immediately in this state, giving the smooth increase in distance with passing time. The region where $\tilde{A}$ and $\tilde{B}$ cross is at about $R = 6.25$ au and the centre of the wavepacket reaches this region at around 80 fs. In fact, looking at the full excited wavepacket in $\tilde{A}$, it is seen that the leading edge enters the region of the crossing around 40 fs after excitation and it has completely passed through after about 120 fs.

The expectation value for state $\tilde{B}$ begins to increase from about 20 fs due to a small amount of excitation from the ground state. This minor wavepacket immediately moves down the diabat towards the dissociation limit. It is clearly moving faster than the wavepacket on $\tilde{A}$. Density from the wavepacket on $\tilde{A}$ crosses to $\tilde{B}$ from about 50 fs onwards as indicated by the change in the diabatic state populations seen in Fig. 4(b). At around 60 fs, the smooth progress of the wavepacket on diabat $\tilde{B}$ is halted as crossing from $\tilde{A}$ becomes more rapid than the rate at which the small density already on the higher state can move away. In effect, it is slowed by interference from the wavepacket on $\tilde{A}$ as it leaves the interaction region. This period lasts until about 100 fs, when the wavepacket on $\tilde{B}$ accelerates away again on the steep dissociative curve. At this point the interference due to the wavepacket on diabat $\tilde{A}$ is reduced due to the passage of the bulk of the wavepacket. After 150 fs the bond has extended by over 3 au.

The diabatic populations show that after passing through the intersection region, the major proportion of the wavepacket remains on $\tilde{A}$. This corresponds to the formation of Br*. In the adiabatic picture, the population transfer from lower to upper adiabatic surfaces is indicated in Fig. 4(c). The populations cross at 80 fs and then steady themselves at around 100 fs after excitation. These plots show that the dynamics is dominated by diabatic behaviour as the diabatic populations do not cross.

4.3 Parameter dependency

Having examined the process which occurs when IBr is excited and the nuclear wavepacket is allowed to propagate towards dissociation, we now examine, in more detail, the features of Fig. 3(a) in order to understand how the control pulse affects the process just described. We do so by isolating elements of the Hamiltonian through which the control pulse exerts an effect, and varying them to see the effect on the shape of the plot in Fig. 3(a). This can be achieved by varying the elements of the dipole and polarisability matrices and once more plotting the ratio of dissociation products to the timing of the control pulse. In this way we can see which interactions are important in the control process.

4.3.1 Diagonal elements of the dipole matrix.
Initially we look at the effects of the dipole terms, specifically the components on the diagonal of the interaction matrix. Fig. 5(a) shows the effect of varying d_{11} from 0.0 au to -1.0 au (0D to -2.54D). All other dipole and polarisability elements are kept as in Table 1. The general trend to note is the flattening of the peaks and troughs by increasing the magnitude of the dipole element. The negative peak at $t = 0$ fs is reduced in magnitude by about a half over the range of increase, but this is not as dramatic as the effect on the positive peak at $t \approx 90$ fs and the negative peak at $t \approx 60$ fs due to the increase in the magnitude of the dipole element; both lose about 90% of their intensity. There is some indication that the effect is non-linear as the reduction in magnitude of the extrema is not of an even ratio as d_{11} moves to the higher values. Having said that, the qualitative shape of the curves is unaltered by the increase in dipole,

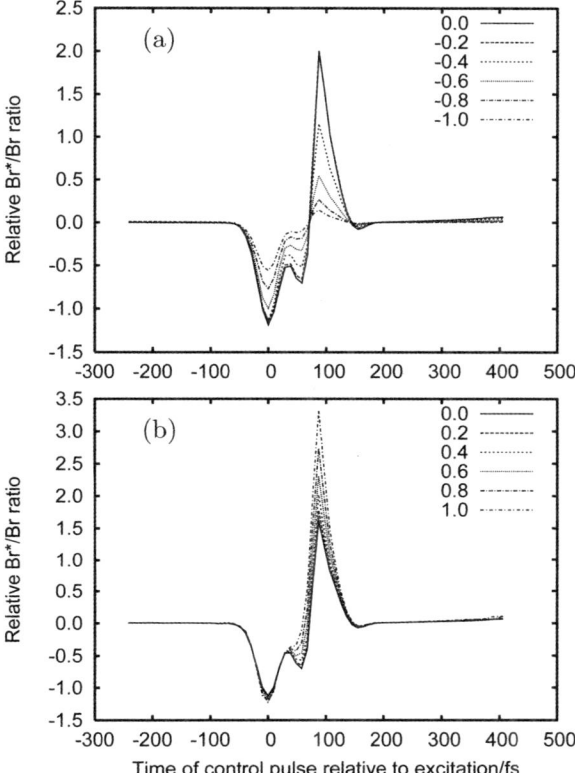

Fig. 5 Ratio of flux through the upper dissociation channel to that through the lower channels as a function of time of application of the control pulse relative to the ratio without control. (a) Each curve represents a different magnitude of dipole element, d_{11}, as indicated in the key. Units are au. (b) Each curve represents a different magnitude of dipole element, d_{22}, as indicated in the key. Units are au.

indicating that nothing fundamental is altered, only the magnitudes of the underlying processes.

Fig. 5(b) shows the effect of d_{22}. Here we see that the negative peak at $t \approx 0$ fs is barely affected by changes in d_{22}. The major effect of the changing dipole component is in the maximum, the magnitude of which increases monotonically as d_{22} increases from 0 au to 1.0 au (0D to 2.54D). As with the changes in d_{11}, the effect of altering the second component appears to be non-linear.

The increase in the height of the peak is accompanied by the reduction and gradual disappearance of the negative peak at $t \approx 60$ fs, meaning that, at greater magnitudes of d_{22}, we are left with a single trough and a single peak in the plots of branching ratio.

The reason for changes, seen above, is the increase in energy of diabat $\tilde{A}$ when the control pulse is applied. This effect is more significant when the interaction of the laser pulse with the energy surface, through d_{22}, is greater. As the energy of the surface increases, the crossing of the energy level with diabat $\tilde{B}$ is changed in character and position. Recalling that the expectation value of the position of the wavepacket reaches the crossing (at $R = 6.3$ au) after about 90 fs, then if the energy level of diabat $\tilde{A}$ is increased, the position of the crossing moves to shorter bond lengths, *i.e.* away from the wavepacket. The likelihood of crossing thus decreases and the wavepacket carries on along the upper dissociation channel.

Calculations involving the variation of d_{33} were also carried out, but they revealed that, at the lower magnitude of d_{33} relative to the other diagonal components of the dipole matrix, changes in the ratio of dissociation through the upper and lower channels were insignificant.

4.3.2 Excited state coupling terms. Moving on from the diagonal elements of the dipole matrix, we look at the coupling between levels $\tilde{A}$ and $\tilde{B}$. Clearly this is of major interest as the branching ratios we are studying are dependent on the ability of the wavepacket on diabat $\tilde{A}$ to cross to diabat $\tilde{B}$.

Fig. 6(a) shows the effect of variation in the off-diagonal dipole element, d_{23}. The polarisability coupling between the two excited states, α_{23}, is set to zero, and the dipole coupling increased from 0.00 au to 0.12 au (0D to 0.305D).

Here we immediately note that the negative peak at $t \approx 60$ fs is not present when both α_{23} and d_{23} are zero. The large peak at $t \approx 90$ fs is reduced in magnitude as d_{23} is increased (and it appears to shift to slightly later times). A negative peak also grows in at $t \approx 40$ fs. This seems to be a different feature to the peak noted above at $t \approx 60$ fs. Thus it appears that d_{23} helps a greater proportion of the wavepacket on $\tilde{A}$ to cross to $\tilde{B}$.

The negative peak at $t \approx 0$ fs appears to be less affected by the change in d_{23}, the part at positive times being subsumed by the growth of the negative peak at $t \approx 40$ fs

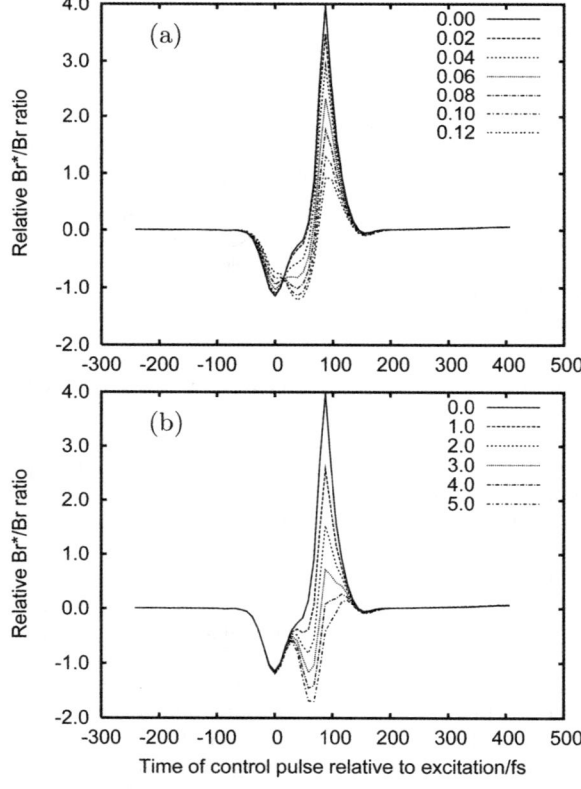

Fig. 6 Ratio of flux through the upper dissociation channel to that through the lower channels as a function of time of application of the control pulse relative to the ratio without control. (a) Each curve represents a different magnitude of the off-diagonal dipole element, d_{23}, as indicated in the key. Units are au. (b) Each curve represents a different magnitude of the off-diagonal polarisability element, α_{23}, as indicated in the key. Units are au.

and the part at negative time gradually flattened into a shoulder of the emerging peak. It ought to be noted that the dipole coupling should not affect the excitation probability.

In Fig. 6(b) we plot the effect of the coupling between the two excited states caused by the control pulse through the polarisability term, α_{23}. As such we set $d_{23} = 0$ au and vary α_{23}.

The key features to note in this plot are that the negative peak at $t \approx 0$ fs is unaffected by the alteration of the polarisability coupling. Recalling the relatively small effect on it due to varying the dipole coupling, we can conclude that this negative peak is only marginally affected by coupling between the two excited states. As this negative peak is coincident with the excitation pulse, and hence appears before the wavepacket has had a chance to reach the region where the two excited surfaces cross, this is perhaps unsurprising.

The plots do, however, change dramatically with increasing α_{23}. A negative peak at $t \approx 60$ fs is seen to appear. It is all but non-existent when $\alpha_{23} = 0$ au (as was also the case in Fig. 6(a)). When the polarisability is between 2.0 au $< \alpha_{23} <$ 3.0 au, it becomes deeper than the negative peak coincident with the excitation. The wavepacket is beginning to pass the crossing region at around the time of this feature, so it makes sense that an increase in the interaction between the two states caused by the coupling, should cause an increased amount of crossing to the lower dissociation channel. This is backed up by the reduction in the magnitude of the peak at $t \approx 90$ fs, which effectively disappears at $\alpha_{23} = 4.0$ au. Clearly the control pulse has a major effect on the ratio of wavepacket passing through the two dissociation routes, forcing the wavepacket to cross to the lower level if a large coupling (dipole and/or polarisability) is available.

The difference in position of the emerging negative peaks, due to the varying d_{23} and α_{23} at $t \approx 40$ fs and $t \approx 60$ fs respectively, leads us to interpret these peaks as due to different mechanisms of control. The control pulse has a frequency of 0.73 eV. The gap between $\tilde{A}$ and $\tilde{B}$ has this value at approximately 6.1 and 6.4 au, either side of the avoided crossing. This is reached by the wavepacket on $\tilde{A}$ at approximately 65 fs. Thus the effect of the d_{23} peak seen in Fig. 6(a) is due to dipole excitation or de-excitation between the states. In contrast, the polarisability peak in Fig. 6(b) at 80 fs coincides with the wavepacket on $\tilde{A}$ reaching the avoided crossing maxima, and so this is due to the increased non-adiabatic coupling flattening out the barrier and so enhancing the adiabatic passage.

Finally, the disappearance of the major peak, at high polarisabilities, reveals a second, small peak at $t \approx 120$ fs, which has been hidden by its dominant neighbour. The peak and trough immediately after at $t \approx 160$ fs are either side of the time when the wavepacket is almost through the intersection region. From Fig. 4(a), at this time the wavepackets on the two surfaces are at the same geometry, $R \approx 6.8$. At this bond length the surfaces are roughly 1.5 eV apart. This effect could thus be due to 2-photon excitation/de-excitation between the states.

4.3.3 Dependency on individual parameters.
Having examined the effect of altering individual dipole and polarisability components, whilst all other elements were present, we now turn to the effects of each component in isolation. Fig. 7(a) and 7(b) show the essential features in the dissociation ratio plots, the former showing the effect of each dipole component in isolation, the latter the contribution of the individual polarisability components. Each line on the plots is associated with a single component being set to the value in Table 1, all other values in the dipole and polarisability matrices being set to zero.

4.3.3.1 Dipole terms. Starting with the dipole components, it is clear that only d_{13} has a significant contribution to Fig. 3(a), this being to the peak at $t \approx 90$ fs. It is not immediately apparent why the interaction between the ground state and diabat $\tilde{B}$ causes a decrease in the amount of the wavepacket crossing down to these levels when the control pulse is applied, however, examination of the populations

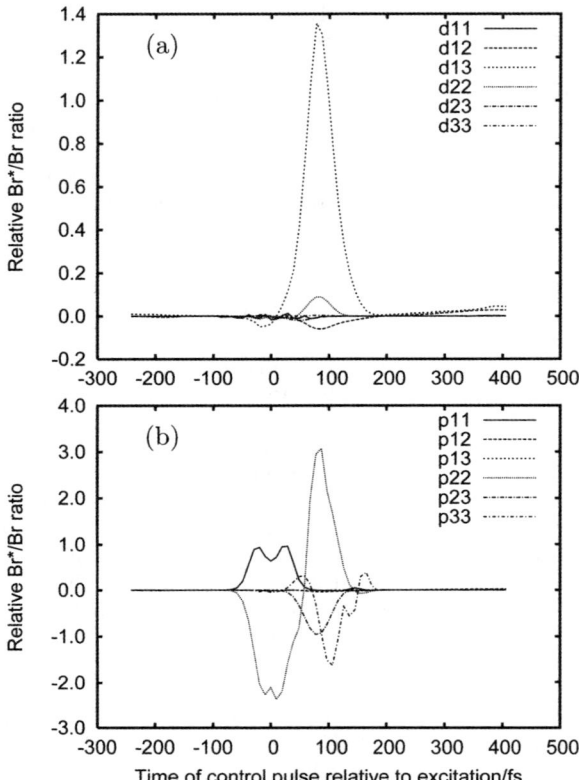

Fig. 7 Ratio of flux through the upper dissociation channel to that through the lower channels as a function of time of application of the control pulse relative to the ratio without control. (a) Each curve represents a different component of the dipole matrix, as indicated in the key. That component is non-zero whilst all others and the polarisability matrix are set to zero. (b) Each curve represents a different component of the polarisability matrix, as indicated in the key. That component is non-zero whilst all others and the dipole matrix are set to zero.

of all of the states provides a clue. Without a control pulse, the population of the ground state stays essentially constant after excitation. When only the d_{13} component is included in a control pulse at 87 fs, however, the population of the ground state increases from 0.9923 to 0.9927 at this time. The population of this state stays constant when no control pulse is applied. It thus appears that the pulse induces a transition of part of the wavepacket down to the ground state, from where it relaxes towards the equilibrium geometry. As such, the flux through the lower channel is reduced, and the proportion exiting *via* the upper channel is increased as seen. As this effect is seen when only the d_{13} component is present, it appears that the transition is due to de-excitation from the $\tilde{B}$ state rather than from $\tilde{A}$. The following mechanism is thus proposed: transfer from $\tilde{A}$ to $\tilde{B}$ occurs as it can do with no control pulse present. At $R = 6.3$ au, the location of the crossing, the energy gap between $\tilde{X}$ and $\tilde{B}$ is about 0.75 ev, in other words, in (near) resonance with the control pulse. The pulse photons thus stimulate emission of photons of the same frequency, causing de-excitation from the upper to the ground state, hence influencing the flux ratios as described above.

Components d_{12} and d_{22} give smaller contributions to the peak, the former increasing the proportion of the wavepacket dissociating through the lower channel. This suggests an interaction between levels $\tilde{X}$ and $\tilde{A}$, giving some crossing to the ground state caused by the control pulse. The latter increases the amount of wavepacket exiting *via* the upper channel. The reason for this was given above when discussing Fig. 5(b) in terms of the shifting of the position of the crossing, when the laser pulse is applied, to a value of the bond length less than the position of the wavepacket on the surface.

At the magnitudes of the dipole components given above, only d_{23} of the remaining members has a significant contribution to the features in Fig. 3(a). This is the beginning of the negative peak seen at $t \approx 40$ fs in Fig. 6(a). It seems that the small value of d_{23} present in the standard calculations means that this feature can be easily overlooked.

4.3.3.2 Polarisability terms. Moving on to the effects of the polarisability components, we immediately note that all components, bar α_{12} and α_{13}, make significant contributions to the control effect. α_{11} and α_{22} are the originators of the negative peak at $t \approx 0$ fs. The former increases the probability of dissociation through the upper channel, while the latter decreases that probability. Taken together, these competing effects give the negative peak seen at this time. This would seem to be an effect due to alteration the energy levels during the excitation process. In contrast, the presence of a non-zero α_{22}, means that if the control pulse is applied as the wavepacket passes the crossing region, then an opposite effect is seen to when the pulse coincides with the excitation. Here, this component increases the proportion of the wavepacket dissociating through the upper channel, giving a major component of the peak at $t \approx 90$ fs. This is a complementary effect to that seen by the presence of non-zero d_{22} in Fig. 7(a), the explanation being the relative positions of the wavepacket and crossing when the control pulse is applied.

For the last diagonal component, α_{33}, small peaks are seen at $t \approx 50$ fs and $t \approx 160$ fs, as is a trough at $t \approx 100$ fs. Thus, the crossing probability is increased as the bulk of the wavepacket passes the intersection of diabats $\tilde{A}$ and $\tilde{B}$, but reduced soon before and after. This is a subtle effect of altering the energy of diabat $\tilde{B}$, thus changing the position of the crossing and hence the probability of transition of the wavepacket from one surface to the other.

Of the off-diagonal elements, only α_{23} makes a significant contribution to the standard plot. The effect is similar to that of d_{23}, in that the control pulse causes an increase in the crossing from the upper to the lower dissociation channel if it occurs as the wavepacket moves through the crossing region. As noted earlier, the effect of the polarisability component is most significant about 20 fs after the dipole makes its contribution. The reason for this has been explained above.

4.3.3.3 Summary. It is apparent that the negative peak at the $t \approx 0$ fs is due to effects on the excitation caused by polarisability induced changes in the states involved in the excitation. The maximum at $t \approx 90$ fs is affected by d_{13}, d_{22} and α_{22}, all of which reduce the probability of crossing to the lower level, and by d_{12} and α_{33} which have the opposite effect. The negative peaks around $t = 40$–60 fs are caused by the 2–3 coupling terms in the dipole and polarisability matrices, interaction between the crossing states being maximised by the presence of the control pulse, causing an increase in the chance of the wavepacket crossing to the lower level. The small features at later times are most likely due to a shift in the energy of level $\tilde{B}$ preventing the tail of the wavepacket from crossing down.

5 Conclusion

Using our 3-state model we support the idea that the NRDSE is able to control the photodissociation of IBr by changing the topography of the PES in the region of non-adiabatic coupling. In addition to this, by varying the parameters of the model and comparing to the simulated wavepacket dynamics we have identified the different control mechanisms available in a 2-pulse NRDSE experiment.

The control is seen by a change in the branching ratio. The first control mechanism occurs when the control pulse overlaps the excitation pulse. This Stark shift of the surfaces affects the states excited, and subsequently the dynamics of the non-adiabatic dissociation. As all polarisabilities are negative the surfaces are shifted up by the field interaction. The $\tilde{A}$ state is shifted up relative to the ground-state. Thus the excitation is to states lower in the well. These have a lower probability of dissociating on the upper channel so cross to the lower with decreased branching ratio.

The second effect is the peak at 40 fs controlled by the transition dipole coupling the surfaces, d_{23}. This coincides with the wavepacket reaching the region where the gap between the surfaces matches the control pulse frequency and is due to 1-photon transfer between the states. In contrast, the peak at 60 fs is controlled by the magnitude of the polarisability coupling α_{23}. It coincides with the wavepacket reaching the centre of the avoided crossing and is due to changing the barrier height on the adiabatic surface. Both of these effects decrease the branching ratio.

The major effect in IBr, $e.g.$ seen the study of the individual parameters, is due to the Stark shift of the $\tilde{A}$ state relative to the $\tilde{B}$ state. As $\tilde{A}$ effectively moves up against $\tilde{B}$, the avoided crossing moves to a shorter distance when the control pulse is on. Thus at short times the crossing moves to meet the wavepacket, staying with it longer on its evolution, increasing the amount of crossing and decreasing the branching ratio. At longer times, the crossing is moved away, behind, the evolving wavepacket, so decreasing the amount of crossing and increasing the branching ratio. The change over point for the 2 regimes is around 80 fs, when the uncontrolled wavepacket reaches the avoided crossing.

Finally, small peaks are seen in the branching ratio change as the wavepacket comes out of the non-adiabatic region. These are probably due to 2-photon population transfer as the gap between the states is approximately twice the energy of the control photon.

These conclusions are in good agreement with, and add to the understanding of the mechanisms behind, the experimental results seen by the Stolow group.[18] The agreement between our model and the simpler Guo–Stolow model is also notable. The main point of disagreement between the theoretical models and the experiment is the timescale of the features described above. In the experiment, the peaks, as seen in the branching ratio plots, correspond with application of the control pulse at much greater times after the excitation than is observed in the theoretical calculations. Assuming our explanatory mechanisms are valid, this implies that the dissociation processes are roughly twice as quick in the theoretical models than in the experiment. The reasons for this are not clear, but certainly would warrant further investigation.

Acknowledgements

CSS would like to thank the Ministry of Education and Science in Spain and its CONSOLIDER-INGENIO program which under the grant CSD2009-00038 is getting the funding to continue with her research.

Notes and references

1. S. A. Rice and M. Zhao, *Optical Control of Molecular Dynamics*, John Wiley, New York, 2000.
2. P. W. Brumer and M. Shapiro, *Principles of the Quantum Control of Molecular Processes*, John Wiley, New York, 2003.
3. G. A. Worth and C. Sanz Sanz, *Phys. Chem. Chem. Phys.*, 2010, **12**, 15570–15579.
4. S. Autler and C. Townes, *Phys. Rev.*, 1955, **100**, 703.
5. M. Quesada, A. Lau, D. Parker and D. Chandler, *Phys. Rev. A: At., Mol., Opt. Phys.*, 1987, **36**, 4107.
6. R. G. Fernandez, A. Ekers, J. Klavins, L. P. Yatsenko, N. N. Bezuglov, B. W. Shore and K. Bergmann, *Phys. Rev. A: At., Mol., Opt. Phys.*, 2005, **71**, 23401.
7. P. Yi, M. Song, Y. Liu, R. Field, L. Li and A. Lyyra, *Opt. Commun.*, 2004, **233**, 131.

8 E. Ahmed, A. Hansson, P. Qi, A. Lazoudis, S. Kotochigova, A. Lyyra, L. Li, J. Qi and S. Magnier, *J. Chem. Phys.*, 2006, **124**, 084308.
9 E. Ahmed, P. Qi, B. Beser, J. Bai, R. Field, J. Huennekens and A. Lyyra, *Phys. Rev. A: At., Mol., Opt. Phys.*, 2008, **77**, 053414.
10 U. Schlöder, T. Deuschle, C. Silber and C. Zimmerman, *Phys. Rev. A: At., Mol., Opt. Phys.*, 2003, **68**, 051403.
11 J. Qi, G. Lazarov, X. Wang, L. Li, L. Narducci, A. Lyyra and F. Spano, *Phys. Rev. Lett.*, 1999, **83**, 288.
12 O. Salihoglu, P. Qi, E. Ahmed, S. Kotochigova, S. Magnier and A. Lyyra, *J. Chem. Phys.*, 2008, **129**, 174301.
13 B. Girard, G. Sitz, R. Zare, N. Billy and J. Vigue, *J. Chem. Phys.*, 1992, **97**, 26.
14 B. Tolra, C. Drag and P. Pillet, *Phys. Rev. A: At., Mol., Opt. Phys.*, 2001, **64**, 061401.
15 S. Sweeney, E. Ahmed, P. Qi, T. Kirova, A. Lyyra and J. Huennekens, *J. Chem. Phys.*, 2008, **129**, 154303.
16 Y. Kim, G. Hall and T. Sears, *Phys. Chem. Chem. Phys.*, 2006, **8**, 2823.
17 B. Sussman, M. Ivanov and A. Stolow, *Phys. Rev. A: At., Mol., Opt. Phys.*, 2005, **71**, 051401.
18 B. Sussman, D. Townsend, M. Ivanov and A. Stolow, *Science*, 2006, **314**, 278.
19 J. ling and K. Wilson, *J. Chem. Phys.*, 1976, **65**, 881.
20 G. Busch, R. Mahoney and K. Wilson, *J. Chem. Phys.*, 1969, **51**, 837.
21 K.-W. Jung, J. Griffiths and M. El-Sayed, *J. Chem. Phys.*, 1995, **103**, 6999.
22 M. V. D. Villeneuve and A. Stolow, *J. Chem. Phys.*, 1996, **105**, 5647.
23 H. Schor and F. M. da Silva, *THEOCHEM*, 1997, **394**, 147.
24 A. Hussain and G. Roberts, *J. Chem. Phys.*, 1999, **110**, 2474.
25 M. Shapiro, M. Vrakking and A. Stolow, *J. Chem. Phys.*, 1999, **110**, 2465.
26 K. Vandana and M. Mishra, *J. Chem. Phys.*, 1999, **110**, 5140.
27 M. Shapiro, *J. Raman Spectrosc.*, 2000, **31**, 59.
28 K. Vandana and M. Mishra, *J. Chem. Phys.*, 2000, **113**, 2336.
29 E. Volkers, A. Wiskerke, R. Mooyman, M. Vrakking and S. Stolte, *PhysChemComm*, 2000, **10**.
30 E. Wrede, S. Laubach, S. Schulenberg, A. Brown, E. Wouters, A. Orr-Ewing and M. Ashfold, *J. Chem. Phys.*, 2001, **114**, 2629.
31 M. A. Robb, F. Bernardi and M. Olivucci, *Pure Appl. Chem.*, 1995, **67**, 783–789.
32 G. A. Worth and L. S. Cederbaum, *Annu. Rev. Phys. Chem.*, 2004, **55**, 127–158.
33 *Conical intersections: Electronic structure, dynamics and spectroscopy*, ed. W. Domcke, D. R. Yarkony and H. Köppel, World Scientific, Singapore, 2004.
34 www.molpro.net.
35 S. Patchkovskii, *Phys. Chem. Chem. Phys.*, 2006, **8**, 926.
36 H. Guo, *J. Chem. Phys.*, 1993, **99**, 1685.
37 M. H. Beck, A. Jägckle, G. A. Worth and H. D. Meyer, *Phys. Rep.*, 2000, **324**, 1.
38 G. A. Worth, M. H. Beck, A. Jackle and H. D. Meyer, *The MCTDH Package, Version 8.4*, University of Heidelberg, Heidelberg, Germany technical report, 2009.

General discussion

Professor Brumer opened the discussion of the paper by Professor Motzkus: These are very interesting results! In earlier papers[1] we described a theory that provides insight into the control of radiationless transitions, such as internal conversion. Specifically, we showed that enhanced control over the rate of internal conversion results from the presence of overlapping resonances in the S_2 manifold to which the system is initially excited. If these zeroth order states in S_2, which are resonances since they decay to S_1, overlap then one can obtain control over the rates of internal conversion by manipulating the relative phases of the states created in S_2 upon excitation. Is this a mechanism that you believe you have seen in your experiments?

1 See, for example: P. S. Christopher, M. Shapiro and P. Brumer, *J. Chem. Phys.*, 2006, **125**, 124310.

Professor Motzkus replied: In principle, if I understood your proposed mechanism correctly, the mechanism may play a role in our experiment. Our current explanation is based on the localization of a wavepacket and therefore can be explained better in the time domain. By manipulating the different vibrational states the wavepacket can be focused at a specific region, *e.g.* the coupling region, or can be delocalized over the entire coordinates of the active potential. Depending on the timescales of the decay channels (open-system) more or less control is possible in the linear regime. This explanation is similar to the earlier comments given by Professor Kosloff on chirped excitation. On the other hand, in order to exploit the mechanism of overlapping resonances the different eigenstates must overlap in frequency which is presumably better described in the frequency domain. Due to the short lifetime of the excited states (of the order of 100–200 fs) this interference should be more pronounced in low-frequency modes, which have been identified as the essential modes not only in our control experiment but also in several other control studies. In this regard the magnitude of overlapping resonances might be of the same importance as the different timescales of system–bath interactions. Therefore it would be very interesting to adapt your mechanism to our molecular system and experimental data in order to explore similarities and develop probably another quantitative description of control.

Professor Weinacht asked: Am I correct in understanding that you control the time-dependent population of different vibrational eigenstates by varying the phase between pulses and since the coupling to the bath can act as a time-dependent 'probe' of the dynamics, one can control the dynamics, even though in the weak field limit and in the absence of coupling to a bath the final vibrational populations are independent of the spectral phase of the shaped pulse?

Professor Motzkus answered: Yes, your understanding is correct. The calculations show that, if the system is a closed system without any coupling to a bath, the dynamics of the different vibrational states can be manipulated by the phase differences of the sub-pulses in the multipulse sequence. For example, for a specific phase parameter c of the sine-phase which is applied to the spatial light modulator the amplitudes of all even vibrational states increase faster compared to the odd states in the first half of the multipulse sequence (Fig. 6 in the paper). During the second half of the pulse this behavior is reversed so that, after the multipulse sequence is over, the population distribution is always the same in a closed system and independent of the phase structure. Therefore this behavior shows some similarities with coherent transients studied by Siberberg, Girard *et al.*[1–3] in alkali atoms. However if the system is an open system and *e.g.* a second decay channel exists like in light harvesting complexes (energy transfer process) or bacteriorhodopsin (isomerization) the additional pathway could act as a time-dependent probe leading to control of the dynamics.

1 N. Dudovich, D. Oron and Y. Silberberg, *Phys. Rev. Lett.*, 2002, **88**, 123004.
2 J. Degert, W. Wohlleben, B. Chatel, M. Motzkus and B. Girard, *Phys. Rev. Lett.*, 2002, **89**, 203003.
3 S. Zamith, J. Degert, S. Stock , B. de Beauvoir, V. Blanchet, M. A. Bouchene and B. Girard, *Phys. Rev. Lett.*, 2001, **87**, 033001.

Professor Miller enquired: You have used an electronic dephasing time of 600 fs in your modeling (T_{2e} = 600 fs). The usual decoherence time for electronic states for such systems is much shorter than this, by nearly an order of magnitude. It is precisely this problem we wrestled with in the case of coherent control of bacteriorhodopsin[1] that we addressed using coherent control 2D spectroscopy in the present *Faraday Discussions* paper. As you state in the paper, there should be little coherence transfer if the pulse interval of the shaped pulse is longer than the decoherence time. How does your modeling depend on the electronic dephasing times and other constants? Can you provide some additional insight into how to reconcile the period of the shaped pulses with the conventional views on the timescale of electronic decoherence for such large molecular systems?

As a follow up question in comparing theory with experiment, the degenerate four wave mixing (DFW) probe is a nice concept in terms of providing higher sensitivity to excited state absorption. However, it is only formally equivalent to transient absorption with respect to tracking excited state dynamics if you heterodyne detect the signal (which gives much higher signal to noise) and separate the real and imaginary components. In your experiment the signal intensity was detected directly (homodyne detection) and will thus contain dispersive contributions that I believe are not taken into account in the modeling. The excited state spectrum is shifting in time and should affect the absorptive and dispersive components of the signal response differently. The dispersive contribution generally dominates the homodyne detected signal. How does this affect your modeling of the signal response? Are you assuming the two terms show the same spectral dependence?

1 V. I. Prokhorenko, A. M. Nagy, S. A. Waschuk, L. S. Brown, R. R. Birge and R. J. D. Miller, *Science*, 2006, **313**(5791), 1257–1261.

Professor Motzkus responded: The time constant of 600 fs describes only the pure electronic dephasing time for the appropriate fundamental transition with the solvent when no coupling to other electronic states is involved. The effective electronic dephasing time (T_2) in our modeling (as observed in the experiment) is limited by the electronic population relaxation time (T_1) of the S_2 state of β-carotene and is, therefore, much shorter than that. Since the T_1 time for polyenes with 10–11 conjugated double bonds is generally between 40–200 fs, the T_2 time will be automatically much shorter than the pure T_2*. Therefore, the effect described in our paper does not strongly depend on the value of T_2* as originally expected. Coherence transfer is indeed of little importance when high frequency modes are considered, and that has been observed in numerous transient absorption experiments. However low frequency modes seem to conserve the coherence during the coupling and this interaction can be seen in the simulations. The effective time constant of 200 fs allows the interaction of at least three pulses (inter-pulse spacing of 56 fs) with the molecule, and the interference of the three wavepackets leads to a corresponding phase behavior which, with respect to the vibrational period, also resembles the work on quantum carpets. A discussion on the simulation and their parameters can be partly found in the work by J. Voll and R. de Vivie-Riedle (ref. 24 in the paper).

In addition, the role of electronic dephasing time and the controllability of such large systems has already been thoroughly investigated by us numerically and experimentally in another publication (ref. 18 in the paper). The degree of control, in that case, of the population transfer between ground and excited states, depended crucially on the electronic dephasing time of the transition. We have seen (Fig. 9 of ref. 18 in the paper) that the population enhancement in the excited state increases with the T_2 time, until it reaches a maximum for T_2 times larger than about the total multipulse duration. Therefore, in our opinion, there is no contradiction in any of these control experiments involving polyenes: The rather small controllability (10–30%) observed in these experiments is due to the short effective T_2 times compared to the sub-pulse separations used.

Regarding your follow up question, the short answer is: All contributions should be automatically taken in account in our simulations since the signal is calculated directly from the density matrix evolution. The transient evolution of the absorption spectra and the interplay between this evolution and the tailored excitation was also investigated numerically in a transient absorption (TA) experiment (again, see ref. 18 in the paper). Figure 8 of that paper shows that the imaginary part of the sum of the electronic coherences, which describes the absorption, can be clearly tailored with different multipulse separations. For an effective absorption process, we observed that the overlap between the real part of the excitation field and the dispersive part of the electronic coherences should be maximal. For pump–DFWM, the interaction of the tailored initial pump generates such transient evolution of the S_0–S_2 absorption in the same manner as the tailored pump pulse in TA. Since the DFWM signal is proportional to the evolution of the squared number of molecules, we simulated the signal by calculating the square of the population in the S_1 state. In other words, there is no modeling of the DFWM signal *per se*.

Professor Miller asked: There now appears to be a real trend observed. The optimal control pulses have a substructure that resonantly drives the low frequency modes around 200 cm^{-1}. These modes appear to be everywhere. Modes in this frequency range generally involve the entire molecule, typically as low frequency torsional motions. In the case of their role in directing chemical reactions, it is surprising as the reaction itself is much more localized. These modes seem to be right in the spectral region where the modes are strongly damped but not overdamped, and delineate the upper frequency cutoff to diffusive type motions. The damping, however, is sufficiently strong that these modes are not good approximations to pure modes but are highly mixed with high frequency modes and probably serve to connect vibrational energy relaxation pathways. Can you provide some physical insight into the special role these modes play in promoting the control process and by implication their role in the photochemistry?

Professor Motzkus replied: In the case of the three control experiments that are mentioned in our presentation (LH2, Dyads and BR) the chemical structure of the initially excited chromophore is similar. The skeletal structure is based on a linear change of alternating C–C single and double bonds. In these molecules it is known[1] that those low frequency modes are, for example, angle-alternating bending modes which break the molecular symmetry and present promoting modes of conical intersections. The pulse sequence selectively addresses these modes and focuses or defocuses the wavepacket so that the path through the region of the conical intersection is modified. In other words the transient tailoring of the vibrational population affects the electronic lifetime of the S_2 state, *i.e.* the population will remain longer in the higher-lying electronic state compared to Fourier-limited excitation in this situation. If there are additional reaction channels which compete with the internal relaxation, like *e.g.* the Förster-type energy transfer in the LH2 case, such a transient evolution can then in principle affect the branching ratio between competing channels in the excited state. Therefore, we think that our results are able to rationalize the pulse shapes found in this whole class of coherent control experiments involving polyenic biochromophores.

1 W. Fuss, Y. Haas and S. Zilberg, *Chem. Phys.*, 2000, **259**, 273.

Professor Matsika commented: Conical intersections are described by two coordinates, the nonadiabatic coupling and the tuning mode. The nonadiabatic coupling has a specific symmetry determined by the symmetry of the two states coupled, as Professor Stolow mentioned, but it is further uniquely defined by the overlap between the wavefunction of one state and the derivative of the wavefunction of the second state with respect to nuclear coordinates. So even though there may be more than one mode with the correct symmetry it does not mean that all of them are important modes for the conical intersection. The contribution of each normal mode to the nonadiabatic coupling is defined by the above overlap integral. It is likely then that low frequency torsional modes are contributing to the nonadiabatic coupling vector, and become very important in the nonadiabatic dynamics of polyenic chromophores.

Professor Miller responded: The role of the low frequency modes seems to be general and I agree with your discussion of the coupling between coordinates. The question, to my mind, is still that these low frequency modes are spatially delocalized over the entire molecule; yet the induced changes or motion through conical intersections (as defined) are dominated by highly localized motions. To reconstruct the observed reaction dynamics through a conical intersection necessarily involves a large superposition of higher frequency modes as well. The fact that these low frequency modes are the most strongly damped may be the key to their predominant role in propagating structural changes.

Professor de Vivie-Riedle remarked: The geometric distortion involved in the process, *i.e.* driving the system towards a conical intersection, defines the time scale. Depending on the system this can be a high frequency mode or a low frequency mode. Both is possible.

Professor Miller replied: I agree completely with this statement. The curious point again is that most processes appear to involve low frequency modes. The anomaly is that such modes involve the entire molecule yet the photoproduct must involve highly localized motions. A highly localized motion would necessarily involve high frequencies. I think at reactive crossings the modes are so highly mixed that a clear distinction of low and high frequency modes becomes tenuous in terms of connecting to the observed times scales. The consideration of the modal character of a reactive crossing may be of most value in assigning coupling coefficients to specific microscopic motions, as you state, however, we should keep in mind that the modes are highly mixed, especially for the very anharmonic nature of far from equilibrium fluctuations involved at reactive crossings.

Professor Ashfold said: The predissociation of phenols following excitation to their $S_1(^1\pi\pi^*)$ state provides another clear example wherein low frequency, non-totally symmetric modes drive the coupling at a conical intersection (CI). A recent re-appraisal of the measured vibrational energy disposal in the phenoxyl products formed following excitation of phenol to its $S_1(v=0)$ level identifies the out-of-plane ring puckering vibration v_{16a} (of a_2 symmetry) as the dominant mode promoting tunnelling through the barrier under the CI between the bound S_1 and dissociative (S_2, $^1\pi\sigma^*$) potential energy surfaces *en route* to O–H bond fission.[1,2]

1 M.G.D. Nix, A.L. Devine, B. Cronin, R.N. Dixon and M.N.R. Ashfold, *J. Chem. Phys.*, 2006, **125**, 133318.
2 R.N. Dixon, T.A.A. Oliver and M.N.R. Ashfold, *J. Chem. Phys.*, 2011, **134**, 194303.

Professor Weinacht queried: I am wondering about how the agreement between experiment and simulation depends on the coordinate dependent potentials that you used. Were the potentials fit to a separate data set (different from the control measurements) meaning that agreement between experiment and simulations provides an independent check of the control model?

Professor Motzkus answered: The potentials used in the simulations are based on normal mode analysis of the ground state potential of β-carotene.[1,2] The displacement and couplings of the excited states have been chosen such that the simulations are matching the spectroscopic findings of ordinary pump–probe dynamics, like absorption and decay rates *etc.* determined in independent experimental studies (see *e.g.* ref. 3 in the paper). Therefore, the excellent agreement of the phase-dependent experimental observations and simulations is an independent check of the control model and in particular allows us to identify the specific mode which is responsible for control. This is also the reason why we prefer to call this way of performing experiments "quantum control spectroscopy" rather than "coherent control of molecules".

1 S. Saito and M. Tasumi, *J. Raman Spectrosc.*, 1983, **14**, 310.
2 A. Requena, J. P. Ceron-Carrasco, A. Bastida, J. Zuniga and B. Miguel, *J. Phys. Chem. A*, 2008, **112**, 4815.

Dr Stolow opened the discussion of the paper by Dr Offerhaus: Can you describe the relationship between your method and the "all optical processing" approach previously described by Oron, Dudovich and Silberberg?[1]

1 D. Oron, N. Dudovich and Y. Silberberg, *Phys. Rev. A*, 2004, **70**, 023415.

Dr Offerhaus responded: In that paper a flat (unshaped) pump and Stokes are used, followed by a probe that consists of multiple frequencies. The different frequencies are chosen such that different resonances (or polarizations from those resonances) interfere constructively at some wavelength in the output spectrum. I would argue that this is partially the same (parts of the probe that are phased correctly to ensure constructive interference) but also different in the sense that the overall (spectrally integrated) response was not optimized, only the signal in a selected spectral region. The matched filter approach aims to optimize the integrated signal.

Professor Rabitz commented: The general principles involved for the control of one species in the presence of similar other ones has been established with optimal dynamic discrimination (ODD). The fact that similar molecules can be controlled implies that the mechanisms will be different for the similar species. This circumstance is important to consider in the general analysis of control mechanisms in any context.

Professor Rabitz then asked: To what degree can your empirical simulations reliably identify pulses for high quality discrimination? In particular, can they avoid experimental optimization when multiple similar species are present?

Dr Offerhaus replied: Let me start by saying that since we do not have a lot of experimental data yet (we have some but only on very simple phase structures), I cannot claim any results in this direction. However, in the simulations we have considered a number of factors that could degrade the selectivity such as phase noise and chirp from propagation. The general approach of using the inverse phase shape of the molecular response is quite a robust approach in that respect. The biggest problem that we have encountered so far is that the vibrational response of some materials vary as a function of their environment (pH, water content, *etc.*) which means that a pulse that is optimal for a species in one condition is not the same in other conditions. If the differences between similar species are small then these changes can cause significant confusion.

Professor Weinacht remarked: I note that in your calculations you demonstrate large discrimination ratios, and I just wanted to comment that I don't think very high control ratios are actually required for large discrimination as long as you can get sufficient signal to noise in your measurements, since one can always subtract images with different pulse shapes to find the independent contributions from different target molecules. I would argue that if one were to plot a histogram of the control ratio for two different pulse shapes aimed at discriminating between two different molecules (where the histograms are constructed for say 1000 measurements for each pulse shape and show the number of measurements on the y-axis and the ratio of signal from one molecule to the other on the x-axis), then it is the separation between the histograms divided by their width which determines the contrast you can achieve in a selective microscopy experiment. So, even for very poor discrimination (say only 20%), you can still get reasonable selectivity if you have a noise value (histogram width) less than 1%.

Dr Offerhaus responded: I completely agree that high selectivity based on multiple pulse shapes can be reached even if the discrimination for a single shape is low. You might want to trade the selectivity of a particular pulse shape against robustness. For example, if the sample causes dispersion which degrades the selectivity then it might be advantageous to image with multiple robust pulses of lower discrimination.

The imaging speed, however, suffers if multiple shapes have to be used to achieve a selective image.

Professor Rabitz commented: Multiple control solutions often exist, and these may be utilized to reduce the chance of false positives in the detection of a particular species in a complex mixture.

Dr Offerhaus responded: The existence of multiple solutions can in fact be used to optimize multiple goals. In particular we have tried to optimize specific species in the presence of others, where the (integrated) signal from the others is rejected. In biological samples this includes the rejection of signal from the ubiquitous water or strong lipid signal near protein bands. We envisage a closed-loop system where optimization of selectivity is performed in a way that incorporates (*a priori*) spectral information with spatial information from the sample.

Professor de Vivie-Riedle asked: Can you elaborate on the Kautz bases in more detail?

Dr Offerhaus answered: The Kautz function is the (current) result in our search for modeling the phase function using a basis set with as few terms as possible, yet covering a diverse set of solutions. Consider the two conventional models:
 1) Fourier series (discrete resonances without damping), and
 2) Fourier transform (continuum of resonances with/without damping)
Three alternative classes of basis sets are:
 1') Sum of discrete damped resonances (=Lorentzian functions),
 2') Sum of multiple orders of a given damped resonance, and
 3') A continuous transform with a Lorentzian kernel.[1]
The Kautz functions used in our study represent the class 2' (multiple orders of a given resonance). They are Laplace transforms of an orthonormal basis set in the time domain.[2] In the light of the partial fraction decomposition theorem, Kautz

functions form a subset of the more general class of rational functions (ratios of two polynomials).

Note 1: We have also briefly addressed the class 1' (See Sec. 6.3 in ref. 3.).[3]

Note 2: In some cases, a Lorentzian function should be considered with a scale factor inversely proportional to the linewidth. As such, the limit of a Lorentzian profile (as the damping vanishes) can be a delta function and not a constant.

Note 3: We are only interested in Kautz functions with complex poles.[4] In filter theory, "Kautz functions" assume real poles and form a generalization of the Laguerre basis. These two different classes of Kautz functions correspond to underdamped and overdamped systems, respectively.

Note 4: Rational functions can also assume orthogonality of some sort (like in the complex plane and in conjunction with z-transform). However, such orthogonality is not (at least directly) applicable to a nonlinear coherent control problem, and may not be meaningful in the first place.

As mentioned in our manuscript, the key feature correlating "the ideal basis of a linear operator" and "the ideal landscape of an optimization problem" is a hypersphere energy landscape (or a manifold/hyper-surface as close to it as possible). So, instead of trying to look for the "modes" or "orthogonality" for a nonlinear operator, we seek to find a basis set that result in a hypersphere energy landscape (with minimal distortion).

1 W. Leidemann, V. D. Efros, G. Orlandini and E. L. Tomusiak, *Few-Body Systems*, 2011, **49**, 71.
2 M. Telescu, N. Tanguy, P. Bréhonnet and P. Vilbé, *Signal Process.*, 2007, **87**, 3234.
3 A. C. W. van Rhijn, H. L. Offerhaus, P. van der Walle, J. L. Herek and A. Jafarpour, *Optics Express*, 2010, **18**, 2695.
4 M. Telescu, N. Tanguy, P. Bréhonnet and P. Vilbé, *Signal Process.*, 2007, **87**, 3234.

Professor de Vivie-Riedle said: The question of similar or non-similar electronic structure of the crossing electronic states at conical intersections depends on the basis used for discussion. In the adiabatic basis the electronic structures of the crossing electronic states are strongly mixed in the vicinity of a conical intersection. In the diabatic basis they keep to their dominant electronic configuration.

Dr Stolow enquired: The interference between the non-resonant background and the resonant signal has consequences for the Gouy phase in the CARS image formation process.[1] Will the Gouy phase shift play a role in your approach to chemically-selective microscopy?

1 K. I. Popov, A.F. Pegoraro, A. Stolow and L. Ramunno, *Optics Express*, 2011, **19**, 5902.

Dr Offerhaus replied: I am not sure if the Gouy phase is really the source of the phase artifacts that you can see at sharp interfaces between resonant and non-resonant materials, but I agree that sharp boundaries in combination with a finite imaging numerical aperture cause imaging artifacts because light with fast (spatial) phase variations propagates at large angles and might be lost during collection. This is really a spatial (phase) problem. In our spectral phase shaping scheme these artifacts will be present in much the same way as in conventional CARS. The problem will not be worse but also will not be less. No amount of spectral phase shaping will solve these spatial problems.

Professor Whitaker opened the discussion of the paper by Professor Ulbricht:†
You showed a figure in your summary presentation depicting the interference pattern after the second grating in a Talbot–Lau interferometer. I am curious to

---

† Professor Ulbricht's paper was presented by Mr Venn, University of Southampton, UK.

know if this was obtained experimentally or by calculation, and how I am supposed to interpret the structures seen as a function of distance from the grating? I expected to see, perhaps naively, the evolution of a fringe pattern as predicted by the Bohm–de Broglie interpretation of quantum mechanics, similar to the observation of average photon trajectories in a two slit interferometer recently reported by Kocsis et al.,[1] but your picture shows a rich structure that I do not immediately understand.

1 S. Kocsis, B. Braverman, S. Ravets, M.J. Stevens, R. P. Mirin, L. K. Shalm and A. M. Steinberg, *Science*, 2011, **332**, 1170.

Mr Venn responded: The complex diffraction pattern has been obtained by simulation of this Talbot–Lau interferometer. It is the so-called quantum carpet known also for near-field light optics. For light diffraction these carpets have been measured for instance in ref. 1. In the case of molecule matter waves such a pattern has not been measured. So far only the high contrast self-image at Talbot distances has been seen. We plan to measure this quantum carpet more systematically in our interferometer at Southampton to then reconstruct the Wigner function of the motional quantum state of the molecule. The complexity of the pattern comes from a mixing of all existing diffraction orders of many coherently illuminated slits in the near-field regime, where the size of the diffraction pattern (interference fringe) is comparable to the grating period. The double slit pattern mentioned is a far field pattern, much larger than the structural size of the diffracting element. To see such a far-field pattern for molecular matter waves the transverse beam coherence must be two to three orders of magnitude higher/larger than in the Talbot–Lau case. More details on quantum carpets can be found for instance in Wolfgang Schleich's book.[2]

1 W. B. Case, M. Tomandl, S. Deachapunya, and M. Arndt, Realization of Optical Carpets in the Talbot and Lau Configurations, *Optics Express*, 2009, **17**, 20966-20974.
2 W. P. Schleich, *Quantum optics in phase space*, Wiley-VCH, 2001.

Professor Rabitz asked: Is the goal of this research to make a better detector or to manipulate the internal molecular dynamics? Can a practical device be realized?

Mr Venn answered: The aim of the research is to manipulate internal molecular dynamics and to link/couple these dynamics to the external motion towards quantum coherent effects and quantum entanglement of external and internal degrees of freedom. A possible practical device would be for the sorting of molecules with different properties at the same mass (isomers). This could be particles in different conformations and chiralities. Also neutral particles with different mass, which can hardly be sorted chemically, such as nanoparticles and nanotubes, could be separated by their difference in the polarisability (susceptibility) to mass ratio. As an example, we have demonstrated the separation of C_{60} and C_{70} fullerenes.[1] There has also been a related attempt to file a patent for a nano-particle sorting machine. The analytics industries need to be interested for that.

1 Ulbricht et al., *Nanotechnology*, 2008, **19**, 045502.

Professor Rabitz asked: Is a yield difference required for seeing a signal?

Mr Venn replied: No, in the ideal experiment we would shift the interference pattern using the deflection electrode and move the third grating to detect the shift for analyzing the polarisability (susceptibility). The total yield (molecules per area and time) would be the same with and without the deflection field. In real experiments other effects play a role. For instance imperfect velocity selection ($\Delta v/v = 15\%$) and the related dispersive effect in the electric deflection field smears out the interference contrast, which means it affects the count rate depending on the position of the third (the scanning) grating (Note: we don't loose molecules, so the total number is still constant, but the spatial distribution is changed). When using material gratings in the interferometer the dispersive van der Waals–Casimir-Polder interactions between the molecule and the grating have an additional visibility-reducing effect.

So, I guess the correct answer to your question is: the total yield is constant, but the spatially dependent yield (interference contrast, depending on the transverse position of the third grating, resolution 10nm) varies.

Professor Rabitz enquired: Is it correct that utilizing larger molecules is favorable due to their generally larger polorizability?

Mr Venn answered: The larger polarisability is only one part of the game. The phase shift for the molecule interacting with the optical grating in the interferometer depends only on polarisability and we gain here from using larger molecules. For the deflection the case is different and the polarisabiltiy to mass ratio is important (it is a gradient force deflection effect) and this ratio is (amazing or not) close to equal for molecules and nanoparticles. So we have to find good candidate molecules to demonstrate the conformation switching experiments (we think the diazobenzene derivative is one). Small molecules have the advantage that some manipulation techniques have been developed over the last decade (such as by Gerard Meijer, *etc.*). In the end it would be very interesting to study this with large molecules, as our general interest is to see macroscopic (quantum) effects and our experimental expertise is with those molecules.

Professor Kosloff commented: Your experiment of matter wave interferometry is based on a standing wave of light field perpendicular to the molecular motion. It is then possible to construct this field from a combination of a one and three-photon field. Then this can be used in the context of 1+3 cw coherent control initiated by Brumer and Shapiro[1] and demonstrated experimentally by Robert Gordon and Valeria Kleiman.[2] I speculate that you will obtain an alternating interference pattern of reaction products. Your detection by ionization could observe this phenomena.

1 P. Brumer and M. Shapiro, *Principles and Applications of the Quantum Control of Molecular Processes*, Wiley Interscience, 2003.
2 V. D. Kleiman, L. Zhu, J. Allen and R. J. Gordon, Coherent Control Over the Photodissociation of CH3I, *J. Chem. Phys.*, 1995, **103**, 10800.

Mr Venn responded: That is a very interesting idea. First it should be possible (while not very easy) to implement the 1+3 photon cw or pulsed laser beams to realize an optical grating. The wavelength would have to match perfectly (they are matched if one is the third harmonic of the other). Then the optical standing lightwave period has to match (be equal to) the periods of the other two Talbot–Lau interferometer (TLI) gratings. It should also be possible to measure the different reaction products directly with the interferometer, if those differ in mass.

One difficulty is that the reaction products are ions. In the TLI the particles have to pass the third grating for detection. This third grating is in our case a material SiN_x structure. Ions will be trapped or deflected here. But there is a recently proposed scheme for a TLI, where all three gratings are realized by light.[1] This experiment will be based on photo-ionisation and photo-neutralisation of molecules. For our Southampton experiment, I think a different coherent controlled molecular process (without ionisation, such as conformation change) would be interesting to be identified for a feasible experiment. Then the control field can be combined with the optical grating. I like the idea very much!

1 S. Nimmrichter, P. Haslinger, K. Hornberger and M. Arndt, *Concept of an ionizing time-domain matter-wave interferometer*, *New J. Phys.*, 2011, **13**, 075002.

Professor Miller asked: This source appears to be extremely coherent. How do you obtain such high transverse coherence in order to resolve fringes for such small de Broglie wavelengths? How sensitive is the alignment of your spatial filters. Also how sensitive is the observation to uniformity in feature size?

Mr Venn answered: The molecule source is a conventional thermal (even effusive) beam source. The transverse coherence is obtained from spatial filtering by pinholes and slits. For a conventional far-field experiment with C_{60} at about 150 m s^{-1} one would need to collimate the molecular beam to about 10 μrad (for gratings of periodicity of 100 nm). As the phase space density of the molecular beam is low, this experiment is basically limited by the detection efficiency.

One nice trick is the use of a Talbot–Lau interferometer (TLI) invented by John Clauser in the 1990s. Here the spatial (transverse) coherence requirement is lowered to 1 mrad before the first grating. The first grating of the TLI is a multi-slit spatial filter. In the three grating ensemble one can basically increase the intensity of the molecular beam by the number of slits. All illuminated first grating slits contribute to the interference pattern. The price we have to pay for the relaxation of the requirement on the spatial coherence is the alignment precision: For typical grating dimensions (250 nm to 1 μm periodicity), the distance between grating one and two and between two and three has to be equal and of the order of 1 μm, and all rotation angles of all gratings (pitch, yaw, roll) have to be aligned to each other on the order of 10 μrad to 100 μrad. This alignment and stabilisation of the three gratings for the duration of the interference makes molecule interference a technically challenging experiment.

The spatial resolution to detect the interference pattern comes from scanning the third TLI grating transverse to the molecular beam propagation direction, over the Talbot self-image of the second grating. This means that only a uniform pattern of integer fractions and multiples of the periodicity of the third grating can be detected with high contrast. It also depends on the opening fraction of the third grating. So if there is a miss-match between the period of the interference pattern and the period of the third grating by half a period (~100 nm) over the molecule beam diameter (~1 mm), that would already wash out the pattern in the integrated detection. The use of a single scanning slit instead of the third grating would again lower the detection count rate by about a factor of 1000. With a very efficient detector (more than 10%) that should be possible.

Professor Weinacht enquired: How can you distinguish between changing the orientation or the configuration of the molecule?

Mr Venn replied: We cannot directly image the orientation of the molecules in the proposed experiments, but we would see a difference in the transverse shift (Δs) if molecules (particles) have different polarisabilities (susceptibilities) along different molecular axes. So we would see an indirect alignment effect if the molecules are perfectly aligned during their passage through the deflection electrode and diffraction grating. Then changing the configuration would give a different change to the polarisability. The two effects will need to have been backed up by computation (*ab initio* such as in Gring *et al.*[1]). The computations need to predict polarisability and molecules (conformations of molecules) where the difference is large enough to be detected. The described diazobenzene derivative seems to be such a molecule.

1 Gring *et al.*, *Phys. Rev. A*, 2010, **81**, 031604(R).

Professor de Vivie-Riedle asked: Can your scheme be extended to use external motion for cooling internal degrees of freedom of a molecule? Coupling of external and internal motions are often used in cavity mediated cooling schemes.

Mr Venn responded: That would be one of the motivations to understand this coupling of external and internal degrees of freedom in our experiment, to use it for cooling—possibly cavity cooling—schemes. At this point we are investigating cavity cooling schemes to cool the centre of mass motion of such large molecules as C_{60}, but experimentally that has not yet been achieved. It would be interesting to see if then, *via* an external internal coupling as proposed for our experiment, internal degrees can also be cooled. But I think for cooling one would need to control the external temperature first. I am aware of your recent paper.[1] I would be very interested to think about this further.

1 G. Morigi, P. W. H. Pinkse, M. Kowalewski, and R. de Vivie-Riedle, *Cavity Cooling of Internal Molecular Motion*, *Phys. Rev. Lett.*, 2007, **99**, 073001.

Dr Leonard continued the discussion of the paper by Professor Motzkus: In Marcus Motzkus' oral presentation (but not in his paper #13), he illustrated very well (with simulations) how different vibrational states of the electronic excited state may be TRANSIENTLY populated in different ways, by using differently shaped pump pulses. However, the same simulations show that in all cases after the end of the light pulse (or pulse train), the population (amplitude) distribution on various vibrational levels is INDEPENDENT of the spectral phase of the pulse, and in particular is the same as for a transform-limited pulse. So, within the linear regime of excitation, after the light interaction is switched back off, the population of various vibrational levels in the excited state is independent of the spectral phase of the pulse. This very fact has been part of the discussion around Dr Miller's Science paper[1] (2006) on coherent control in bacteriorhodopsin in the linear regime of excitation: Is it or not possible to achieve coherent control (that is to have a relaxation pathway which depends on the spectral phase of the optical electric field) within the linear regime of excitation?

Apparently, the answer seems to be "yes in an open system" (*e.g.* retinal in bacteriorhodopsin), "no in a closed system" (*e.g.* the simulation I am referring to in Professor Motzkus' oral presentation). In Professor Motzkus' paper, significant coupling occurs to a lower lying electronic state, BEFORE the end of the overall pump pulse (or pulse train). This is thought to explain why even in the linear regime, the spectral phase of the pump beam may influence the outcome of a photoreaction. Hence the second question is: In order to achieve coherent control in the linear regime of excitation, should we really have an open system (*i.e.* interaction with a bath, in an incoherent manner) or is the coupling to a limited number states and/or degrees of freedom enough? In other words, is the (incoherent) interaction with a bath required to achieve quantum control in the linear regime of excitation? (Which seems paradoxical, as Dr Stolow was arguing earlier.)

1 V. I. Prokhorenko, A. M. Nagy, S. A. Waschuk, L. S. Brown, R. R. Birge and R. J. D. Miller, *Science*, 2006, **313**, 1257.

Professor Brumer replied: We published a paper last year[1] that proves that one photon phase control is possible in open quantum systems, and discusses the isolated system as well. This is an "existence proof", but the exact conditions under which such open system control is achieved is not yet clear.

1 M. Spanner, C. Arango and P. Brumer, *Conditions for One-Photon Coherent Phase Control in Isolated and Open Systems*, *J. Chem. Phys.*, 2010, **133**, 151101.

Professor Motzkus replied: Please refer to the comments already given in the discussion above. The molecular system studied in the present case (β-carotene in solution) has to be considered in a pure sense as a closed system since all population initially excited to S_2 will finally end up in S_1. Therefore in our experiment we see a transient control effect, or in other words we manipulate the path of a wavepacket passing a conical intersection by pure phase shaping. (The corresponding experimental results are shown in Fig. 4 of the paper, where the arrival of population in S_1 depends on the phase parameter c, and the shape of the wavepacket is represented by the strength of the oscillations.) This situation may be completely changed if new relaxation channels are added to the system. In the case of LH_2 an additional energy relaxation pathway, which is incoherent Förster-type energy transfer, opens up due to the presence of bacteriochlorophylls.

Professor Miller replied: To have coherent control within the weak field, linear, regime requires an open quantum system in which the stochastic fluctuations of the bath lead to real decoherence onto transient populations. Subsequent dissipation into the bath leads to the formation of barriers and trapping of the coherently transferred population to one of two or more possible states for observation. The actual measurement, if you will, is made by the bath in which the system–bath coupling leads to the measurement or observable. The key is that the decoherence must involve dissipation to the bath (T_1 process) and the time scales for this dissipative coupling must occur on similar timescales to the collisionally induced decoherence. As stated earlier, the biggest question in my view was the requirement for long lived coherences, on the order of 100–400 fs, in order for the observed optimal pulse shapes to exert any degree of control. Previously, for bacteriorhodopsin, three-pulse photon echo measurements indicated that the decoherence time is on the order of a few tens of femtoseconds (<50 fs) or an order of magnitude shorter than would be required to explain the observed pulse shaping effects. This timescale is faster than any known bath fluctuation, especially within a protein environment. We now know based on the coherent control 2D experiments that this previous measurement of coherence is really dephasing of the initially launched wavepacket as it quickly moves out of the Franck–Condon region by virtue of the strongly repulsive surface of the excited state in this system (see additional earlier discussion by Professor Weinacht and Professor Miller on this point).

In the case of a closed quantum system, if the closed quantum system of interest was large enough, one could see similar effects within linear response if the observation of the quantum state distribution were made on timescales shorter than quantum recurrences. However, if one waited long enough you would eventually see that there was no effect of the shaped pulses on the state distribution (Dr Stolow's earlier point). However, for open quantum systems, one is dealing not with eigenstates but with true dynamics. The system is coupled to a continuum of states and there are no recurrences, even in the t goes to infinity limit (as the size of the bath can similarly be scaled to infinite dimensions.) It is the very incoherent aspects of the bath and coupling to a continuum that make such observations possible only with open quantum systems. This is an important fundamental distinction for coherent control of open and closed quantum systems.

Professor Shapiro said: There has been much discussion concerning the possibility of weak field control when interactions with a bath in an open system are involved. The issue I would like to raise is how can we tell that we are in the truly weak field regime? The usual linear dependence of the product yield on the laser intensity may not be enough because there are situations, such as when a manifold of bound states is coupled to a "flat" continuum (or quasi-continuum) of states, that rate processes dominate the dynamics. Under these circumstances the decay of the initial state $\psi_i(t)$ is given as:

$$\psi_i(t) = \psi_i(0)\exp\left(-\mu \int_0^t dt'\, \epsilon^2(t')\right) \qquad (1)$$

where μ is an average transition dipole moment squared and $\varepsilon(t)$ is the electric field of the pulse as a function of time, e.g.,

$$\varepsilon(t) = \varepsilon_0 \exp(-\alpha t^2) \exp(-i\omega t). \tag{2}$$

The time dependence of the products' wavefunction $\psi_p(t)$ then assumes the form

$$\psi_p(t) = \mu_{p,i} \int_0^t dt' \, \epsilon(t') \psi_i(t'), \tag{3}$$

where $\mu_{p,i}$ is the dipole matrix element coupling the continuum (that leads to the formation of the product p) to the initial state i.

Substituting eqn 1 in eqn 3 and integrating $|\psi_p(t)|^2$ over the short time durations of $\varepsilon(t)$, we often find that the yield is proportional (until the depletion of the ground state becomes noticeable, at which point the yield begins to saturate and flatten out) to ε_0^2, *i.e.*, it is linearly dependent on the peak intensity. This may happen regardless of whether we are in the weak field regime or not. A more complete exposition of this topic is being prepared.[1]

1 A. Han and M. Shapiro, *to be published*.

Professor Miller responded: It is difficult to comment to this point without seeing more details of the treatment. However, you state that the process is linear up until you see ground state saturation when the observable of interest is coupled to a continuum and rate processes dominate. If you are not saturating or significantly depleting the ground state, you are not in the strong field limit. There may be nonlinear field (higher order) contributions but these have to be detected within the background of the truly linear one-photon responses that dominate the observable before the onset of saturation of the ground state transition. From an experimental standpoint, the linear, truly weak field, observables will dominate higher order nonlinear contributions, even if the nonlinear process scales linearly in intensity. The cross sections are orders of magnitude different. In order to isolate higher order field interactions that happen to show linear dependence on intensity, one would have to have some well defined difference in the observable between the one-photon process and the higher order process. These two can then be separated in this manner.

Professor Kosloff commented: The issue of weak field coherent control has attracted much attention.

Experimentally weak field is defined by the linearity of the observed phenomena.

Nevertheless this is not a sufficient criteria. Calculations show that a linear relation can be measured for a limited class of observables and other observables show a non linear relation.

If for example we check a chirp effect in the weak field, the linearity of population transfer is not a sufficient criteria. The whole range of chirps measured should be checked if they scale linearly with the field. A good criterion is to refer to the intensity level where only one photon could be present in the sample. We have studied weak field control theoretically for excited state observables (Fig. 1–4).[1] We found, according to theory, that control is induced by the influence of the bath. Our model was composed of a ground electronic surface and bright and dark excited electronic surfaces. We found a strong chirp effect. We proposed the mechanism in the time domain. Without the bath in the weak field we prepared a superposition of excited state eigenstates. This means that no phase control is possible. With the bath a new timescale emerges. We can imagine a control mechanism that the initial impulsively produced wavepaket is focused on the crossing point. As a result the nonadiabatic transfer from the bright to the dark state is enhanced. Then the bath takes over and stabilizes the product dark state by dissipating energy below the barrier. If the transfer takes longer the bath will stabilize the bright state. The evidence for focusing is the enhancement by negative chirp. If one wants to consider the phenomena from the energy point of view, in this model there is a very long timescale where tunneling will equillibrate the bright and the dark excited states.

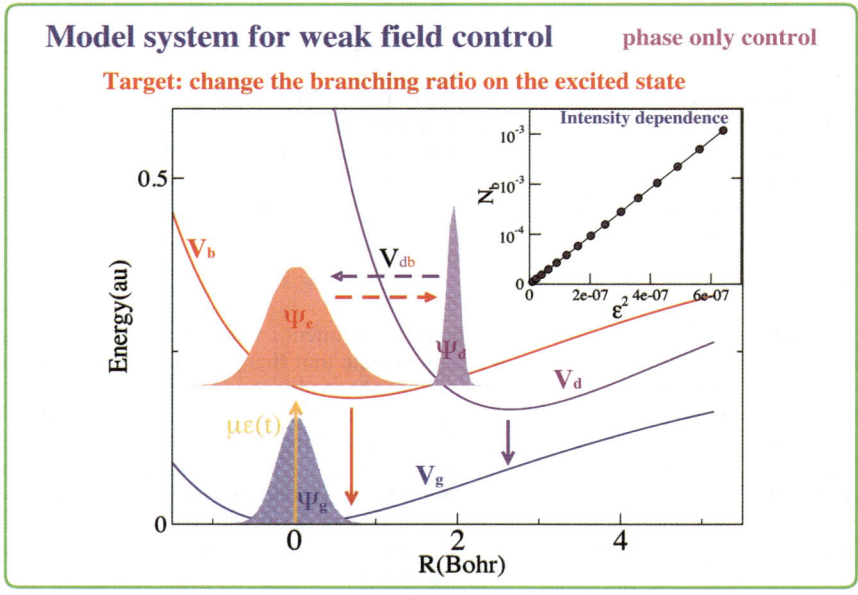

Fig. 1 Model System for weak field control. The wavefunctions are superimposed on the ground bright excited state and the dark excited state. The inset shows the linearity of the population transfer with the field intensity.

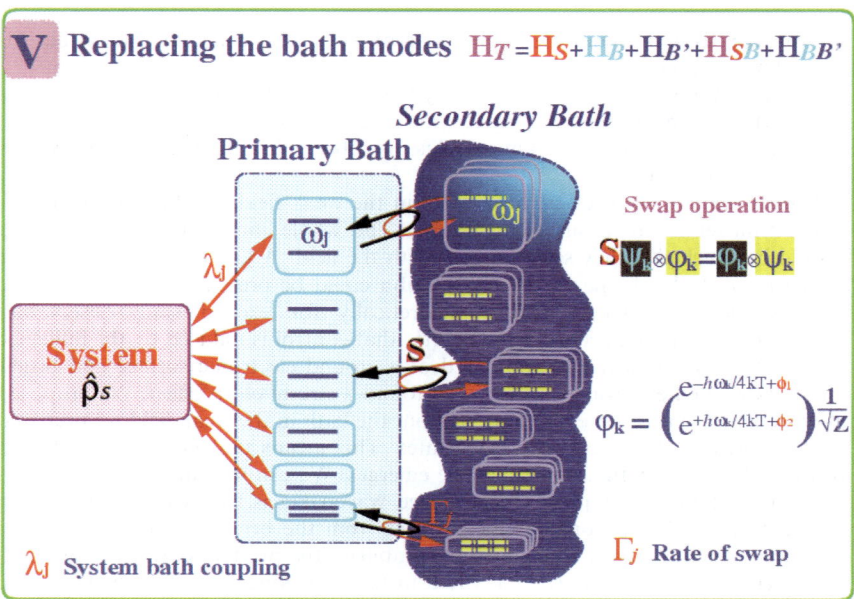

Fig. 2 The general scheme of the stochastic surrogate Hamiltonian method.

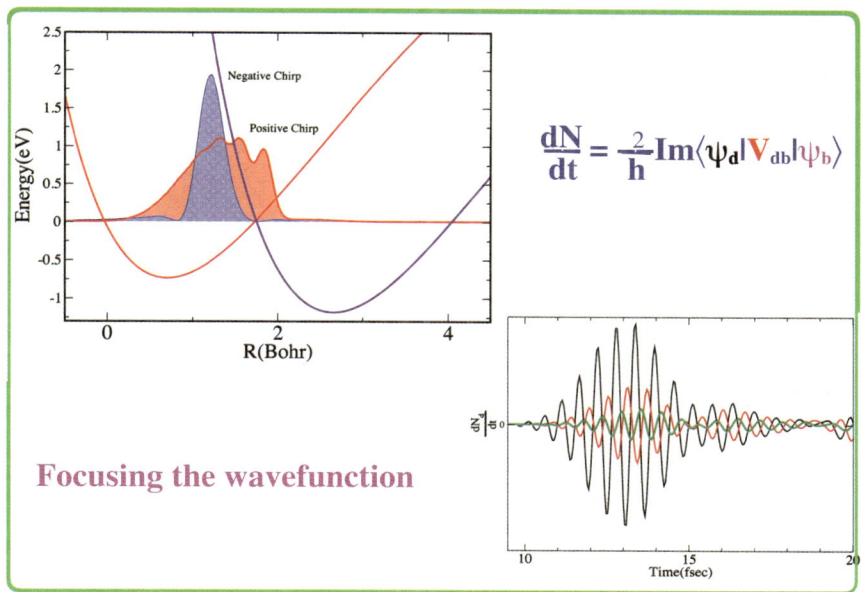

Fig. 3 The mechanism of phase only control: Focusing the wavefunction on the nonadiabatic transition point. The rate of population transfer as a function of time for a transform limited pulse and a positive and negative chirped pulse.

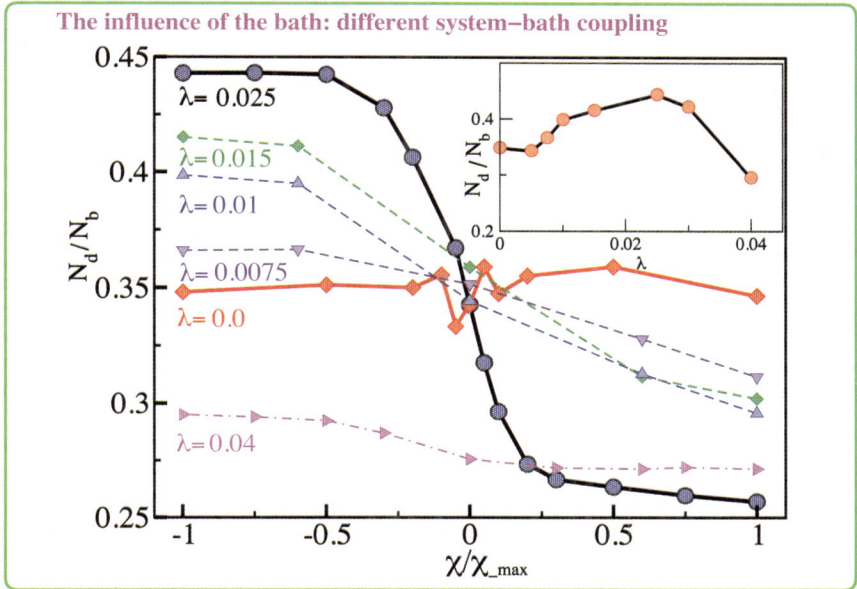

Fig. 4 The final branching ratio between the bright and dark excited state population as a function of chirp for different system bath coupling parameters. The inset shows the maximum effect as a function of system bath coupling exhibiting a turnover.

Our model of the system–bath interaction is based on the stochastic Surrogate Hamiltonian scheme.[1] This is a non-Markovian description which can deal explicitly with initial system bath correlation. We find that some weak field control scenarios do not work in the Markovian limit.

1 G. Katz, M. Ratner and R. Kosloff, *Control by decoherence: weak field control of an excited state objective*, New J. Phys., 2010, **12**, 015003.

Professor Miller noted: The confusion arises when one considers the observable to be an eigenstate of the system. For a closed quantum system, one can use an eigenstate basis. In open quantum systems, the number of degrees of freedom is so large that there is no reasonable timescale on which one could define a strictly stationary state, or eigenstate, even though in principle one could define an orthonormal basis *etc.* One has to take into consideration the dynamics of the situation in open quantum systems in which we are effectively dealing with a continuum of states. A dynamics or time domain picture is most appropriate in this case. The problem is generally broken down into a system and a bath, where there is a clear distinction in the density of states for the two energetic and spatially distinct regions. In principle, the system could be expanded to include the bath and try to recast the problem within an eigenstate basis, however, the bath as defined leads to a continuum of states with no clearly defined observable. The target molecule or system of interest has optically accessible states that can be selectively excited, manipulated, and followed in time, so this delineation of the system and bath is consistent with the actual experimental conditions. Within this basis, the coupling of the system to the bath leads to dissipation and collapse of the coherently prepared state within the short time limit of observation. The subsequent redistribution (dissipation) of energy into the various degrees of freedom leads to a barrier between the two possible pathways in coherently controlled process. Depending on the barrier, the system is now trapped in one of two possible non-stationary states that are defined globally as the observables. In principle, if one waited long enough all the states, system + bath, would return to the initial state and you would have no control. However, this would take an infinitely long time for any reasonable system of interest, especially in the condensed phase. Basically, for an open quantum system, the coupling to the bath truncates the observable as defined in the short time limit of observation—a real observation within real experimental time frames. An eigenstate basis of discussion is not a useful construct as the observations are not involving strictly stationary states as one discusses for closed or isolated quantum systems.

This issue can be perhaps better appreciated by considering a model problem in which a coherence can be manipulated to control population on one of two spatially distinct sites. For two states coupled and with the initial state preparation on one site, the coherence will oscillate back and forth with the inverse of the electronic coupling. Within this time domain picture, if more states are added, it will take longer and longer for the coherence to re-phase at the initial site or reconstruct the initial state preparation, *i.e.*, as the number of states coupled to the two spatially distinct sites is increased the recurrence time gets longer and longer (see for example ref. 1 for the case of electron transfer). In the limit that dissipation to uncoupled (bath) states occurs faster than the recurrence time of the coupled states, the system will collapse into one of the two possible sites. Even for closed quantum systems, if the observations are made within the short time limit, the system response is dependent on the phase of the optical excitation in the state preparation. In the limit of an open quantum system, the number of degrees of freedom is so large that the system never experiences recurrences. The system collapses onto the short time window of observation. I should add that a second pulse or field is not needed in principle to observe this product state as it is long-lived. Any means of distinguishing the two possible outcomes (such as chromatography) can be used and the process is strictly linear and depends only on the initial, time sensitive, state preparation. This is the point we specifically made in the first reports of coherent control in the weak field limit.[2] There are now more detailed theoretical analyses that bear this point out by the groups of Professors Brumer[3] and Kosloff,[4] as discussed above.

1 J. Lanzafame and R. J. D. Miller, in *Ultrafast Dynamics of Chemical Systems*, ed. J. Simon, Kluwer Academic Publishers, Netherlands, 1994, pp. 163–204; J. Lanzafame, D. Wang, S. Palese, A. A. Muenter and R. J. D. Miller, *J. Phys. Chem. B*, 1994, **98**, 11020–11033.
2 V. I. Prokhorenko, A. M. Nagy, S. A. Waschuk, L. S. Brown, R. R. Birge and R. J. D. Miller, *Science*, 2006, **313**(5791), 1257–1261; V. I. Prokhorenko, A. M. Nagy, L. S. Brown and R. J. D. Miller, *Chem. Phys.*, 2007, **341**(1–3), 296–309.
3 M. Spanner, C. Arango and P. Brumer, *J. Chem. Phys.*, 2010, **133**, 151101.
4 G. Katz, M. Ratner and R. Kosloff, *New J. Phys.*, 2010, **12**, 015003.

Professor Tannor commented: Sometimes a particular physical effect is much more transparent in one representation than in another. In this case, to understand the effect of weak field pulse shaping on final product distributions, I think it is much more fruitful to think in the time domain rather than the energy domain. In the time domain, the phase of the weak field pulse can, *e.g.* affect the direction of the momentum of the wavepacket, *i.e.* two different pulses with the same amplitudes but different phases can prepare wavepackets with the same energy distribution but with different momentum. It then seems clear that the relaxation could trap one of these wavepackets on one side of a double well while trapping the other wavepacket on the other side of the double well.

Professor Kosloff said: When a quantum primary system is embedded in a bath there is a drastic change in the prospects for control. A bath is contracting, eventually leading the system to thermal equilibrium. It would seem that control is lost. But on the intermediate timescale the bath opens new opportunities for control. For example, cooling the primary system is only possible if a bath exists. We studied weak field control and found that the bath enables phase-only control when the isolated system is non controllable. For example, if the target is population transfer from a ground to an excited electronic surface, we find a chirp effect which is due to the interplay of the pulse and the bath dynamics influencing the population of the transient.

Professor Sension commented: It is not clear what thermal equilibrium has to do with weak field control when dissipation of energy into the bath can (and does) stabilize the products into a non-equilibrium distribution. Even without control the distribution of products in condensed phase photochemistry is generally out of thermal equilibrium. It is all a matter of timescales. If phase control of an excitation pulse can produce an excited state wavepacket that modifies the conformational distribution on a time scale comparable to the inherent dephasing and dissipation in the system, then the dissipation of energy into the bath may then stabilize and trap this perturbed distribution.

Professor Rabitz opened the discussion of the paper by Professor Matsika: What conclusions may be drawn from the control experiments that were performed?

Professor Matsika answered: The control experiments that were performed were not successful in altering the ratio of fragments except for a trivial case where the pump and probe pulses overlapped. So our main conclusion is that we could not influence the dynamics of the wavepacket on the neutral potential energy surfaces over the different photophysical pathways present. We are not certain why. We suspect that our limited pump pulse bandwidth does not allow for adequate manipulation of the dynamics of the wavepacket. This is consistent with previous theoretical studies by Regina de Vivie-Rielde on cyclohexene, where it was found that very short pulses in the deep UV, with durations as short as 5–10 fs, were required to control the wavepacket dynamics around a conical intersection.[1]

1 R. de Vivie-Rielde, *Appl. Phys. B*, 2004, **79**, 987.

Professor de Vivie-Riedle enquired: The need for extremely short laser pulses to control the photoinduced electrocyclic ring opening of cyclohexadiene was partly due to the vanishing transition dipole the moment the system left the Franck–Condon region. Thus the time window for control is very small. How does the transition dipole moment change for the DNA bases beyond the Franck–Condon area towards the conical intersections, which are responsible for the photophysical deactivation?

Professor Matsika replied: The situation can be different for the different DNA bases. In uracil, which is the one we tried the control experiments, there is a conical intersection between the bright absorbing state and a dark state near the Franck–Condon region. In this case the transition dipole moment is expected to vanish very fast. The exact timescale of the encounter with the conical intersection is not established and thus how fast the transition dipole moment vanishes is not clear, but it is expected to be fast. It is very possible that extremely short laser pulses are needed to control the photophysical behavior in uracil similarly to the previous cyclohexadiene control studies.

Professor Rabitz said: Are the simulations consistent with the experiments?

Professor Matsika responded: Our theoretical calculations do not attempt to mimic the experiments directly, since it will be very difficult to do time-dependent calculations on the dynamics of this polyatomic molecule in the presence of an intense laser field. So, there are no direct observables that we calculate and are also measured experimentally that can be compared. Instead we perform time-independent electronic structure calculations for both the neutral and the ion which help interpret the experiments, and the calculations in that sense offer a qualitative description of the experiment.

Dr Stolow asked: In strong laser fields, there may be multiple pathways to a given ionic fragmentation channel. For example, dynamical evolution in a neutral state followed by ionization may produce a certain fragmentation pattern. Ionization followed by dynamical evolution may produce an overlapping fragmentation pattern but one which depends on phase, time delay, and intensity in quite different ways. Experimentally, your method records the coherent and incoherent sum (populations) of all paths to a given channel. How then can one unambiguously assign a given fragmentation pattern to a given dynamical pathway?

Professor Matsika answered: It is true that fragments observed after interaction of molecules with strong fields can be produced by a variety of mechanisms, and interpretation of the observed signals is far from trivial. For example ionization may lead to many ionic states, or the dynamical evolution we observe may be either from the neutral or the ion. For this reason we are very careful in our analysis and the fragments we choose to interpret, and we don't try to understand all the events that occur. On the contrary, we first try to identify fragments that are easier to analyze, and can make clean assignments. For example, in our cytosine studies we chose to use the parent ion as one of the signature fragments that was used to label one of the pathways for neutral dynamics. Any parent ion signal observed has to come from the ground ionic state D_0 since ionization to higher ionic states will provide enough energy for fragmentation. Of course, as the wavepacket moves along the neutral excited state and interacts with the probe pulse, ionization can lead to higher ionic states in addition to D_0, and we don't assume that it does not. However, the parent ion signal will only come from the fraction of the population that is on D_0. Based on the dissociation barriers in the cytosine cation and the energies of the states we can further eliminate all pathways but one in which the parent ion can survive. Based on this elimination we can use the parent ion as a signature of this pathway. In summary, even though there are many fragments observed in the time of flight, we only choose the ones that can provide simple interpretations. This may be restrictive in the amount of information we can extract, but we can still obtain useful information.

Professor Miller remarked: There was a great deal of discussion in the early days of studying the photophysics of DNA that there were extremely short lived excited states within double stranded DNA and that this fast nonradiative relaxation was key to providing inherent UV photoprotection. The idea back then was that there were low frequency modes or numerous other nonradiative relaxation channels that led to a form of built in photoprotection of DNA. The recent work of Berne Kohler and his group[1] showed that the nucleotides themselves had very short-lived excited states in contrast to expectations based on the energy gap law for nonradiative relaxation. This fast nonradiative relaxation for the nucleotide bases is now understood in terms of a conical intersection connecting the upper and lower electronic surfaces. We now find, paradoxically, that double stranded DNA actually has long-lived photoexcited states, most likely involving photoinduced charge separated electron–hole states. These intramolecular charge transfer states do not fluoresce, and they explain the lack of fluorescence observed in the earlier studies. It does bring into question how to connect the observed photophysics to any role of biological significance. This issue aside, can you comment on how to connect the photophysics of model nucleotides to DNA photophysics as the role of the DNA structure clear affects the photophyics and even photochemistry?

1 . B. Kohler, *J. Phys. Chem. Lett.*, 2010, **1** (13), 2047.

Professor Matsika replied: It is true that studying the photophysics of the individual nucleobases and nucleotides does not lead to an immediate understanding of the photophysics of the double stranded DNA. In single and double stranded DNA there are additional excited states and additional pathways that play an important role in the photophysics and photochemistry of DNA. As an example, there are the charge transfer states that you mentioned which have been found to contribute to longer lifetimes. Nevertheless, the pathways found in the individual bases are still present (perturbed by the environment) even in single and double strand DNA, and the overall dynamics will be determined by the competition between all the pathways present. So, even though the pathways found for the individual bases cannot be used to explain the dynamics in DNA as a whole, they are one piece of the puzzle that should be included in an overall picture. Understanding the more complicated picture when several bases are interacting is a much more challenging task, and several groups, including ours, are currently working on it. Based on Kohler's[1] and others'[2] work the π-stacking interactions even within a single strand are crucial for the presence of the charge transfer long lived states, so studying the excited states of π-stacked bases is an important next step we are pursuing.

1 C. E. Crespo-Hernandez, B. Cohen and B. Kohler, *Nature*, 2005, **436**, 1141.
2 I. Vaya, F.-A. Miannay, T. Gustavsson, and D. Markovitsi, ChemPhysChem, 2010, **11**, 987.

Professor Shapiro commented: The extra stability of DNA discussed may be due to the fact that the separation between nucleotides, hence charges, is much larger in DNA compared to gas phase dimers.

Professor Miller added: This point further enforces my comment. The further apart the nucleotides are, the greater the distance between the photoinduced charge separated states. The resulting dipole and medium repolarization are larger and the barrier to the back electron transfer process is increased. These reactive radical ion states will actually be longer-lived with this larger separation and should be more susceptible to reactions that lead to damage of the DNA. I don't believe we can extrapolate the photophysical properties of nucleotides to evolutionary features of DNA or infer any property of inherent UV photostability of DNA based on the short time dynamics of excited states even within DNA. We need additional information on the chemical reaction dynamics within the cell for both the deleterious photoreactions and repair mechanisms. These are the relevant details that are missing at the present time.

Professor Rabitz enquired: Did you attempt to perform photo-fragmentation and ionization at 800 nm?

Professor Matsika answered: The probe pulse is around 800 nm (780 nm), so after the pump pulse excites the molecules to the first neutral excited state the intense probe pulse ionizes and photofragments them. So what we do is photofragmentation and ionization at 800 nm. Another interesting experiment would be to use the intense IR probe to ionize the molecules and then use a weak one-photon IR probe to probe the ionic dynamics. We have done a similar experiment for other molecules (*i.e.* substituted acetones), but it may be harder to do it for the nucleobases.

Professor Rabitz noted: There are many organic-based molecules that are stable under charge separation. Such cases deserve consideration.

Professor Miller added: I agree. The overall issue of photostability of DNA requires knowledge of the kinetics of the deleterious reactions and the repair mechanisms. These kinetics are much slower than the photoexcited state dynamics and dominate the situation with respect to the balance between UV protection and loss of integrity of the gene sequence. It is generally only the photoinduced thymine dimer that is involved in photoinduced DNA damage and lives long enough to cause problems. The other charge separated states undergo geminate recombination (ground state recovery) much faster and aren't a problem as you suggest. Again, my point was that we should not try to extrapolate excited state dynamics to system properties of the cell that occur on much longer length and time scales.

Dr Marquetand queried: In your simulations, you calculated several excited states whose ultrafast dynamics you assume to play a role in the ionization pathways. Did you consider triplet states in your calculations? The background of my question is the following: We have preliminary results employing our newly developed SHARC (Surface Hopping in the Adiabatic Representation including arbitrary Couplings) method showing very fast singlet–triplet transitions in cytosine. The timescale might be on the order of hundreds of femtoseconds.

Professor Fielding also commented: Triplet states may well play an important role in the excited state dynamics and it would be interesting to include them in the calculations. In recent work[1,2] we have reported femtosecond time-resolved photoelectron spectroscopy experiments and calculations investigating the ultrafast dynamics of benzene at energies above the onset of channel 3. These experiments and calculations demonstrated that the ultrafast decay that was previously attributed to internal conversion through a conical intersection is actually due to competing internal conversion and intersystem crossing, and that both processes occur on similar femtosecond timescales. The key is the near degeneracy of the S^1 and T^2 states over a wide range of geometries traversed by the wavepacket as it moves away from the Franck–Condon region. This allows for efficient intersystem crossing, despite a weak spin–orbit coupling, at the expense of internal conversion at the more distant S^1/S^0 conical intersection. We suspect that such ultrafast intersystem crossing may well be widespread.

1 H. H. Fielding *et al.*, *Chem. Phys. Lett.*, 2009, **469** 43–47.
2 H. H. Fielding *et al.*, *Phys. Chem. Chem. Phys.*, 2010, **12**, 15607–15615.

Professor Matsika responded to Dr Marquetand: We have not included triplet states in our calculations. It is true that these states are present and they may play an important role. However, we assumed that intersystem crossing through a singlet–triplet transition will in general be slower than a singlet–singlet transition through a conical intersection, and that is why we focused on the singlet states. The problem is quite complicated even when ignoring the triplet states, so we decided to understand the singlet manifold before making the problem even more complicated. Some recent work,[1,2] including yours, seems to indicate that triplet states are participating in the radiationless decay, so a complete picture should include these states. Professor Fielding's comment also points to the possibility that an intersystem crossing may be even more efficient than internal conversion in some cases.

1 J. Jose Serrano-Perez, R. Gonzalez-Luque, M. Merchan, L. Serrano-Andres, *J. Phys. Chem. B*, 2007, **111**, 11880.
2 M. Merchan, L. Serrano-Andres, M. A. Robb, L. Blancafort, *J. Am. Chem. Soc.*, 2005, **127**, 1820.

Professor Matsika responded to Professor Fielding: This is a quite interesting observation, and the results on benzene should challenge our intuitive assumptions that intersystem crossing is not as important when conical intersections are present. As my reply to Dr Marquetand's earlier question indicates, we did not include triplet states in our work since we did not think they would be as important, and they would make the problem even more complicated. There is some work by others where triplet states are included in the studies of the nucleobases, but more extended work is needed to establish the importance of triplet states in the photophysics of the nucleobases.

Professor de Vivie-Riedle enquired: Fig. 1(a) of your paper shows the energetic positions of critical geometries relevant to two S_1 relaxation pathways in cytosin. The parent ion, which can be formed from the butterfly minimum is selected as signature of one of the pathways along the S_1. However, the scheme in Fig. 1(a) shows that both pathways lead through the butterfly minimum. How in this case can the signal of the parent ion be useful to discriminate between both pathways?

Professor Matsika answered: It is true that according to the minimum energy paths shown in Fig. 1 the butterfly minimum appears to be present in both pathways. However, based on our joined experimental–theoretical analysis described in detail elsewhere,[1] this picture is revised so that the butterfly minimum is only involved in one pathway, and that is why we can use the ions produced from it as a signature for that pathway. In summary we believe the molecule does not follow the minimum energy path shown in Fig. 1.

1 S. Matsika *et al.*, *J. Chem. Phys.*, 2011, **134**, 184309.

Professor de Vivie-Riedle asked: Input from quantum chemistry can be used to decide which of the two S_1 relaxation pathways might be preferred after photoexcitation. Did you check on the excited state gradients towards the different S_1 relaxation channels?

Professor Matsika replied: The gradients initially at the Franck–Condon region lead to the butterfly minimum. However, as was pointed out earlier, the minimum energy path may not be enough to explain the dynamics here. Other dynamical studies[1] have also found that multiple minima are accessed simultaneously.

1 H. R. Hudock and T. J. Martínez, *ChemPhysChem.*, 2008, **9**, 2486—2490.

Professor Tannor opened the discussion of the paper by Dr Marquetand: We have developed a method, still unpublished, for solving nonadiabatic dynamics using complex classical trajectories (the trajectories evolve in complex position and momentum). The trajectories carry a complex phase and thus in principle bear the imprint of the phase of an external laser field and will undergo interference. The method can be derived rigorously from the time-dependent Schrödinger equation, but we are still struggling with numerical difficulties, so at this point I can't advertise it as a practical method for calculating non-adiabatic dynamics in large molecules.

Professor Kleiman responded: We are currently working on developing methods that will allow us to calculate the effect of phase modulation in these very large systems in solution.

Through a collaboration with Fernandez-Alberti, Roitberg and Tretiak[1,2] we started to look at on-the-fly calculations of non-adiabatic transitions leading to energy transfer processes in ssimilar molecular systems. We are still on the process of including the phase sensitive component and we hope to be able to simulate these systems very soon. We need to keep in mind that these systems are quite large, and their dynamics behavior coupled to the fact that we do ensemble measurements greatly affects the electronic coupling and the measured parameters.

1 S. Fernandez-Alberti, V. D. Kleiman, S. Tretiak and A. E. Roitberg, *J. Phys. Chem. A*, 2009, **113**, 7535.
2 S. Fernandez-Alberti, V. D. Kleiman, S. Tretiak, and A. E. Roitberg, *J. Phys. Chem. Lett.*, 2010, **1**, 2699.

Dr Leibscher opened the discussion of the paper by Dr Sanz-Sanz: Did you consider the possible influence of rotational dynamics on the non-resonant dynamic Stark effect (NRDSE) in IBr? I assume that the non-resonant control pulse (with an intensity of approx. 10^{13} W cm^{-2}) also excites rotational states of IBr and thus induces molecular alignment. If the control pulse is applied before the excitation pulse, the molecules are likely to be partially aligned during excitation. If the control pulse is applied after the interaction, randomly oriented molecules are excited. Since the absorption of light depends on the orientation of the molecules, the excitation process will be different in both cases. Moreover, rotationally excited molecules have—due to the centrifugal force—effective potential energies depending on the rotational quantum number. This might also influence the nuclear dynamics and control *via* the NRDSE.

Dr Marquetand replied: So far, we have not considered the rotation of the molecule in the laser field but we are currently working on the implementation of the rotational degree of freedom. We also expect to see an influence on the orientational dynamics and on the Stark control effect.

Dr Sussman replied: Application of a non-resonant control field should produce rotational and vibrational influence simultaneously. The mutual interaction of these two effects will depend on the various system couplings. There are many interesting applications where the two degrees of freedom are simultaneously excited, including modelling of vibrational decoherence due to a rotational "bath". Experimental efforts to observe alignment during vibrational control in IBr were not successful, suggesting, in this case, that the rotational influence is weaker. However, this warrants further investigation and confirmation.

Dr Sanz-Sanz replied: We have not yet considered the influence of the rotational dynamics in the process. Of course, the control pulse and its high intensity will affect the system differently when it is applied either before or after the excitation, however the shifting and distortion of states due to non-resonant Stark effect is expected to be stronger than the other effects. The non-resonant character of the field is supposed to produce little excitation of molecules within the excited states but high distortion of the potential curves. With respect to the possible orientation of the molecules, this can only happen while the pulse is on and no other field is applied to produce any extra process, therefore when the control pulse ends molecules are field-free and move along the dissociation with no external influences. All possible effects should of course be included to make the simulation more realistic, but we don't consider the rotational dynamics as being a decisive effect in this photodissociation process.

Professor de Vivie-Riedle said: Summarizing the discussion: Can theory and experiment agree that the confusion regarding the sign of the polariazbilty and how the corresponding matrix elements are treated in the models of different degree in complexity is solved now?

Dr Sussman responded: Despite initial confusion, the time-dependent simulations presented by Dr Sanz-Sanz and our simulation and experimental work previously published[1] provide similar results for branching ratios control. Polarizabilities and dipole moments, however, were not known at the time of the original time-dependent simulations, and fits of the Stark shifts were made to the experimental data. A typical molecular polarizability may be on the order of 10 $Å^3$. Such a molecule receives a 0.01 eV shift in a control field of 7×10^{11} W cm^{-2}, suggesting that a significant level shift is possible in common systems.

In the application of a control field to IBr during excitation, the ground state is shifted down with respect to the upper state. This is equivalent to spectrally red-shifting the excitation and results in a decreased wavepacket velocity and hence more adiabatic behaviour (larger Br production). Control during propagation lowers the crossing point and the wavepacket velocity is increased at the intersection resulting in a more diabatic transition (larger Br* production). In the 2006 simulations, the maximal Stark shifts used in were +0.125 eV for the diabatic $B^3\Pi(O^+)$ state and −0.022 eV for the diabatic $Y^3\Sigma a^-(O^+)$ state. These shift amplitudes are proportional to the polarizability and control intensity. The intensity was approaching 10^{13} W cm^{-2} (strong, but non-ionizing), though it was not accurately characterized and so obtaining precise estimates for the polarizabilities was not possible at that time. The newer simulations by Dr Sanz-Sanz now use *ab initio* calculated dipole moments and polarizabilities.

1 Sussman *et al.*, *Science*, 2006, 314, 278–281.

Professor Weinacht continued the discussion about the paper by Dr Marquetand: This is a question about the number of non-resonant states required to get an accurate calculation of the dynamic Stark shift. In calculations of the polarizability or dynamic Stark shift that we have performed for atoms, we perform an essential states calculation, including as many off-resonant states as needed in the expression for the polarizability as required for a given level of error. The further off resonance a given state is and the lower the dipole moment, the lower the contribution to the Stark effect. We found that including up to 10 levels in atomic sodium allowed us to include all states which contributed more than 1% to the total Stark shift. It sounds to me that the case here is much worse—that many more states have to be included. If one were to rank the states as contributing from most to least, how many states would one need to include to all states contributing more than 1%? Another way of stating the question is to ask how much the total Stark shift (or perhaps differential Stark shift) varies in going from say 20 to 30 states.

Dr Marquetand answered: At that high level of theory which we are applying, it is not feasible for us to calculate more states than the 36 shown also by Dr Sanz-Sanz. Therefore, we are not able to predict how many states you would actually need here to achieve convergence. All we can say is that 36 states are not enough in this model to account for the dynamic polarizabilities. As Professor Kosloff pointed out earlier, you would need to also include ionic states to get the correct polarizability. Since we need all 36 states to obtain the correct potential energy curves, we did not look at the effect of fewer states on the Stark shift.

Professor Kosloff continued the discussion of the paper by Dr Sanz-Sanz: Did you include ion pair states in these calculations? I would expect that the states I^+Br^- and I^-Br^+ play an important role in excited state Stark dynamics.

Dr Sanz-Sanz replied: We have not included in the calculations the ion pair states. For the kind of process we are trying to study, which is the effect of the Stark shifting near a curve crossing, we did not consider these states of enough importance and they would complicate even more the calculations. There is another thing to point out, the laser strengths we are using are considered strong enough to produce Stark shifting but not ionization.

Professor Shapiro continued the discussion of the paper by Dr Marquetand: The polarizability expression you used is only valid for off-resonance transitions. In the case of on-resonance transitions one must compute the full "Light Induced Field Dressed Potential".

Mr von den Hoff asked: We recently showed, that it is possible to control the branching ratio of a photochemical reaction mediated by a local conical intersection, by preparing a superposition state between the two coupled electronic states with a phase stable few cycle laser pulse shortly before the intersection is reached. Changing the carrier envelope phase of the coupling laser changes the passage of the superposition state through the local conical intersection and thereby the branching ratio. Is this kind of control describable with your approach to introduce various kinds of couplings into on-the-fly dynamics?

Dr Marquetand answered: As we have shown, laser control can be modelled in general with our approach. However, for the case you are referring to, the phase of the wavepacket is important. Classical trajectories, on the contrary, usually do not have a phase and the single trajectories of a swarm representing a wavepacket are uncoupled. Therefore, it is not possible with the current implementation to describe this kind of control. However, it might be possible to add the necessary features, as Professor Tannor pointed out earlier, by using complex-valued trajectories.

Professor de Vivie-Riedle remarked: Roland Mitric from Berlin developed together with Vlasta Bonacic-Koutecky and coworkers[1] a similar approach for on-the-fly dynamics including laser excitation. In their *ansatz* the electronic coherence is included and is the only control knob for any molecular reaction. Is your algorithm also sensitive to the coherence induced in an electronic superposition state?

1 R. Mitric, J. Petersen and V. Bonacic-Koutecky, *Phys. Rev. A.*, 2009, **79**, 053416.

Dr Marquetand responded: Already in the original approach of surface hopping by Tully,[1] the time-dependent Schrödinger equation is solved for the electronic part, *i.e.* quantum effects like coherence are considered. Both methods, the one of Roland Mitric and coworkers as well as ours, are based on Tully's surface hopping and therefore are sensitive to electronic coherence. However, as a classical *ansatz* is chosen for the nuclei, the overall coherence of the total system is not accessible in these semiclassical methods. The dynamics will most probably be well reproduced as long as a wavepacket mimicked by a swarm of trajectories remains localized.[2]

1 J. C. Tully, *J. Chem. Phys.*, 1990, **93**, 1061–1071.
2 see *e.g.* S. Gräfe, P. Marquetand, V. Engel, *J. Photochem. Photobiol. A: Chem.*, 2006, **180**, 271–276.

Professor Tannor continued the discussion of the paper by Dr Sanz-Sanz: The application we heard about now was for a diatomic. Can the method be used for selective control of bond breaking in polyatomics? If so, can you give us some insight into the different knobs available and the resulting mechanism for controlling the motion within a 3N-6 dimensional coordinate space?

Dr Stolow replied: The 'knobs' available for control using the second order nonresonant dynamic Stark effect (NRDSE) are the (i) strength, (ii) duration, (iii) pulse envelope shape and (iv) timing of the applied Stark field. The coordinate and field strength dependence of the second order NRDSE are molecular properties given by nature. As such, the nature of the control will be system dependent. It may or may not be possible to control a 'specific bond breaking' pathway. The use of computer-controlled feedback on the NRDSE interaction is certainly feasible but has not yet been implemented. Another 'knob' worth considering is theinterferencee control of lifetimes *via* Stark modification of the phase difference (*i.e.* via the classical action) between paths, as discussed by Sussman, Ivanov and Stolow.[1] Experimental investigations of branching ratio control in real polyatomic systems, we think, will add to our understanding of how these various 'knobs' apply.

1 B. J. Sussman, M. Yu. Ivanov and A. Stolow, *Phys. Rev. A*, 2005, **71**, 051401R.

Professor Ashfold commented: We have previously reported a comprehensive ion imaging study of the I, Br and Br* atoms resulting from IBr photolysis throughout the wavelength range 440–685 nm.[1] Analysis of the measured product branching ratios and recoil anisotropies allowed decomposition of the total parent absorption spectrum into partial cross-sections for $A^3\Pi_1$–$X^1\Sigma^+$, $B^3\Pi_{0+}$–$X^1\Sigma^+$ and $C^1\Pi_1$–$X^1\Sigma^+$ excitation (if we assumed that the respective transition dipole moments were each independent of R_{I-Br}). Combining these data with the results of earlier spectroscopic studies of the bound v' vibrational levels supported by the A and B state potentials[2,3] allowed us to propose analytic expressions for these four potentials, and for the $1^3\Sigma^-$ (0^+) potential—three of which potentials underpin the model calculations reported in the papers by Marquetand *et al.*[4] and Sanz-Sanz *et al.*[5] Vrakking and co-workers later demonstrated large variations in the predissociation rates of high v' levels of the B state of $I^{79}Br$.[6] We note that our spectral decomposition implies that A–X and C–X absorption (neither of which are included in the recent modelling) accounts for $\sim$10% of the total dissociation yield at the excitation energy of interest, and are interested to know how well the field-free potentials used in the papers by Marquetand *et al.*[4] and Sanz-Sanz *et al.*[5] reproduce those derived from inverting the various experimental data.

1 E. Wrede, S. Laubach, S. Schulenburg, A. Brown, E. R. Wouters, A. J. Orr-Ewing and M. N. R. Ashfold, *J. Chem. Phys.*, 2001, **114**, 2629.
2 M. S. Child, *Mol. Phys.*, 1976, **32**, 1495.
3 D. R. T. Appadoo, P. F. Bernath and R. J. Le Roy, *Can. J. Phys.*, 1994, **72**, 1265.
4 P. Marquetand, M. Richter, J. González-Vázquez, I. Sola and Leticia González, *Faraday Discuss.*, 2011, DOI: 10.1039/c1fd00055a.

5 C. Sanz-Sanz, G. W. Richings and G. A. Worth, *Faraday Discuss.*, 2011, DOI: 10.1039/c1fd00039j.
6 E. A. Volkers, A. E. Wiskerke, R. Mooyman, M. J. J. Vrakking and S. Stolte, *PhysChemComm*, 2000, **3**, 56–60.

Dr Sanz-Sanz responded: We thank Professor Ashfold for bringing this work to our attention and we will make a comparison between our calculated field-free potentials and the experimentally derived one.

Dr Richings said: Further to the work described in this paper, we have recently been working on extending the concept of the NRDSE control of photo-dissociation to polyatomics. In particular, we have constructed a two-state, two-dimensional model of ammonia. Using as coordinates the distance of the dissociating hydrogen atom from the centre of mass of the NH_2 fragment, and the angle of this vector to the bisector of the fragment, we have calculated surfaces for the energies, dipoles and polarisabilities (including the transition components of the latter two).

Immediately prior to this series of discussions, we completed a preliminary calculation on the control of the dissociation of ammonia through a conical intersection, using these surfaces and the same methodology as presented in the paper under discussion. The results of this are shown in Fig. 5. We note the change in the relative branching ratio of the high and low energy paths as a function of delay time of the intermediate field, 150 fs duration control pulse with respect to the initial excitation. A negative peak, concurrent with excitation and similar to that seen in IBr, is followed by an extended feature at later control times corresponding to an increased proportion of the wavepacket exiting *via* the lower energy channel.

With this early result, we get evidence for control of branching ratios through a single conical intersection, an extension of the idea from the avoided crossings, seen in diatomics, to the more complex features seen in polyatomics. As such there is hope that the idea can be taken onwards to even more extended systems.

Professor de Vivie-Riedle enquired: When the dynamic stark control scheme is used for a polyatomic system, will there be a conflict due to unavoidable resonant transitions?

Dr Sussman answered: There are two limits of dynamic Stark control: dipole or Raman dominant. In the former case, when the control field is resonant with dipole coupled states the control will follow the instantaneous electric field and result in population oscillations. In the later case, when the control field is not resonant with the relevant states the control field will follow pulse envelope. Both limits provide opportunities for control.

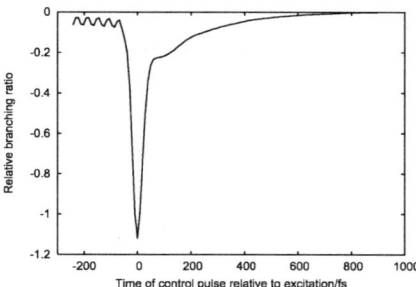

Fig. 5 Plot showing the ratio of the flux through the excited and ground state dissociation channels of ammonia as a function of control pulse timing (excitation at $t=0$), with the dissociation ratio, seen without a control pulse, subtracted.

Professor Weinacht asked: In comparing applications of the non-resonant dynamic Stark shift to rotational dynamics *vs.* vibrational dynamics my sense is that the forces that one can apply as a function of angle are much larger than the forces one can apply as a function of vibrational coordinate, particularly for linear diatomic molecules because the polarizability varies more with angle than it does with vibrational coordinate. I think this is why the original experiment on Cl_2 was able to spin up the molecule until it dissociated, making more than 100 rotational Raman transitions, while experiments using impulsive stimulated Raman scattering have not succeeded in dissociating a molecule from the vibrational ground state as far as I know. Do you know of any cases where the polarizability can vary with vibrational coordinate more than with angle—other than C_{60}?

Dr Sussman responded: The depth of the alignment potential created by the dynamic Stark effect depends on the differences between the principle polarization axes of a molecule being large. In molecules with large polarization difference the rotational alignment potential can be significant. There is, also, a shift in the rotational zero point that is usually neglected. This zero point change can be of use in the vibrational degrees of freedom where it can generate a relative shift between potential surfaces. As well, if the surfaces intersect, the crossing point will move as the two surfaces shift relatively. This can have a dramatic coordinate-dependent effect on the potential energy surface. The higher order effects will result in position-dependent alterations of the potential surfaces as well, though these are typically less dramatic in the regime discussed. While it is not required for the crossed-curve case, there are many molecules where the polarizability can vary more with vibrational coordinate more than with angle. For example, all spherical tops.

Dr Stolow commented: Within the electric dipole approximation, which certainly applies here, the external field applies uniformly to the whole molecule. Therefore, it is only the nuclear coordinate dependence of the dynamic Stark effect—an innate molecular property—which is relevant. By changing the timing, duration and intensity of the dynamic Stark interaction, control may be achieved.

Professor Miller added: I agree. The problem will be scaling to larger molecular systems. Using the analogy of a catalyst, the changes in polarizability and charge density induced by a typical molecular catalyst to control barrier heights are very localized. Here I am thinking of typical enzyme–substrate interactions as an example. The dynamic Stark effect uniformly affects the whole molecule, as you state, and it may not be possible to introduce sufficient differential polarizability to modify the effective barrier to the reaction coordinate of interest. It will be interesting to see how far this approach can be scaled to control molecular reaction dynamics.

PAPER

From molecular control to quantum technology with the dynamic Stark effect

Philip J. Bustard,[ab] Guorong Wu,[a] Rune Lausten,[a] Dave Townsend,[c] Ian A. Walmsley,[b] Albert Stolow[a] and Benjamin J. Sussman*[a]

Received 15th April 2011, Accepted 6th June 2011
DOI: 10.1039/c1fd00067e

The non-resonant dynamic Stark effect is a powerful and general way of manipulating ultrafast processes in atoms, molecules, and solids with exquisite precision. We discuss the physics behind this effect, and demonstrate its efficacy as a method of control in a variety of systems. These applications range from the control of molecular rotational dynamics to the manipulation of chemical reaction dynamics, and from the suppression of vacuum fluctuation effects in coherent preparation of matter, to the dynamic generation of bandwidth for storage of broadband quantum states of light.

1. Introduction

There are deep connections between the quantum control of chemical processes and the burgeoning development of technologies based on quantum information. Both utilize electromagnetic forces which govern the static and dynamic behaviour of quantum systems containing electrons. The first and most famous example of external field control over quantum states is the Stern-Gerlach experiment where a static but spatially inhomogeneous magnetic field was used to modify the translational states of free silver atoms. More generally, time-dependent electromagnetic interactions may be applied to the manipulation of quantum systems, perhaps ideally using lasers fields.[1,2] The electric field strengths generated by modern short pulse lasers can exceed those which bind electrons to matter. Furthermore, these electric interactions can be rapidly varied, on time scales comparable to those of internal motions within molecules. Ultrafast lasers make an ideal tool for the manipulation of quantum systems using the these electrodynamic interactions. As is discussed in more detail below, we will be particularly interested in the second order (quadratic) non-resonant dynamic Stark effect (NRDSE). The first order (linear) Stark effect directly induces oscillations in the target quantum system at the frequency of light which, particularly for visible light fields, may be too high for certain purposes. The NRDSE, by contrast, exploits the response of a quantum system to the intensity envelope of a laser pulse rather than the oscillating electric field. As the envelope frequencies in a femtosecond laser pulse are on the time scale of molecular vibrations and rotations, the NRDSE is well suited to the control of quantum dynamics in molecules. That the interaction is non-resonant means that it can be described by use of a polarizability and, hence, the only *a priori* requirement of the quantum system is that it have non-zero polarizability. We will also make use of the non-perturbative nature of the matter-field interaction wherein the usual

[a]*Steacie Institute for Molecular Sciences, National Research Council of Canada, 100 Sussex Drive, Ottawa, Ontario, K1A 0R6, Canada. E-mail: ben.sussman@nrc.ca*
[b]*Clarendon Laboratory, University of Oxford, Parks Road, Oxford, OX1 3PU, UK*
[c]*School of Engineering and Physical Sciences, Heriot-Watt University, Edinburgh, EH14 4AS, UK*

spectroscopic picture of transitions will fail. The latter is based on a converging power series expansion in electric field strength, where the complex coefficient of each term is the appropriate order of electric susceptibility. For the NRDSE under consideration here, the non-perturbative interaction extends to all orders but in a specific way, as is discussed in more detail below. Importantly, this means that the quantum transition probabilities in the field cannot be understood by consideration of the linear (or non-linear) Fourier power spectrum of the laser pulse. Rather, the NRDSE interaction can be thought of as the Rabi cycling of stimulated Raman scattering. Using this point of view, we will see that both internal (vibration, rotation, electronic) and external (translation) degrees of freedom may be controlled.

Important to our work is that the NRDSE increases only gradually as the field intensity is increased. Therefore, we can choose non-perturbative intensities at which the quantum system experiences Stark shifts of sufficient magnitude so as to exert an influence over the dynamics, yet produce negligible contributions from other strong field effects such as multi-photon ionization. For non-resonant near-infrared fields, this is in an intensity regime on the order of 10^{12} W cm^{-12}. As a relevant example, this means that we can control the evolution of a chemical reaction on a *given* potential energy surface using strong non-resonant laser fields which do not induce *any* transitions to other potential energy surfaces (*i.e.* other electronic states). In this chemical example, the control is exerted *via* the coordinate dependence of the Stark effect. To make an analogy with skiers going down a hill, the NRDSE interactions modifies the shape of the ski hill (potential energy surface) while the skiers are skiing, but leaves the hill unchanged at the end of the run. Furthermore, the skiers always remain on the same hill throughout.

Closely associated with the field of quantum control is quantum information, computation, and communication (QICC). QICC makes use of the intrinsic uncertainty in quantum mechanics, providing fundamental advantages over operations with classical systems. For these advantages to emerge, however, we will require the accurate articulation and observation of quantum systems. Here too we will see that the NRDSE will play a role. An example discussed in more detail below is a quantum memory which must write a photon to matter and then read the memory to produce an output photon. This procedure is analogous to the well known control protocol STIRAP,[3] which can be used to move population between states in a three-level system. The dynamic bandwidth associated with the NRDSE interaction permits a dramatic speed up of quantum memory devices.

In the following, we will develop the background necessary for understanding the NRDSE interaction and our implementation of it. Various uses of the dynamic Stark effect are then presented and discussed; these diverse applications emphasize the great utility of the dynamic Stark effect in the control and articulation of seemingly disparate quantum systems. We will illustrate its ability to control molecular dynamics such as axis alignment, and chemical dynamics such as branching ratio control in non-adiabatic photodissociation. We then consider the extension of NRDSE control to the non-adiabatic photodissociation of polyatomic molecules where the Stark interaction modifies dynamics at conical intersections.[4] Moving on to quantum technologies, we consider the preparation and amplification of quantum coherence where the NRDSE plays a key role: amplification of Raman coherences. The amplification will utilize strong fields that produce stimulated emission, but also a weak field whose amplitude is designed to control and suppress the effect of vacuum fluctuations. Finally we discuss the role of the dynamic Stark effect in high speed quantum memories and efforts towards the creation of entangled states in solid state materials such as bulk diamond.

2. Chemical control

The interaction of a control laser field with a molecule can give rise to a quasistatic shift of the energy levels, much like the Stark effect for a system in a static field. The level shift can be interpreted as being due to an effective potential energy term created by the control laser. For molecules, the effective potential varies with the direction of the principal axis with respect to the field and can be used to align their molecular frame. The potential also varies with nuclear coordinate and electronic state; it can therefore be used to control vibrational motion. When the control laser intensity is increased sufficiently, significant amounts of stimulated Raman emission may be produced. This creates a feedback mechanism in which the Raman emission may also interact to create a potential. Classically, one might not expect this effect because the optical field oscillations are bipolar and therefore may appear to average to zero. However, the interaction energy is generally nonlinear and in this case rectification can occur in the quadratic term. Quantum mechanically the origin of the quadratic potential energy term is the effect of non-resonant states that do not directly participate in the nuclear dynamics, but nevertheless influence the system.

In this section of the paper, we consider the non-resonant dynamic Stark effect (NRDSE)[5,6] and its use as a method of chemical control. In section 2.1 we develop a classical model and an equivalent quantum mechanical model to derive the generalized Raman interaction Hamiltonian for the NRDSE. Some potential applications of this Hamiltonian as a way of controlling rotational motion are then presented in section 2.2. In section 2.3 we consider the use of the NRDSE as a technique for control of vibrational dynamics in molecules.

2.1 Non-resonant dynamic Stark effect

2.1.1 Classical approach.
The classical interaction of a scalar electric field with a static dipole is given by

$$V^{dipole}(t) = -\mu E(t). \tag{1}$$

However, if the charges are not fixed, the applied field can alter or induce the dipole moment, even if there is no static dipole. The dipole can then be written as a Taylor series in the field

$$\mu(E) = \mu_0 + \alpha E + \ldots \tag{2}$$

where μ_0 is a possible static dipole moment and $\alpha = \frac{d}{dE}\mu(E)$. The energy dV required to move a dipole through an electric field variation dE is $dV = -\mu(E)dE$, so that the total induced dipole moment energy is just the sum

$$V(t) = -\int_E \mu(E)dE = -\mu_0 E - \frac{1}{2}\alpha E^2 + \cdots. \tag{3}$$

If we assume the applied laser has a slowly varying envelope $\varepsilon(t)$ the electric field can be written as

$$E(t) = \frac{1}{2}\varepsilon(t)e^{-i\omega t} + c.c. \tag{4}$$

where ε is allowed to be complex to incorporate an optical phase. We can then write out the interaction (3) explicitly using the oscillating field and envelope quantities:

$$V = -\mu_0 E - \frac{1}{8}\alpha\left(\varepsilon(t)e^{-i\omega t} + \varepsilon^*(t)e^{-i\omega t}\right)^2 + \cdots. \tag{5}$$

Although the field here is completely classical, the quantum processes of photon absorption and emission can still be pictured. The terms oscillating as $e^{-i\omega t}$ correspond to absorption (depicted as an upward arrow ↑) and the terms oscillating as

$e^{i\omega t}$ correspond to emission ($\downarrow$).[7] When the product is expanded, three quadratic terms are formed.

$$V = -\mu_0 E - \frac{1}{8}\alpha\left(2|\varepsilon(t)|^2 + \varepsilon^2(t)e^{-i2\omega t} + \varepsilon^{*2}(t)e^{i2\omega t}\right) + \cdots. \quad (6)$$

A two-photon process is then the product of two single photon processes: two-photon absorption ($\uparrow\uparrow$) arises from the term oscillating at $-2\omega t$ and the reverse two-photon emission ($\downarrow\downarrow$) arises from the term oscillating at $2\omega t$. Raman type excitations ($\uparrow\downarrow$, $\downarrow\uparrow$) arise from the two quasi-static cross-terms (hence the factor of $2|\varepsilon(t)|^2$), which are a product of two terms oscillating at ωt and $-\omega t$. The Raman excitations are quasi-static in the sense that the only time dependence they exhibit is due to the envelope.

While the light electrons can respond to the fast optical oscillations, nuclei are too heavy and cannot. Therefore the fast oscillation terms at $\pm 2\omega t$ can be neglected and only the portion responding to the pulse envelope remains. This is equivalent to saying that two-photon absorption or emission are very weak processes when far from resonance with an electronic transition. The remaining slowly evolving portion represents the Stark induced potential

$$V \approx -\mu_0 E(t) - \frac{1}{4}\alpha|\varepsilon(t)|^2. \quad (7)$$

In the presence of a three dimensional field $\mathbf{E}(t) = \frac{1}{2}\boldsymbol{\varepsilon}(t)e^{-i\omega t} + c.c.$ the above formalism yields an interaction given by:

$$V \approx -\boldsymbol{\mu}_0 \cdot \mathbf{E}(t) - \frac{1}{4}\boldsymbol{\varepsilon}^*(t) \cdot \boldsymbol{\alpha} \cdot \boldsymbol{\varepsilon}(t). \quad (8)$$

In summary, the classical approach indicates that, in the presence of an oscillating electric field, a dipole moment may be induced that interacts with the field to produce a quadratic potential. The original dipole interaction remains, but it is augmented by a term quadratic in the field strength. Importantly, it should be noted that while the interaction is quadratic, the interaction appears directly in the Hamiltonian: in terms of quantum mechanical perturbation theory, the influence will be to all orders 2,4,6,.... There are two limits for this general interaction potential: dipole dominated or Raman dominated. Which term dominates depends on the optical frequency and the system matrix elements and energy levels. For purposes of rotational and vibrational control, the optical fields are usually chosen to be non-resonant such that the Raman term dominates.

2.1.2 Quantum approach. In the transition to semi-classical quantum mechanics, the classical induced Hamiltonians (7) or (8) could be inserted directly into Schrödinger's equation. However, this is not formally correct since it replaces the operator μ with its expectation value. Only the true dipole interaction $V(t) = -\mu E(t)$ should be used and a fully quantum approximation should be investigated. There is no reason to expect identical results from the two approaches although it turns out that the quantum approach gives approximately the same answer, under a more stringent set of limits.

Consider a non-degenerate two-level system starting with a fully populated ground state a and empty excited state b which are coupled by an oscillating field.

$$i\dot{a} = \omega_a a - \mu_{ab}E(t)b \quad (9)$$

$$i\dot{a} = \omega_b b - \mu_{ba}E(t)a \quad (10)$$

The equation for b can be formally integrated† and inserted into the equation for a. In this case, the equation for a is:

$$i\dot{a} = \omega_a a - i\mu_{ab}E(t)e - i\omega_b t \int_0^t \mu_{ba} aE(t')e^{i\omega_b t'} dt' \qquad (11)$$

Any dependence on b has been eliminated, but a new term replaces its effect on state-a. The most obvious point is that the new term is quadratic in $E(t)$, like the quadratic Stark effect. However, the amplitude a appears within the integral, suggesting that this is not yet a simple potential energy term.

To demonstrate that this is a potential term, the next step is to approximate the integral and see if the classical approach can be recreated. The integral is rewritten to have the product of a slow part which cannot be analytically integrated and fast oscillating part which can:

$$\int_0^t \left(e^{i\omega_b t}(\tfrac{1}{2}\mu_{ba}\varepsilon(t)e^{-i\omega t} + c.c.)e^{-i\omega_a t} \right) \left((e^{i\omega_a t}a) dt \right) \qquad (12)$$

Since a is time dependent it will have some spectral content. The spectral content will consist of a portion at its eigenvalue ω_a and another portion at other frequencies induced by the applied field and resulting dynamics. If the spectral content at ω_a dominates, $e^{i\omega_a t}a$ will be slowly varying (*i.e.*, its time derivative is appropriately small), then the integral can be integrated by parts and the surface term dropped due to small time variations:

$$\frac{1}{2i}\frac{\mu_{ba}\varepsilon(t)}{\omega_b - \omega_a - \omega}e^{i(\omega_b-\omega)t}a + \frac{1}{2i}\frac{\mu_{ab}\varepsilon^*(t)}{\omega_b - \omega_a + \omega}e^{i(\omega_b+\omega)t}a \qquad (13)$$

When the laser frequency ω is near resonance with the level spacing $\omega_b - \omega_a$, the second term dominates and the first term may be neglected. Here, both terms are kept so that the off-resonance case is included. The integral is now inserted back into (14) and the rotating wave approximation is taken to eliminate all fast oscillating terms:

$$i\dot{a} = \omega_a a - \frac{1}{4}\left(\frac{\mu_{ab}\varepsilon(t)\mu_{ba}\varepsilon^*(t)}{\omega_b - \omega_a - \omega} + \frac{\mu_{ba}\varepsilon^*(t)\mu_{ab}\varepsilon(t)}{\omega_b - \omega_a + \omega}\right)a \qquad (14)$$

or

$$\dot{a} = \omega_a a - \frac{1}{4}\varepsilon^*(t)\alpha\varepsilon(t)a \qquad (15)$$

where the dynamic polarizability is

$$\alpha(\omega) = \left(\frac{\mu_{ab}\mu_{ba}}{\omega_b - \omega_a - \omega} + \frac{\mu_{ba}\mu_{ab}}{\omega_b - \omega_a + \omega}\right). \qquad (16)$$

This form recreates the classical induced potential (7) when there are two states.

To extend beyond the two-state approximation is relatively straightforward. When interested in the dynamics in state a and multiple other non-resonant states are included, the calculation is essentially the same (although, strictly speaking, the approximations are stronger[5]). In this case, the polarizability in a state a depends on the sum over all non-resonant states b:

---

† The most convenient form is achieved by transforming to the interaction picture, eliminating b, and then transforming back.

$$\alpha_a^{ij}(\omega) = \sum_b \left(\frac{\mu_{ab}^i \mu_{ba}^j}{\omega_b - \omega_a - \omega} + \frac{\mu_{ba}^i \mu_{ab}^j}{\omega_b - \omega_a + \omega} \right) \qquad (17)$$

where the Cartesian indices i,j have been included for the vector case. The Schrödinger equation is now written as:

$$i\dot{a} = Ha - \frac{1}{4}\boldsymbol{\varepsilon}^*(t) \cdot \boldsymbol{\alpha} \cdot \boldsymbol{\varepsilon}(t) a \qquad (18)$$

and the induced potential, as in the classical case

$$V_{NRDSE} = -\frac{1}{4}\boldsymbol{\varepsilon}^*(t) \cdot \boldsymbol{\alpha} \cdot \boldsymbol{\varepsilon}(t) \qquad (19)$$

is recovered. For the case of multiple spectrally distinct fields centered at frequencies ω_p, the field may be written

$$\mathbf{E}(t) = \frac{1}{2}\sum_p [\boldsymbol{\varepsilon}_p(t)e^{-i\omega_p t} + c.c.] \qquad (20)$$

so the potential generalizes to,

$$V_{NRDSE} = -\frac{1}{4}\sum_{p'p} \boldsymbol{\varepsilon}_{p'}^*(t) \cdot \boldsymbol{\alpha} \cdot \boldsymbol{\varepsilon}_p(t) e^{-i(\omega_p - \omega_{p'})t}. \qquad (21)$$

This Hamiltonian is responsible for all of the dynamics induced by the NRDSE, including the well-known molecular frame alignment.

2.2 Alignment with the NRDSE: adiabatic and non-adiabatic techniques

As with the evolution of many other quantum systems, NRDSE control interactions may be separated into one of two categories: those which are *adiabatic*, and those which are *non-adiabatic*. For the control dynamics to be adiabatic, the interaction Hamiltonian must only undergo a small fractional change in a typical period of the unperturbed system. A system starting in an eigenstate of the initial Hamiltonian will then always remain in the corresponding eigenstate of the perturbed Hamiltonian. To violate the condition for adiabatic evolution, the perturbation amplitude or time variation may be increased, leading to transitions between the perturbed states. An examination of (21) reveals that the degree of adiabaticity in a NRDSE control experiment is determined by the applied field strength and its temporal profile, as compared to the timescale of the nuclear rotational, or vibrational motion. Vastly different control outcomes can be achieved in molecules by varying these two parameters. In this section we discuss adiabatic and non-adiabatic control by the NRDSE for the specific case of molecular alignment;[8–11] molecules are said to be aligned when an applied field defines order with respect to a space-fixed axis. This discussion involves a more detailed consideration of Raman scattering, and so we briefly introduce this phenomenon before proceeding.

2.2.1 Raman scattering.
In a Raman process, an input pump field scatters off a molecule and a new field is produced at a longer Stokes wavelength (fig. 1). One photon of the pump is destroyed and a longer wavelength photon at the Stokes frequency is created. Energy is conserved and so the deficit is deposited in the system as a rotational excitation at the pump-Stokes energy differences ω_{pS}, depicted in fig. 1 as between states 0 and 2. This figure shows the specific case of spontaneous Raman scattering, where only a pump field is incident on the medium and emission to the Stokes field is stimulated by the vacuum fluctuations. A theoretical description of spontaneous Raman scattering requires quantization of the Stokes field, which is

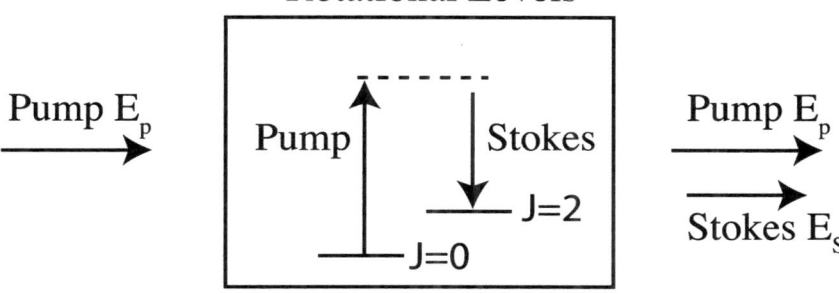

Fig. 1 A spontaneous rotational Raman scattering process. An input field excites the medium and a Stokes field is generated to conserve energy and momentum. In the photon perspective a pump photon is destroyed and a Stokes photon and rotational excitation quantum are created. The energy difference between pump and Stokes is the energy difference between states $J = 2$ and $J = 0$. If the pump is short, it may have sufficient bandwidth so that spectrum is present at the Stokes wavelength, thus stimulating the emission of Stokes light.

beyond the scope of the semi-classical model discussed herein. We note, however, that in the absence of an input Stokes field, the phase of the output Stokes field is random due to its quantum mechanical origin.[12,13] Stimulated Raman scattering occurs when the Stokes field, and the Raman excitation, are made to grow nonlinearly by increasing the Raman gain of the medium such that the Stokes field can stimulate the emission of further Stokes photons. For transient phenomena, where the Raman excited molecules maintain a well-defined phase relationship with the Stokes field, the Raman gain is enhanced by increasing the number of scattering molecules, or by increasing the fluence of the pump field.[13]

Should the population in the upper energy level become large enough, the process can be reversed, with the Stokes field producing anti-Stokes light at the shorter pump wavelength. Because the cross sections are so small, the generation of Raman scattered light is often ignored in alignment experiments, however its mechanism is intimately related to non-adiabatic alignment approaches.

In order to discuss potential mechanisms of alignment by the NRDSE, it is helpful to explicitly write the control field as an interaction of two spectrally distinct input fields $E = E_p + E_S$, the pump p and the scattered Stokes S fields, both assumed to be slowly varying. Eqn (21) then has the following form:

$$V_{NRDSE} = V_p + V_S + V_{pS} + V_{Sp}. \qquad (22)$$

where

$$V_p = -\frac{1}{4}\varepsilon_p^* \cdot \boldsymbol{\alpha} \cdot \varepsilon_p \quad \text{standard alignment } c.f., (8, 19) \qquad (23)$$

$$V_S = -\frac{1}{4}\varepsilon_S^* \cdot \boldsymbol{\alpha} \cdot \varepsilon_S \quad \text{Stokes alignment} \qquad (24)$$

$$V_{pS} = -\frac{1}{4}\varepsilon_p^* \cdot \boldsymbol{\alpha} \cdot \varepsilon_S e^{-i\omega_{pS}t} \quad \text{Stokes scattering} \qquad (25)$$

$$V_{Sp} = -\frac{1}{4}\varepsilon_S^* \cdot \boldsymbol{\alpha} \cdot \varepsilon_p e^{i\omega_{pS}t} \quad \text{anti} - \text{Stokes scattering}. \qquad (26)$$

The four terms each result in different modification of the molecular motion. The first term V_p contains no rapid time-dependence since the field is assumed to be slowly varying. As a result, the frequency spectrum of the V_p term is not resonant with any non-degenerate rotational levels and the V_p term is therefore responsible for adiabatic alignment; this is discussed further in section 2.2.2. Similarly, the second term V_S represents the alignment potential due to the Stokes field, which we also assume to be slowly-varying. In the absence of a large input Stokes field, or stimulated emission due to large Raman gain, V_S is usually small compared to V_p and can be neglected. Importantly, the inclusion of the pump and Stokes fields has resulted in the appearance of an oscillation at the beat frequency at ω_{pS} in the V_{pS} (Stokes) and V_{Sp} (anti-Stokes) scattering terms. This beat frequency may efficiently drive the internuclear dynamics when ω_{pS} is tuned close to resonance with a rotational energy level splitting. Physically, this is because the electrons respond to the fast oscillations of the field but the heavier nuclei cannot. The electrostatic coupling of the electrons to the nuclei means that modulating the control field intensity at ω_{pS} will parametrically drive the nuclear motion at this frequency, thereby transferring population between the rotational levels by Stokes scattering. The V_{Sp} term can similarly drive transitions in the opposite direction by anti-Stokes scattering.

2.2.2 Adiabatic and non-adiabatic alignment. The critical molecular timescale in determining the adiabaticity of alignment is the timescale of rotational motion τ_{rot}. For an applied field to adiabatically induce alignment, the control pulse intensity must vary slowly in time, compared with τ_{rot}. This is typically achieved by using a smoothly-varying control pulse of duration τ satisfying[8] $\tau \gg \tau_{rot}$. As the field is applied it 'dresses' the system and, taking advantage of the adiabatic theorem, slowly transports an unaligned state to an aligned state. If the field is slowly turned back off, the aligned state will adiabatically evolve to its original unaligned state, up to a global phase shift. In this adiabatic limit, only the V_p term is significant, and emission to the Stokes field can be neglected.

The essential feature of an adiabatic aligning potential V_p is that it contains no oscillating exponential component. It is basically a DC field that follows the pump pulse envelope; *i.e.* the frequency spectrum of V_p is not resonant with different states. Since it is not resonant, following the interaction the state populations remain unchanged. This is the mechanism of adiabatic alignment. To observe this explicitly, Fermi's Golden rule can be used to determine whether or not a transition occurs. From the time-domain version of Fermi's Golden rule, the amplitude for state $J = 2$ following excitation from state $J = 0$ can be written in first order perturbation theory for V_P as

$$c_2(t) \approx \frac{i}{4}\int_{-\infty}^{t} e^{-i\omega_{02}t} \langle 2|, \varepsilon_p^* \cdot \alpha \cdot \varepsilon_p|0\rangle dt. \qquad (27)$$

This is a finite time Fourier transform, with a slow varying argument $\langle 2|, \varepsilon_p^* \cdot \alpha \cdot \varepsilon_p| 0\rangle$. At late times there is no population excitation since $c_2(\infty) = 0$. At early times, while the pump ε_p is on, amplitude c_2 is excited (the molecule is aligned) due to apparent bandwidth introduced by truncating the integral at time t, instead of at infinity. That is, the finite time integral is physically equivalent to the case of a pump ε_p that is suddenly switched off at time t. A sudden turn off generates large resonant bandwidth. This leads to the notion of 'switched wave-packets', discussed below. At later times, when the pulse is over, the Fourier transform samples the full-time excitation and finds no resonant excitation which would leave population in the excited state: at first the system experiences a large resonant bandwidth, and does not discover that the excitation is really non-resonant until the pulse is over. This demonstrates that an adiabatic pulse aligns a molecule even if there is no resonant excitation, because while the pulse is applied the molecule experiences greater apparent bandwidth then is actually present, and population is excited, thus generating adiabatic alignment. The adiabatic alignment effect may also be understood by diagonalizing the full Hamiltonian to introduce pendular states which are hybrids of the field-free rotational eigenstates.[9]

Alignment *via* non-adiabatic means can be achieved by increasing the rate of change of the interaction Hamiltonian V_{NRDSE} with respect to the frequency-splittings of the molecule rotational levels. In this case, the frequency spectrum of the interaction Hamiltonian will come into resonance with the rotational level splittings such that population can be transferred from one rotational level to another. A simple way to achieve this[14] is to apply a Stokes field in addition to the pump field such that $\omega_{pS} \approx (\omega_{J+2} - \omega_J)$ where ω_J is the frequency of the rotational level with quantum number J. If the lower energy level is populated more than the upper level, the V_{pS} term will resonantly transfer population from the lower level to the upper level; if the population inversion is reversed, the V_{Sp} term will dominate and transfer population in the opposite direction. The phase of the coherent rotational motion induced by this process is set by the phase difference of the pump and Stokes fields. In the absence of decoherence mechanisms such as collisional dephasing, non-adiabatic excitation of the coherent rotational motion may result in field-free superpositions of the initial rotational eigenstates after the control field is turned off. The generation of field-free molecular alignment is key theme in emerging methods for the study of ultrafast molecular dynamics.[8]

Instead of applying an external Stokes field, in some instances significant alignment may be generated by simply increasing either the fluence of a slowly-varying pump, or the number of scattering molecules, so that spontaneous emission to the Stokes field can reach the regime of stimulated Raman scattering. As the pump field propagates, positive feedback from the Stokes field stimulates further Stokes emission. In this way, energy is transferred from the pump field to the medium as the pump and Stokes fields drive transitions among the rotational levels. A key point to note, however, is that the phase of such a Raman excitation will be random on any given laser shot due to the quantum mechanical origin of the Stokes light. It is therefore not possible to probe any ultrafast rotational dynamics using short probe pulses as there is no phase relationship between the molecules and the probe. Quantum coherence amplification is discussed in section 2.4 as a solution to this problem.

An extreme example of non-adiabatic alignment is *impulsive* alignment, where a short pulse with duration at or below the timescale of typical rotational motion $\tau \leq \tau_{rot}$ is used to induce alignment by impulsive stimulated Raman scattering.[15] The bandwidth of a pulse $\Delta\omega$ is inversely related to its duration as $\Delta\omega \sim 1/\tau$. Therefore, impulsive pulses have sufficient bandwidth to overlap with an initial state and a target state. As the short pulse consists of a broad coherent phase-locked spectrum, the subsequent phases of the target molecules' states are locked by the impulsive excitation and are therefore left in a superposition. The energy deposited in the molecules by the Raman scattering interaction results in a red-shifting of the pulse spectrum. The broad bandwidth of short pulses makes a clear separation of V_{NRDSE} into distinct frequency components difficult. Typically this is dealt with theoretically by considering only the V_P term in V_{NRDSE} while relaxing the condition that the pulse envelope be slowly-varying.

2.2.3 Alignment potentials.
The type of order induced by an alignment potential depends principally on the molecular structure, the control field polarization, and the control field temporal profile. In systems where sufficient numbers of molecules are excited, feedback from the molecules can significantly alter the control field. Propagation then becomes a significant factor, resulting in spatial variation of the interaction Hamiltonian V_{NRDSE} and of the aligned sample. For example, in the case where a pump field and a Stokes field are applied in two-photon

resonance with a rotational transition, the Stokes field grows at the expense of the pump field as it propagates through the medium, resulting in growth of the rotational coherence along the direction of propagation. In this section, we consider the effect of the control field polarization and the control field temporal profile on the alignment. We then use experimental results to highlight the effect of the control field temporal profile, and contrast adiabatic with non-adiabatic alignment.

From the general form of the non-resonant Stark potential (8) or (19), the angular dependence of the aligning force can be demonstrated. We use the coordinate system shown in Fig. 2 for the ensuing discussion. For molecules symmetric about the major axis, the polarizabilities on the two minor axis are equal such that $\alpha_1 = \alpha_2$. When an electric field vector is polarized along the z direction such that

$$\varepsilon = \varepsilon_0(t) \begin{bmatrix} 0 \\ 0 \\ 1 \end{bmatrix} \quad (28)$$

the interaction (21) then simplifies to

$$V(t) = -\frac{1}{4}|\varepsilon_0(t)|^2 \big(\alpha_1 + (\alpha_3 - \alpha_1)\cos^2(\theta)\big). \quad (29)$$

Therefore the molecules experience a Stark shift which is equivalent to adding or *shaping* a potential energy surface so as to create a minimum that varies as $\cos^2(\theta)$. The potential well minimum draws the principal axis of the molecule toward the z coordinate axis and aligns the system. Note that since the interaction pulls toward both $\theta = 0, \pi$ the molecules do not orient themselves with respect to one end or the other; for orientation this interaction cannot be used alone.

For consideration of circular polarized fields, the equations can be simplified by choosing a different orientation for the aligning field. The equations are simplified for circular polarization in the xy-plane:

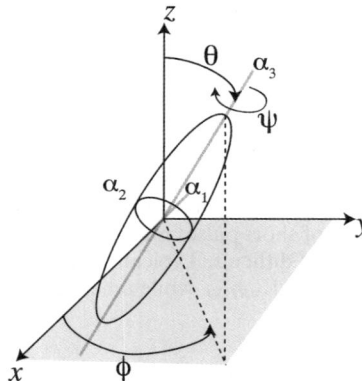

Fig. 2 A molecule (black ellipsoid) and its principle axes. The Euler angles θ and ϕ relate the principle axis of the molecule to the space-fixed frame. The principal axes of the polarizabilities are shown in grey, aligned here with the (moment of inertia) principle axes of the molecule. The Euler angle ψ is a rotation about the main principal axis. The polarizabilities α_i are shown along the grey axes. For example, application of a linear laser field polarized along z creates an aligning potential that aligns the principal axis toward the z axis.

$$\varepsilon = \frac{\varepsilon_0(t)}{\sqrt{2}} \begin{bmatrix} i \\ 1 \\ 0 \end{bmatrix} \quad (30)$$

The interaction then simplifies to

$$V(t) = -\frac{1}{4}|\varepsilon_0(t)|^2 \left(\alpha_3 + \frac{1}{2}(\alpha_3 - \alpha_1)\sin^2(\theta) \right). \quad (31)$$

In contrast to the linear polarized case, now the potential energy is minimized when $\theta = \pi/2$. The effect of circularly polarized light is thus to pull the molecule to the xy-plane in which the light field rotates.

2.2.4 Field-free molecular alignment.
The application of nonresonant fields can be used to align molecules in induced potentials. However, while the aligning potentials are on, the molecular states are subject to level shifts. This field-induced restructuring will significantly hinder the interpretation of any spectroscopic or dynamics studies in terms of field-free behaviour, the general goal. For dynamical studies, molecular alignment methods should be field-free. In this section, we discuss two alternative methods for the preparation of field-free molecular alignment using linearly-polarized pump fields: switched wave-packets,[16,17] and impulsive stimulated Raman scattering.[10] To induce field-free alignment, each of these techniques are necessarily non-adiabatic. The relevant potential for these alignment fields is (29). The degree of molecular alignment was measured by detecting the transient birefringence of the sample as a function of time using a pump–probe technique. In the absence of an aligning potential, the molecules are randomly aligned and so no birefringence is detected. By contrast, alignment of the sample along an axis induces a directional dependence in the polarizability, and therefore net birefringence. The probe pulse was linearly-polarized at 45° to the aligning field polarization; measurement of the polarization-rotated component through a high-extinction ratio polarizer as part of a Kerr cell is then a sensitive way to sample the birefringence, and thus alignment, of the sample.[11]

In the switched wave-packet (SWP) technique, a 100 ps, 1.064 μm pump pulse is truncated using a plasma shutter to provide a control pulse with a slowly increasing intensity to its peak, followed by a sudden drop of intensity to zero in a few tens of femtoseconds. Fig. 3 (top) shows the resulting polarization-rotated component of an 80 fs, 800 nm probe pulse detected using a photo-multiplier tube (PMT) after propagation through a sample of CO_2 molecules at 300 Torr. It should be noted that when the pump and probe pulses overlap in time, the measured birefringence includes a component due to the extremely transient (field-following) electronic response and the slower nuclear alignment response; only the latter is present at positive probe delays however. At negative probe delays, the birefringence smoothly increases as the sample is adiabatically driven to align with the pump polarization by the slowly increasing intensity. The birefringence signal drops sharply at zero-delay as the pump pulse is truncated, however periodic revivals are seen in the birefringence signal at larger delays. Coherence is left among the rotational levels because the pump pulse is turned off much more rapidly than the time-scale of rotational motion in CO_2, thus projecting the field-perturbed state on to the field-free eigenstates. The periodic revivals occur due to rephasings of the rotational wave-packet as it freely evolves in the absence of the control field. The revivals correspond to times at which the molecules are aligned but no external field is present. These revivals are very useful for the dynamical study of molecular frame properties.

In the second experiment, a 55 fs, 800 nm pump pulse was used to induce alignment in N_2 by impulsive stimulated Raman scattering. The time-dependent

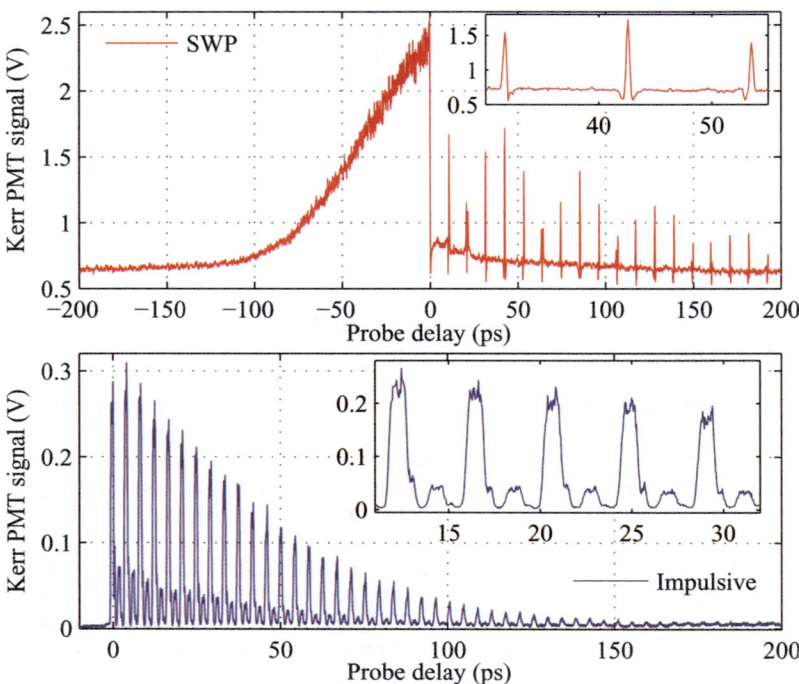

Fig. 3 Kerr cell detection of field free alignment for two contrasting control fields. Top: A switched wave-packet (SWP) technique is used to adiabatically align CO_2 molecules with the smoothly increasing linearly-polarized pump pulse envelope. The pump pulse is rapidly turned off, projecting the molecules onto a superposition of the field-free rotational eigenstates. Revivals in the molecular alignment are seen at positive probe delay times due to periodic rephasings of the non-adiabatically prepared rotational wave-packet. The inset shows the structure of the revival signal in detail. Bottom: A sample of N_2 molecules is impulsively aligned by a linearly polarized 55 fs pump pulse. Revivals in the alignment signal are again apparent due to the periodic rephasing of the non-adiabatically prepared rotational wave-packet.

birefringence signal is plotted in Fig. 3 (bottom). At negative probe delay times, no birefringence is detected using the 400 nm probe pulse, however prompt birefringence is induced within tens of femtoseconds of zero-delay. As with the SWP experiment, due to the non-adiabatic nature of the control field potential, coherence is left among the rotational levels. The resulting rotational wave-packet periodically rephases resulting in revivals in the birefringence signal. In both experiments, the birefringence signal decays as a function of time due to collisional dephasing of the rotational coherence, and intrinsic dephasing of the wave-packet caused by centrifugal distortions to the simple rotor Hamiltonian.[18]

2.3 Vibrational control

Many chemical reactions can be thought of as trajectories on vibrational potential energy surfaces which start as reagents, proceed over a barrier, and end as products. Catalysts often enhance reactions by exerting multi-polar electric forces that reduce the barrier. In a similar manner, the non-resonant dynamic Stark effect (NRDSE) can be used to modify the barriers of a chemical reaction on ultrafast timescales in order to control the outcome of a reaction.

Vibrational control using the NRDSE has been applied to an important class of photochemical reactions: nonadiabatic processes. These processes, such as internal conversion or intersystem crossing, entail charge rearrangements that occur along a reaction path at the intersections of potential energy surfaces and act as triggers of the ensuing chemistry. Chemical branching ratios in nonadiabatic processes are very sensitive to the intersection geometry, and therefore the dynamic modification of these processes is an important application of the NRDSE. So far, control on chemical branching ratios using the NRDSE has only been demonstrated on a diatomic molecule. The specific example was the nonadiabatic photodisociation of IBr.[19]

In IBr, there is an avoided crossing between 2(B) and 3(Y) excited states, which leads to two chemically distinct, neutral atomic channels: IBr $\rightarrow$ I + Br($^2P_{3/2}$) and I + Br*($^2P_{1/2}$).[20] An infrared NRDSE field can be used to modify the curve-crossing barrier at a specific time, thus promoting the yield of one chosen product over another. In the experiment, the reaction was initiated by a 100 fs laser pulse centered at 520 nm, above the dissociation limit for both channels. A strong, time-delayed non-resonant 1.7 µm infrared pulse, 150 fs in duration, was used as the control field. About 60 ps after the pump and control pulses, a third weak pulse at 304.5 nm was used to spectroscopically detect free, neutral ground state iodine atom products *via* (2 + 1) resonance-enhanced multiphoton ionization (REMPI). Measurement of the iodine atom kinetic energy distribution permits unambiguous determination of the Br*/Br product branching ratio, *via* conservation of kinetic energy and linear momentum. A dramatic variation of the overall integrated Br*/Br branching ratio as a function of control pulse delay was observed. At early and late times, the branching ratio was identical to that of the molecule under field-free conditions. Although the product branching ratios were controlled by the NRDSE field, the asymptotic recoil velocities of I atom fragments remained unmodified—they always matched those of the field-free dissociation. These results demonstrate that there is no significant net absorption of photons during the NRDSE interaction: there are no real electronic or ionizing transitions due to the application of the control pulse. There are two time delays when the reaction outcome is critically sensitive to the control field: (i) during initiation, and (ii) during traversal of the crossing point. If the NRDSE pulse is applied simultaneously with the initiation pulse, the reaction begins on a Stark-shifted potential energy surface. The Stark lowering of the ground state has an effect equivalent to spectrally red-shifting the pump laser wavelength. The result is that, after the control pulse turns off, the wave-packet velocity is reduced at the crossing point, decreasing the 'hopping' probability and enhancing the Br channel. An enhancement of 60 percent of Br yield at the expense of the Br* channel was observed. Application of the control pulse during traversal of the crossing point shifts and lowers the adiabatic barrier, thereby increasing the diabatic Landau–Zener hopping probability[21] and thus production of Br*. The enhancement of the Br* channel was more than 30 percent. The fractional change of peak-to-valley contrast of the branching ratio was over 90 percent and, importantly, due to the non-perturbative nature of the interaction, control was exerted on 100 percent of the reacting population.

An extension of the NRDSE control scheme to non-adiabatic dynamics at conical intersections in polyatomic molecules is a longstanding goal. However, the intrinsic multidimensionality of polyatomic systems makes this extension challenging. In diatomic molecules, the branching of chemical processes at an avoided crossing is determined by the velocity of wave-packet, the relative slopes of the two coupled potential energy curves, and their coupling strengths. In polyatomic molecules, the potential energy surfaces are multidimensional hypersurfaces, and the zero dimensional avoided crossing of diatomic molecules becomes a N^{int}-2dimensional seam of conical intersection (CI), where N^{int} is the number of internal degrees of freedom.[22] The branching between adiabatic and diabatic dynamics at the CI is largely determined by the local topography. However, there is evidence that the speed and direction of approach of the wave-packet to the CI may also affect the diabatic *versus* adiabatic branching.[23] Wave-packets on excited state potential energy surfaces may also decay through more than one type of CI. The routing of the wave-packet to different CIs is certainly dependent on details of the potential. Therefore, modification of potentials due to the NRDSE would be expected to change this routing and, hence, control the relative contributions of the electronic channels involved in the dynamics. In addition, modification of potentials using the NRDSE may also lead to different state distribution amongst the products within each channel, a phenomenon which is absent in diatomic cases. NRDSE control of chemical processes in polyatomic molecules is expected to be much more challenging and requiring of more detailed information on dynamical pathways. In return, the studies of NRDSE control over polyatomic molecule electronic branching ratios may also provide new insights into the dynamics of these systems. These developments will require considerable experimental and theoretical effort.

In order to extend the control of diatomics to a polyatomic system exhibiting electronic branching control *via* the NRDSE, we must begin by experimentally characterizing the field-free dynamics. As a candidate system, we have chosen to investigate the UV photodissociation of N-methylpyrrole (NMP), which has a number of appealing features. Experiments demonstrated the existence of methyl radical elimination channels with differing kinetic energy release (termed here 'fast' and 'slow').[24,25] The relative contribution of these channels varies with excitation energy. A sudden change in the slow *versus* fast methyl radical branching ratios occurs around 239 nm excitation. At wavelengths longer than 239 nm, there are two clear maxima in the CH_3 radical kinetic energy distribution, corresponding to slow and fast products. When the pump wavelength is tuned to 238 nm or below, the fast CH_3 radical channel apparently disappears. At these wavelengths, the molecule is vibronically excited to the $S_1(^1A_2, \pi\sigma^*)$ state. At even shorter wavelengths, the initial excitation is to the dipole allowed $\pi\pi^*$ state, which has a maximum around 217 nm.[26] However, the kinetic energy distribution of CH_3 has a very similar profile to the one at 238 nm. It was suggested that the slow *versus* fast CH_3 branching is due to the diabatic *versus* adiabatic splitting of the excited state wave-packet upon traversal of the CI connecting the first excited state (S_1) with the ground state (S_0), as shown in Fig. 4. One wave-packet component decays along the repulsive $\pi\sigma^*$ surface (which is the S_1 state in the Franck–Condon (FC) region and has 3s

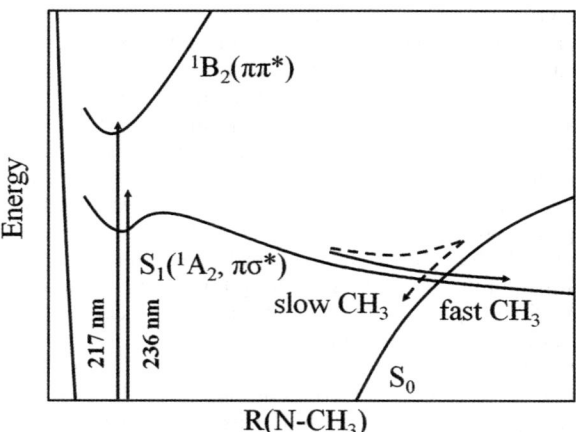

Fig. 4 Schematic potential energy curves for the ground state and two singlet excited states of NMP plotted as a function of R(N-CH₃).[48] This figure also illustrates the branching of photodissociation at the S_0/S_1 conical intersection. The direct dissociation along the repulsive $\pi\sigma^*$ surface leads to fast CH_3 radicals, while the dissociation of the hot ground state following internal conversion from the S_1 state leads to slow CH_3 radicals. The vertical arrows indicate the pump wavelengths.

Rydberg character), generating the fast CH_3 radical elimination. The other wave-packet component decays to the ground state *via* internal conversion: the hot ground state molecule subsequently undergoes statistical unimolecular dissociation to yield the slow CH_3 radical channel. In order that control be effective, the time scale of the dynamics must be accessible with available ultrafast control lasers. For these reasons, the photodissociation results and theoretical calculations[27,28] on the relevant potential energy surfaces make NMP a promising candidate for NRDSE control. However, the nonadiabatic excited state dynamics of NMP, especially the nature of the dependence of the slow *versus* fast branching ratio upon excitation energy are not well understood and require further study. Time-resolved measurements allow us to directly follow the wave-packet evolution on the excited state potential energy surfaces, and will provide complementary information. Due to the lack of previous experimental research, we performed femtosecond time-resolved photoelectron spectroscopy (TRPES)[29] studies. This is very important in understanding the dynamics of the branching and its excitation energy dependence, which is critical to the implementation of coherent control of NMP photodissociation. Measurements were made at three different pump laser wavelengths: (i) 242 nm, where both slow and fast CH_3 elimination channels operate; (ii) 236 nm, where only the slow channel exists; and (iii) 217 nm where NMP has strong absorption to the dipole-allowed $\pi\pi^*$ state. The excited state molecules were photoionized by a time-delayed femtosecond probe pulse of 267 nm. Dramatic changes in the form of the TRPES spectra were observed for these three different excitation energies, as shown in Fig. 5.

Global 2D least squares fits at all time delays and photoelectron kinetic energies were obtaining using Levenburg-Marquart algorithm.[30] For both 236 nm and 242 nm, a parallel kinetic model with two time constants represents the experimental data very well. Addition of more components only leads to a minor improvement in fit quality. The two time constants are 5.2/87.3 ps for 236 nm, and 5.7/952 ps for 242 nm. Decay associated spectra (DAS) of the two time constants at each wavelength are also derived from the fitting (not shown here). At each wavelength, both DAS have Rydberg-state-like features: a single sharp peak due to $\Delta v = 0$ FC overlap between the neutral Rydberg state and the cationic state; their peaks are also at the same energetic position. All these observations suggest that the two components for each wavelength may come from the same electronic state: the $S_1(^1A_2,\pi\sigma^*)$ state (see Fig. 4). For a pump pulse of 217 nm, a sequential kinetic model with two time constants fits the data very well. The two time constants are $\tau_1 = 101$ fs and $\tau_2 = 353$ fs. The implications of these time constants and fit models are not yet clear. High level theoretical calculations and analysis are currently in progress.[31]

In future experiments, we will measure the kinetic energy distribution of methyl radical to determine the fast *versus* slow branching ratio using the velocity map imaging technique. By studying these as a function of the delay, intensity and polarization of the NRDSE field, we will investigate control over electronic branching at the S_1/S_0 CI. The details of the control experiments will be dependent on the outcomes of current field-free dynamics studies, which continue.

2.4 Quantum coherence amplification

The potential effect of vacuum fluctuations on molecular quantum control was discussed in section 2.2.2. In particular, if spontaneous Raman emission influences a control process then the excited material dynamics will have a fluctuating phase from shot to shot. This poses a general problem in coherent preparation of matter: how to generate large amplitude vibrational or rotational excitations with a predetermined phase such that the dynamics are phase locked to an ultrashort pulse source. Typically, this is achieved *via* impulsive stimulated Raman scattering (ISRS) using a high power ultrashort pulse source, either in the form of a single ultrashort pulse,[10,32,33] or sequence of pulses.[34] In quantum coherence

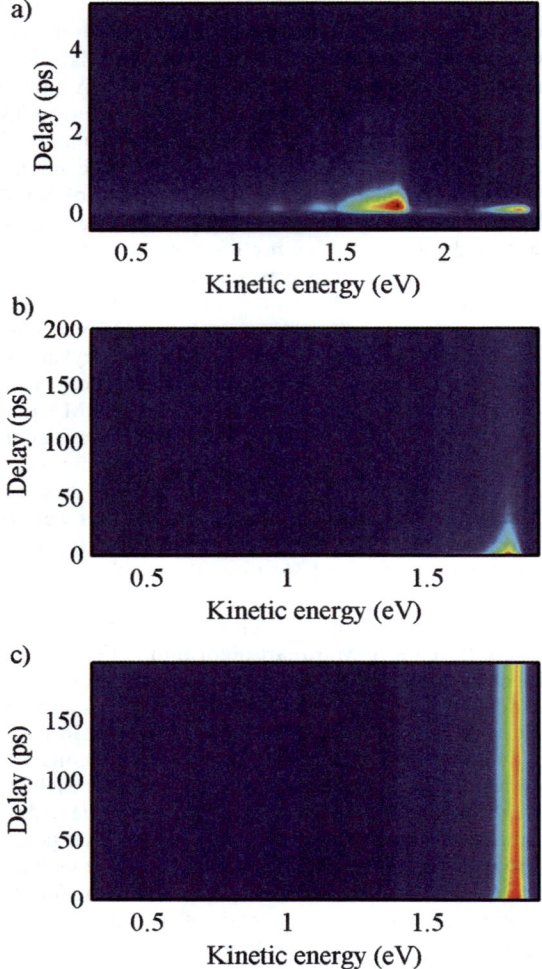

Fig. 5 TRPES spectra of N-methylpyrrole with pump wavelength of a) 217 nm, b) 236 nm, and c) 242 nm are shown (after subtracting background photoelectrons generated from single color multiphoton ionization). The probe wavelength is 267 nm.

amplification,[35] this preparation is achieved in two steps. Initially, a small fraction of energy from the ultrashort pulse is used as a 'seed', to pre-excite the rotational or vibrational levels of the system by ISRS, and dominate the effect of vacuum fluctuations. For ISRS to occur, the ultrashort pulse must have duration shorter than, or comparable to, the ro-vibrational dynamics it is to excite; in the frequency domain this means that the pulse spectrum contains both the pump and Stokes frequencies necessary for Raman scattering. In the next step, a subsequent pump pulse can then amplify the coherence by stimulated Raman scattering. While the pump pulse drives the electrons at optical frequencies, their dynamics are modulated by the pre-existing coherent nuclear motion excited by the seed pulse. As a result, the electronic motion consists of a frequency component at the Stokes frequency, red-shifted from the pump, and Stokes light is emitted. This feedback process continues as the Stokes light stimulates the emission of further Stokes photons, in the process transferring energy from the pump field to the rotational, or vibrational, coherence. Crucially, because the pre-existing coherence in the medium was generated by the seed, the amplified excitation will be phase-correlated with the ultrashort pulse source rather than with the random vacuum phase. In the absence of the seed, this is not true: spontaneous emission stimulated by the vacuum fluctuations would initiate stimulated Raman scattering and the phase of the excitation would be random on any given laser shot.

A significant advantage of quantum coherence amplification is that the pump pulse source can be completely phase-independent of the ultrashort pulse source, while still preserving phase coherence of the excitation with respect to the latter. While the phase information is provided by the ultrashort pulse source, the majority of the excitation energy is provided by the pump pulse source. In quantum coherence amplification, the pump pulse may be significantly longer than the ultrashort seed pulse, thereby allowing much higher pulse energies to be used while mitigating deleterious nonlinear effects such as field-ionization; such effects can result from coherence preparation using a single high energy, ultrashort excitation pulse.

We have demonstrated quantum coherence amplification on a rotational transition in hydrogen. This implementation may be thought of as an alignment experiment: rather than inducing order along the polarization axis (as in Fig. 3), here the order is induced along the direction of light propagation. In Fig. 6 we show quantum coherence amplification in hydrogen gas, with rotational coherence seeded using a 12 μJ, 55 fs pulse centered at 798 nm, and amplified using a 60 mJ, 10 ns pump pulse at 532 nm. A strong rotational Stokes sideband, shifted from the pump pulse, was detected at the cell output; this indicates that a significant amount of the pump energy had been deposited in the hydrogen molecules as rotational energy. The phase-stability of the rotational coherence with respect to the ultrashort pulse source was tested by scattering a 55 fs probe pulse from the excited molecules. The spectrum of the scattered pulse was measured as a function of delay from an initial starting delay of 3.9 ns. A rotational Stokes sideband, and a rotational anti-Stokes sideband were measured, in addition to the depleted input probe spectrum; energy is transferred from the central probe spectrum into new sidebands by molecular phase modulation.[36,37] Fig. 6 (top) shows a lineout from the centre of the probe spectrum at $\lambda = 798$ nm as a function of delay, while the lower plot shows a corresponding lineout from the rotational sideband at $\lambda = 837$ nm (a shift in wavelength of one rotational quantum). Delay-dependent oscillations are

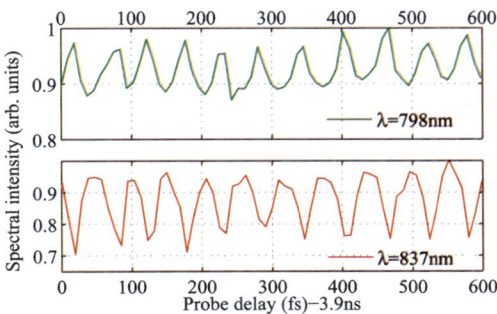

Fig. 6 Plot of delay-dependent oscillations in the spectrum of a probe pulse after phase modulation on propagating through an ensemble of H_2 molecules prepared in a rotationally excited state by quantum coherence amplification. Top: A line-out from the probe spectrum at the central probe wavelength of 798 nm, plotted from an initial delay of 3.9 ns; bottom: the corresponding line-out from the rotational Stokes sideband at 837 nm. The delay-dependent oscillations in the spectral intensity are at the period of the excited rotational transition, and indicate that the coherence is phase-stable with respect to the ultrashort pulse source. Oscillations in the Stokes sideband are π-radians out of phase with those at the central probe wavelength.

apparent in each of the plots, at the period of the excited rotational transition ($\tau_{rot} =$ 57fs). These oscillations demonstrate the phase-stability of the rotational motion with respect to the ultrashort pulse source: when the seed pulse is not present, no oscillations are seen and so the source is not phase-stable. The oscillations in the Stokes sideband are π-radians out of phase with those at the central probe wavelength, with dips in the central probe spectrum resulting from energy transfer into the rotational Stokes sideband.

3. Quantum technologies

In this section, we now move on from the use of the NRDSE in control of chemical systems to discuss its use in the burgeoning domain of quantum technologies. In the two implementations discussed, the main tool for articulating control in the target system is again the laser induced dynamic Stark effect. The application of the effect in these diverse systems indicates its broad utility, not just in quantum control, but also in the development of quantum technologies.

Along with the development of Quantum Theory came the realization that the probabilities associated with quantum systems exhibit features not expected from classical probability theory. Over time it became clear that these violations of classical probability theory actually imply that certain devices may be built with characteristics superior to their classical analogues, including improved algorithm processing and communications protocols. One experimental challenge in building these devices is finding suitable quantum systems and to interact with them in an efficient and precise manner. Various systems have been investigated including propagating photons,[38,39] ions,[40] nuclear spins,[41] high Q-cavities and others.[42] All have appealing features and challenges. To bridge the discussion of quantum control with quantum technologies we briefly conclude by discussing two quantum systems that we have investigated that, when subjected to the off-resonant dynamic Stark effect, can be utilized in the development of quantum technology devices. The first system is a quantum memory in caesium vapour and the second is a room-temperature solid state lambda system in bulk diamond.

3.1. Ensemble quantum memory in caesium

The ability to store quantum information will be a key enabling capability for quantum information processing and communication.[43] In quantum communication, designs for quantum networks require the ability to store and controllably re-emit photons, while preserving their initial quantum characteristics with high fidelity and efficiency. Quantum information processing will utilize 'buffers' to store quantum information deterministically while other information is processed and manipulated, just as in classical computation. Storage of quantum information is more problematic than its classical analogue because the no-cloning theorem prohibits the copying of quantum states. Due to their high speeds and low interaction strengths with the environment, photons are an ideal carrier of quantum information. However, these same features render photons challenging for storage of quantum information: a different physical manifestation of the information is required, such as an atomic or solid-state system. A key challenge is therefore to transfer quantum information from photons to such a system, acting as a storage device or 'quantum memory', with high fidelity.

The storage of quantum light states in an atomic ensemble,[44] for example, requires precise control of the internal atomic dynamics to enable efficient read and write operations on the memory. Here again the dynamic control of energy levels in the storage medium *via* non-resonant fields is a key enabling feature. Recently a high-speed Raman memory protocol has been developed, in which input photons are mapped into a collective atomic exciation by Raman interaction with a much stronger control field.[45] Application of the control field dynamically generates

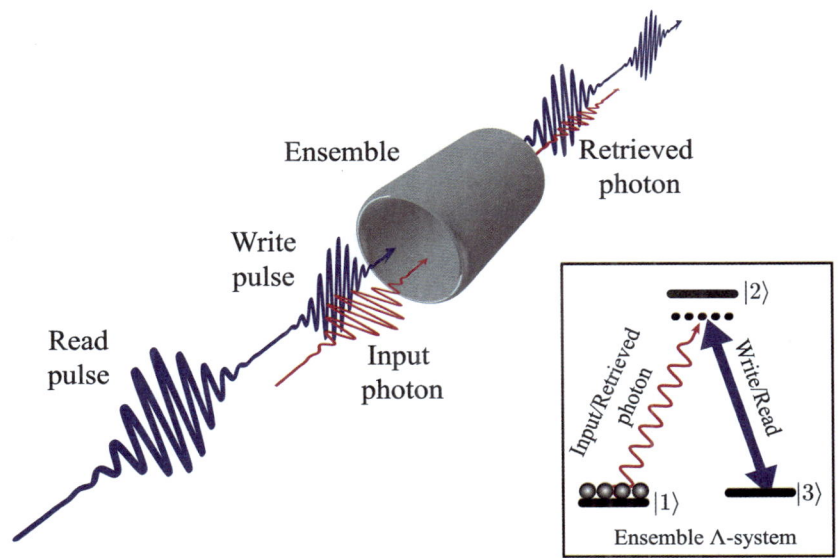

Fig. 7 Ensemble three-level system. The input photon is directed into the ensemble along with a strong write pulse and is stored. A subsequent strong read pulse extracts the stored excitation, mapping it to an output photon. *Inset*: Under appropriate circumstances an ensemble can behave as an individual quantum system, greatly reducing the complexity that would be required to couple to an individual atomic Λ-level structure. In the caesium quantum memory the $F = 3,4$ 'clock states' of the $6^2S_{1/2}$ ground-level hyperfine manifold are the states $|1\rangle$ and $|3\rangle$, and are linked to the excited state $|2\rangle$ ($6^2P_{3/2}$ manifold) through the D2 line at 852 nm. In bulk diamond the ground state $|1\rangle$ is the acoustic phonon and the storage state $|3\rangle$ represents an optical phonon. The transition is then linked *via* the far off resonance exciton state $|2\rangle$.

bandwidth from a narrow transition, enabling the input photons to be absorbed. Subsequent application of a second control field then enables the process to be reversed, and the stored information can be read out.

In the experimental implementation (fig. 7), a 300 ps write pulse and a weak 300 ps input photons pulse are overlapped and directed into a caesium vapour cell. The input photons are mapped *via* a Raman transition into a collective atomic excitation on the hyperfine levels of the caesium vapour. After a delay, a read pulse is sent into the vapour cell and converts the excitation into an optical output signal. Due to the use of an off-resonant Raman interaction which generates bandwidth for storage and read-out of the input photons, the technique has the potential for GHz data rates, a hundred-fold increase over other techniques.

3.2. Ensemble Λ-system in bulk diamond

Due to their intrinsic stability and compactness, efforts continue to develop easily accessed quantum systems with solids. In particular, bulk diamond phonons represent a simple and stable solid-state system in which to investigate quantum phenomena.[46,47] The extraordinary physical and optical properties of diamond make it a model quantum workbench. The workbench can be considered as a room temperature, solid-state, ultra-broadband 'lambda' system where control pulses can be used to write and read photons into and from the optical phonon state. Quantum features are difficult to observe in room temperature solids. Decoherence usually destroys any quantum effects before it can be measured or utilized. However, due to the large splitting between storage states (1332 cm^{-1} between the states $|1\rangle$ and $|3\rangle$), bulk diamond phonons can accommodate femtosecond pulses and therefore demonstrate an unusually high number of operational time bins.

As in the caesium quantum memory, the write process is performed by a control pulse that dynamically generates bandwidth (fig. 7). Unfortunately phonon coherence lifetimes in diamond are 7–11 ps, so the stored phonon rapidly decays. However, although the decay is rapid, the use of femtosecond pulses still permits a significant number of read/write operations (~100) to be performed before coherence is lost. In addition, the loss of system coherence and population in memory can be used to quantify the characteristic decay times T_1 and T_2.[46,47] Performing read and write sequences is a prerequisite to the measurement based entanglement approaches under development. The use of control fields to dynamically generate bandwidth may become a useful tool on the quantum information workbench.

4. Conclusion

Externally applied electric forces provide many strategies for control. The use of the non-resonant dynamic Stark effect (NRDSE), induced by strong near infrared laser fields, is a powerful tool within this class. For example, the NRDSE can be used to align molecules (in both 1D and 3D), in order to make improved (i.e. molecular frame) dynamical observations and potentially alter chemical interactions. The NRDSE may also be used to control product branching yields in neutral molecule photodissocation. This is achieved through the application of laser fields that are of sufficient intensity to modify potential energy barriers during the dissociation event. Importantly, this is done without inducing further electronic transitions (i. e. the NRDSE is a non-peturbative but non-ionizing—or non-destructive—control strategy). From this point of view, the NRDSE interaction may be viewed as a photonic catalyst rather than a photonic reagent. The extension of NRDSE control of product branching during polyatomic photodissociation is in progress. In many polyatomic systems, multiple conical intersections may be traversed as the initially prepared state evolves towards the final product state distribution and one can perhaps envisage the use of multiple control pulses, each precisely timed, to effect control at several different points along the reaction coordinate.

The links between quantum control and quantum information based technology continue to grow. One example is optical quantum memories which can store and controllably re-emit photons, preserving their initial quantum characteristics with high efficiency and fidelity. These need to be both written and read, and are therefore related to control approaches. For example, the high-speed Raman protocol in caesium uses dynamic Stark shifts to sweep an absorption transition during the storage of a photon, generating bandwidth over an otherwise a narrow transition. This increased bandwidth leads to GHz data rates—a hundred-fold increase over other techniques. Another area is the solid state phonon system in bulk diamond. Again, the dynamically increased bandwidth due to the control pulse interaction provides a novel route to high speed quantum storage, but this time in a compact solid state device that operates at room temperature.

The NRDSE is a general interaction between strong non-resonant light fields and all quantum systems containing charged particles. It is tacitly present in all strong field experiments, but can be utilized as a unique tool in different systems (e.g. alignment, product control, quantum memories). We anticipate that further development in control based on NRDSE will continue to open new areas for control and coherence in chemistry, physics and, quantum information science.

Acknowledgements

The authors would like to acknowledge the support of NSERC. We gratefully acknowledge valuable discussions with M.Yu. Ivanov, J.G. Underwood, K. Reim, K.C. Lee, J. Nunn and B.A. Jr. McCabe.

References

1 S. Rice and M. Zhao, *Optical control of molecular dynamics*, John Wiley, 2000.
2 M. Shapiro and P. Brumer, *Principles of the quantum control of molecular processes*, Wiley-Interscience, 2003.
3 K. Bergmann, H. Theuer and B. W. Shore, *Rev. Mod. Phys.*, 1998, **70**, 1003–1025.
4 *Conical Intersection: Electronic Structure, Dynamics & Spectroscopy*, ed. W. Domcke, D. R. Yarkony and H. Koppel, World Scientific Publishing Company, 2004.
5 B. J. Sussman, *Am. J. Phys.*, 2011, **79**, 477–484.
6 D. Townsend, B. J. Sussman and A. Stolow, *J. Phys. Chem. A*, 2011, **115**, 357–373.
7 R. Loudon, *The Quantum Theory of Light*, Oxford University press, USA, 2000.
8 H. Stapelfeldt and T. Seideman, *Rev. Mod. Phys.*, 2003, **75**, 543–557.
9 B. Friedrich and D. Herschbach, *Phys. Rev. Lett.*, 1995, **74**, 4623–4626.
10 M. Morgen, W. Price, L. Hunziker, P. Ludowise, M. Blackwell and Y. Chen, *Chem. Phys. Lett.*, 1993, **209**, 1–9.
11 J. P. Heritage, T. K. Gustafson and C. H. Lin, *Phys. Rev. Lett.*, 1975, **34**, 1299–1302.
12 M. G. Raymer and J. Mostowski, *Phys. Rev. A*, 1981, **24**, 1980–1993.
13 M. Raymer and I. A. Walmsley, *Prog. Opt.*, 1990, **28**, 181.
14 D. D. Yavuz, D. R. Walker, G. Y. Yin and S. E. Harris, *Opt. Lett.*, 2002, **27**, 769.
15 Y. Yan, E. Gamble Jr and K. Nelson, *J. Chem. Phys.*, 1985, **83**, 5391–5399.
16 J. G. Underwood, M. Spanner, M. Y. Ivanov, J. Mottershead, B. J. Sussman and A. Stolow, *Phys. Rev. Lett.*, 2003, **90**, 223001.
17 B. J. Sussman, M. Y. Ivanov and A. Stolow, *Phys. Rev. A*, 2005, **71**, 051401.
18 G. Herzberg, *Molecular Spectra and Molecular Structure: 1. Spectra of diatomic molecules*, Van Nostrand, Princeton, 1950.
19 B. J. Sussman, D. Townsend, M. Y. Ivanov and A. Stolow, *Science*, 2006, **314**, 278–281.
20 M. S. Child, *Mol. Phys.*, 1976, **32**, 1495–1510.
21 C. Zhu, Y. Teranishi and H. Nakamura, in *Adv. Chem. Phys.*, ed. I. Prigogine and S. A. Rice, John Wiley & Sons, Inc., 2007, vol. 117, ch. Nonadiabatic transitions due to curve crossings: complete solutions of the Landau–Zener-stueckelberg problems and their applications, pp. 127–233.
22 D. R. Yarkony, *Rev. Mod. Phys.*, 1996, **68**, 985.
23 A. M. D. Lee, J. D. Coe, S. Ullrich, M.-L. Ho, S.-J. Lee, B.-M. Cheng, M. Z. Zgierski, I.-C. Chen, T. J. Martinez and A. Stolow, *J. Phys. Chem. A*, 2007, **111**, 11948–11960.
24 A. G. Sage, M. G. Nix and M. N. R. Ashfold, *Chem. Phys.*, 2008, **347**, 300–308.
25 G. Piani, L. Rubio-Lago, M. A. Collier, T. N. Kitsopoulos and M. Becucci, *J. Phys. Chem. A*, 2009, **113**, 14554–14558.
26 R. McDiarmid and X. Xing, *J. Chem. Phys.*, 1996, **105**, 867–873.
27 A. L. Sobolewski and W. Domcke, *Chem. Phys.*, 2000, **259**, 181–191.
28 M. Vazdar, M. Eckert-Maksic, M. Barbatti and H. Lischka, *Mol. Phys.*, 2009, **107**, 845–854.
29 A. Stolow and J. G. Underwood, in *Adv. Chem. Phys.*, ed. S. A. Rice, John Wiley & Sons, Inc., 2008, vol. 139, ch. Time-Resolved Photoelectron Spectroscopy of Nonadiabatic Dynamics in Polyatomic Molecules, pp. 497–584.
30 K. Levenberg, *Quart. Appl. Math.*, 1944, **2**, 164–168.
31 G. Worth and M. N. R. Ashfold, personal communication.
32 S. D. Silvestri, J. G. Fujimoto, E. P. Ippen, E. B. Gamble, L. R. Williams and K. A. Nelson, *Chem. Phys. Lett.*, 1985, **116**, 146–152.
33 R. Bartels, S. Backus, M. Murnane and H. Kapteyn, *Chem. Phys. Lett.*, 2003, **374**, 326–333.
34 J. P. Cryan, P. H. Bucksbaum and R. N. Coffee, *Phys. Rev. A*, 2009, **80**, 063412.
35 P. J. Bustard, B. J. Sussman and I. A. Walmsley, *Phys. Rev. Lett.*, 2010, **104**, 193902.
36 M. Wittmann, A. Nazarkin and G. Korn, *Phys. Rev. Lett.*, 2000, **84**, 5508.
37 A. V. Sokolov and S. E. Harris, *Journal of Optics B: Quantum and Semiclassical Optics*, 2003, **5**, R1.
38 P. Kok, W. Munro, K. Nemoto, T. Ralph, J. Dowling and G. Milburn, *Rev. Mod. Phys.*, 2007, **79**, 135–174.
39 C. Perez-Delgado and P. Kok, *Arxiv preprint arXiv:0906.4344*, 2009.
40 J. Cirac and P. Zoller, *Phys. Rev. Lett.*, 1995, **74**, 4091–4094.
41 Kane, *Nature*, 1998, **393**, 133–138.
42 H. Mabuchi and A. Doherty, *Science*, 2002, **298**, 1372.
43 A. Lvovsky, B. Sanders and W. Tittel, *Nature Photonics*, 2009, **3**, 706–714.
44 K. Hammerer, A. S. Sørensen and E. S. Polzik, *Rev. Mod. Phys.*, 2010, **82**, 1041–1093.
45 K. Reim, J. Nunn, V. Lorenz, B. Sussman, K. Lee, N. Langford, D. Jaksch and I. Walmsley, *Nature Photonics*, 2010, **4**, 218–221.

46 F. Waldermann, B. Sussman, J. Nunn, V. Lorenz, K. Lee, K. Surmacz, K. Lee, D. Jaksch, I. Walmsley and P. Spizziri, *et al.*, *Phys. Rev. B*, 2008, **78**, 155201.
47 K. Lee, B. J. Sussman, J. Nunn, V. Lorenz, K. Reim, D. Jaksch, I. Walmsley, P. Spizzirri and S. Prawer, *Diamond and Related Materials*, 2010, **19**, 1289–1295.
48 O. Schalk, Private communication (2011).

PAPER

Controlled redistribution of vibrational population by few-cycle strong-field laser pulses

William A. Bryan,*[ad] C. R. Calvert,[b] R. B. King,[b] J. B. Greenwood,[b] W. R. Newell[c] and I. D. Williams[b]

Received 14th March 2011, Accepted 27th April 2011
DOI: 10.1039/c1fd00042j

The use of strong-field (*i.e.* intensities in excess of 10^{13} Wcm^{-2}) few-cycle ultrafast (durations of 10 femtoseconds or less) laser pulses to create, manipulate and image vibrational wavepackets is investigated. Quasi-classical modelling of the initial superposition through tunnel ionization, wavepacket modification by nonadiabatically altering the nuclear environment *via* the transition dipole and the Stark effect, and measuring the control outcome by fragmenting the molecule is detailed. The influence of the laser intensity on strong-field ultrafast wavepacket control is discussed in detail: by modifying the distribution of laser intensities imaged, we show that focal conditions can be created that give preference to this three-pulse technique above processes induced by the pulses alone. An experimental demonstration is presented, and the nuclear dynamics inferred by the quasi-classical model discussed. Finally, we present the results of a systematic investigation of a dual-control pulse scheme, indicating that single vibrational states should be observable with high fidelity, and the populated state defined by varying the arrival time of the two control pulses. The relevance of such strong-field coherent control methods to the manipulation of electron localization and attosecond science is discussed.

1 Introduction

Quantum mechanics describes internal molecular dynamics in terms of the amplitude and phase of vibrational wavepackets, which are coherent superpositions of states and contain all information about the associated populations and phases. In the energy or frequency domain, the quantum beating is observable by resonant photonic processes; in the temporal domain, this interference results in characteristic time-varying motion. Ultrafast laser systems generating near-infrared (NIR) pulses with durations of hundreds of femtoseconds allowed the first observation of such wavepackets[1,2] with significant applications in chemical dynamics, opening up the field of femtochemistry. Vibrational wavepackets have been observed in a range of systems, often initiated by optical pumping with an ultrafast laser pump pulse and observed by fragmenting the molecule with a similar probe or dump pulse. Recent advances have allowed such wavepacket motion to be resolved in individual molecules[3] through the application of single-molecule detection schemes.

---

[a]*Department of Physics, Swansea University, Swansea, SA2 8PP, UK. E-mail: w.a.bryan@swansea.ac.uk; Fax: +44 (0)1792 295324; Tel: +44 (0)1792 295301*
[b]*Centre for Plasma Physics, School of Mathematics and Physics, Queen's University Belfast, Belfast, BT7 1NN, UK*
[c]*Department of Physics and Astronomy, University College London, WC1E 6BT, UK*
[d]*STFC Rutherford Appleton Laboratory, Harwell Science and Innovation Campus, Didcot, Oxon, OX11 0QX, UK*

A range of coherent control strategies have been demonstrated whereby temporal or spectral shaping is applied to laser pulse to modify the launch or evolution of vibrational wavepacket motion, thus altering internal states of the molecular system under study.[4] Generally, such modifications require optical coupling between states, and complex spectral or temporal shaping is required to populate pre-defined final state.[5]

State-of-the-art ultrafast laser technology allows access to the few-cycle regime, where typically the near-infrared (NIR) photon energies correspond to a electric field period of around 2.7 fs. Ultrabroadband Ti:sapphire oscillators with dispersion control optics and chirped pulse amplification in temperature-controlled crystals produce (NIR) pulses with a duration of the order 20–30 fs over a bandwidth of >30 nm.[6] By spectrally broadening through self-phase modulation in a gas-filled hollow fibre,[7–9] the bandwidth is extended to hundreds of nanometers which allows subsequent compression durations as short as 3.3 fs[10] with few-mJ pulse energies and kHz repetition rates. Such pulses are basis for attosecond XUV pulse generation via high-order harmonic generation.[11]

The vibrational periods of the lightest and simplest molecules H_2^+ and D_2^+ are 13 and 20 fs respectively. NIR few-cycle pulses are therefore perfectly suited for imaging vibrational wavepackets as a pulse duration shorter than the vibrational period is readily achievable. Using an interferometrically stable pump–probe configuration and reflection focusing tight enough to generate an intensity of the order 10^{14} Wcm^{-2} allows ultrafast strong-field imaging of vibrational motion. The wavepacket is generated by tunnel ionization of the neutral molecules,[12,13] projecting the ground state wavefunction in the neutral molecule onto all vibrational states in the molecular ion. The ionization rate varies as a function of internuclear separation,[14] hence in long (i.e. >50 fs) laser pulses the subsequent ionization and fragmentation dynamics is dominated by enhanced ionization at large internuclear separation,[12] facilitated by dissociative wavepacket dispersion. In a few-cycle pulse, the wavepacket initiated by tunnel ionization is well localized at a small internuclear bond, a result of the favorable temporal conditions.

Following the launch of the vibrational wavepacket by a few-cycle pump, a similar strong-field probe pulse is applied to the H_2^+ or D_2^+ molecular ions, initiating fragmentation via photodissociation or Coulomb explosion.[15,16] Measuring the distribution of kinetic energy release allows the wavepacket shape to be imaged via the well-known potential energy surfaces. As is frequently the case in molecules, the vibrational wavepacket in H_2^+ or D_2^+ was predicted to dephase due to anharmonicity of the potential surface, and rephases or revives when all components interfere in phase.[17] Such vibrational wavepackets were observed in hydrogenic molecular ions by a number of groups: Cocke and co-workers observed the initial dephasing of hydrogenic wavepacket,[18] Moshammer, Ullrich and co-workers then studied D_2^+ observed dephasing and revival.[19] The authors and co-workers imaged similar behaviour in HD$^+$, observing dephasing and revival[20,21] and explored the coherence of a D_2^+ wavepacket by stretching the pump and probe pulses.[22]

At the high intensities employed in ultrafast studies of vibrational wavepackets using strong-field pulses, a significant rotational impulse is applied by the field-induced torque on the molecule.[23] The time the torque is applied for is far shorter than the characteristic period of rotational motion, hence such intense pulses launch a rotational wavepacket through impulsive alignment, generating a coherent superposition that continues to evolve once the initial pulse is turned off. Rotational wavepackets have been observed in hydrogenic molecules[24,25] by observing fragmentation which is modulated by the revival of the wavepacket, and are well described by recent theory.[26]

It might intuitively appear that the conditions associated with generating vibrational and rotational wavepackets are very similar if not identical, however recent investigations by the authors indicate that either a rotational wavepacket in D_2 or a vibrational wavepacket in D_2^+ is generated.[22,25] This is not to say that the processes are mutually exclusive, however in recent experiments there is no evidence for coherent rovibrational excitation. We suspect this is the result of interactions occurring in different volumes: at focus, the ionization rate is high, hence a vibrational wavepacket is generated before rotation gets underway. Off focus, the ionization rate is much lower hence impulsive alignment can cause rotation. Alternatively, the frequency resolution is limited so as not to be able to resolve rovibrational components which in the future could be rectified by extending the pump–probe delay to the picosecond timescale. As detailed in ref. 25, rotational and vibrational contributions can be isolated by bandpass filtering. Also, see ref. 48 for a recent review of studies of vibrational and rotational wavepackets in hydrogenic molecular ions.

Time- and frequency-domain measurements of wavepackets in molecules facilitated by techniques such as two-dimensional Fourier-transform spectroscopy[27] allow the inter- and intranuclear transfer of energy to be resolved on femtosecond and nanometer scales, which have proven to be short enough to resolve complex biochemical reactions. A very promising alternative has been proposed for small molecules exposed to strong-field few-cycle pulses: quantum beat imaging employs a pump–probe configuration identical to earlier systems, with the pump causing ionization to launch a (ro)vibrational wavepacket and the probe fragments *via* Coulomb explosion.[12] By measuring the pump–probe delay dependent kinetic energy release from D_2^+ and applying a filtered Fourier transform, a section of the laser-dressed potential energy surface has been recovered from the experimental wavepacket.[28] This proof-of-principle measurement led to the theoretical extension of quantum beat imaging to a range of pulse durations and intensities[29] and multiple spatial dimensions.[30–32] Recent experimental observations of vibrational wavepackets in electronically complex diatomics[33,34] will lead to this method being extended to larger systems. The real potential of this technique lies in the ability to resolve the laser-modified potential energy surfaces without prior knowledge.

As with coherent control investigations, once it has been established that a vibrational (or indeed rovibrational) wavepacket has been generated and imaged, it is natural to try to modify its evolution. Spectral shaping unavoidably leads to an increase in pulse duration, which defeats the purpose of employing few-cycle pulses. Recent theoretical predictions indicate a strong-field few-cycle pulse applied at the correct time can heavily perturb a bound electron orbital leading to the transfer of vibrational population as the nuclei adjust nonadiabatically to the rapidly-varying electronic environment.[35–37] This process can be treated as a dynamic Raman or Stark process.[38] In the former, multiphoton coupling between ground and excited states as wavepacket oscillates causes a bond length (hence time) dependent redistribution; the latter is a polarization of the molecular orbital by the dipole force leading to a time-varying distortion of potential surfaces. The nuclear wavepacket then propagates on the modified potential, and the diabaticity of the process causes population transfer. In both cases, changing intensity, wavelength and intensity of control pulse influences population transfer. Impressive experimental evidence for time-dependent manipulation of a vibrational wavepacket has been presented by Niikura and co-workers,[39,40] however only a limited portion of the time-series is reported.

In the rest of this paper, we briefly review a recent theoretical model which predicts the strong-field modification of a vibrational wavepacket. This efficient quasi-classical model allows the influence of the distribution of intensity within a focal volume to be explored. We then discuss in detail the experimental facilitation of control of a vibrational wavepacket in D_2^+. Finally, we extend the quasi-classical model to a dual-control pulse scheme, again investigating how changing relative delays and intensity allows range of vibrational states to be populated with varying fidelity. Finally, the prospect for manipulating electron localization as a nuclear wavepacket evolves under external control is discussed.

2 Theoretical model

The scheme for vibrational wavepacket control by a strong-field few-cycle pulse is sketched in Fig. 1, and is comparable to those discussed theoretically. The pump pulse initiates tunnel ionization of D_2, populating a coherent superposition of vibrational states in the electronic ground state ($1s\sigma_g$) of the D_2^+ molecular ion. The manipulation pulse which induces a dipole force is applied after some delay during which the vibrational wavepacket has evolved in time, as shown in Fig. 1b for the solution of the time-dependent Schrödinger equation and the equivalent quasi-classical trajectory model (QCM), Fig. 1c. Finally, the subsequent evolution of the modified wavepacket is mapped by a probe pulse, *via* photodissociation (PD) of the molecular ion.

In the QCM (see Bryan *et al.*[41] for details), the $1s\sigma$ $v = 0$ ground state of the neutral D_2 precursor is projected onto the available electronic states in D_2^+ ($1s\sigma_g$ ground, $2p\sigma_u$ dissociative) and $D^+ + D^+$ (Coulomb explosion) states by tunnel ionization[42] and the relative populations found by numerically solving the resulting coupled differential rate equations. Stable D_2^+ ions are generated over a narrow pump intensity range of $4 \times 10^{13} < I_{pump} < 1.1 \times 10^{14}$ Wcm^{-2}, and a range of vibrational states ($0 \leq v \leq 24$) populated. The wavepacket is approximated as a classical ensemble of particles moving on the $1s\sigma_g$ potential, and allowed to evolve in a Newtonian manner.

As the orbital configuration defines the potential surface constraining the nuclear motion, by polarizing the electrons with a strong field, the nuclear dynamics can be controlled. In the present work, the key mechanism is nonadiabatically altering the nuclear environment by coupling the electric field associated with the control pulse

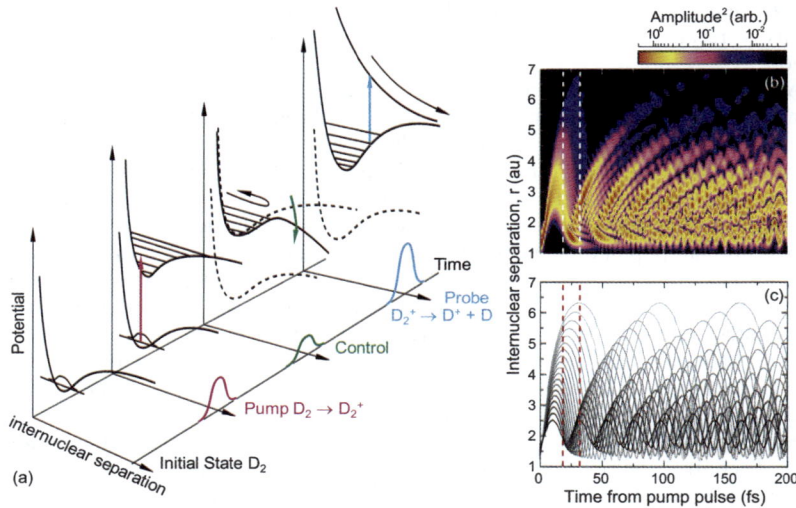

Fig. 1 Schematic of the pump–control–probe scheme and unperturbed vibrational wavepacket motion. (a) The pump pulse (intensity 4×10^{13} Wcm^{-2} to 1.1×10^{14} Wcm^{-2}) ionizes at room temperature $D_2 \rightarrow D_2^+$ creating a coherent superposition of states. Some time later the control pulse (intensity 1.3×10^{13} Wcm^{-2} to 3.7×10^{13} Wcm^{-2}) distorts the molecular potential energy surface *via* the polarization of the electronic orbital by the laser field, causing the wavepacket to rapidly adjust. The population redistribution and phase shift caused by the control pulse is imaged by photodissociating the $D_2^+ \rightarrow D^+ + D$ in the probe pulse (intensity 6×10^{13} Wcm^{-2} to 2×10^{14} Wcm^{-2}). (Right) The unperturbed vibrational wavepacket created in the D_2^+ molecule: the solution to the time-dependent Schrödinger equation (top, b) and current quasi-classical model[41] (bottom, c) are compared. Both exhibit the same return to the inner turning point, the region over which the control pulse is applied is indicated by vertical dashed lines.

to the nuclear motion *via* the transition dipole and the Stark effect. The former is a transient dipolar polarization created by the interaction of the electromagnetic field of the control pulse with the molecular orbital (most influential at small internuclear separations), the latter is the direct shifting and distortion of the molecular potential by the applied field (dominant at large internuclear separations). It is assumed that the mass of the nuclei makes the polarization of these positive particles negligible.

We model the time-varying modification of the classical ensemble trajectories during the control pulse, which adjust in amplitude, frequency (hence *v*-state) and phase offset. The control pulse is applied around the time of the first return of the wavepacket to its inner turning point, between the dashed lines on Fig. 1(b and c) as the wavepacket is still well localized at this time. Comparing the perturbed and unperturbed ensemble trajectories allows the final *v*-state populations and relative phases to be deduced. Rather than having to quantify the coupling between the $1s\sigma_g$ and $2p\sigma_u$ states during photodissociation in the probe pulse, we use the critical R-cutoff method, which has been proven to be a highly efficient approximation.[43] It is this predicted PD yield as a function of pump–probe and pump–control delay that will be compared directly to experimental yields. Following photodissociation which releases a kinetic energy dependent on the vibrational states populated, the fragment ions (D^+ and D) travel along the polarization direction of the probe, with the charged fragment recorded in an ion spectrometer.

A vital consideration for strong-field control is that any attempt to manipulate the wavepacket should not destroy either the coherence of the system or photodissociate the molecule. The photon energy and bandwidth of the control pulse could introduce complex inter-state couplings which must be quantified. For a spectrally well-defined control the manipulation operation should also be carefully tailored: readily ionized molecules are not able to withstand the same level of polarization by the field as a molecule with a high ionization potential, hence the intensity and duration of the control pulse is a key experimental parameter.

Fig. 2 illustrates the importance of having a well-quantified distribution of laser intensity and imaging a subsection of the focal volume. The Huygens–Fresnel diffraction integral is numerically solved for an experimental system, which has typical parameters of $\lambda = 790$ nm, hollow-fibre internal diameter 250 μm, recollimated 1 metre from exit, focused into the spectrometer with $f = 50$ mm spherical mirror, focusing mirror focal length $f = 50$ mm, $1/e^2$ beam radius $D = 1.5$ mm, total propagation distance from source to focal volume $L = 5$ m, and is comparable to a number of experimental systems world-wide. The laser beam profile on exit of our hollow fibre (see later) is reasonably described by the predicted EH_{11} hybrid mode. Propagation through our optical system unavoidably introduces diffraction which makes the focal volume non-Gaussian, the relevance of which to atomic ionization studies has been well documented,[44,45] hence is expected to be highly influential when experimentally realizing strong-field control. Taking $\log_{10}I(x,y,z)$ of the resulting three-dimensional intensity distribution, then calculating the histogram of relative volume contributions allows a straightforward application of the QCM results. This 'focal histogram' (Fig. 2a) also allows the relative radius of the PD-yield detector to be investigated.

The beam waist at focus is approximately 10 μm, generating a pump pulse peak intensity approaching 10^{15} Wcm^{-2}. If the ion detector radius is 1 mm, the full variation of intensity over the focal volume is seen, with significantly more volume at low intensities: over the intensity range of interest the focal histogram varies by two orders. As the radius of the detector is reduced, the subsection of the focal volume available exhibits a massive change in distribution of relative intensity. As the detector radius approaches 125 μm, the focal histogram is predicted to be constant over the intensity range of interest.

The result of operating the QCM at a peak pump intensity between 10^{13} Wcm^{-2} and 1.58×10^{14} Wcm^{-2}, a pump–control pulse intensity ratio of 3 : 1, and durations

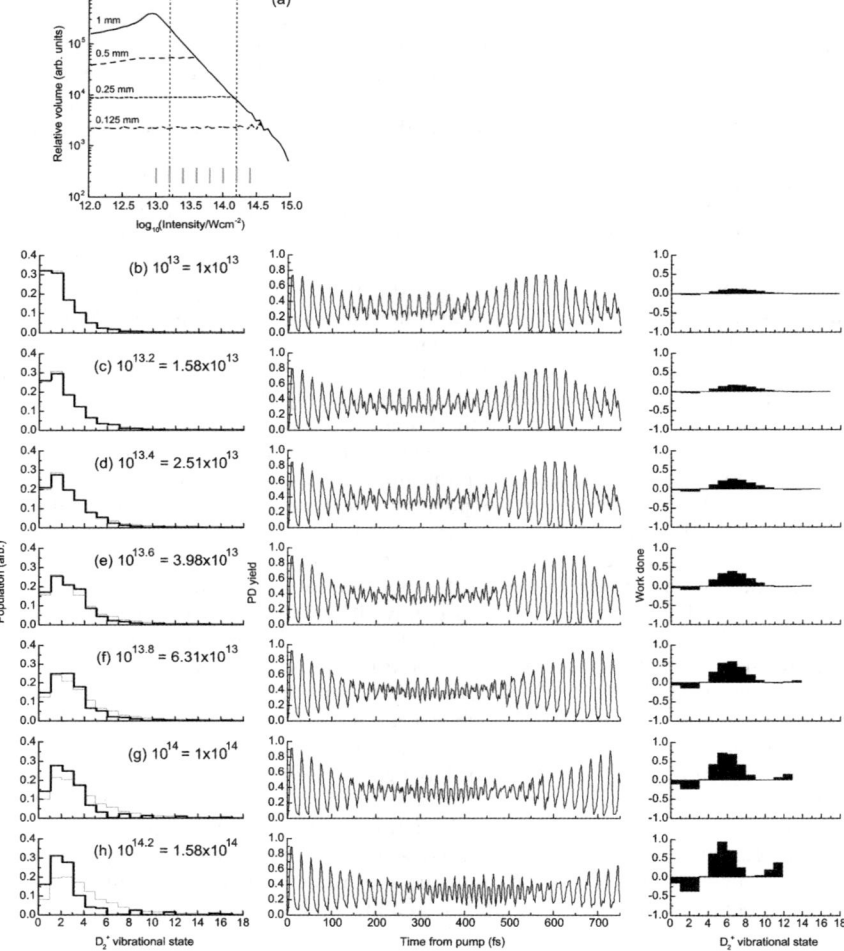

Fig. 2 (a) The focal volume of a typical ultrafast laser focus ($\lambda = 790$ nm, focal length $f = 50$ mm, $1/e^2$ beam radius $D = 1.5$ mm, total propagation distance from source to focal volume $L = 5$ m) characterized with a relative-volume histogram. The distribution of intensity is found by numerically solving the Huygens–Fresnel diffraction integral. By reducing the radius of the circular aperture through which the PD fragments are detected, a subsection of the focal volume is imaged. The red dashed lines indicate the intensity range over which stable D_2^+ ions are produced, and the green ticks correspond intensities for which the calculations are provided below. (b) To (h) demonstration of the QCM for a pump–control intensity ratio of 3 : 1, where the pump, control and probe have a duration of 6 fs. (Left) Initial (green line) and final (black line) vibrational population distributions for a varying pump pulse intensity. (Centre) Photodissociation yield as a function of pump–probe delay. (Right) Work done on or by the vibrational wavepacket as a function of vibrational state. The modification of the vibrational population can be seen to be the result of significant work done.

of 6 fs FWHM is presented in Fig. 2(b to h). The QCM returns the initial and final vibrational populations, predicted PD yield as a function of pump–probe delay and work done as a function of vibrational population. Note that the results are not normalized for relative volume. At low intensity, Fig. 2(b to e), there is minimal evidence for the influence of the control pulse, as the predicted PD yield is essentially that of the unperturbed wavepacket, *i.e.* a pump–probe configuration without a control pulse. As the intensity of the pump is further increased for the same pump–control ratio, Fig. 2(f to h), the initial population distribution is observed to shift to a higher v-state and the distribution of final vibrational states increasingly modified. There is also a clear shift in the predicted PD yield structure, and the work done or by the control pulse increases. If an experimental measurement was attempted with a 2 mm detector, the essentiality unperturbed wavepacket motion shown in Fig. 2(b to e) would swamp the control-modified motion in Fig. 2(f to h), and would most likely be unobserved.

By reducing the diameter of the detector, the focal volume imaging is altered, as evidenced from the focal histogram modification. As the detector diameter is reduced, the subset of the focal volume imaged shifts in favour of the higher intensities. As the pump–control intensities are set by the relative energies of the two pulses, a more prominent variation of the predicted photodissociation yield is seen as more work is done on the wavepacket. At a detector radius of 125 µm, the intensity histogram is flat, and as we've taken $\log_{10} I(x,y,z)$, this results in a heavy bias in favor of highest intensity. As demonstrated in the following section, this makes strong-field control of the vibrational wavepacket experimentally observable. Furthermore, as the photodissociation of H_2^+ and D_2^+ is strongly peaked around the probe polarization direction,[12] we further increase our ability to observe the influence of the control pulse.

3 Experimental demonstration

The recently published first demonstration of strong-field control of a vibrational wavepacket (see Bryan *et al.*[46]) requires three precisely-timed intense laser pulses to interact with an hydrogenic molecule in the gas phase. Linearly polarized pulses from a Ti:sapphire FemtoLasers CompactPro HP (30 fs, 800 nm, 1 mJ at 1 kHz) were focused into a $\sim$1 m long, 250 µm internal diameter hollow fibre with a pressure gradient of argon ($\sim 10^{-5}$ to 2 bar), increasing the bandwidth from 30 nm FWHM to $\sim$140 nm. Eight chirped mirror reflections minimized the group-delay dispersion (GDD) producing a 10.2 fs pulse as measured in an all-reflective FROG with an error of $G = 0.009$.

The pump, control and probe pulses are interferometrically derived from the output of the hollow fibre as illustrated in Fig. 3; all beam splitters (Femtolasers low dispersion parts OA135, OA037 and OA200) introduce minimal GDD. The input energy into the nested Mach–Zehnder interferometer was 240 µJ; following optical losses, the P:C:P pulse energies were 32 : 11 : 35 µJ, polarized perpendicular:parallel:parallel with respect to the ion spectrometer axis. The relative delay between the three pulses is independently controllable with two high-precision translation stages (Newport MFA-CC). To establish P:C:P pulse synchronization, an autocorrelation signal was monitored as the P-C and P-P delays are independently scanned and temporal satellites observed from P-C and P-P cross correlation. We estimate the temporal overlap (hence P-C-P delay) uncertainty is $\sim$300 attoseconds, derived from the translation state resolution and reproducibility and pixel size in the autocorrelator CCD camera.

An effusive jet of room temperature D_2 is introduced into the interaction region of an ion time-of-flight mass spectrometer (TOFMS)[20,22,25,46–48] containing an aperture of radius 125 µm (cf Fig. 2), which essentially reduces this to a 1D problem given the prevalence for photodissociation along the laser polarization direction, hence the spectrometer is only sensitive to molecules aligned to within a few degrees of the probe polarization direction. Following precompensated transmission through a fused silica window, the P-C-P pulse sequence was reflection focused into the interaction region of the TOFMS by an spherical $f = 50$ mm silver mirror. A room temperature beam of D_2 was crossed effusively with the laser focus, and the generated D_2^+ and D^+ ions electrostatically separated; the kinetic energy of fragmentation was also measured for the D^+ ions. Fig. 4 shows the focal histogram for the

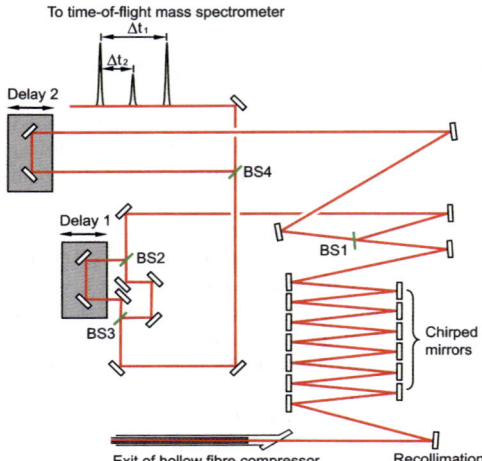

Fig. 3 Schematic of the ultrafast pulse manipulation allowing the observation of strong-field vibrational wavepacket control. Two nested Mach–Zehnder interferometers constructed using low dispersion silver mirrors and thin dielectric beamsplitters produce the pump–control–probe sequence with independently variable delays.

experimental set-up, confirmed by measuring the spatial distribution of atomic ionization in argon;[44] differences between Fig. 2a and 4 are due to the finite diameter of the optics and vacuum entrance window being considered sequentially. This distribution of intensities is employed when integrating the QCM over the focal volume, thus allowing a very accurate comparison with the measured photodissociation yields. Ions arriving at the end of the TOFMS drift tube struck a pair of microchannel plates, and the resulting electron cascade collected on a solid anode. The induced voltage was monitored by a digital phosphor oscilloscope (DPO, Tektronix DPO-7254B), sampled at 2.5 GS/s; the DPO also drives both delay lines.

Following data collection, the D^+ fragmentation yield was integrated over 0–1 eV, the full range of photodissociation energies.[12] A Butterworth bandpass filter was applied to the PD yield as a function of pump–probe delay to suppress high-frequency noise and the low-frequency contribution from the D_2 rotational wavepackets.[22,25] The parameters of this filter (high and low pass frequencies and sampling frequency) were optimized by superimposing two representative sine waves of varying frequencies and adding random noise of variable amplitude, then filtering to recover the higher frequency wave. While there is a frequency overlap at the extremes of the vibrational (16 to 47 THz, peaking at 42 THz) and rotational (5 to 20 THz, peaking at 12 THz) spectra, the asymmetric distribution of population

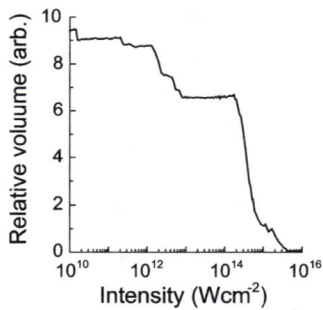

Fig. 4 Focal histogram for experimental demonstration.

across the vibrational and rotational states[25] allowed the Butterworth filter to accurately recover the high frequency component even with 100% noise modulation.

The post-filtering PD yields associated with the evolution of the vibrational wavepacket as a function of probe delay for different control delays are presented in Fig. 5, where the control and probe delays are always referenced to the pump pulse at $t = 0$. As the control delay is varied between 18 fs and 32 fs, the experimental PD yield exhibits subtle but statistically significant variations: the oscillatory structure is observed to increase in amplitude, cycle-averaged yield level and periodicity. Qualitatively, this demonstrates that the wavepacket amplitude in the higher vibrational states has increased, indicative that the molecular orbital distortion by the control pulse is redistributing vibrational population. A variation of oscillation shape is also observed, from saw-tooth (18 fs) to more sinusoidal (22 fs) to almost plateau (28 fs), finally to large amplitude features well-separated from $t = 0$ (32 fs). Considering that the average period of D_2^+ is around 24 fs, such rapid variations in structure reveals that shifting the P-C delay by 2 fs steps dramatically changes the bound wavepacket.

After isolating the vibrational signatures in the experimental data, the volume-integrated QCM results are overlapped. The similarities in shape, amplitude, cycle-average yield offset and periodicity of the theoretical prediction and measured PD yield indicates the coherence of the vibrational wavepacket is retained throughout the control pulse. This is the first experimental demonstration of not only the subsequent evolution of a modified D_2^+ vibrational wavepacket in an ultrashort strong-field pulse, but also the first evidence that the vibrational population and phases of the superposition can be modified in the ultrafast strong-field regime in a predictable manner.

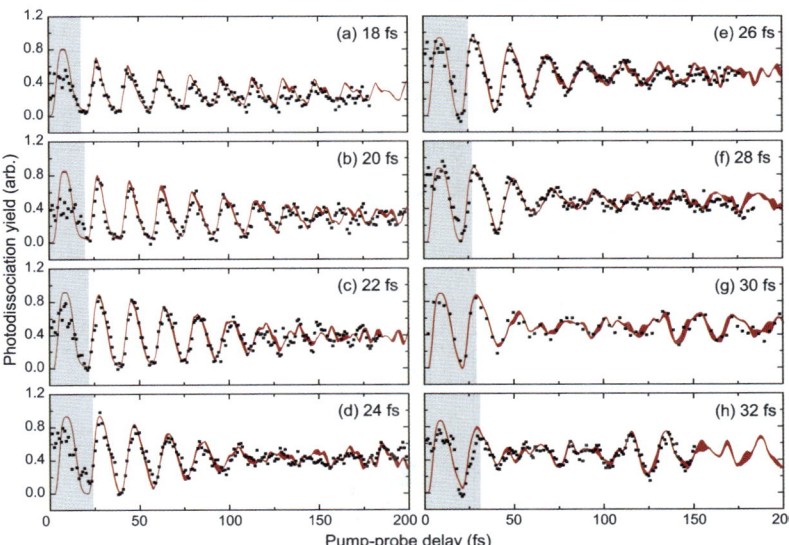

Fig. 5 Experimentally measured photodissociation yield[46] and predicted focal–volume–integrated yield derived from the quasi-classical model.[41] The agreement between theory (red line) and experiment (black squares) demonstrates that the control pulse is manipulating the wavepacket evolution in a quantifiable manner and the molecular PES returns to the field-free state following the control pulse. The varying vertical thickness of the predicted yield indicates the uncertainty in fitting the experimental results. The grey shading indicates the presence of the control pulse, hence the QCM is not expected to describe the experimental data accurately in this region.

The vibrational populations corresponding to the best-fit signal-volume-integrated PD yields (Fig. 5) are presented in Fig. 6. It should be noted that these results are not derived directly from the data presented in Fig. 5, rather are simulations using the known experimental parameters, hence the phrase 'predicted' is employed. Sources of uncertainty are defining the zero delay time (estimated as 300 as from a linear delay calibration) and the range of active intensities (estimated as better than 8×10^{12} Wcm^{-2} for all three pulses). These uncertainties are combined in the QCM to produce upper and lower bounds indicated by the thickness of the theoretical prediction. Clearly, uncertainty in intensity and either delay will result in a significant variation of the QCM outcome, indicated in Fig. 6.

The initial vibrational population caused by the pump pulse is indicated (green line). The final vibrational population distribution (black line) is observed to vary significantly as the control delay is scanned. When the control pulse is applied at 18 fs, the final state distribution is driven coherently towards the lowest v-states: $v = 0$–3 contain almost all the population. As the control delay is increased, the lowest v-states are depopulated relative to the 18 fs distribution, particularly the $v = 0, 1$ states. At a control delay of around 24 to 26 fs a significant population exists in v

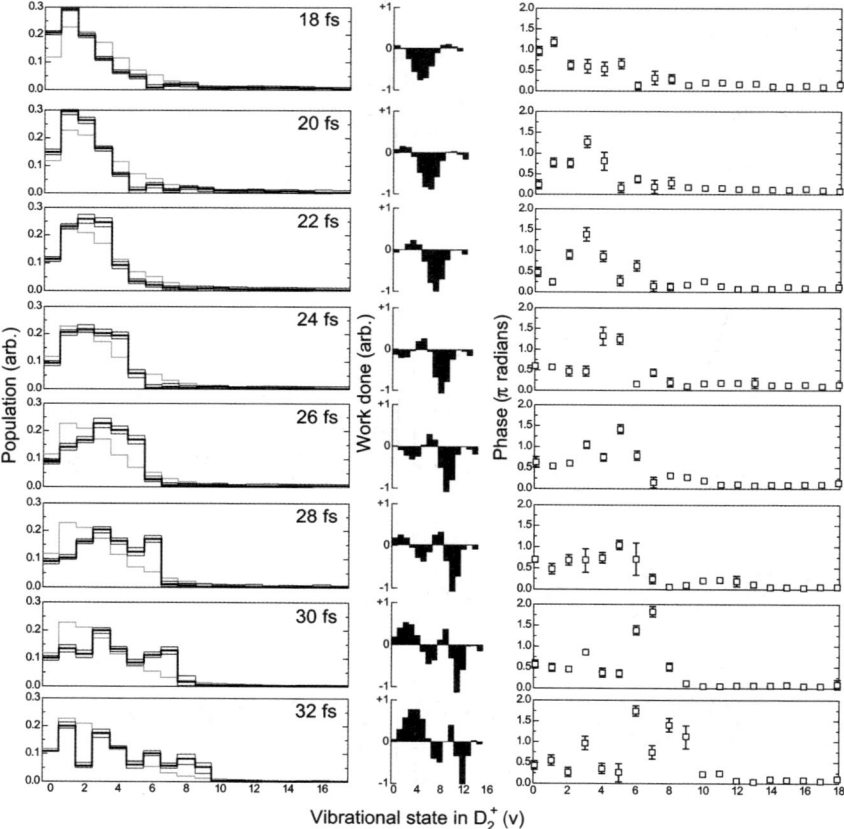

Fig. 6 The best-fit predictions of vibrational population, work done on or by the control pulse, and classical phase distributions as the pump–control delay is varied. As the control delay is sequentially changed from 18 to 32 fs, a significant shifting in the vibrational population (black) with respect to the initial distribution of states (green) is observed, and is a result of the work done. The largest change in classical phase follows the maximal population shift. The uncertainty in population (grey bars) and classical phase (small markers) is derived from the uncertainty when fitting the experimental PD yield (Fig. 5).

= 3 and 4, and this population shoulder is observed to move to a higher v-state as the control delay increases up to ~30 fs, at which point the distribution of v-states becomes structured. At a control delay of greater than 26 fs, the lowest v-states are depopulated with respect to the initial conditions. The shifting of the vibrational population from the low to high v-states as the control delay is increased is compatible with our earlier qualitative discussions of the periodicity and oscillatory structure.

By integrating the product of the force applied and distance travelled by each ensemble element during the control pulse, the work done on the wavepacket can be predicted as a function of vibrational state. As shown in Fig. 6, at a P:C delay of 18 fs, the majority of vibrational states work on the field-modified potential, which is responsible for the skewing of the population distribution to lower v. As the P:C delay is increased, the distribution of work done shifts to higher v, as the control pulse acts more efficiently on more highly lying v-states. For P:C delays above 26 fs, the work done is seen to oscillate rapidly which shifting to higher vibrational state, causing the evolution in the structure of the vibrational population.

The phases of the QCM trajectories are distorted by the control pulse; while this classical phase is not strictly a quantum phase, it is an intuitive indicator of the ensemble motion. For all final v-states in D_2^+, the distorted trajectories are compared with the unperturbed trajectories (Fig. 1c), and a population-weighted mean calculated, as presented in Fig. 6. The maximum phase distortion as a function of final vibrational state follows the maximum population change, with phase changes of the order π radians exhibited. Importantly, a large phase change is possible irrespective of the state population, opening the possibility for storing quantum information in the vibrational phase.

Fig. 7 shows volume-integrated vibrational population and phase matrices illustrating the redistributive action of the control pulse. During the operation of the QCM, all perturbed trajectories are compared to the unperturbed trajectories, resulting in an $n \times n$ matrix between 0 and 1 indicating the best fit to the final v-state and an $n \times n$ phase matrix between 0 and 2π. In both cases, n is the number of vibrational states considered, here $n = 24$. The transfer function is converted to a vibrational population matrix by scaling by each row by the initial vibrational population. Were no control pulse present, the population matrices would be a diagonal line; population above the diagonal is up-shifted in vibrational state. Depending on the control delay and v-state, up- and down-shifting is clear: summing the matrix horizontally results in the initial distribution of states; summing vertically generates Fig. 6. The asymptotic tendency at high v is a result of the strongest influence of the control pulse at the largest internuclear separations. The empty area at a final state of $v > 13$ indicates that all trajectories in these states photodissociate. The opposite effect is responsible for the lowest-lying v-states being minimally influenced. A population above 10^{-5} is the threshold for generating the phase matrices. Note that the phase is represented in terms of fractions of an oscillation in the final v-state.

4 Modelling multiple control pulses

To demonstrate the feasibility of more extensive control of the vibrational state of a molecule with strong-field pulses, we now extend the QCM to two control pulses with independent delays. The proposed technique is distinct from traditional coherent control methods as, rather than applying one shaped pulse, we propose that multiple few-cycle impulsive actions on the wavepacket at well-timed intervals can achieve a useful fidelity of final state. This approach has been discussed for solution to the TDSE for unique double control pulse scenario,[36] however here we further the state of the art by systematically investigating the control landscape. Such a study is a demonstration of the efficiency of the QCM, as solving the TDSE repeatedly would take a prohibitive long time.

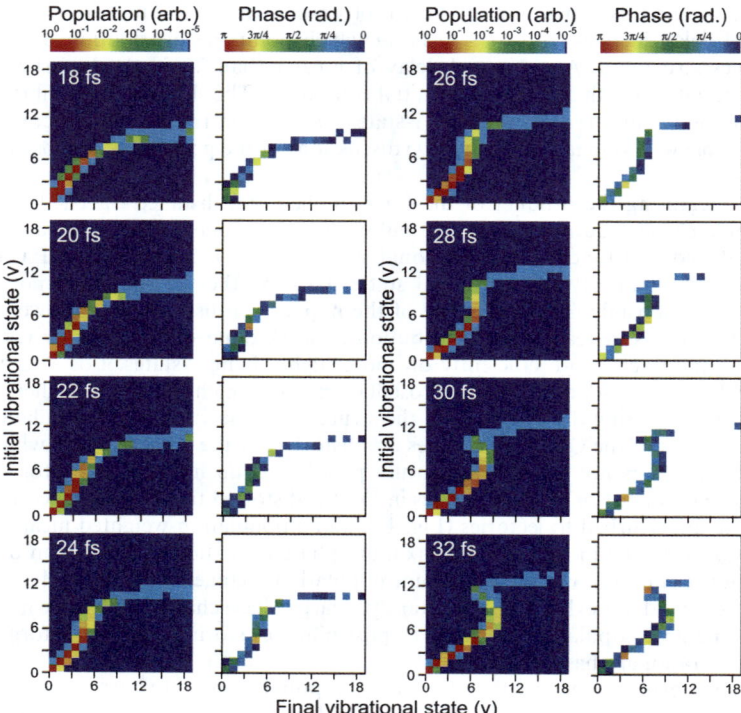

Fig. 7 Volume-integrated vibrational population and phase matrices as the pump–control delay is scanned. The QCM compares all initial and final trajectories allowing arbitrary transfer of population. The amplitude of the initial state is defined by the tunnel ionization step; when a match to the trajectory of the final state is found, this population is transferred to the final state. To facilitate a best-fit, it is necessary to include a phase offset, summarized in the phase matrix. This process is repeated across the focal histogram then summed vertically to produce Fig. 6.

The QCM is readily modified to include an additional control pulse as the ensemble trajectory motion is modelled at each time-step, allowing a control field of arbitrary complexity. As with the single-control case, the QCM is run out to 750 fs in 50 as steps, and returns the photodissociation yield, phase and population matrices and work done as a function of vibrational state. Operating the QCM 20 000 times requires 42 h on a standard-performance PC (Intel 2.4 GHz Core 2 Duo P8600, 4 GB RAM, Microsoft Windows 7). Scaling to more complex molecules will require an increase in computing power: each populated electronic state will have to be included, and for tri- and polyatomics, the QCM will have to be extended to cover each active degree of freedom. If resonant control pulses are applied, interstate coupling will have to be introduced and a more complete representation of rovibrational dynamics is necessary. Finally, as discussed earlier, volume integration is clearly necessary. While such modelling is significantly more advanced than the current discussion, the problem is perfectly suited to parallelization methods.

Experimentally, deriving additional control pulses from one ultrafast pulse is nontrivial as repeated interferometric splitting of the laser output leads to significant loses, leading to a drop in the peak intensities available. An alternative to additional nested interferometers is the use of a focusing optic divided into independently movable annuli, however for large delays, the spatial overlap of the resulting focal volumes would degrade. Furthermore, such optics introduce significant diffraction,[44,45] however the resulting distortion to the pulse wavefronts may compensate for focal walk-off. Another method for introducing independent delays is the azimuthal rotation of glass plates; again, by dividing into annuli, different spatial elements of the pulse can be delayed but this introduces a delay-dependent group-delay dispersion, temporally distorting the pulse. These difficulties could potentially be overcome through the use of a noncollinear geometry, whereby the pump and probe are derived from one interferometer and multiple control pulses from another. While the requirement for tight focusing may make noncollinear propagation difficult, the necessity for the detector to image a restricted section of the focal volume implies that even tight (tens of micron) waists could be overlapped at angles of tens of degrees and still form a common interaction region, and could be improved with diffractive shaping of the foci.

The result of applying dual control pulses (C1 and C2) to the vibrational wavepacket in D_2^+ is presented in Fig. 8. The delay between the pump ($t = 0$) and probe

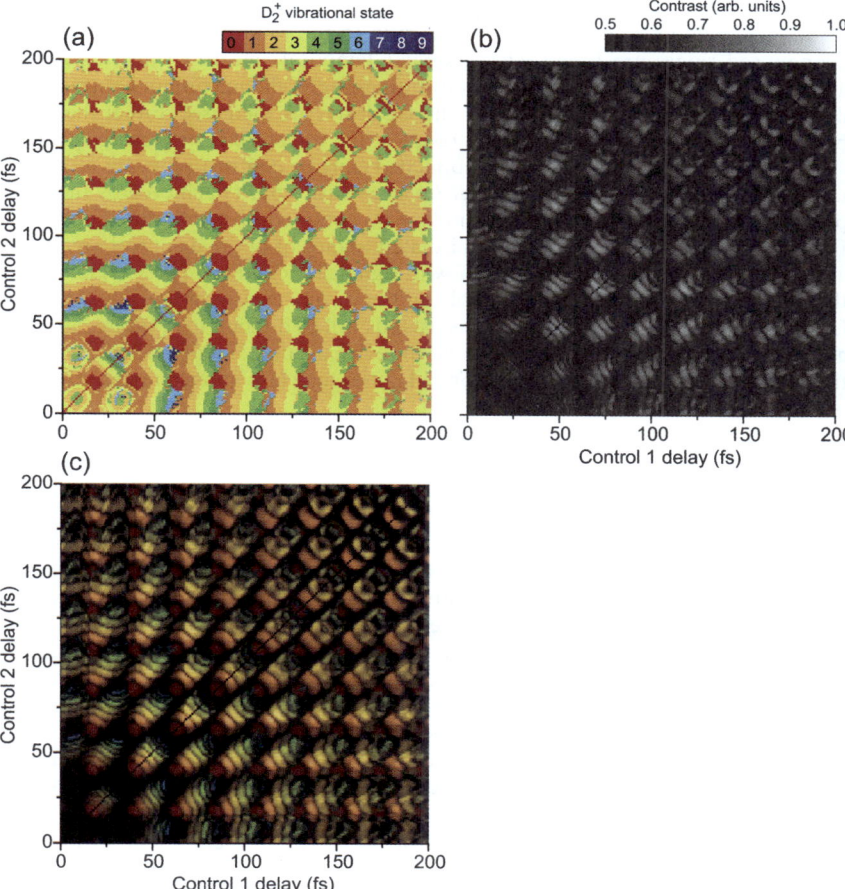

Fig. 8 Modelling the outcome of applying two control pulses to the D_2^+ ensemble. Pump intensity $= 10^{14}$ Wcm^{-2}, pump–control intensity ratio $= 3:1$, duration of pump, control and probe $= 6$ fs. (a) Colour map of the most populated vibrational state as control pulses C1 and C2 are scanned from 0 to 200 fs. As pulses C1 and C2 are identical, the region below the diagonal is reflected. (b) Vibrational state contrast, C as a greyscale map, where C = (pop$_{max}$ − pop$_{min}$)/(pop$_{max}$ + pop$_{min}$), and pop$_{min}$ is the mean of the remaining populations not equal to pop$_{max}$. (c) Most populated state and contrast map overlayed to illustrate the final state fidelity that can be achieved, so the ridged colour indicates vibrational state and the luminosity indicates the purity of state.

is scanned for $0 \leq$ (C1, C2) ≤ 200 fs in 1 fs steps. A pump intensity of 10^{14} Wcm^{-2} is employed to launch the wavepacket and a pump–control intensity ratio of 3 : 1 is defined for both C1 and C2. To elucidate the transfer of population, the most populated vibrational state as a function of C1, C2 delay is presented as a colour map in Fig. 8a. A regular modulation of the most populated state is predicted: at small temporal separations from $t = 0$, vibrational states up to $v = 9$ are populated with a periodicity defined by the average period of oscillation of the D_2^+ molecular ion. At larger delays at constant C1 or C2 delay, an overlap of the repetitive structure is found, the result of the control pulses acting on a more spatially dispersed wavepacket. This effect is magnified along the C1 $\simeq$ C2 diagonal, resulting in a suppressed level of control.

Clearly, the purity of the vibrational population is of interest to coherent control applications and the most populated state is only part of the story, hence the contrast of the population as a function of C1 and C2 is presented in Fig. 8b. The contrast is calculated by taking the difference of the peak population to the average of other populations and calculating the ratio to the total population. Groups of pronounced ridges are found with a similar temporal smearing effect as the C1 and C2 delays are increased from $t = 0$. Interestingly, the maximum contrast only degrades by a small amount as the C1 and C2 delays are increased, rather the ridges blur into each other, which is again a result of the wavepacket dispersing spatially.

Taken in isolation, the maximum populated state or contrast plots are of limited use. By overlaying the two results (Fig. 8c), the relative purity achieved by the dual-pulse control scheme is revealed. Around $t = 0$, a poor control outcome is observed. For C1 or C2 delays up to 100 fs, well isolated islands of optimal control are found, allowing single state access up to $v = 7$ with significant contrast. These regions offer the best chance of experimentally resolving the wavepacket modification. At C1 or C2 delays above 100 fs, the high contrast islands will be difficult to separate, however as the unperturbed wavepacket is known to revive around 580 fs, further investigation is required. Nonequal intensity control pulses may also allow an interesting mix of final states to be populated.

As demonstrated in the single control pulse theoretical and experimental results, variation of the laser intensity has a dramatic influence on the observability of the control operation. In Fig. 9, we demonstrate the same is even more so the case for the dual-control pulse scheme. For a fixed pump intensity of 10^{14} Wcm^{-2}, the pump–control intensity is varied from 3 : 1 (as in Fig. 8) to 2 : 1 and 1 : 1 and a subsection of the C1–C2 variation landscape is presented where the clearest manipulation is found in Fig. 8. A pump–control ratio of 2 : 1 significantly improves the contrast of the final state populations over a ratio of 3 : 1, particularly for the lowest lying states. This is the result of the more intense control pulse distorting the potential more severely, thus driving all states including the lowest lying. Despite the increase in control intensity from 3.33×10^{13} to 5×10^{13} Wcm^{-2}, the high-contrast maximal population ridges do not shift significantly with C1 or C2 delay.

As the pump–control ratio is further increased to 1 : 1 the population and contrast of the $v = 0$ state is observed to be enhanced even further than the 2 : 1 or 3 : 1 cases, however the disruption of the ensemble is now so large that all other vibrational states are erratically populated. The loss of the regular structure seen in the 2 : 1 and 3 : 1 cases is therefore indicative of the upper useful limit of the dual-control scheme. To further improve the contrast or to populate a pre-defined distribution of states will require additional control pulses.

5 Conclusions and outlook

We have demonstrated that controlling the motion of a bound vibrational wavepacket in D_2^+ by an ultrashort control pulse can be experimentally implemented and quantified. The redistribution of vibrational population can be recovered using a relatively simple quasi-classical model that incorporates tunnel ionization and

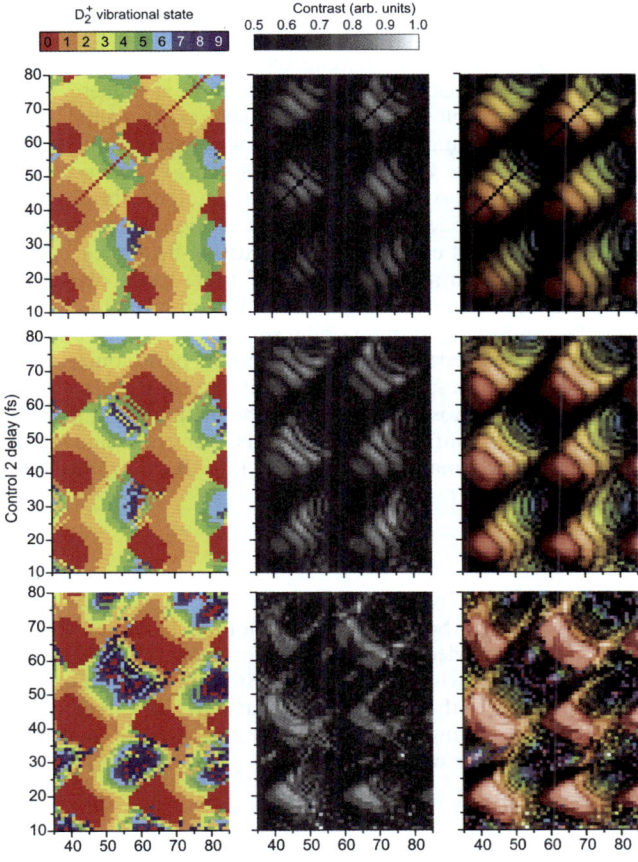

Fig. 9 Section of the two control pulse landscape as the pump–control ratio is varied from 3 : 1 (top), 2 : 1 (middle) and 1 : 1 (bottom). The pump intensity = 10^{14} Wcm^{-2} and the duration of pump, control and probe = 6 fs. The most populated vibrational state, contrast and final state fidelity as in Fig. 8.

dynamic Stark-shift deformation of the potential surface. The simplicity of this model allows it to be integrated over the focal volume, making realistic comparisons with experiment possible. While this model is approximate, it demonstrates the validity of applying a strong-field treatments to a simple system. The sensitivity of the control process to the intensity of all pulses employed is explored, and the requirement for accurate quantification of the focal intensity distribution is established.

A novel application of the quasi-classical model has been presented, allowing a systematic study of the application of two intense few-cycle control pulses. High fidelity population transfer to individual vibrational states is predicted, and following earlier discussions of experimental feasibility, we demonstrate how the range of available states and the transfer contrast depends heavily on the intensity of the control pulses, establishing an upper limit. With access to more computational power, this systematic approach could be improved to search for an optimal outcome to a pre-defined final state using genetic algorithms, with applications in quantum information.[49–51]

The manipulation of a vibrational wavepacket by a strong-field control pulse has very interesting applications to attosecond science. It has recently been demonstrated that the localization of the electron in D_2^+ can be externally manipulated by applying a carefully defined light field.[52–54] By varying the relative phase of the carrier and envelope of a few-cycle pulse (referred to as the carrier-envelope phase, CEP), the electron is observed to be driven from one nuclei to the other. This manipulation of the electron wavepacket is evidenced from the asymmetry of the photodissociation or Coulomb explosion process.[55] Such experimental demonstrations naturally point to controlling the vibrational wavepacket while simultaneously driving the electron motion: Calvert et al.[56] theoretically showed that significant asymmetry should be observable while modifying the vibrational population. Such methods allowing the nuclear and electronic motions to be selectively directed, allowing additional coherent control routes for strong-field science.

Rather than using the controlled light field of a few-cycle pulse, it has also been shown recently that attosecond XUV pulses can be employed to launch vibrational wavepackets in D_2^+, improving the temporal resolution by around an order of magnitude.[57] Launching the wavepacket by single-photon photoionization rather than tunnel ionization allows the populated rovibrational and electronic states to be selected to some degree. Attosecond pump–probe methods will make a significant contribution to coherent control, as coupling the electronic and nuclear motions allow dynamic wavepacket manipulation beyond the Born–Oppenheimer approximation.

Acknowledgements

This work was supported by the Engineering and Physical Sciences Research Council (EPSRC) and the Science and Technology Facilities Council (STFC), UK. CRC and RBK acknowledge financial support from the Department of Education and Learning (NI). We wish to thank Chris Froud, Edmond Turcu, Cephise Cacho and Emma Springate of the Artemis Laser Facility, STFC Rutherford Appleton Laboratory, UK for operating the laser system employed to generate the experimental measurements in this work.

References

1 H. Zewail, *Science*, 1988, **242**, 1645–1653.
2 M. J. Rosker, T. S. Rose and A. H. Zewail, *Chem. Phys. Lett.*, 1988, **146**, 175–179.
3 D. Brinks, F. D. Stefani, F. Kulzer, R. Hildner, T. H. Taminiau, Y. Avlasevich, K. Müllen and N. F. van Hulst, *Nature*, 2010, **465**, 905–908.
4 K. Ohmori, *Annual Review of Physical Chemistry*, 2009, **60**, 487–511.
5 C. Brif, R. Chakrabarti and H. Rabitz, *New J. Phys.*, 2010, **12**, 075008.
6 S. Backus, C. G. D. III, G. Mourou, H. C. Kapteyn and M. M. Murnane, *Opt. Lett.*, 1997, **22**, 1256–1258.
7 M. Nisoli, S. de Silvestri and O. Svelto, *Appl. Phys. Lett.*, 1996, **68**, 2793–2795.
8 M. Nisoli, S. de Silvestri, O. Svelto, R. Szipöcs, K. Ferencz, C. Spielmann, S. Sartania and F. Krausz, *Opt. Lett.*, 1997, **22**, 522–524.
9 M. Nisoli, S. Stagira, S. de Silvestri, O. Svelto, S. Sartania, Z. Cheng, M. Lenzner, C. Spielmann and F. Krausz, *Applied Physics B: Lasers and Optics*, 1997, **65**, 189–196.
10 A. L. Cavalieri, E. Goulielmakis, B. Horvath, W. Helml, M. Schultze, M. Fieβ, V. Pervak, L. Veisz, V. S. Yakovlev, M. Uiberacker, A. Apolonski, F. Krausz and R. Kienberger, *New J. Phys.*, 2007, **9**, 242.
11 F. Krausz and M. Ivanov, *Reviews of Modern Physics*, 2009, **81**, 163–234.
12 J. H. Posthumus, *Reports on Progress in Physics*, 2004, **67**, 623.
13 T. K. Kjeldsen and L. B. Madsen, *Phys. Rev. Lett.*, 2005, **95**, 073004.
14 A. Saenz, *J. Phys. B*, 2000, **33**, 4365.
15 I. D. Williams, P. McKenna, B. Srigengan, I. M. G. Johnston, W. A. Bryan, J. H. Sanderson, A. El-Zein, T. R. J. Goodworth, W. R. Newell, P. F. Taday and A. J. Langley, *J. Phys. B*, 2000, **33**, 2743–2752.
16 K. Sändig, H. Figger and T. W. Hänsch, *Phys. Rev. Lett.*, 2000, **85**, 4876–4879.
17 B. Feuerstein and U. Thumm, *Phys. Rev. A*, 2003, **67**, 063408.
18 A. S. Alnaser, B. Ulrich, X. M. Tong, I. V. Litvinyuk, C. M. Maharjan, P. Ranitovic, T. Osipov, R. Ali, S. Ghimire, Z. Chang, C. D. Lin and C. L. Cocke, *Phys. Rev. A*, 2005, **72**, 030702.

19 T. Ergler, A. Rudenko, B. Feuerstein, K. Zrost, C. D. Schröter, R. Moshammer and J. Ullrich, *Phys. Rev. Lett.*, 2006, **97**, 193001.
20 J. McKenna, W. A. Bryan, C. R. Calvert, E. M. L. English, J. Wood, D. S. Murphy, I. C. E. Turcu, J. M. Smith, K. G. Ertel, O. Chekhlov, E. J. Divall, J. F. McCann, W. R. Newell and I. D. Williams, *J. Mod. Opt.*, 2007, **54**, 1127–1138.
21 D. S. Murphy, J. McKenna, C. R. Calvert, W. A. Bryan, E. M. L. English, J. Wood, I. C. E. Turcu, W. R. Newell, I. D. Williams and J. F. McCann, *J. Phys. B*, 2007, **40**, 359.
22 W. A. Bryan, J. McKenna, E. M. L. English, J. Wood, C. R. Calvert, R. Torres, D. S. Murphy, I. C. E. Turcu, J. L. Collier, J. F. McCann, I. D. Williams and W. R. Newell, *Phys. Rev. A*, 2007, **76**, 053402.
23 H. Stapelfeldt and T. Seideman, *Rev. Mod. Phys.*, 2003, **75**, 543–557.
24 F. Lee, K. F. Légaré, D. M. Villeneuve and P. B. Corkum, *Journal of Physics B: Atomic, Molecular and Optical Physics*, 2006, **39**, 4081.
25 W. A. Bryan, E. M. L. English, J. McKenna, J. Wood, C. R. Calvert, I. C. E. Turcu, R. Torres, J. L. Collier, I. D. Williams and W. R. Newell, *Phys. Rev. A*, 2007, **76**, 023414.
26 R. Torres, R. de Nalda and J. P. Marangos, *Phys. Rev. A*, 2005, **72**, 023420.
27 D. M. Jonas, *Annual Review of Physical Chemistry*, 2003, **54**, 425–463.
28 B. Feuerstein, T. Ergler, A. Rudenko, K. Zrost, C. D. Schröter, R. Moshammer, J. Ullrich, T. Niederhausen and U. Thumm, *Phys. Rev. Lett.*, 2007, **99**, 153002.
29 U. Thumm, T. Niederhausen and B. Feuerstein, *Phys. Rev. A*, 2008, **77**, 063401.
30 M. Winter, R. Schmidt and U. Thumm, *Phys. Rev. A*, 2009, **80**, 031401.
31 M. Winter, R. Schmidt and U. Thumm, *New J. Phys.*, 2010, **12**, 023020.
32 M. Magrakvelidze, F. He, T. Niederhausen, I. V. Litvinyuk and U. Thumm, *Phys. Rev. A*, 2009, **79**, 033410.
33 S. De, I. A. Bocharova, M. Magrakvelidze, D. Ray, W. Cao, B. Bergues, U. Thumm, M. F. Kling, I. V. Litvinyuk and C. L. Cocke, *Phys. Rev. A*, 2010, **82**, 013408.
34 I. A. Bocharova, A. S. Alnaser, U. Thumm, T. Niederhausen, D. Ray, C. L. Cocke and I. V. Litvinyuk, *Phys. Rev. A*, 2011, **83**, 013417.
35 H. Niikura, D. M. Villeneuve and P. B. Corkum, *Phys. Rev. Lett.*, 2004, **92**, 133002.
36 T. Niederhausen and U. Thumm, *Phys. Rev. A*, 2008, **77**, 013407.
37 D. S. Murphy, J. McKenna, C. R. Calvert, I. D. Williams and J. F. McCann, *New Journal of Physics*, 2007, **9**, 260.
38 C. R. Calvert, T. Birkeland, R. B. King, I. D. Williams and J. F. McCann, *Journal of Physics B: Atomic, Molecular and Optical Physics*, 2008, **41**, 205504.
39 H. Niikura, P. B. Corkum and D. M. Villeneuve, *Phys. Rev. Lett.*, 2003, **90**, 203601.
40 H. Niikura, D. M. Villeneuve and P. B. Corkum, *Phys. Rev. A*, 2006, **73**, 021402.
41 W. A. Bryan, C. R. Calvert, R. B. King, G. R. A. J. Nemeth, J. B. Greenwood, I. D. Williams and W. R. Newell, *New J. Phys.*, 2010, **12**, 073019.
42 G. L. Yudin and M. Y. Ivanov, *Phys. Rev. A*, 2001, **64**, 013409.
43 C. R. Calvert, R. B. King, T. Birkeland, J. D. Alexander, J. B. Greenwood, W. A. Bryan, W. R. Newell, D. S. Murphy, J. F. McCann and I. D. Williams, *J. Mod. Opt.*, 2009, **56**, 1060–1069.
44 W. A. Bryan, S. L. Stebbings, E. M. L. English, T. R. J. Goodworth, W. R. Newell, J. McKenna, M. Suresh, B. Srigengan, I. D. Williams, I. C. E. Turcu, J. M. Smith, E. J. Divall, C. J. Hooker and A. J. Langley, *Phys. Rev. A*, 2006, **73**, 013407.
45 W. A. Bryan, S. L. Stebbings, J. McKenna, E. M. L. English, M. Suresh, J. Wood, B. Srigengan, I. C. E. Turcu, J. M. Smith, E. J. Divall, C. J. Hooker, A. J. Langley, J. L. Collier, I. D. Williams and W. R. Newell, *Nature Physics*, 2006, **2**, 379–383.
46 W. A. Bryan, C. R. Calvert, R. B. King, G. R. A. J. Nemeth, J. D. Alexander, J. B. Greenwood, C. A. Froud, I. C. E. Turcu, E. Springate, W. R. Newell and I. D. Williams, *Phys. Rev. A*, 2011, **83**, 021406.
47 W. A. Bryan, S. L. Stebbings, J. McKenna, E. M. L. English, M. Suresh, J. Wood, B. Srigengan, I. C. E. Turcu, I. D. Williams and W. R. Newell, *J. Phys. B*, 2006, **39**, S349.
48 C. R. Calvert, W. A. Bryan, W. R. Newell and I. D. Williams, *Phys. Rep.*, 2010, **491**, 1–28.
49 E. A. Shapiro, M. Spanner and M. Y. Ivanov, *Phys. Rev. Lett.*, 2003, **91**, 237901.
50 E. Persson, J. Burgdörfer and S. Gräfe, *New J. Phys.*, 2009, **11**, 105035.
51 J. Voll and R. de Vivie-Riedle, *New J. Phys.*, 2009, **11**, 105036.
52 M. F. Kling, C. Siedschlag, A. J. Verhoef, J. I. Khan, M. Schultze, T. Uphues, Y. Ni, M. Uiberacker, M. Drescher, F. Krausz and M. J. J. Vrakking, *Science*, 2006, **312**, 246–248.
53 M. Kremer, B. Fischer, B. Feuerstein, V. L. B. de Jesus, V. Sharma, C. Hofrichter, A. Rudenko, U. Thumm, C. D. Schröter, R. Moshammer and J. Ullrich, *Phys. Rev. Lett.*, 2009, **103**, 213003.
54 B. Fischer, M. Kremer, T. Pfeifer, B. Feuerstein, V. Sharma, U. Thumm, C. D. Schröter, R. Moshammer and J. Ullrich, *Phys. Rev. Lett.*, 2010, **105**, 223001.
55 Z.-T. Liu, K.-J. Yuan, C.-C. Shu, W.-H. Hu and S.-L. Cong, *J. Phys. B*, 2010, **43**, 055601.

56 C. R. Calvert, R. B. King, W. A. Bryan, W. R. Newell, J. F. McCann, J. B. Greenwood and I. D. Williams, *J. Phys. B*, 2010, **43**, 011001.
57 G. Sansone, F. Kelkensberg, J. F. Pérez-Torres, F. Morales, M. F. Kling, W. Siu, O. Ghafur, P. Johnsson, M. Swoboda, E. Benedetti, F. Ferrari, F. Lépine, J. L. Sanz-Vicario, S. Zherebtsov, I. Znakovskaya, A. L'Huillier, M. Y. Ivanov, M. Nisoli, F. Martín and M. J. J. Vrakking, *Nature*, 2010, **465**, 763–766.

Control of coherent excitation of neon in the extreme ultraviolet regime

Jürgen Plenge,* Andreas Wirsing, Christopher Raschpichler, Bernhard Wassermann and Eckart Rühl

Received 3rd March 2011, Accepted 27th April 2011
DOI: 10.1039/c1fd00032b

Coherent excitation of a superposition of Rydberg states in neon by the 13th harmonic of an intense 804 nm pulse and the formation of a wave packet is reported. Pump–probe experiments are performed, where the 3d-manifold of the $2p^6 \rightarrow 2p^5\ (^2P_{3/2})\ 3d\ [1/2]_1$- and $2p^6 \rightarrow 2p^5\ (^2P_{3/2})\ 3d\ [3/2]_1$-transitions are excited by an extreme ultraviolet (XUV) radiation pulse, which is centered at 20.05 eV photon energy. The temporal evolution of the excited state population is probed by ionization with a time-delayed 804 nm pulse. Control of coherent transient excitation and wave packet dynamics in the XUV-regime is demonstrated, where the spectral phase of the 13th harmonic is used as a control parameter. Modulation of the phase is achieved by propagation of the XUV-pulse through neon of variable gas density. The experimental results indicate that phase-shaped high-order harmonics can be used to control fundamental coherent excitation processes in the XUV-regime.

1 Introduction

Coherent excitation of atomic transitions by ultrashort laser pulses is a general way to excite wave packets and coherent transients, which have been the subject of several investigations.[1–13] Often, wave packet dynamics is studied by using pump–probe techniques. In a typical pump–probe experiment, the pump laser initiates the dynamics by exciting a superposition of states. The temporal evolution is probed by a time-delayed pulse exciting an atomic or molecular system to a final state, by producing a signal, which reveals the dynamics as a function of the pump–probe delay. This final state is subsequently detected by time-independent methods, such as electron or ion yields. The detection of the dynamics of the system requires that the probe step probability varies significantly with the time evolution of the wave packet. This approach led to the investigation of numerous phenomena that are not accessible using standard spectroscopic techniques such as the precession of orbital and spin angular momentum.[14,15] However, if the probe step shows no time dependence, an alternative approach for probing wave packet dynamics relies on producing interferences between two wave packets created by two identical time-delayed laser pulses.[16] This approach provides a useful alternative to pump–probe techniques when these cannot be applied. In addition, this allows one to coherently control the dynamics of the excited wave packet, where the time delay of the two pulses is used as a control parameter.[17] Also, it has been shown that an adjustment of the excitation wavelength[18] or changes in the polarization of the probe pulse[15] in wave packet dynamics studies can give access to different excitation pathways. Furthermore, pulse shaping techniques have led to novel results in coherent control

---

Physikalische und Theoretische Chemie, Institut für Chemie und Biochemie, Freie Universität Berlin, Takustr. 3, 14195 Berlin, Germany. E-mail: j.plenge@fu-berlin.de; Fax: +49 30 8385 2717; Tel: +49 30 8385 5354

of photon induced processes. This includes the possibility to control the wave packet dynamics using shaped laser pulses.[6]

The excitation of atomic two-level systems with chirped femtosecond laser pulses has been studied in great detail before.[1–3] This leads to excitation of coherent transients that are a signature of an interference between resonant and non-resonant excitation pathways and result in oscillations of the excited state population.[3] This fundamental process has been investigated by using ultrashort laser pulses in the near infrared regime, where atomic rubidium has been used as a model system.[2,3] The manipulation of the interaction process between laser light and the atomic system and coherent control of the excited state population has been demonstrated by applying pulse shaping techniques. Zamith et al. studied the effects of coherent excitation of a two-level system with linearly shaped pulses, where the 5s–5p transition in Rb was excited.[3] Silberberg and coworkers applied pulse shaping techniques to enhance the transient excited level population in a two-level atomic transition compared to that achievable from a transform-limited pulse.[2] In addition, the authors demonstrated how the dispersion induced by the absorption line itself leads to rapidly oscillating transients in the excited state population. Recently, Weber et al. investigated the influence of chirping the probe pulse in coherent transient experiments.[1]

Most of the studies on the excitation of wave packets and coherent transients have been performed in the near infrared regime, where Ti:sapphire laser systems have been used. This often limits the investigation of such fundamental processes to those model systems, which have accessible states in the near infrared regime. Alternatively, multi-photon processes can be employed. However, these can be accompanied by ac Stark effects so that the analysis of the results can be difficult.[10,11] An extension of the investigation of these fundamental excitation processes from the near infrared to the extreme ultraviolet (XUV) regime has become feasible with the development of ultrashort XUV-radiation sources. High-order harmonic generation (HHG) of intense infrared laser radiation is a well-established approach leading to coherent XUV- as well as soft X-ray-sources.[19] These sources have been applied to investigations on photoionization of atoms, where combining high-order harmonics with infrared photons from the same femtosecond laser gives access to two-color ionization of atoms.[20] Two-color photoionization of such systems is of fundamental interest and is also a powerful tool to probe properties of high-order harmonics.[21] Recently, Swoboda et al. investigated two-color two-photon ionization of helium via the 1s3p $^1P_1$ state using the 15th harmonic to excite this Rydberg state.[22] Haber et al. combined high-order harmonics and velocity-map imaging to measure partial photoionization cross sections of the 1s3p and 1s4p excited states of helium.[23] Photoionization of $2p^53d$ Rydberg states in neon has been recently investigated by combining synchrotron radiation and femtosecond infrared laser pulses, where the photoelectron distribution was investigated.[24] However, no time-resolved experiments have been performed on neon. In addition, high-order harmonics have been used to probe the dynamics of dissociating molecules by photoionization.[25,26] Further, spectral phase modulation of the 15th harmonic of an intense 805 nm pulse and coherent transient excitation of the 1s3p state in helium has been demonstrated.[27]

We report in this study experimental results on the coherent excitation of Rydberg states in neon by ultrashort XUV-pulses. We demonstrate that the 13th harmonics of an intense 804 nm pulse can be utilized to excite a two-state wave packet, which is composed of a superposition of two 3d Rydberg states that are probed by photoionization. We demonstrate the manipulation of the spectral phase of the 13th harmonic by propagating the pulse through a variable gas density of neon. The spectral phase modulation is probed by measuring coherent transients. In addition, it is shown that the modulation of the spectral phase can be used to manipulate the dynamics of the two-state wave packet. Therefore, this work goes beyond previous investigations on the coherent excitation of a single Rydberg state in helium by the phase-shaped 15th harmonic.[27]

2 Experimental

A Ti:sapphire laser system containing a Kerr lens mode-locked oscillator (Vitesse-800, Coherent) and a regenerative amplifier (Hidra-25, Coherent) is used to generate femtosecond laser pulses with a center wavelength of 804 nm and a bandwidth of 11.4 nm at full width half maximum (FWHM). The experimental pulse duration of the amplified pulses is typically 85 fs and the pulse energy is 1 mJ at a repetition rate of 1 kHz.

Fig. 1 shows the experimental pump–probe setup used in this work. The output of the laser system is sent to a beamsplitter directing 80% of the 804 nm laser light to a high-order harmonic generation beam path, where the XUV-pump beam is generated. High-order harmonics are generated by focusing the 804 nm beam onto a gas cell filled with argon. The gas pressure and the laser focal parameters are varied to optimize the HHG yield. The HHG beam and the remaining 804 nm beam exit the gas cell. Two silicon plates that are rotated to the Brewster angle of the 804 nm laser wavelength are used to remove the remaining 804 nm fundamental. A spherical, gold-coated mirror (focal length = 500 mm) is used to focus the divergent HHG beam into the ionization region of a magnetic-bottle photoelectron spectrometer.

The other 20% of the femtosecond laser system output are used for the probe beam that is sent through a computer-controlled delay stage. It is focused by a lens (focal length = 1000 mm) into the ionization region of the photoelectron spectrometer where it spatially overlaps with the XUV-pump beam. The pump and the probe beam intersect at a small angle in the ionization region of the photoelectron spectrometer, where both beams are crossed by an effusive neon beam. A magnetic-bottle type photoelectron spectrometer is used to detect the photoelectrons. Further details on this spectrometer have been described before.[28] Photoelectron spectroscopy allows us to distinguish between contributions of different high-order harmonics when all generated harmonics are simultaneously incident on the sample since adjacent high-order harmonics are separated by 3.08 eV in photon energy. The polarization vectors of the XUV-pump beam and the 804 nm

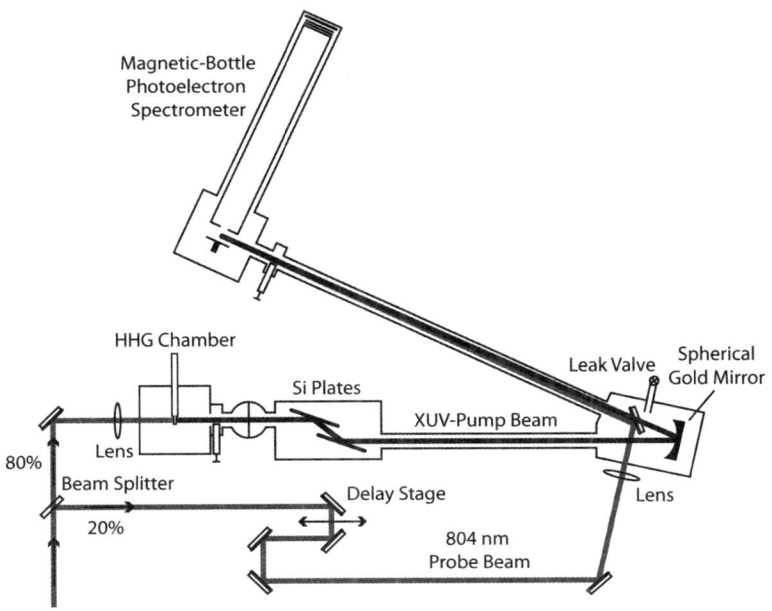

Fig. 1 Schematic diagram of the pump–probe setup.

probe beam are set to be parallel to each other. The relative delay between the pump and the probe beams is varied using a computer-controlled delay stage (Physik Instrumente), where the probe pulse is typically delayed by up to 100 ps.[29]

The output signals from a microchannel plate detector used for photoelectron detection are averaged using a digitizer (Agilent, Acqiris DC271) typically over 1000 laser shots for each delay time, where the delay between the pump and the probe beam is systematically varied. The time resolution of the pump–probe setup has been established to be of the order of 100 fs by measuring the time-dependence of side bands in the photoelectron spectra of rare gases.[21]

The measurements are performed with an effusive gas beam with low backing pressure in order to maintain a background pressure below 10^{-5} mbar during the measurements. A leak valve is used to introduce gaseous neon along the optical path of the XUV-pump beam between the refocusing mirror and the ionization region of the photoelectron spectrometer. The spectral phase of the XUV-pump beam is modified by changing the optical thickness of the resonant medium due to the dispersion near the absorption resonances of neon.

3 Results and discussion

The excitation scheme that is used in the pump–probe experiment is shown in Fig. 2. Rydberg states of neon are excited by an XUV-pump pulse, where the 13th harmonic centered near 20.05 eV is used. The intermediate states Ne $2p^5(^2P_{3/2})3d$ $[1/2]_1$ and Ne $2p^5(^2P_{3/2})3d[3/2]_1$ are within the bandwidth of the pump pulse and can be excited from the ground state by using photon energies of 20.0264 eV and 20.0404 eV, respectively.[30] The excited states are described within the J_cK-coupling scheme,[31] which is used to take into account the coupling of the 3d-electron to the core with the electron configuration $1s^22s^22p^5$. The term symbol of the excited states can be written as $[K]_J$ in the J_cK-coupling scheme, where the orbital momentum of the excited electron is coupled to the angular momentum J_c of the core yielding the quantum number K. The angular momentum J_c is obtained from the LS-coupling scheme. The quantum number K and the electron spin S determine the total angular momentum J. The dynamics of the excited states is probed by one-photon ionization

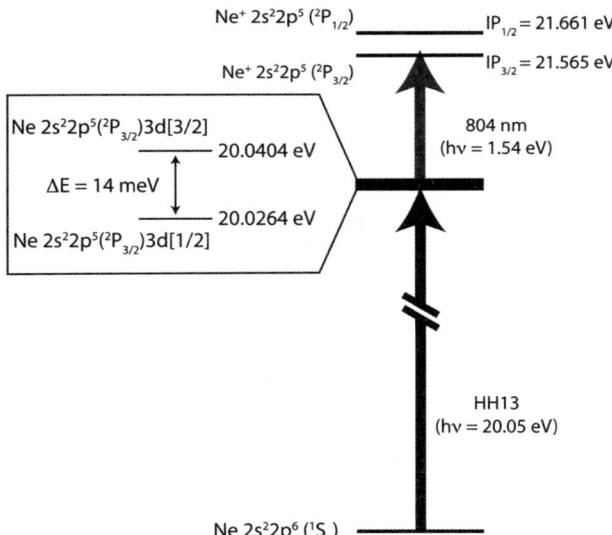

Fig. 2 Energy level diagram and excitation scheme of atomic neon.

with a time-delayed 804 nm femtosecond laser pulse ($\hbar\omega_L = 1.54$ eV). The ionization energy of neon is 21.565 eV and 21.661 eV, where the cation is formed in the $^2P_{3/2}$ or $^2P_{1/2}$ state, respectively.[30] The summed photon energy of the XUV-pump pulse and the 804 nm probe pulse is 21.59 eV, so that one-photon ionization occurs only below the $^2P_{1/2}$ threshold. The kinetic energy of the ejected electrons is calculated to be 25 meV. Therefore, the investigated pump–probe scheme can be summarized as follows:

$$\text{Ne } 2p^6 \xrightarrow{\hbar\omega_{HH13}} \text{Ne } 2p^5 3d[1/2]_1 \xrightarrow{\hbar\omega_L} \text{Ne}^+ 2p^5(^2P_{3/2}) + e^- \quad (1)$$

and

$$\text{Ne } 2p^6 \xrightarrow{\hbar\omega_{HH13}} \text{Ne } 2p^5 3d[3/2]_1 \xrightarrow{\hbar\omega_L} \text{Ne}^+ 2p^5(^2P_{3/2}) + e^- \quad (2)$$

Here, $\hbar\omega_{HH13}$ and $\hbar\omega_L$ denote the photon energy of the 13th harmonic and the 804 nm probe pulse, respectively. It should be noted that the 15th harmonic and higher orders are capable of direct ionization of neon without the need of an 804 nm probe pulse. Photoionization of neon by the 15th harmonic ($\hbar\omega_{HH15} = 23.13$ eV) results in the emission of photoelectrons with kinetic energies of 1.565 eV and 1.469 eV for the $^2P_{3/2}$ and $^2P_{1/2}$ ionization channel, respectively. Therefore, this ionization channel can be easily separated from the pump–probe signal of the 13th harmonic by the kinetic energies of the emitted photoelectrons. Furthermore, the 11th harmonic ($\hbar\omega_{HH11} = 16.96$ eV) and lower orders are not capable of direct ionization of neon. Also, excitation of Rydberg states of neon by the 3rd–11th harmonics can be excluded since the respective photon energy of these harmonics does not correspond to any electronic transition in neon.[32] Therefore, a contribution of the 11th harmonic and lower orders to the pump–probe signal can be excluded. The bandwidth of the 13th harmonics is calculated to be 45 meV using a procedure given in Ref. 33. Also, previous experimental work indicates that the spectral bandwidth of high-order harmonics can be on the order of 100 meV[34] so that the two transitions given by eqn (1) and eqn (2) can be excited by the coherent XUV-pulse. Therefore, it is expected to excite a two-state wave packet in neon that can be described as:[11]

$$|\Psi(t)\rangle = C_1|3d[1/2]_1, M = 0\rangle + C_2 e^{-i\Delta\omega t}|3d[3/2]_1, M = 0\rangle \quad (3)$$

Here, $C_{1,2}$ denotes complex amplitudes determined by the excitation process and $\Delta\omega$ corresponds to the energy difference between the two intermediate states.

Fig. 3(a) shows the time-dependent photoelectron yield of neon as a function of the XUV-pump pulse and the 804 nm probe pulse. The time traces are obtained from the integrated photoelectron signal, which is accumulated in the $^2P_{3/2}$ ionization channel. The photoelectron yield shows a step-like onset at a time delay of 0 fs, and there is no significant decay of the signal shown in Fig. 3(a), as has been measured for time delays reaching up to 100 ps. This is in accordance with the lifetime of the excited 3d states, which is estimated from tabulated transition probabilities to be 30.3 ns for the $3d[1/2]_1$ state and 10.8 ns for the $3d[3/2]_1$ state, respectively.[32] These values are two orders of magnitude higher than the temporal regime covered by this study.

The photoelectron yield shows a regular and pronounced oscillation with a period of 296 ± 0.5 fs. These oscillations are typical for the coherent excitation of a superposition of states and result from quantum beats between different excitation paths sharing the same final state.[11,14] The amplitude of the oscillation can be quantified by using $C = 2(I_{max} - I_{min})/(I_{max} + I_{min})$, where I_{max} and I_{min} are the photoelectron yields at the maximum and minimum of the oscillations. From the photoelectron yield shown in Fig. 3(a) we derive a contrast of $C = 0.50 \pm 0.01$. Further, oscillations up to 100 ps are observed with only a slight reduction in contrast. This indicates that no significant dephasing occurs. It should be noted that the experimental value of the

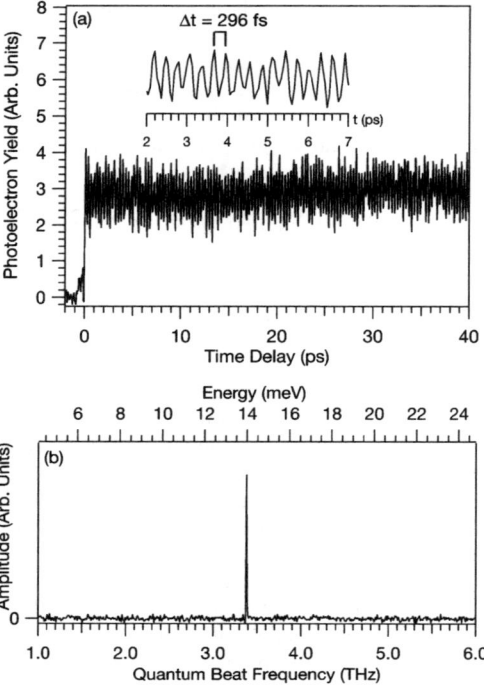

Fig. 3 (a) Time-dependent photoelectron yield as a function of the delay between the 13th harmonic and the 804 nm probe pulse; (b) Fourier-transform of the time-dependent photoelectron signal.

contrast are expected to be a lower limit because the limited time resolution of the pump–probe set-up can smear out the oscillations of the photoelectron yield in Fig. 3(a) leading to a reduction in contrast. The quantum beat frequency is extracted by a Fourier transform of the time-dependent photoelectron signal shown in Fig. 3 (a). The result of the Fourier transformation is shown in Fig. 3(b) and reveals a single signal at a frequency of 3379.0 ± 4.6 GHz. This corresponds to an energy difference between both Rydberg states of 13.97 ± 0.02 meV. This is in full agreement with the energy difference between the 3d[1/2]$_1$ and the 3d[3/2]$_1$ states of 14 meV, which is known from earlier spectroscopic work.[30] The present results indicate that a coherent excitation of the two target states has been achieved. It is noted that any Stark shifting of the energy levels is avoided by using one-photon excitations by the weak XUV-pump pulse. This is unlike the approach, where the excitation of high-lying Rydberg states requires intense laser pulses to populate these states by multiphoton absorption.[10,11]

A formalism for the description of wave packet dynamics probed by photoionization in pump–probe experiments is given for example in Ref. 14, where a stationary state and a bright-state/dark-state formalism are used to describe the wave packet dynamics in the 4p-manifold of atomic potassium. In the framework of the stationary state formalism the photoelectron signal can be described by eqn (4):

$$P(\tau) \propto \sum_f \mu_{kg} \mu_{k'g}^* \mu_{fk} \mu_{fk'}^* \, e^{-i(\omega_k - \omega_{k'})\tau} \qquad (4)$$

Here, μ_{kg} and $\mu_{k'g}$ represent the matrix elements of the dipole moment operator μ for the transitions between the ground state $|g\rangle$ and the two intermediate states $|k\rangle$

and $|k'\rangle$, respectively. The dipole matrix elements μ_{fk} and $\mu_{fk'}$ are the matrix elements for the probe step by photoionization into the final state $|f\rangle$. Using $|g\rangle = {}^1S$, $|k\rangle = 3d[3/2]_1$ and $|k'\rangle = 3d[1/2]_1$ it follows that the contrast is given by eqn (5)

$$C = 4 \frac{\sum_f \mu_{fk}\mu_{fk'}}{\sum_f \mu_{fk}^2 + \sum_f \mu_{fk'}^2} \quad (5)$$

The calculation of the contrast following eqn (5) requires knowledge of the dipole matrix elements for photoionization of the intermediate 3d states. Photoionization of the intermediate 3d states results in the emission of photoelectrons that can have either p- or f-character, according to the ionization process $2p^5({}^2P_{3/2})3d \rightarrow 2p^5({}^2P_{3/2}) + \varepsilon p, \varepsilon f$. We do not consider here the $2p^53d \rightarrow 2p^5({}^2P_{1/2}) + \varepsilon p, \varepsilon f$ ionization channel, since it is assumed that photoionization with a change in J_c is of negligible importance. In addition, the total photon energy of the 13th harmonic and the 804 nm probe pulse is just below the ${}^2P_{1/2}$ ionization threshold. Coupling of the outgoing photoelectron with the ionic core leads to five continuum states. These are described by the J_cK-coupling scheme as summarized in Fig. 4. The continuum states $[1/2]_{0,1}$ and $[3/2]_{1,2}$ are accessible from two intermediate states $3d[1/2]_1$ and $3d[3/2]_1$, according to the selection rule $\Delta J = 0, \pm 1$, whereas the $[5/2]_2$ continuum state is only accessible from the $[3/2]_1$ intermediate state (cf. Fig. 4). Therefore, the $[5/2]_2$ continuum state cannot act as a common final state to probe the wave packet dynamics. The dipole matrix elements for the transitions shown in Fig. 4 are not accessible in the literature to the best of our knowledge. Therefore, we calculated the dipole matrix elements for the relevant ionization processes (cf. Fig. 4) using the procedure outlined in Ref. 35. The coupling matrix elements for the transitions from the intermediate states into the ionization continua are evaluated by using the expression given in Ref. 35 for J_cK-coupling wave functions. The radial integrals for the transitions were calculated by using the Wentzel–Kramers–Brillouin (WKB) approximation for phase shifts. The results are summarized in Table 1. Using the values of the dipole matrix elements given in Table 1 we derive a contrast of $C = 0$ for a pure J_cK-coupling scheme indicating that no wave packet dynamics should be observed. However, one has to take into account the propensity rule predicting that upon absorption by one photon those transitions should be favored which lead to an increase of the angular momentum.[14,36] These are transitions with $\ell \rightarrow \ell + 1$ and $J \rightarrow J + 1$. Therefore, ionization in the $2p^5({}^2P_{3/2}) + \varepsilon f$ channel should be favored, as compared to the $2p^5({}^2P_{3/2}) + \varepsilon p$ channel. If we consider ionization exclusively into the f-continuum one obtains a contrast of $C = 0.4$. Alternatively, ionization exclusively into the p-continua would lead to $C = 1.6$. The experimental value of $C = 0.50 \pm 0.01$ is close to the contrast predicted for the

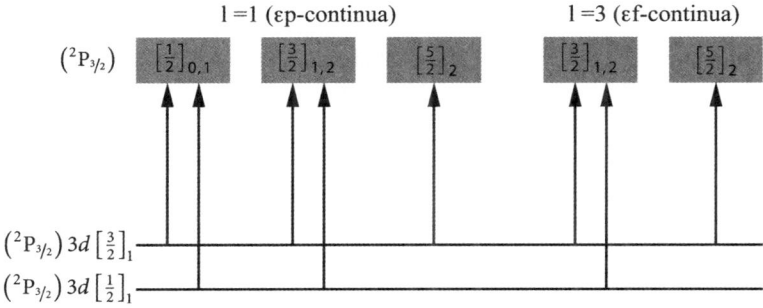

Fig. 4 Ionization scheme for neon showing the intermediate states $2p^5({}^2P_{3/2})3d[1/2]_1$ and $2p^5({}^2P_{3/2})3d[3/2]_1$ and the possible continuum states for the $2p^5(P_{3/2}) + \varepsilon p, \varepsilon f$ ionization channel.

Table 1 Calculated matrix elements of the dipole moment operator for the ionization of the intermediate Rydberg states [3/2]₁ and [1/2]₁ into different ionization continua using J$_c$K-coupling wave functions (in atomic units)

		εp-continua			εf-continua	
		$\left[\frac{1}{2}\right]_{0,1}$	$\left[\frac{3}{2}\right]_{1,2}$	$\left[\frac{5}{2}\right]_2$	$\left[\frac{3}{2}\right]_{1,2}$	$\left[\frac{5}{2}\right]_2$
$\Delta J = -1$	$\left[\frac{3}{2}\right]_1$	11.83	—	—	—	—
	$\left[\frac{1}{2}\right]_1$	11.83	—	—	—	—
$\Delta J = 0$	$\left[\frac{3}{2}\right]_1$	8.37	-14.79	—	-11.22	—
	$\left[\frac{1}{2}\right]_1$	-16.74	-3.74	—	11.22	—
$\Delta J = +1$	$\left[\frac{3}{2}\right]_1$	—	-6.70	5.02	-5.02	-24.60
	$\left[\frac{1}{2}\right]_1$	—	8.40	—	-25.10	—

ionization into the f-continuum. This indicates that this ionization process gives the main contribution to detection of the wave packet, as anticipated from the propensity rule for photoionization.

After having established that the coherent excitation of neon by the 13th harmonic of an 804 nm pulse creates a two-state wave packet, we discuss possible ways to control the dynamics of the excited states. It has been shown before that excitation by chirped laser pulses in the weak field regime leads to coherent transients which result in an oscillation of the excited state population.[3] Control mechanisms in both the time and frequency domain have been exploited to control this dynamics using laser pulses in the near infrared regime, where pulse shapers can be used.[2,3] These techniques cannot be employed for investigating excitations in the XUV-regime. However, it has been shown before that the spectral phase of the 15th harmonic of an intense 805 nm pulse can be manipulated by propagation of the pulse through a variable density gas phase containing helium.[27]

In the present work we introduce instead of helium gaseous neon into the optical path of the 13th harmonic to achieve a spectral phase modulation. Fig. 5 shows a series of time-dependent photoelectron yields as function of the delay between the XUV-pump pulse and the 804 nm probe pulse where the gas density of neon in the propagation path of the 13th harmonic is systematically varied. The photoelectron signal drops significantly above $\Delta t = 0$ fs and there are distinct and pressure dependent oscillations in the picosecond time regime (see Fig. 5), similar to earlier work.[27] Specifically, slow oscillations occurring on the scale of tens of picoseconds are analyzed in greater detail. These oscillations arise from the excitation of coherent transients. They are the result of an interference between the resonant and non-resonant excitation paths induced by the phase-shaped 13th harmonic. This is unlike the photoelectron yield shown in Fig. 3(a), where a step-like onset at a time delay of 0 fs is observed that does not significantly decay within the time window of 40 ps. The result is rationalized in the following way: The 13th harmonic propagates through the resonant neon gas filter. Neon exhibits the two absorption resonances discussed above, occurring in the photon energy regime of the 13th harmonics at 20.0264 eV and 20.0404 eV.[30] These transitions are due the excitation of 3d electrons as described in the excitation scheme in eqn (1) and (2). The line widths of these resonances at room temperature are dominated by Doppler broadening[31] resulting in a line width of 13 GHz. This is much smaller than the bandwidth of the 13th harmonic, which is estimated to be at least 45 meV,[32] so that the amplitude of the pulse will not be significantly changed, except at the resonance frequencies, where a depletion of the amplitude of the XUV-pulse occurs as a result of absorption. However, the refractive index of the medium changes significantly in the vicinity of the absorption line. Therefore, the dispersion near the absorption resonance adds either a positive or negative chirp to the frequencies of the XUV-pulse below and above the resonances, respectively. As a result, reshaping of the pulse occurs during propagation of the XUV-pulse through the resonant medium. It should be noted that these effects have been studied before for the propagation of an optical pulse in a resonant medium.[37,38] We also note that the propagation of intense X-ray pulses was investigated in theoretical studies, as a suitable way for pulse compression.[39,40]

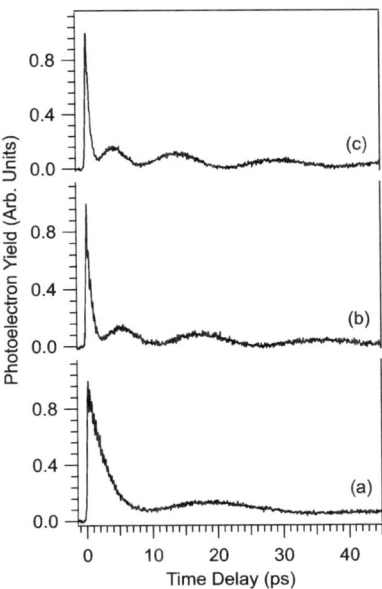

Fig. 5 Photoelectron yields as a function of the delay between the 13th harmonic and the 804 nm probe pulse recorded at different neon partial pressures along the propagation path of the 13th harmonic: (a) $p = 3 \times 10^{-5}$ mbar; (b) $p = 1 \times 10^{-4}$ mbar; (c) $p = 3 \times 10^{-4}$ mbar.

After propagating through the resonant neon gas filter the phase-shaped XUV pulse excites the intermediate 3d states of neon. The amplitude of an excited state can be written in first-order perturbation theory as:[5]

$$a_e(t) = \frac{\mu_{eg}}{\hbar}\left[\frac{1}{2}E_{HH13}(\omega_{eg})e^{i\varphi(\omega_{eg})} + \frac{i}{2\pi}\wp\int_{-\infty}^{\infty}d\omega\frac{e^{-i[(\omega-\omega_{eg})t-\varphi(\omega)]}}{\omega-\omega_{eg}}E_{HH13}(\omega)\right] \quad (6)$$

Here, $\wp$ is the principal value of Cauchy. $E_{HH13}(\omega)$ and $\phi(\omega)$ describe the amplitude and the phase of the 13th harmonic and ω_{eg} is the transition frequency between the ground state and the intermediate state. The first term in eqn (6) describes the excitation of the intermediate state at the resonance frequency. The second term has contributions from all the other non-resonant frequencies of the XUV-pulse. This term vanishes for a transform limited pulse, but can contribute significantly if the phase function $\phi(\omega)$ of the XUV-pulse is non symmetric with respect to the transition frequency ω_{eg}. A non-symmetric phase function for the XUV-pulse is formed after propagating the pulse through the resonant gas filter leading to chirped XUV-pulses. For the chirped pulses the frequency of the pulse sweeps with time and crosses the resonance ω_{eg}. Most of the population transfer occurs at this resonance, but a small fraction of the excited state amplitude is created by the frequency components of the XUV-pulse after the resonance and leads to strong oscillations in the excited state population due to the interference between the oscillating atomic dipole and the exciting XUV-field.[4]

The occurrence of coherent transients shown in Fig. 5(a)–(c) indicates that the spectral phase of the 13th harmonic is shaped by the interaction with neon, which is similar to previous work on helium.[27] However, it should be noted that in contrast to the work on helium (cf. Ref. 27), where only a single transition (He 1s → 3p) in the intermediate state occurs, in the present work we have to take into account two transitions into the intermediate states (Ne $2p^5(^2P_{3/2})3d[1/2]_1$ and Ne $2p^5(^2P_{3/2})3d[3/2]_1$). These contribute to the reshaping mechanism of the 13th harmonic in neon resulting in a more complex phase. The results shown in Fig. 5 already demonstrate that the phase modulation of the 13th harmonic can induce and control coherent transient excitation of the two intermediate states.

The influence of the phase shaped 13th harmonic on the excitation of the two-state wave packet is shown in Fig. 6. The photoelectron yield depicted in Fig. 6(a) has been recorded at a neon pressure (p) of 1×10^{-4} mbar along the propagation path of the 13th harmonic. This spectrum is equivalent to the conditions of the trace shown in Fig. 5(b), where the enhanced temporal resolution shows more details. Two distinct features can be seen: (i) a slow oscillation occurs on the scale of tens of ps that can be assigned to the excitation of coherent transients, as discussed above and (ii) a fast oscillation on the scale of a few hundred femtoseconds which is due to the dynamics of the two-state wave packet composed of the intermediate states $2p^5(^2P_{3/2})3d[1/2]_1$ and Ne $2p^5(^2P_{3/2})3d[3/2]_1$. The latter component is visualized by subtracting the contribution to the photoelectron yield on the ps timescale from the temporal evolution of the photoelectron signal shown in Fig. 6(a) by using an analytical function to describe the slow oscillation on the ps time scale. The result is depicted in Fig. 6(b). It shows the contribution of the wave packet dynamics to the modulation of the photoelectron signal induced by the resonant neon gas filter. This result can be compared to the photoelectron yield shown in Fig. 3(a), which has been obtained from excitation of neon by the 13th harmonic without using a resonant gas filter. There, a pronounced oscillation of the photoelectron signal without any significant reduction of the contrast is observed. In contrast, the time trace shown in Fig. 6(b) gives evidence for a strong modulation of the fast wave packet oscillation amplitude on the timescale of tens of ps. This clearly indicates that the phase manipulation of the 13th harmonic can control the coherent transient excitation of the intermediate 3d states in neon and at the same time the temporal

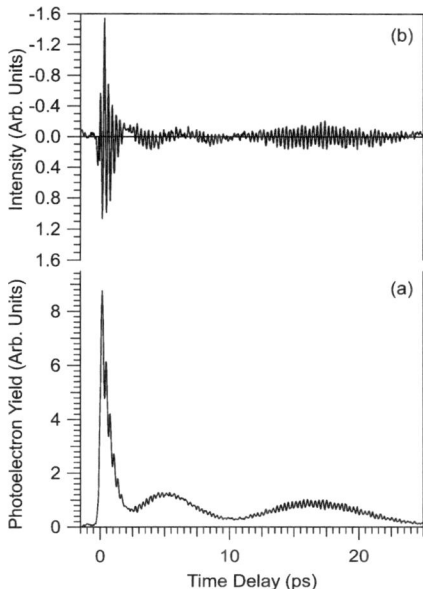

Fig. 6 (a) Photoelectron yield as a function of delay between the 13th harmonic and the 804 nm probe pulse recorded at a neon pressure (p) of 1×10^{-4} mbar along the propagation path of the 13th harmonic. (b) Time trace obtained from subtracting the contribution to the photoelectron yield with an oscillation on the ps time scale from the trace shown in Fig. 5(a) (see text for details).

evolution of the wave packet, as evidenced from the modulation of oscillation amplitude (see Fig. 6(b)). The shape of time traces shown in Fig. 6 appears to be similar to time-resolved fluorescence yields recorded after two-pulse excitation of rubidium atoms in the near infrared regime.[41] However, to the best of our knowledge this is for the first time that such processes have been experimentally investigated in the XUV excitation regime. The present approach allows for studying a broad range of atomic and molecular systems in the entire XUV-regime. For example, using an optical parametric amplifier instead of the fundamental of a Ti:sapphire laser for generating high-order harmonics gives access to tuneable XUV-pump pulses that enable to investigate the selective, coherent excitation of high-lying states, such as autoionizing states in atoms or molecules. Phase-shaped high-order harmonics can give access to coherent control of their ultrafast decay dynamics.

4 Conclusions

In conclusion, we have presented pump–probe experiments performed on the $2p^5 3d$ states of neon atoms. The 13th harmonic of an intense 804 nm pulse has been used to excite a superposition of the two Rydberg states $2p^5$ ($^2P_{3/2}$)$3d[1/2]$ and $2p^5$ ($^2P_{3/2}$) $3d$ [3/2]. The excited states are probed by one-photon ionization using an 804 nm probe pulse. The time-dependent photoelectron signal reveals the dynamics of the excited wave packet, where the dynamics is followed up to 100 ps.

The spectral phase of the 13th harmonic has been shaped by gaseous neon, where the dispersion near the absorption resonances changes the chirp of the XUV-pulse. The amount of chirp has been controlled by the atomic density along the propagation path, yielding time-dependent photoelectron yields with distinct features. A slow oscillation occurs on a scale of tens of ps that results from the excitation of coherent transients and depends strongly on the gas density, whereas a fast oscillation with a period of 296 ± 0.5 fs is due the wave packet dynamics. The experimental results clearly indicate that the phase manipulation of the 13th harmonic can control the coherent transient excitation of the intermediate 3d states in neon and at the same time the dynamics of the wave packet.

The observed control of coherent excitation of neon in the XUV-regime will facilitate further investigations using coherent short-wavelength radiation sources, such as free electron lasers. The spectral phase modulation, which has been exemplified to control the dynamics of a two-state wave packet in neon, can be explored to control inner-shell excitations processes of atoms, molecules, and clusters.

Acknowledgements

Financial support by the DFG (Sonderforschungsbereich 450 (TP B10) and SPP 1391) is gratefully acknowledged.

References

1 S. Weber, B. Girard and B. Chatel, *Phys. Rev. A*, 2010, **81**, 023415.
2 N. Dudovich, D. Oron and Y. Silberberg, *Phys. Rev. Lett.*, 2002, **88**, 123004.
3 S. Zamith, J. Degert, S. Stock, B. de Beauvoir, V. Blanchet, M. A. Bouchene and B. Girard, *Phys. Rev. Lett.*, 2001, **87**, 033001.
4 A. Monmayrant, B. Chatel and B. Girard, *Phys. Rev. Lett.*, 2006, **96**, 103002.
5 J. Degert, W. Wohlleben, B. Chatel, M. Motzkus and B. Girard, *Phys. Rev. Lett.*, 2002, **89**, 203003.
6 B. Chatel, D. Bigourd, S. Weber and B. Girard, *J. Phys. B: At., Mol. Opt. Phys.*, 2008, **41**, 074023.
7 C. Nicole, M. A. Bouchene and B. Girard, *J. Mod. Opt.*, 2002, **49**, 183.
8 J. Parker and C. R. Stroud, Jr, *Phys. Rev. Lett.*, 1986, **56**, 716.
9 G. Alber and P. Zoller, *Phys. Rep.*, 1991, **199**, 231.
10 S. Gilb, V. Nestorov, S. R. Leone, J. C. Keske, L. Nugent-Glandorf and E. R. Grant, *Phys. Rev. A*, 2005, **71**, 042709.
11 S. Gilb, E. A. Torres and S. R. Leone, *J. Phys. B: At., Mol. Opt. Phys.*, 2006, **39**, 4231.
12 H. H. Fielding, *Annu. Rev. Phys. Chem.*, 2005, **56**, 91.
13 J. R. R. Verlet, V. G. Stavros, R. S. Minns and H. H. Fielding, *Phys. Rev. Lett.*, 2002, **89**, 263004.
14 S. Zamith, M. A. Bouchene, E. Sokell, C. Nicole, V. Blanchet and B. Girard, *Eur. Phys. J. D*, 2000, **12**, 255.
15 E. Sokell, S. Zamith, M. A. Bouchene and B. Girard, *J. Phys. B: At., Mol. Opt. Phys.*, 2000, **33**, 2005.
16 C. Nicole, M. A. Bouchene, S. Zamith, N. Melikechi and B. Girard, *Phys. Rev. A*, 1999, **60**, R1755.
17 V. Blanchet, C. Nicole, M. A. Bouchene and B. Girard, *Phys. Rev. Lett.*, 1997, **78**, 2716.
18 C. Nicole, M. A. Bouchene, C. Meier, S. Magnier, E. Schreiber and B. Girard, *J. Chem. Phys.*, 1999, **111**, 7857.
19 T. Pfeifer, C. Spielmann and G. Gerber, *Rep. Prog. Phys.*, 2006, **69**, 443.
20 P. O'Keeffe, R. López-Martens, J. Mauritsson, A. Johansson, A. L'Hullier, V. Véniard, R. Taïeb, A. Maquet and M. Meyer, *Phys. Rev. A*, 2004, **69**, 051401(R).
21 A. Bouhal, R. Evans, G. Grillon, A. Mysyrowicz, P. Breger, P. Agostini, R. C. Constantinescu, H. G. Muller and D. von der Linde, *J. Opt. Soc. Am. B*, 1997, **14**, 950.
22 M. Swoboda, T. Fordell, K. Klünder, J. M. Dahlström, M. Miranda, C. Buth, K. J. Schafer, J. Mauritsson, A. L'Huillier and M. Gisselbrecht, *Phys. Rev. Lett.*, 2010, **104**, 103003.
23 L. H. Haber, B. Doughty and S. R. Leone, *Phys. Rev. A*, 2009, **79**, 031401(R).
24 P. O'Keeffe, P. Bolognesi, A. Mihelič, A. Moise, R. Richter, G. Cautero, L. Stebel, R. Sergo, L. Pravica, E. Ovcharenko, P. Decleva and L. Avaldi, *Phys. Rev. A*, 2010, **82**, 052522.
25 L. Nugent-Glandorf, M. Scheer, D. A. Samuels, A. M. Mulhisen, E. R. Grant, X. Yang, V. M. Bierbaum and S. R. Leone, *Phys. Rev. Lett.*, 2001, **87**, 193002.
26 Ph. Wernet, M. Odelius, K. Godehusen, J. Gaudin, O. Schwarzkopf and W. Eberhardt, *Phys. Rev. Lett.*, 2009, **103**, 013001.
27 D. Strasser, T. Pfeifer, B. J. Hom, A. M. Müller, J. Plenge and S. R. Leone, *Phys. Rev. A*, 2006, **73**, 021805(R).

28 C. G. Eisenhardt, M. Oppel and H. Baumgärtel, *J. Electron Spectrosc. Relat. Phenom.*, 2000, **108**, 141.
29 J. Plenge, A. Wirsing, I. Wagner-Drebenstedt, I. Halfpap, B. Kieling, B. Wassermann and E. Rühl, *Phys. Chem. Chem. Phys.*, 2011, **13**, 8705.
30 M. A. Baig and J. P. Connerade, *J. Phys. B: At. Mol. Phys.*, 1984, **17**, 1785.
31 R. D. Cowan, *The Theory of Atomic Structure and Spectra*, University of California, Berkeley, 1981.
32 J. R. Fuhr and W. L. Wiese, *CRC Handbook of Chemistry and Physics*, 79th Edition, D. R. Lide (Editor), CRC Press, Boca Raton, 1998.
33 X. He, M. Miranda, J. Schwenke, O. Guilbaud, T. Ruchon, C. Heyl, E. Georgadiou, R. Rakowski, A. Persson, M. B. Gaarde and A. L'Huillier, *Phys. Rev. A: At., Mol., Opt. Phys.*, 2009, **79**, 063829.
34 L. Nugent-Glandorf, M. Scheer, M. Krishnamurthy, J. W. Odom and S. R. Leone, *Phys. Rev. A: At., Mol., Opt. Phys.*, 2000, **62**, 023812.
35 R. D. Cowan and K. L. Andrew, *J. Opt. Soc. Am.*, 1965, **55**, 502.
36 U. Fano, *Phys. Rev. A: At., Mol., Opt. Phys.*, 1985, **32**, 617.
37 M. A. Bouchene, *Phys. Rev. A: At., Mol., Opt. Phys.*, 2002, **66**, 065801.
38 R. Netz, T. Feurer and J. A. Fülöp, *Phys. Rev. A: At., Mol., Opt. Phys.*, 2001, **64**, 043808.
39 Y.-P. Sun, J.-C. Liu, C.-K. Wang and F. Gel'mukhanov, *Phys. Rev. A: At., Mol., Opt. Phys.*, 2010, **81**, 013812.
40 Y.-P. Sun, J.-C. Liu and F. Gel'mukhanov, *J. Phys. B: At., Mol. Opt. Phys.*, 2009, **42**, 201001.
41 D. Felinto, L. H. Acioli and S. S. Vianna, *Opt. Lett.*, 2000, **25**, 917.

Optical manipulation of coherent phonons in superconducting $YBa_2Cu_3O_{7-\delta}$ thin films

Yasuaki Okano,[a] Hiroyuki Katsuki,[ab] Yoshihiro Nakagawa,[abc] Hiroshi Takahashi,[d] Kazutaka G. Nakamura[cd] and Kenji Ohmori*[abc]

Received 18th April 2011, Accepted 24th May 2011
DOI: 10.1039/c1fd00070e

The coherent phonons of $YBa_2Cu_3O_{7-\delta}$ are believed to be strongly coupled to its superconductivity. Controlling the phonons below its transition temperature, therefore, may serve as a promising scheme of the control of superconductivity. Here we demonstrate optical manipulation of the Ba–O and Cu–O vibrations in a thin-film $YBa_2Cu_3O_{7-\delta}$ below its transition temperature using a pair of femtosecond laser pulses. The interpulse delay is tuned to integral and half-integral multiples of the oscillation period of a specific phonon mode (Ba–O or Cu–O vibration) to enhance and suppress its amplitude, respectively.

Introduction

Optical manipulation of atomic motions in quantum mechanical systems is known as "coherent control." This technique is a promising way of controlling a variety of quantum systems, ranging from isolated atoms to bulk solids, and is relevant to novel quantum technologies such as bond-selective chemistry and quantum information processing.[1–4] Coherent control has been developed mostly with isolated atoms and molecules in the gas phase.[5–11] We have recently succeeded in visualizing and controlling vibrational motion of an isolated gas-phase molecule on the picometre and femtosecond spatiotemporal scales.[7,8] Moreover this high-precision control has been applied to ultrafast computing with molecular wave functions.[10] With these developments we believe that the coherent control of isolated atoms and molecules in the gas phase has matured enough for us to proceed to the next stage, in which our high-precision control will be applied to the solid state where many-body interactions among nuclei and electrons may severely degrade its controllability.

Atomic motion in the solid state can be triggered with optical pulses. If the pulse duration is shorter than the periods of lattice vibrations, the atoms start oscillating collectively. This collective oscillation is referred to as coherent phonons. Optical manipulation of coherent phonons has been studied with motivations to optically induce phase transitions or novel functionalities of solid-state materials.[12–15] We have recently carried out such a phonon-control experiment with an $YBa_2Cu_3O_{7-\delta}$ (YBCO) film at room temperature by using a pair of femtosecond (fs) laser pulses,[16] whose delay is tuned to integral and half-integral multiples of the oscillation period of a specific phonon mode (Ba–O or Cu–O vibration) to enhance and suppress its amplitude, respectively. It is suggested from previous experimental studies on

[a]*Institute for Molecular Science, National Institutes of Natural Sciences, Okazaki, 444-8585, Japan. E-mail: ohmori@ims.ac.jp*
[b]*SOKENDAI (The Graduate University for Advanced Studies), Okazaki, 444-8585, Japan*
[c]*CREST, Japan Science and Technology Agency, Kawaguchi, Saitama, 332-0012, Japan*
[d]*Materials and Structures Laboratory, Tokyo Institute of Technology, Yokohama, 226-8503, Japan*

YBCO films over a wide temperature range that a correlation exists between phonon motions and superconductivity;[17–19] the amplitude of Ba–O vibration is substantially enhanced below the transition temperature. These findings suggest a strong electron–phonon coupling below the transition temperature.[18] Controlling the phonons below the transition temperature, therefore, may lead to the control of superconductivity. As a first step toward the test of such novel controllability, here we demonstrate the double-pulse control of the Ba–O and Cu–O vibrations of YBCO below its transition temperature.

Experimental

Fig. 1 schematically shows our pump–control–probe setup. The output of a Ti: sapphire oscillator (VENTEON PULSE: ONE PE, center wavelength: ~800 nm) was split with a partial beam splitter into two beams with 9 : 1 ratio, and the smaller portion was used as the probe pulse. The other portion was input to a Michelson-type interferometer to produce a pair of identical laser pulses, which travelled collinearly to be used as the pump and control pulses. The dispersion arising from air and optical elements between the oscillator and a YBCO film was pre-compensated by wedge plates, chirp mirrors, and a quartz plate.

The sample was a YBCO single-crystal film with its 500 nm thickness deposited on a bulk MgO substrate (THEVA Co.). It was installed in an optical cryostat to be cooled below the transition temperature $T_c = 88.1$ K with liquid nitrogen or helium.

The pump, control, and probe pulses were focused with a concave mirror ($f = 250$ mm) to be overlapped on the sample surface. The incident angle of the pump and control pulses was ~0 degree, and that of the probe pulse was ~3 degrees to the surface normal. The energies of the pump and control pulses were ~0.3 nJ pulse^{-1} (~10^{-5} J cm^{-2}), and that of the probe pulse was ~0.1 nJ pulse^{-1}. The reflectivity change of the probe pulse induced by the pump and control pulses were monitored with a pair of balanced PIN photodiodes (PD1 and PD2). The differential signal from the PDs was amplified with a current preamplifier (SRS, SR570) and averaged in a digital oscilloscope (Tektronics, TDS5054B). A bandpass filter from 3 kHz to 300 kHz was used in all of the measurements reported in the present paper. The temporal evolution of the reflectivity change was measured by scanning the probe delay τ_{probe} repetitively at 20 Hz with a fast-scan mechanical stage (APE GmbH, ScanDelay), which was inserted on the optical path of the pump and control pulses. The displacement of this stage was proportional to the voltage applied, and the voltage was swept sinusoidally in each scan. This temporal variation of the

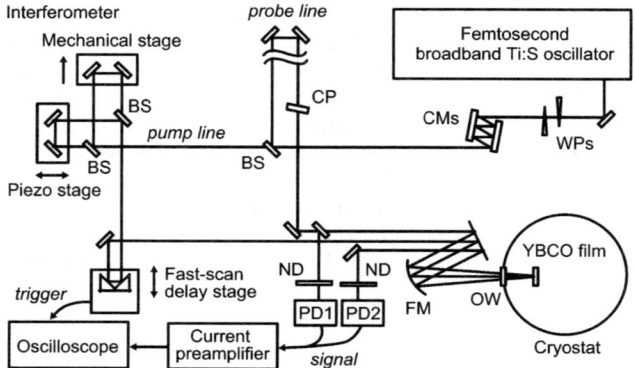

Fig. 1 Schematic of the pump–control–probe setup. WPs: wedge plates; CMs: chirp mirrors; BS: beam splitter; CP: compensation plate (quartz); ND: variable neutral density filter; FM: focusing mirror; OW: Optical window; PD1, 2: biased PIN photo diode.

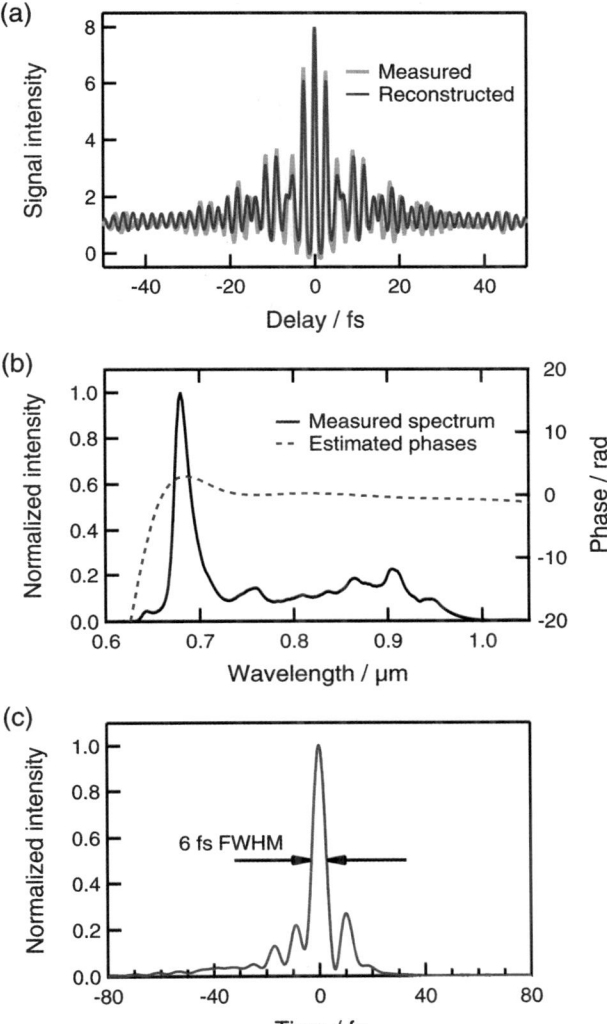

Fig. 2 Laser pulse characterization. (a) Measured (gray) and reconstructed (blue) FRAC traces of the pump and control pulses. (b) Measured spectrum and estimated phases of the pump pulse. (c) Reconstructed temporal profile of the pump pulse. See text for further details.

voltage was associated with electrical noise, and was fitted by a sine function to calibrate τ_{probe}. The data points thus sampled along the τ_{probe} axis were not equidistantly distributed. We linearly interpolated these measured data points to give equidistant points in steps of 1 fs to be fast-Fourier-transformed later. These interpolated reflectivity changes are shown in the present paper.

The duration of the pump and control pulses at the sample was evaluated with a fringe-resolved autocorrelation (FRAC) technique. In this measurement we put another mirror between the focusing mirror and the cryostat to introduce the pump and control pulses onto a BBO crystal with its thickness of 10 μm, which was placed at the same distance from the focusing mirror as that of the YBCO sample in the cryostat. An antireflection-coated fused-silica plate, whose width is the same as that of the optical window of the cryostat, was placed in front of the

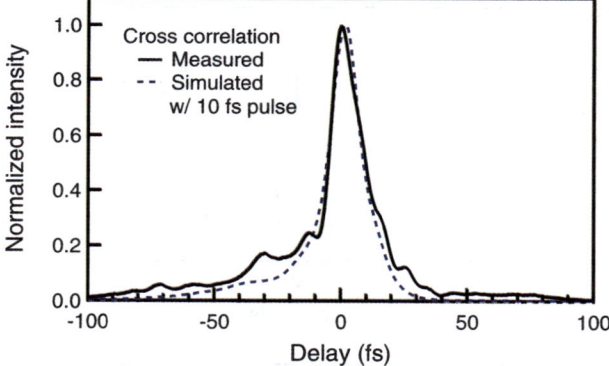

Fig. 3 Measured and simulated cross-correlation traces of the pump and probe pulses. See text for further details.

BBO crystal to reproduce the same amount of dispersion as for the pulses introduced into the cryostat. The intensity of the second harmonic was measured as a function of the delay between the pump and control pulses scanned with a piezo stage within the interferometer. An example of the measured FRAC traces is shown in a gray solid line in Fig. 2(a). Fig. 2(b) shows the spectrum of the pump pulse in a black solid line. The blue broken line in Fig. 2(b) represents the phases estimated for the pump pulse, and combined with its spectrum to give the FRAC trace shown in the blue solid line in Fig. 2(a). It is seen that this FRAC trace is in fairly good agreement with the measured one, so that we conclude that the phases shown in Fig. 2(b) have been reasonably estimated for the current pump pulse. These estimated phases and the measured spectrum in Fig. 2(b) are combined to give the temporal profile of the pump pulse shown in Fig. 2(c). Its main component has a 6 fs FWHM and the other sub-components are attributable to high-order dispersions unable to be compensated with the current wedge plates, chirp mirrors, and quartz plate.

Fig. 3 shows a measured cross-correlation of the pump and probe pulses as a black solid line. The blue broken line represents the cross correlation simulated with the

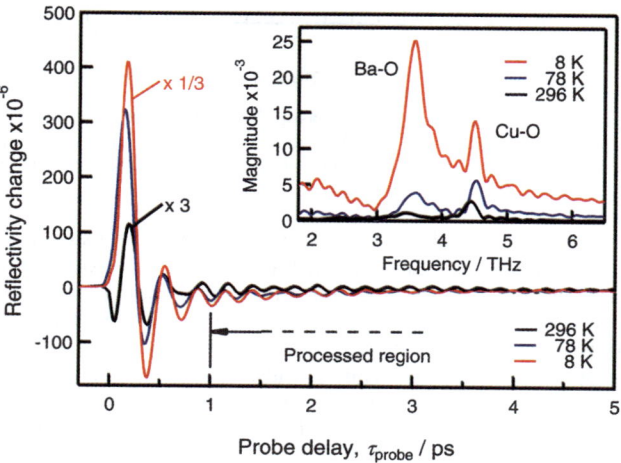

Fig. 4 Temporal evolution of the reflectivity change induced by the pump pulse measured at 296 K (black), 78 K (blue) and 8 K (red). Inset: fast-Fourier-transform of the temporal evolutions. See text for further details.

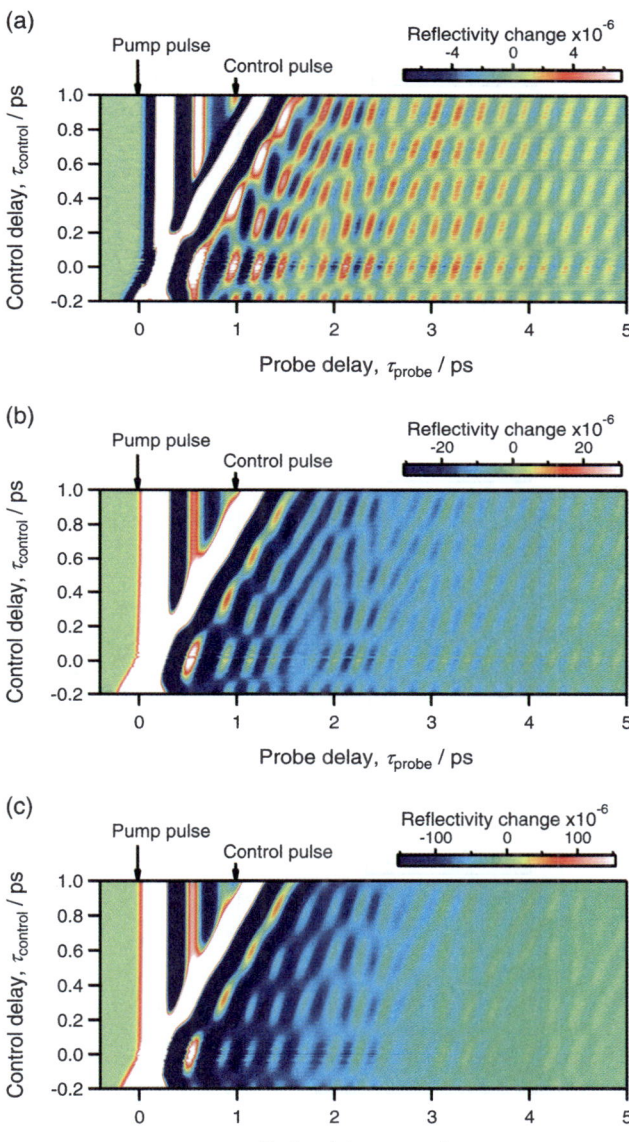

Fig. 5 Temporal evolution of the reflectivity change induced by the pump and control pulses. (a) 296 K, (b) 78 K and (c) 8 K.

temporal profile of the pump pulse taken from Fig. 2(c) and that of the probe pulse assumed to be Gaussian with its FWHM to be 10 fs. The simulated cross correlation is in reasonable agreement with the measured one.

Results and discussion

Fig. 4 shows the temporal evolution of the reflectivity change measured without the control pulse at 296 K, 78 K and 8 K. The origin of the probe delay was determined by the cross correlation measurement. Each trace is an average of 2,000 scans. In

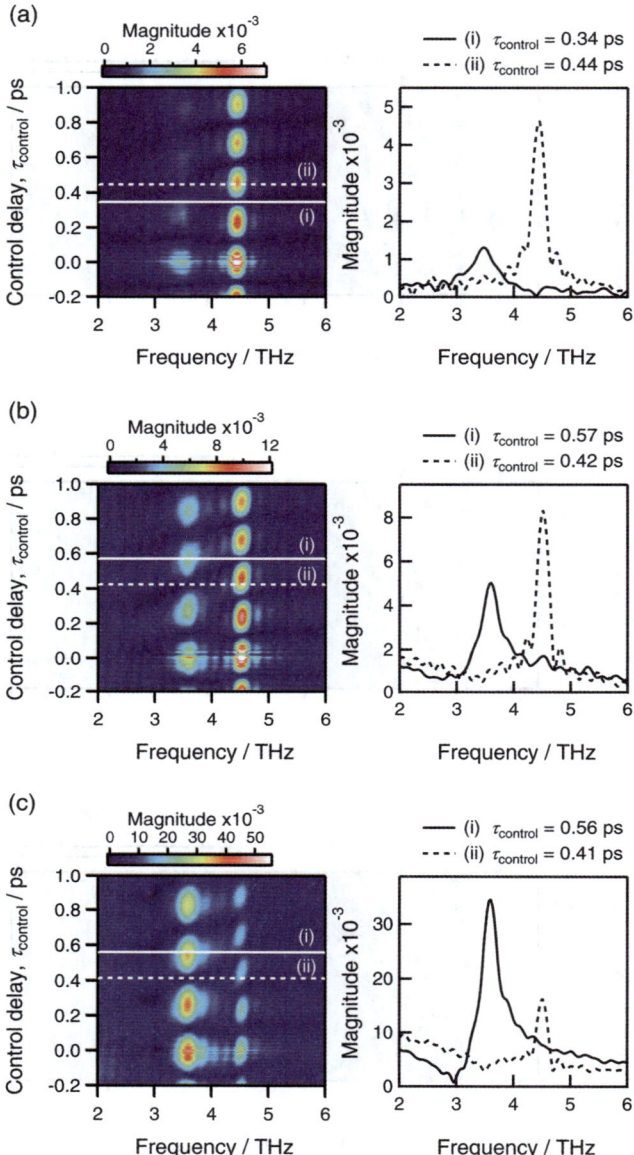

Fig. 6 Fast-Fourier-transform of the temporal evolutions shown in Fig. 5. (a) 2976 K, (b) 78 K and (c) 8 K. The right column shows cross sections along the solid (i) and broken (ii) lines in the left column at each temperature.

each trace a periodic oscillation is seen after the initial impulsive response around the zero delay, which is attributable to photo-carrier generation. Hereafter we focus on the region after the initial photo-carrier response with $\tau_{probe} \geq 1$ ps. This region has been fast-Fourier-transformed with zero-padding to give a spectrum shown in the inset of Fig. 4 for each temperature. In each spectrum two frequency components are seen around 3.5 THz and 4.5 THz and are assigned to Ba–O and Cu–O vibrations, respectively. It is clearly seen that Ba–O vibration is drastically enhanced at

8 K. This is in consistent with previous observations below T_c.[18,19] The backgrounds seen in the current fast-Fourier-transform (FFT) spectra at 78 K and 8 K arise from the slowly decaying components seen in Fig. 4. This component is attributable to carrier relaxation.[18,20]

Fig. 5 shows the temporal evolution of the reflectivity change induced by the pump and control pulses at 296 K, 78 K and 8 K. The ordinate represents the pump-control delay $\tau_{control}$ tuned with the mechanical stage within the interferometer shown in Fig. 1. At each control delay $\tau_{control}$, the region in this temporal evolution after the control pulse with $\tau_{probe} \geq \tau_{control} + 1$ ps is fast-Fourier-transformed with zero-padding to give a spectrum as a function of the control delay $\tau_{control}$ shown in the left column of Fig. 6 for each temperature. The cross sections along the solid (i) and broken (ii) lines are shown in the right column of Fig. 6. It is clearly seen that the relative amplitudes of Ba–O and Cu–O vibrations have been optically manipulated even below the transition temperature T_c.

It is seen in Fig. 6 that the enhancement and suppression of each vibrational mode appear alternately with its own vibrational period along the $\tau_{control}$ axis. This period is different from the period close to the optical cycle of the pump and control pulses that we have previously observed for the constructive and destructive interferences of the vibrational wave-packets in the iodine molecule.[6] It is thus understood that the relative phase between the pump and control electric fields does not play a role in the present enhancement or suppression of each vibrational mode, whereas it is selectively induced by the envelope of the whole electric field composed of the pump and control pulses.

In summary, we have succeeded in manipulating the coherent phonons of $YBa_2Cu_3O_{7-\delta}$ in its superconducting state with a pair of femtosecond laser pulses. The Ba–O vibration, which is believed to be strongly coupled to the superconductivity, has been selectively enhanced and suppressed below the transition temperature, with the interpulse delay tuned to integral and half-integral multiples of its oscillation period, respectively. The present technique could be combined with electric-current measurements to further investigate the coupling between a particular phonon mode and superconductivity.

Acknowledgements

This work was partly supported by Grant–in-Aid for scientific research by JSPS and Photon-Frontier-Consortium Project by MEXT of Japan, Collaborative Research of IMS, and Collaborative Research of MSL.

References

1 P. Brumer and M. Shapiro, *Chem. Phys. Lett.*, 1986, **126**, 541.
2 D. J. Tannor, R. Kosloff and S. A. Rice, *J. Chem. Phys.*, 1986, **85**, 5805.
3 T. C. Weinacht, J. Ahn and P. H. Bucksbaum, *Phys. Rev. Lett.*, 1998, **80**, 5508; T. C. Weinacht, J. Ahn and P. H. Bucksbaum, *Nature*, 1999, **397**, 233.
4 D. Meshulach and Y. Silberberg, *Nature*, 1998, **396**, 239.
5 K. Ohmori, Y. Sato, E. E. Nikitin and S. A. Rice, *Phys. Rev. Lett.*, 2003, **91**, 243003.
6 K. Ohmori, H. Katsuki, H. Chiba, M. Honda, Y. Hagihara, K. Fujiwara, Y. Sato and K. Ueda, *Phys. Rev. Lett.*, 2006, **96**, 093002.
7 H. Katsuki, H. Chiba, B. Girard, C. Meier and K. Ohmori, *Science*, 2006, **311**, 1589.
8 H. Katsuki, H. Chiba, C. Meier, B. Girard and K. Ohmori, *Phys. Rev. Lett.*, 2009, **102**, 103602.
9 K. Ohmori, *Annu. Rev. Phys. Chem.*, 2009, **60**, 487.
10 K. Hosaka, H. Shimada, H. Chiba, H. Katsuki, Y. Teranishi, Y. Ohtsuki and K. Ohmori, *Phys. Rev. Lett.*, 2010, **104**, 180501.
11 H. Goto, H. Katsuki, H. Ibrahim, H. Chiba and K. Ohmori, *Nat. Phys.*, 2011, **7**, 383.
12 K. Sokolowski-Tinten, C. Blome, J. Blums, A. Cavalleri, C. Dietrich, A. Tarasevitch, I. Uschmann, E. Forster, M. Kammler, M. H. Hoegen and D. Van der Linde, *Nature*, 2003, **422**, 287.

13 D. M. Fritz, D. A. Reis, B. Adams, R. A. Akre, J. Arthur, C. Blome, P. H. Bucksbaum, A. L. Cavalieri, S. Engemann, S. Fahy, R. W. Falcone, P. H. Fuoss, K. J. Gaffney, M. J. George, J. Hajdu, M. P. Hertlein, P. B. Hillyard, M. Horn-von Hoegen, M. Kammler, J. Kaspar, R. Kienberger, P. Krejcik, S. H. Lee, A. M. Lindenberg, B. McFarland, D. Meyer, T. Montagne, É. D. Murray, A. J. Nelson, M. Nicoul, R. Pahl, J. Rudati, H. Schlarb, D. P. Siddons, K. Sokolowski-Tinten, Th. Tschentscher, D. von der Linde and J. B. Hastings, *Science*, 2007, **315**, 633.
14 M. Rini, R. Tobey, N. Dean, J. Itatani, Y. Tomioka, Y. Tokura, R. W. Schoenlein and A. Cavalleri, *Nature*, 2007, **449**, 72.
15 K. Makino, J. Tominaga and M. Hase, *Opt. Express*, 2011, **19**, 1260.
16 H. Takahashi, K. Kato, H. Nakano, M. Kitajima, K. Ohmori and K. G. Nakamura, *Solid State Commun.*, 2009, **149**, 1955.
17 J. M. Chwalek, C. Uher, J. F. Whitaker, G. A. Mourou and J. A. Agostinelli, *Appl. Phys. Lett.*, 1991, **58**, 980.
18 W. Albrecht, Th. Kruse and H. Kruz, *Phys. Rev. Lett.*, 1992, **69**, 1451.
19 O. V. Misochko, K. Kisoda, K. Sakai and S. Nakashima, *Phys. Rev. B: Condens. Matter*, 2000, **61**, 4305.
20 S. D. Brorson, A. Kazeroonian, D. W. Face, T. K. Cheng, G. L. Doll, M. S. Dresselhaus, G. Dresselhaus, E. P. Ippen, T. Venkatesan, X. D. Wu and A. Inam, *Solid State Commun.*, 1990, **74**, 1305.

Femtosecond coherent control of thermal photoassociation of magnesium atoms

Leonid Rybak,[a] Zohar Amitay,*[a] Saieswari Amaran,[b] Ronnie Kosloff,*[b] Michał Tomza,[c] Robert Moszynski[c] and Christiane P. Koch[d]

Received 1st April 2011, Accepted 12th May 2011
DOI: 10.1039/c1fd00052g

We investigate femtosecond photoassociation of thermally hot atoms in the gas phase and its coherent control. In the photoassociation process, formation of a chemical bond is facilitated by light in a free-to-bound optical transition. Here, we study free-to-bound photoassociation of a diatomic molecule induced by femtosecond pulses exciting a pair of scattering atoms interacting via the van-der-Waals-type electronic ground state potential into bound levels of an electronically excited state. The thermal gas of reactants is at temperatures in the range of hundreds of degrees. Despite this incoherent initial state, rotational and vibrational coherences are observed in the probing of the created Mg_2 molecules.

1 Introduction

Coherent control was initiated as a scheme to steer a chemical reaction to its desired outcome.[1] In a nutshell, the quantum interference is employed to constructively enhance the selected product and destructively suppress the other possible channels.[2] A prerequisite for such control is coherent dynamics from the initial state to the final product.[3,4] The control is obtained by an external field which can alter the dynamics, typically by temporarily altering the generator—the Hamiltonian. For the standard situation describing single-photon transition, it reads

$$\hat{H} = \hat{H}_0 - \hat{\mu} \cdot \varepsilon(t), \tag{1}$$

where $\hat{H}_0$ is the bare molecular Hamiltonian, $\hat{\mu}$ the electric dipole moment operator, and $\varepsilon(t)$ the time-dependent external electric field. In a two-photon setting employing adiabatic elimination of all the off-resonant single-photon transitions, a Hamiltonian with a structure similar to that of eqn (1) is obtained except for the molecule-light coupling where the external field enters quadratically.[5]

Coherent control of chemical reactions has been achieved experimentally.[6,7] Almost exclusively, all demonstrations were unimolecular reactions such as photodissociation or photoisomerization. Coherent control of binary reactions with photoassociation as one prominent example has proven to be a much harder task.[8,9] The difficulty originates from the initial state of the reaction. In a unimolecular reaction the final fragments are initially entangled. As a result a coherent pathway connects

[a]The Shirlee Jacobs Femtosecond Laser Research Laboratory, Schulich Faculty of Chemistry, Technion - Israel Institute of Technology, Haifa, 32000, Israel. E-mail: amitayz@tx.technion.ac.il
[b]Institute of Chemistry and The Fritz Haber Research Center, The Hebrew University, Jerusalem, 91904, Israel. E-mail: ronnie@fh.huji.ac.il
[c]Department of Chemistry, University of Warsaw, Pasteura 1, 02-093 Warsaw, Poland
[d]Theoretische Physik, Universität Kassel, Heinrich-Plett-Str. 40, 34132 Kassel, Germany

the initial state to the final outcome. To control binary chemical reactions a prerequisite is to entangle the initial state.[10] In ultracold conditions, initial entanglement is present due to threshold effects such that only a few partial waves of the colliding pair can lead to association.[11,12] In thermal conditions, many partial waves participate and the colliding pair becomes unentangled.

Is coherent control of a binary reaction in a thermal state possible? The answer depends on the ability to filter out an entangled sub-ensemble of binary states. In the present study we demonstrate that a binary sub-ensemble with high purity is filtered out from a thermal ensemble of hot atoms. Moreover molecular coherence is generated. This is the prerequisite for coherent control of binary reactions.

The experiment is based on two-photon photoassociation of hot magnesium atoms to form a bound molecule. Magnesium in its electronic ground state is a closed shell atom. Therefore only a weak van der Waals attractive well can be found in the ground electronic potential of Mg_2 ($X^1\Sigma_g^+$). Upon promoting an electron to the π orbital, a strong chemical bound is formed in the $(1)^1\Pi_g$ state with a binding energy of $\sim$1.8 eV or, equivalently, $\sim$14500 cm^{-1}. This transition is induced by a femtosecond pulse of 70 fs transform-limited duration with a central wavelength of 840 nm. The energy difference corresponds to a two-photon transition. Two photon transition selection rules differ from the common one-photon photoassociation. As a result hidden potentials become accessible. Photoassociation of ultracold atoms takes place at comparatively long inter-atomic distances[13] where the two-photon transition matrix elements between the electronic ground state and the $(1)^1\Pi_g$ excited state vanish. Different from the ultracold regime, photoassociation of hot atoms occurs close to the repulsive wall of the ground state potential due to thermal population of high partial waves. The initial conditions for the experiment are a hot vapor of Mg atoms ($T \sim$ 1000 K) in Ar buffer gas. The hot conditions imply a broad distribution of linear and angular momentum of the randomly colliding Mg atoms. A major task of the excitation pulse is to preselect a sub-ensemble with increased purity and coherence.

A time-delayed femtosecond probe pulse interrogates the excited dimer *via* a one-photon excitation to a higher excited electronic state ($(1)^1\Pi_u$). This state has a strong one-photon transition back to the ground state. The corresponding experimental observable is the intensity of the resulting UV fluorescence ($\sim$290 nm) measured as a function of the pump–probe time delay. An oscillating time-dependent signal is a manifestation of coherent dynamics in the $(1)^1\Pi_g$ state. The present experiment is the first two-photon photoassociation experiment where the creation of molecular coherence out of an incoherent atomic ensemble is demonstrated. It is also the first hot femtosecond photoassociation experiment where the observed coherence is of both vibrational and rotational nature.

2 Experimental setup and theoretical model

The sequence of events that constitute the experiment is described in Fig. 1. A two-photon broadband transition promotes a pair of Mg atoms to an electronically excited state ($(1)^1\Pi_g$), forming Mg_2 molecules. After a time delay, a probe photon is absorbed transferring the molecules to a higher excited state ($(1)^1\Pi_u$). Emission from the $(1)^1\Pi_u$ state back to the ground state is detected. This signal as a function of the time delay between pump and probe pulse, yielding a pump–probe signal trace, is the main observable in the present study.

2.1 The experiment

The experiments are conducted in a static cell with Ar buffer gas heated to 1000 K, at which the pressure of the Mg vapor is around 5 Torr. The 840 nm 70 fs linearly-polarized pump and probe femtosecond pulses irradiate the sample at 1 kHz repetition rate in a collinear configuration. The measured signal results from the UV

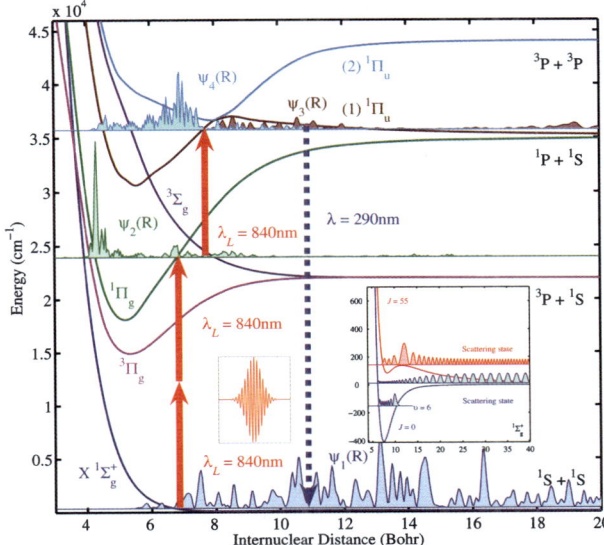

Fig. 1 Two-photon photoassociation of magnesium atoms followed by absorption of a third photon: The relevant potential energy curves are shown with the two-photon and single-photon transitions indicated by red arrows. The experimentally observed fluorescence at ∼290 nm is attributed to the dipole-allowed $(1)^1\Pi_u \rightarrow X^1\Sigma_g^+$ transition (blue arrow). A single realization of the random-phase thermal wavefunction used as initial state is displayed on the ground state potential. The wavefunctions on the electronically excited states correspond to times $t = 320$ fs on the $^1\Pi_g$ potential and $t = 640$ fs on the $^1\Pi_u$ potentials. The inset shows a zoom of the ground state potential with vibrational eigenfunction $v = 6$, $J = 0$ and some representative scattering wavefunctions for $J = 0$ and $J = 55$. Note that for high partial waves such as $J = 55$, the potential $V_g(R) + \dfrac{J(J+1)\hbar^2}{2\mu R^2}$ does not contain any bound levels.

fluorescence at ∼290 nm that is emitted (with a lifetime of a few ns) toward the direction of the beam entrance and collected there at a small angle from the beam axis using a proper optical setup. The final pump–probe trace results from an average over several scans, which averages out most of the interferometric signature existing in a single trace due to the collinear beam configuration.

2.2 Initial state

The thermal initial state is composed of Mg atoms at a temperature of 1000 K. Magnesium atoms in the ground state have a closed electronic shell. As a result, the $X^1\Sigma_g^+$ ground state potential has no chemical bond. The attraction on this surface is due to a shallow van der Walls well of a depth of approximately 400 cm^{-1}. At the high temperature of the experiment, scattering states up to ∼1000 cm^{-1} above threshold are thermally populated. For a Franck–Condon window at an interatomic distance of ∼7 bohr, corresponding to a two-photon excitation at 840 nm, angular momentum of up to $J = 100$ can contribute to the process. The centrifugal barrier at this angular momentum eliminates completely the small potential well of the ground surface (*cf.* inset in Fig. 1). For high values of angular momentum, existence of molecules before photoassociation can therefore be excluded. This is another difference compared to photoassociation of ultracold atoms where a few molecules are already produced by the trapping lasers.[14,15] Some exemplary bound states and scattering states of the ground potential surface are shown in the inset of Fig. 1.

Theoretically the initial state is modeled by a thermal density operator composed of bound and scattering states.[11] We eliminate the center of mass motion and consider the density operator of the relative motion of a pair of atoms,

$$\hat{\rho}_T(t=0) = \frac{1}{Z} e^{-\frac{\hat{H}_g}{k_B T}}, \quad (2)$$

where Z denotes the partition function, $Z = \mathbf{Tr}\left\{e^{-\frac{\hat{H}_g}{k_B T}}\right\}$, and $\hat{H}_g = \frac{\hat{P}^2}{2\mu} + \hat{V}_g$ is the Hamiltonian for nuclear motion in the electronic ground state. k_B is the Boltzmann constant and T is the temperature. We decompose this state into partial waves leading to

$$\hat{\rho}_T(t=0) = \sum_J (2J+1)\hat{\rho}_{T,J}, \quad (3)$$

where $\hat{\rho}_{T,J} = \frac{1}{z_J} e^{-\frac{\hat{H}_J}{k_B T}}$ and $\hat{H}_J = -\frac{\hbar^2}{2\mu}\frac{\partial^2}{\partial R^2} + V_g(R) + \frac{J(J+1)\hbar^2}{2\mu R^2}$ with $V_g(R)$ is the rotationless electronic ground state potential. z_J is the partition function for a given J. Each initial angular momentum component is modeled separately and the final results are summed. We employ an initial random phase thermal wavefunction to model the dynamics. More details on the method of calculation are reported elsewhere.[16]

2.3 Potential energy curves and transition matrix elements

The first step in simulating the dynamics is to generate the potential energy surfaces described in Fig. 1. In addition, two-photon transition matrix elements and dynamical Stark shifts are required. State-of-the-art *ab initio* techniques have been applied to compute the potential energy curves of the magnesium dimer in the Born–Oppenheimer approximation. In all calculations the aug-cc-pVQZ basis set of quadruple zeta quality as the atomic basis for Mg was used. This basis set was augmented by the set of bond functions consisting of $[3s3p2d2f1g1h]$ functions placed in the middle of Mg_2 dimer bond.

The ground $X^1\Sigma_g^+$ state potential was computed with the coupled cluster method restricted to single, double, and noniterative triple excitations, CCSD(T). For the excited $(1)^1\Pi_g$ and $(1)^1\Pi_u$ states we used the linear response theory (equation of motion approach) within the coupled-cluster singles and doubles framework, LRCCSD. The potential energy curve of the excited $(2)^1\Pi_u$ state in the region of the minimum of the potential was also obtained with the LRCCSD method, whereas at larger internuclear distances this potential energy curve was represented by the multipole expansion with electrostatic and dispersion terms C_n/R^n up to and including $n = 10$. The calculated potentials are shown in Fig. 1. More details on the *ab initio* calculations and spectroscopic characteristics of various electronic states of Mg_2 will be reported elsewhere.[17]

The electric transition dipole moments between states i and f, $\mu_j = \langle\Psi_i|\hat{\mu}_j|\Psi_f\rangle$, where the electric dipole operator, $\hat{\mu}_j = r_j$ is given by the jth component of the position vector and $\Psi_{i/f}$ are the wave functions for the initial and final states. These terms were computed as the first residue of the LRCCSD linear response function for $X^1\Sigma_g^+$, $(1)^1\Pi_g$, and $(1)^1\Pi_u$ states, whereas for transitions to $(2)^1\Pi_u$ state MRCI method was applied. Nonadiabatic radial and angular coupling matrix elements, as well as the spin-orbit coupling matrix elements have been evaluated with the MRCI method. The adiabatic $(1)^1\Pi_u$ and $(2)^1\Pi_u$ states are strongly coupled by the nuclear momentum operator $\hat{P}_R = i\nabla_R/\mu$, where μ is the reduced mass of Mg_2. We include this non-adiabatic coupling in the model by switching to the diabatic representation.

We adiabatically eliminate intermediate states that are accessed by single-photon transitions and calculate an effective two-photon coupling,

$$\chi(R) = \frac{1}{4}|S(t)|^2 \sum_{i,j} E_i E_j M_{ij}^{f \leftarrow 0}(\omega_L, R) , \qquad (4)$$

with $S(t)$ being the envelope of the electric field of the laser pulse and E_i being its maximal amplitude, and with ω_L being the pulse carrier frequency in energy units. It is given in terms of the tensor elements of the two-photon transition moment between the ground 0 and excited f state,

$$M_{ij}^{f \leftarrow 0}(\omega_L, R) = -\sum_{n \neq 0}\left[\frac{\langle f|\hat{\mu}_i|n\rangle\langle n|\hat{\mu}_j|0\rangle}{\omega_{n0} - \omega_L} + \frac{\langle f|\hat{\mu}_j|n\rangle\langle n|\hat{\mu}_i|0\rangle}{\omega_{nf} + \omega_L}\right], \qquad (5)$$

where $\omega_{nk} = \omega_n - \omega_k$ is the energy difference between states n and k.

In practice, it can be obtained as a residue of the cubic response function.[18] For transitions between the $X^1\Sigma_g^+$ and $(1)^1\Pi_g$ states, $M_{ij}^{f \leftarrow 0}(\omega_L, R)$ was computed as a residue of the coupled cluster cubic response function with electric dipole operators and wave functions within the CCSD framework.[19,20]

Strong field conditions require to include dynamical Stark shifts. The Stark shift of electronic state k can be expressed through the elements of the dynamic electric dipole polarizability tensor,

$$\omega_k^S(t, R) = -\frac{1}{4}|S(t)|^2 \sum_{i,j} E_i E_j \alpha_{ij}^k(\omega_L, R), \qquad (6)$$

where the tensor elements of the dynamic polarizability are given by

$$\alpha_{ij}^k(\omega_L, R) = \sum_{n \neq k}\left[\frac{\langle k|\hat{\mu}_i|n\rangle\langle n|\hat{\mu}_j|k\rangle}{\omega_{nk} - \omega_L} + \frac{\langle k|\hat{\mu}_j|n\rangle\langle n|\hat{\mu}_i|k\rangle}{\omega_{nk} + \omega_L}\right]. \qquad (7)$$

The tensor elements of the polarizability of the ground $^1\Sigma_g^+$ state were obtained as a coupled cluster linear response function with electric dipole operators and wave functions within CCSD framework.[21]

Dynamic polarizabilities of the exited states were computed as double residues of the coupled cluster cubic response function with electric dipole operators and wave functions within CCSD framework.[22,23]

2.4 Two-photon excitation

The photoassociation step is carried out by a two-photon transition from the $X^1\Sigma_g^+$ electronic ground state to the $(1)^1\Pi_g$ excited state. The photoassociation is induced by a femtosecond pulse of 70 fs transform-limited duration, 840 nm central wavelength, and a transform-limited peak intensity of approximately 5×10^{12} W cm^{-2}.

The transition takes place in a small Franck–Condon window at $R_L \sim 7$ bohr. This two-photon excitation is impulsive on the timescale of vibrational and rotational periods on the excited $(1)^1\Pi_g$ surface ($\sim$175 fs and $\sim$0.5–1 ps, respectively). As a result, coherent transient dynamics is generated on this potential. Indirect experimental evidence for populating the $(1)^1\Pi_g$ state comes from emission lines terminating on the excited triplet ^3P atomic Mg state. The origin of these triplet atoms is most probably a nonadiabatic spin–orbit induced transfer from the singlet $(1)^1\Pi_g$ state to the triplet $(1)^3\Pi_g$ close to the inner turning point and to the $(1)^3\Sigma_g$. We calculated the lifetime on the surface to be $\sim$11 ps. The population on this state will dissociate into ^3P atoms (cf. Fig. 1).

The excitation dynamics was simulated by solving the time-dependent Schrödinger equation,

$$i\hbar \frac{\partial \Psi}{\partial t} = \hat{\mathbf{H}}_{pump}(t)\Psi, \qquad (8)$$

where Ψ is a multi-surface wavefunction. The dynamics are generated by the time-dependent Hamiltonian,

$$\hat{\mathbf{H}}_{pump}(t) = \begin{pmatrix} \hat{H}_g & \chi(t)e^{i\phi(t)} & 0 & 0 \\ \chi(t)e^{-i\phi(t)} & \hat{H}_e & V_{SO1} & V_{SO2} \\ 0 & V_{SO1} & \hat{H}_{im1} & V_{SO3} \\ 0 & V_{SO2} & V_{SO3} & \hat{H}_{im2} \end{pmatrix}, \qquad (9)$$

where $\hat{H}_i$ is the nuclear Hamiltonian, given by $\hat{H}_{g(e)} = \hat{T}_{g(e)} + V_{g(e)}(R) + \omega^S_{g(e)}(t,R)$. $\hat{T}_{g(e)}$ is the vibrational and rotational kinetic energy, $V_{g(e)}(R)$ the potential energy, and $\omega^S_{g(e)}(t,R)$ the dynamic Stark shift. Here, g corresponds to the $X^1\Sigma_g^+$ state and e to the $^1\Pi_g$ state. $im1$ and $im2$ are the two intermediate triplet states, $(1)^3\Sigma_g$ and $(1)^3\Pi_g$, respectively. The Hamiltonian of the intermediate states is given by $\hat{H}_{im1(im2)} = \hat{T}_{im1(im2)} + V_{im1(im2)}(R)$. $V_{SO1(SO2, SO3)}$ are the spin–orbit couplings. The transition is driven by the effective two-photon coupling $\chi(R)$, cf. eqn (4), and $\phi(t)$ is the time-dependent two-photon phase which is zero for transform-limited pulses.

The Hamiltonian is represented on a grid of length $R_{min} = 3.0$ bohr to $R_{max} = 40.0$ bohr, with 1024 grid points. The photoassociation is studied numerically by solving the time-dependent Schrödinger equation for the effective Hamiltonian in eqn (9) with a Chebychev propagator.[24,25] We simulate the dynamics with a pulse duration of 100 fs full-width at half-maximum of (FWHM, with respect to the field) for the transform-limited pulse and a central wavelength of 840 nm. Intensity values were calculated in the range of 10^{10} W cm^{-2} to 10^{13} W cm^{-2}, cf. Fig. 2. This pulse corresponds to the experimental 70 fs pulse (where the FWHM is taken with respect to the intensity profile).

We can estimate the purity of the sub-ensemble in the electronically excited state by constructing the corresponding density operator,

$$\hat{\rho}_e(t) = \hat{\mathbf{P}}_e \hat{\rho}_T(t) \hat{\mathbf{P}}_e, \qquad (10)$$

where $\hat{\mathbf{P}}_e$ is the projector onto the $(1)^1\Pi_g$ excited state. The normalized excited state purity becomes

$$\mathscr{P}_e = \frac{\text{Tr}[\hat{\rho}_e^2]}{\text{Tr}[\hat{\mathbf{P}}_e \hat{\rho}_T(t)]}. \qquad (11)$$

The coherence measure C_e in the excited state allows one to characterize the dynamical contribution to the purity.[26] It is obtained by substracting from $\hat{\rho}_e$ the diagonal elements in the energy representation, squaring and taking the trace. Details of the computational details will be described elsewhere.[16]

2.5 The probe step

The main experimental observable is an emission signal at $\sim$290 nm that is emitted following the $(2\oplus 1)$ absorption of 840 nm photons, i.e., two-photon absorption followed by a time-delayed one-photon absorption. This emission is measured as a function of the time delay between the two-photon and single-photon transitions. The transient signal is associated with the dynamics in the $(1)^1\Pi_g$ excited state. The Fourier transform of this signal should unravel the vibrational and rotational periods on this potential.

We simulate the experimental signal assuming that the population on the $^1\Pi_u$ states is proportional to the emission signal. The pulse parameters remain the same as those of the two-photon pulse. The simulation involves a total of five potential energy surfaces. The initial state for the probe step is $(1)^1\Pi_g$ (denoted as 'g'), the target states, $(1)^1\Pi_u$ and $(2)^1\Pi_u$, are denoted by 'e1' and 'e2', respectively. The intermediate states $(1)^3\Sigma_g^+$ (denoted as $im1$) and $(1)^3\Pi_g$ (denoted as $im2$) are coupled to the 'initial' state $(1)^1\Pi_g$ state by spin–orbit coupling. As discussed in Sec. 2.3, the two $^1\Pi_u$ states exhibit a strong non-adiabatic coupling around the region $R = 8.1$ bohr. Hence, an adiabatic-to-diabatic transformation is performed over these two states and the corresponding diabatic potentials and the diabatic couplings are employed in the simulations.

The Hamiltonian for the probe-pulse simulations is given by

$$\hat{H}_{probe}(t) = \begin{pmatrix} \hat{H}_g & \mu_d^1(R)\varepsilon(t) & \mu_d^2(R)\varepsilon(t) & V_{SO1} & V_{SO2} \\ \mu_d^1(R)\varepsilon^*(t) & \hat{H}_{e1} & V_{12}^d & 0 & 0 \\ \mu_d^2(R)\varepsilon^*(t) & V_{12}^d & \hat{H}_{e2} & 0 & 0 \\ V_{SO1} & 0 & 0 & \hat{H}_{im1} & V_{SO3} \\ V_{SO2} & 0 & 0 & V_{SO3} & \hat{H}_{im1} \end{pmatrix}, \quad (12)$$

where $\mu_d^{1,2}$ are the transition dipole operators in the diabatic basis between the $^1\Pi_u$ states and the $(1)^1\Pi_g$ state, and $\varepsilon(t)$ is the electric field of the laser pulse. V_{12}^d denotes the diabatic coupling between the $(1)^1\Pi_u$ and $(2)^1\Pi_u$ states. The simulation includes only vibrational dynamics, i.e., J is kept fixed.

3 Evidence of coherence

The experimental pump–probe trace for the observed fluorescence and its Fourier transform spectra are shown in Fig. 4 for two pulse bandwidth cases: 70 fs and 100 fs transform-limited pulse duration. Even though the theoretical calculations have been conducted with 70 fs pulse, the 100 fs case is shown here for completion. Both pump–probe traces exhibits time dependence long after the pump (photo-associating) pulse is over, hence they reveal molecular rovibrational coherence of the photoassociated Mg_2 molecules. The Fourier spectra of the pump–probe traces show broad peaks at 5.8 THz and 1.2 THz. The former fits a probed vibrational dynamics in the $(1)^1\Pi_g$ state with vibrational levels around $v \sim 30$–50, while the latter fits a probed rotational dynamics in the $(1)^1\Pi_g$ state with rotations around $J \sim 40$–100. As seen, the vibrational coherence is attenuated for the 100 fs pulse as compared to the 70 fs pulse.

In the theoretical description, it is the autocorrelation function, cf. Fig. 3, that reflects the dynamics in the $(1)^1\Pi_g$ excited state after the pump pulse. We compare the autocorrelation functions obtained with a single realization of a random-phase thermal wavefunction and an average over 17 realizations of random-phase thermal wavefunctions. An oscillatory behavior indicating coherent vibrational dynamics is evident. This coherent dynamics survives the angular momentum averaging as can be seen in Fig. 3(a) (blue line). These results are an indication that coherent dynamics is distilled from an initial thermal ensemble. The probing of this coherent vibrational dynamics into the Π_u states (see above) produces a corresponding oscillatory pump–probe trace that its Fourier spectrum is shown in Fig. 4(b) (purple line) together with the experimental results. It shows a prominent peak at a frequency of 7.2 THz. As described above, the theoretical calculations account here only for the vibrational dynamics on the $(1)^1\Pi_g$ state. As seen in Fig. 4(b), the calculated vibrational Fourier peak is located at the same region of the experimental values, however it is much narrower. We attribute the difference between the experimental and theoretical results to the large uncertainty with regard to the calculated Π_u states due to their corresponding strong non-adiabatic coupling (see above).

4 Prospects for coherent control

The experimental and theoretical evidence for coherent dynamics distilled out of the initial thermal state means that coherent control should be possible. The most

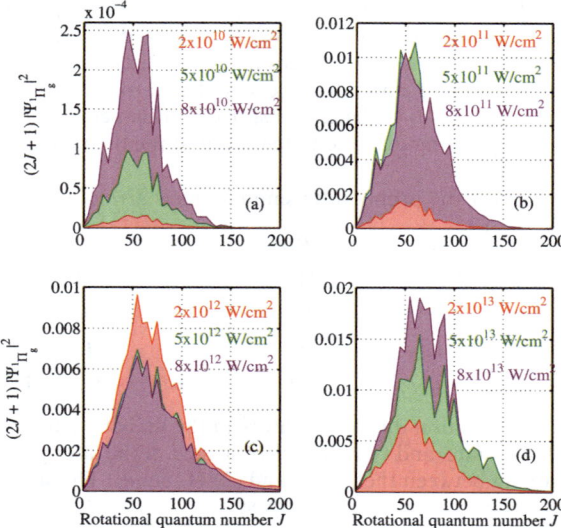

Fig. 2 Thermally averaged population transferred by the 70 fs 840 nm photoassociating pulse *via* two-photon transition to the $(1)^1\Pi_g$ excited state as a function of the initial rotational quantum number J (initial angular momentum) for different intensities. Notice that each panel has a different scale. The highest excitation probability calculated for the experimental intensity of 5×10^{12} W cm^{-2} is at $J = 55$. In general, the peak of the distribution moves to higher J as the intensity increases.

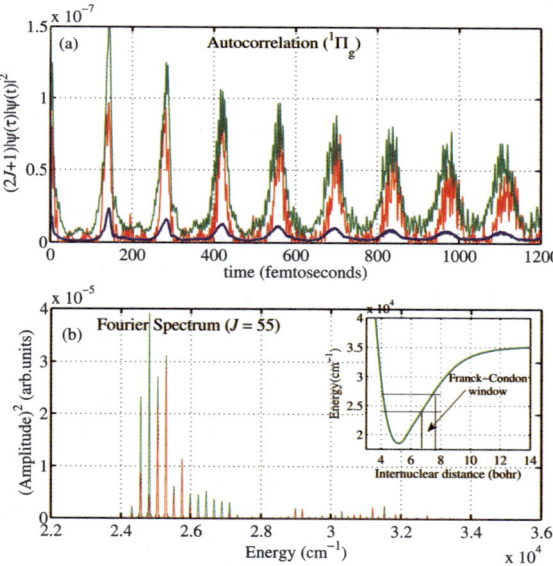

Fig. 3 Autocorrelation function of the transient molecular dynamics on the $(1)^1\Pi_g$ excited state potential for a fixed rotational quantum number $J = 55$ and $T = 1000$ K. (a) The autocorrelation as a function of time. The red curve represents a single realization of random phases, while the green curve has been obtained by averaging over 17 realizations. The blue curve results from an averaging over all the initial rotational quantum numbers (initial angular momentum). (b) The Fourier transform of the autocorrelation function. The inset shows the corresponding range of state energies that are populated on the $(1)^1\Pi_g$ state.

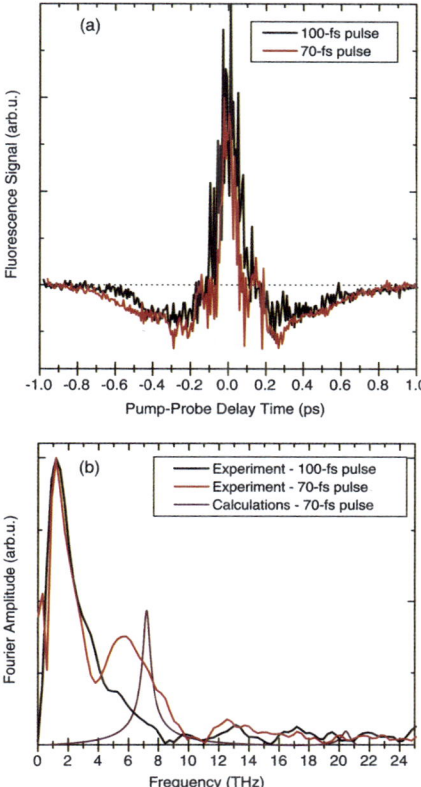

Fig. 4 Bandwidth dependence of the experimentally observed fluorescence signal after excitation with transform-limited femtosecond pulses: (a) the pump–probe signal trace and (b) Fourier spectrum of the trace. The two pulse durations are 70 fs and 100 fs. The probed vibrational coherence is attenuated for the excitation with the 100 fs pulse as compared to the excitation with the 70 fs pulse. The calculated Fourier spectrum for the case of 70 fs pulse is also shown. Note that the theoretical calculations account only for the vibrational dynamics on the excited state (see text).

obvious target is to enhance the production of photoassociated Mg_2 molecules. Other targets could be control of the emission signal.[27] Experimentally, such a target could also be optimized by random feedback techniques.[28] In the present context we demonstrate phase-only control as could be achieved by a pulse shaper. We also show intensity effects.

4.1 Chirp effects

The most simple phase control is a linear chirp, *i.e.*, a time-dependent frequency change. Experimentally it can be obtained by adding a dispersive element to the pulse generator. More elaborate phase control is obtained in the frequency domain by a pulse shaper. In strong fields positive chirp enhances population transfer.[29,30] This is because the dynamics on the excited state does not interfere with the population transfer. In a two-photon transition the chirp effect adds up. We simulate the effect by employing the following pulse form in the time domain *cf.* eqn (4):

$$S_c^2(t) = \frac{E_0^2 \sigma_t}{\sqrt{\sigma_t^4 + 16\chi_{II}^2}} e^{-\frac{t^2 \sigma_t^2}{\sigma_t^4 + 16\chi_{II}^2}} \quad (13)$$

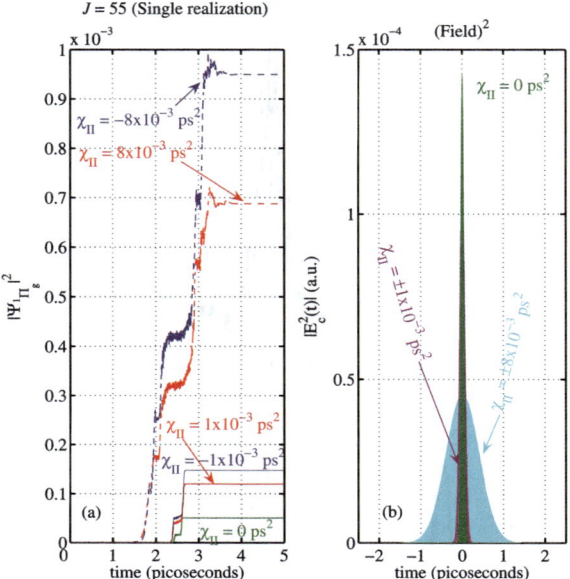

Fig. 5 The effect of the pulse chirp on the population in the $(1)^1\Pi_g$ excited state as calculated for the 70 fs transform-limited pulse and several cases of positive and negative chirps rates. The pulse fluence is kept constant in all the presented cases. The intensity of the (unchirped) transform-limited pulse is 5×10^{12} W cm^{-2}.

and the phase becomes *cf.* eqn (9):

$$\phi(t) = \frac{\chi_{II} 4t^2}{\sigma_t^4 + 16\chi_{II}^2} \qquad (14)$$

where χ_{II} is the two-photon chirp rate in the frequency domain and σ_t is the transform limited pulse width. Eqn (13) reflects the fact that the chirped pulse is stretched in time with respect to the transform-limited pulse with peak intensity attenuated to keep the pulse fluence constant. The transform limited intensity is obtained by $I_0 = \frac{1}{2} c \varepsilon_0 E_0^2$, where c is the speed of light and ε_0 is the free space permittivity.

Fig. 5 shows the influence of a positively and negatively chirped pulse relative to the transform-limited pulse. In all cases the chirp enhances the population transfer with more than an order of magnitude enhancement for $\chi_{II} = \pm 8 \times 10^{-3}$ ps^2. For large chirp rates the positive and negative chirps lead to different enhancements, which is consistent with the theoretical predictions.[30] We expect the chirp enhancement to survive thermal averaging.

4.2 Intensity effects

The two-photon population transfer to the excited $(1)^1\Pi_g$ state is shown as a function of the initial angular momentum J for different intensities in Fig. 2. As the intensity increases, the peak of the distribution shifts to larger J values. For the intensity value used in the experiment, $\sim 10^{12}$ W cm^{-2}, the peak has its maximum at $J = 55$. This represents a compromise between a favorable Franck–Condon window and increasing phase space with J. Other minor peaks could be associated with rotational resonances.

5 Summary and conclusions

We have investigated femtosecond photoassociation of thermally hot atoms in the gas phase and its coherent control. In the photoassociation process, formation of a chemical bond is facilitated by light in a free-to-bound optical transition. We have demonstrated photoassociation of hot atoms experimentally by measuring fluorescence resulting from the absorption of $(2 \oplus 1)$ photons. The Fourier transform of the experimental signal reveals molecular coherence of both vibrational and rotational nature in the intermediate electronically excited state that is accessed by the two-photon photoassociating transition. Our simulations also show molecular coherence of the photoassociated molecules. The theoretical model is based on a state-of-the-art treatment of the electronic structure of Mg_2 including *ab initio* two-photon couplings and dynamic Stark shifts and a novel quantum molecular dynamics approach to modeling a thermal ensemble subject to coherent excitation.

The significance of coherently controlling femtosecond photoassociation is twofold. On one hand, it addresses the fundamental problem of coherently controlling a system having an initial state that is an incoherent thermal mixture of different scattering states. In thermal equilibrium the scattering states correspond to eigenstates of the ground state Hamiltonian. On the other hand, success of femtosecond control of free-to-bound photoassociation is an example of coherent control of a photo-induced bimolecular chemical reaction. Chemical reaction control is a long-standing goal from the early days of the field of coherent control as a possible means for inducing new chemical reactions. Despite its importance, femtosecond control of the photoassociation of 'hot' thermal atoms has previously been investigated experimentally by only a single pioneering study more than a decade ago,[8,31–37] demonstrating only the basic feasibility of spectroscopy of hot atom-atom photoassociation with subsequent rotational coherence using unshaped femtosecond pulses.

Acknowledgements

Financial support from the Deutsche Forschungsgemeinschaft through the Emmy Noether programme and SFB 450 is gratefully acknowledged. This research was supported by The Israel Science Foundation (Grant No. 1450/10) and by The James Franck Program in Laser Matter Interaction. RM and MT would like to thank the Polish Ministry of Science and Higher Education for the financial support through the project N N204 215539. MT was supported by the project operated within the Foundation for Polish Science MPD Programme co-financed by the EU European Regional Development Fund.

References

1 D. J. Tannor and S. A. Rice, *J. Chem. Phys.*, 1985, **83**, 5013.
2 S. A. Rice, *Science*, 1992, **258**(5081), 412.
3 S. A. Rice and M. Zhao, *Optical control of molecular dynamics*, John Wiley & Sons, 2000.
4 P. Brumer and M. Shapiro,*Principles and Applications of the Quantum Control of Molecular Processes*,Wiley Interscience, 2003.
5 C. P. Koch, M. Ndong and R. Kosloff, *Faraday Discuss.*, 2009, **142**, 389.
6 T. Brixner and G. Gerber, *ChemPhysChem*, 2003, 4(5), 418.
7 O. Kühn and L. Wöste, editors, *Analysis and control of ultrafast photoinduced reactions*, Springer, Berlin, 2007.
8 U. Marvet and M. Dantus, *Chem. Phys. Lett.*, 1995, **245**(4-5), 393.
9 W. Salzmann, T. Mullins, J. Eng, M. Albert, R. Wester, M. Weidemüller, A. Merli, S. M. Weber, F. Sauer, M. Plewicki, F. Weise, L. Wöste and A. Lindinger, *Phys. Rev. Lett.*, 2008, **100**, 233003.
10 V. Zeman, M. Shapiro and P. Brumer, *Phys. Rev. Lett.*, 2004, **92**, 133204.
11 C. P. Koch, R. Kosloff, E. Luc-Koenig, F. Masnou-Seeuws and A. Crubellier, *J. Phys. B: At., Mol. Opt. Phys.*, 2006, **39**, S1017.
12 C. P. Koch and R. Kosloff, *Phys. Rev. Lett.*, 2009, **103**, 260401.
13 C. P. Koch, R. Kosloff and F. Masnou-Seeuws, *Phys. Rev. A*, 2006, **73**, 043409.
14 W. Salzmann, U. Poschinger, R. Wester, M. Weidemüller, A. Merli, S. M. Weber, F. Sauer, M. Plewicki, F. Weise, A. Mirabal Esparza, L. Wöste and A. Lindinger, *Phys. Rev. A*, 2006, **73**, 023414.
15 B. L. Brown, A. J. Dicks and I. A. Walmsley, *Phys. Rev. Lett.*, 2006, **96**, 173002.

16 S. Amaran, R. Kosloff, M. Tomza, R. Moszyński, L. Rybak, Z. Amitay, and C. P. Koch, *in preparation*.
17 W. Skomorowski *et al.*, *in preparation*.
18 J. Olsen and P. Jørgensen, *J. Chem. Phys.*, 1985, **82**, 3235.
19 C. Hättig, O. Christiansen and P. Jørgensen, *J. Chem. Phys.*, 1998, **108**, 8331.
20 C. Hättig, O. Christiansen and P. Jørgensen, *J. Chem. Phys.*, 1998, **108**, 8355.
21 O. Christiansen, A. Halkier, H. Koch, P. Jørgensen and T. Helgaker, *J. Chem. Phys.*, 1998, **108**, 2801.
22 C. Hättig, O. Christiansen, S. Coriani and P. Jørgensen, *J. Chem. Phys.*, 1998, **109**, 9237.
23 C. Hättig, O. Christiansen and J. Gauss, *J. Chem. Phys.*, 1998, **109**, 4745.
24 R. Kosloff, *Annu. Rev. Phys. Chem.*, 1994, **45**, 145.
25 R. Kosloff, Time Dependent Methods in Molecular Dynamics, *J. Phys. Chem.*, 1988, **92**, 2087–2100.
26 U. Banin, R. Kosloff and S. Ruhman, *Chem. Phys.*, 1994, **183**, 289.
27 L. Rybak, L. Chuntonov, A. Gandman, N. Shakour and Z. Amitay, *Opt. Express*, 2008, **16**, 21738.
28 R. S. Judson and H. Rabitz, *Phys. Rev. Lett.*, 1992, **68**(10), 1500.
29 S. Ruhman and R. Kosloff, *J. Opt. Soc. Am. B*, 1990, **7**, 1748.
30 J. Vala and R. Kosloff, *Opt. Express*, 2001, **8**, 238.
31 P. Backhaus and B. Schmidt, *Chem. Phys. Lett.*, 1997, **217**, 131.
32 P. Gross and M. Dantus, *J. Chem. Phys.*, 1997, **106**, 8013.
33 U. Marvet, Q. Zhang and M. Dantus, *J. Phys. Chem. A*, 1998, **102**, 4111.
34 P. Backhaus, B. Schmidt and M. Dantus, *Chem. Phys. Lett.*, 1999, **306**(1-2), 18.
35 R. de Vivie-Riedle, K. Sundermann and M. Motzkus, *Faraday Discuss.*, 1999, **113**, 303.
36 P. Marquetand and V. Engel, *J. Chem. Phys.*, 2007, **127**, 084115.
37 E. F. de Lima, TS Ho and H. Rabitz, *Phys. Rev. A*, 2008, **78**, 063417.

General discussion

Professor Weinacht opened the discussion of the paper by Dr Sussman: In some optimal control calculations aimed at controlling dynamics through conical intersections performed by Regina de Vivie-Riedle,[1] she and her colleagues found that in many cases it was more favorable to couple the two potential energy surfaces of interest before the wave packet reached the conical intersection (CI) rather than trying to influence the branching of the wave packet through the conical intersection by shaping it beforehand. This seems to suggest that modifying the ski hill (in her case by resonant coupling between potentials) is more effective than preparing the skier with the appropriate momentum. In the calculations it seemed that the most effective location to modify the ski hill was before the CI. Can you comment on this in light of your experiments?

1 P. von den Hoff, M. Kowalewski, R. Siemering and R. de Vivie-Riedle, *IEEE J. Sel. Top. Quantum Electron.*, 2011, **99**, 1.

Dr Stolow replied: Even in the simplest case of modifying an avoided crossing in a diatomic, we recognized that there will be different scenarios under which non-resonant dynamic Stark effect (NRDSE) control can operate, for example modifying the potential **before** *versus* **during** the traversal of a wavepacket through the region of non-adiabatic coupling. In principle, the physics of the resonant versus the non-resonant case is the same: in the former one matrix element is dominant whereas in the latter, it is a sum over many (*i.e.* the polarizability). However, resonant coupling means dipole allowed transitions are dominant and therefore it is the first order and *not* the second order NRDSE which is operating. This is a very different scenario in which the excited state potentials oscillate at the carrier frequency rather than follow the envelope.

Professor Tannor asked: Consider a 2D potential energy surface where the objective is to steer the wavepacket uphill in the coordinate where the slope is positive. Further, imagine that there is a nearby potential that can be coupled by the light. Although in general, the coupling to this second surface could change the sign of the slope, imagine that for any value of the coupling parameter the sign of the slope the reaction coordinate remains unchanged but that the sign of the slope in the perpendicular coordinate can be changed by the coupling. We studied a case like this in the early 1990s (although we never published it). We called it "sailing into the wind". Our conclusion was that except in singular types of situations, it was always possible to go uphill by a succession of zigzags. This is a somewhat different interaction than the one you are considering (resonant *versus* non-resonant contributions to the polarizability) but the controllability considerations may be analogous.

Professor Baumert enquired: The nonlinear dynamic Stark effect is a powerful tool to alter the dynamics on potential energy surfaces as you convincingly demonstrated in your beautiful experiment on IBr. In your oral contribution you stressed the point, that the strong non resonant interaction is leaving the system without excitation after the laser pulse has finished. Considering the success of impulsive alignment with short laser pulses (where rotational ladder climbing is involved) I wonder to what extent this statement needs to be that strict in terms of further applications of this effect in quantum engineering.

Dr Stolow answered: As you point out, the NRDSE can also lead to ladder climbing, as in the case of molecular axis alignment via control of the rotational degree of freedom. I agree that in general there is no need for such a strict constraint.

Dr Leonard addressed Dr Stolow and Dr Sussman: Is it possible to do quantum control experiments using the dynamic Stark effect (which requires high laser intensities) in dense media?

Dr Stolow replied: NRDSE control may indeed be obtained in condensed media such as, for example, the phonon system in bulk diamond.[1] In general, however, propagation effects due to self-phase and cross-phase modulation, and self-focussing are issues which must be considered in each case.

1 F. C. Waldermann, B. J. Sussman, J. Nunn, V. O. Lorenz, K. C. Lee, K. Surmacz, K. H. Lee, D. Jaksch, I. A. Walmsley, P. Spizziri, P. Olivero and S. Prawer, *Phys. Rev. B*, 2008, **78**, 155201.

Dr Sussman replied: There are a number of challenges in performing quantum control in dense media, particularly using ultrafast pulses, which experience significant dispersion and nonlinear propagation effects. As well, the dephasing of a target system in dense media is often very rapid, particularly in ambient conditions. However, under appropriate circumstances these effects can be mitigated. We have recently been able to demonstrate an example of dense media control in bulk diamond in the context of quantum processing. There, the Stark effect was used to transiently shift phonon levels for the purposes of dynamically generating bandwidth in a quantum memory.

Professor Tannor asked: I am curious about the multidimensional extension of your scheme for control via the dynamical Stark shift. It seems as if you may only have a one-parameter type of control, in that the envelope of the pulse amplitude controls the energy separation between neighboring potential energy surfaces. In multidimensions, do you expect to be able to steer the wavepacket selectively along coordinate 1 or along coordinate 2? This is obviously a key question if one would like to achieve selective bond breaking in polyatomic molecules.

Dr Stolow responded: NRDSE control depends on the coordinate and field strength dependence of the second order dynamic Stark effect. The strength, timing, pulse envelope shape and duration are all independent 'knobs' available for control applications and, therefore, it is not a "one parameter" type of control. We hope to be able to steer wavepackets in multiple dimensions. The details on how this is achieved will be system specific.

Professor Fielding said: In section 2.3 of your paper, N-methylpyrrole is proposed as a candidate system for extending the application of the non-resonant dynamic Stark effect for molecular control from a diatomic to a polyatomic. Schematic potential energy curves are presented in Fig. 4; however, it should be remembered that there is a conical intersection seam along which loss of the CH_3 radical can occur. The strong, non-resonant laser field will modify the shapes of the potential energy surfaces around the seam and so it is likely that the wavepacket will end up accessing a totally different part of the seam. So whilst it may well be possible to control the branching ratio experimentally, do you think there is any possibility that we will understand the control mechanism in any detail?

Dr Stolow answered: In general, strong field control involves a multitude of coherent and incoherent laser–matter interactions, of which the dynamic Stark effect will necessarily be one. The non-resonant dynamic Stark effect (NRDSE) is the simplest nonperturbative strong field effect, since it does not involve any net absorption of photons and therefore does not involve propagation on several potential energy surfaces. It may indeed be challenging to calculate NRDSE control for a polyatomic systems (requiring the excited state polarizability as a function of nuclear coordinates), however it is greatly simpler than the general case of strong field control.

Professor de Vivie-Riedle asked: In case of polyatomic systems, low-lying seams of conical intersections are reached during the reaction. Their topology often defines the outcome of the photoreaction. Is it possible to use the dynamic Stark effect also to direct the reaction towards a certain region of the conical intersection seam? If yes, this would imply control in multi-dimensions. How high do you judge the probability of finding these requirements for flexiblility in polarizability in a molecular system.

Dr Stolow replied: The second order NRDSE operates *via* the molecular polarizability. The coordinate dependence of the polarizability is a molecular property. As such, it is a constraint that must be applied to any NRDSE-based control scheme. However, the timing, duration and strength (within limits) of the NRDSE interaction are under the control of the experimentalist. The details of how the NRDSE may or may not "direct" trajectories in certain directions will be system specific.

Professor Weinacht enquired: Have you considered using interference to measure a dynamic Stark shift? Another way of phrasing the question is to perhaps consider using a dynamic Stark shift to demonstrate interference—*i.e.* use the dynamic Stark shift to vary the phase of one arm of the molecular interferometer.

Dr Stolow responded: This has indeed been considered and was discussed by Sussman, Ivanov and Stolow[1] where NRDSE interference effects were shown to control the predissociation lifetime of a molecule.

1 B. J. Sussman, M. Yu. Ivanov and A. Stolow, *Phys. Rev. A*, 2005, **71**, 051401R.

Dr Sage asked: Some ambiguity remains in the dynamics of N–CH$_3$ bond cleavage in N-methylpyrrole (NMP), specifically in whether the channel forming low velocity methyl fragments comes to dominate as the photolysis wavelength is decreased owing to the increased efficiency of the S$_0$/S$_1$ conical intersection at extended N–CH$_3$ bond length, or whether this is due to some other channel, perhaps an alternative intersection in some other dimension of nuclear motion. Do you anticipate being able to shed light on this problem with your current dynamics experiments, and does this uncertainty affect the suitability of NMP as a candidate for NRDSE studies?

Dr Stolow answered: We hope that the combination of femtosecond time-resolved photoelectron spectroscopy[1] with *ab initio* molecular dynamics calculations will help to shed light on the excited state pathways of NMP. The outcome of these studies will address the suitability of NMP as a candidate for a pure second order NRDSE experiment. Of course, if one relaxes the constraint of implementing solely the second order NRDSE, then there will be many candidate molecules.

1 See, for example, J. G. Underwood and A. Stolow, *Adv. Chem. Phys.*, 2008, **139**, 497.

Professor Ohmori addressed Dr Sussman and Dr Stolow: You claim that this approach based on strong laser pulses gives high data rates with your Cs quantum memory. I consider that this quantum memory might be used as a quantum repeater, in which you should have coherence or entanglement distillation. Then it seems more important to have a longer storage time than a higher bit rate. How long is the storage time? Do you have any feasible idea for the entanglement distillation that can be implemented during your storage time? Can you think of another usage of your quantum memory?

Dr Sussman replied: Quantum memories are an essential component of quantum networks. Depending on their specific characteristics, quantum memories can be used in quantum repeaters, as timing devices, as on-demand single photon-sources, or for high precision metrology applications. There are two important metrics for memory storage times: the total duration and the number of operational time bins before the system dephases. The current principle storage time constraint in caesium is diffusion out of the beam, which limits the current configuration to microseconds before the stored state decays. The caesium quantum memory is capable of operating at GHz data rates, implying a significant number of operational time bins before coherence is lost. For both metrics, these parameters are highly competitive with current alternatives.

Professor Shapiro remarked: Regarding the issue of the effect of phase on the IBr photodissociation to yield $I^+Br(2P3/2)$ *versus* $I^+Br(2P1/2)$, we have shown[1] that weak field phase control can bring about great selectivity of these reaction products while obviating the need to synchronize pulses or to know the exact potential curves involved. One may argue that strong is "beautiful" because strong fields bring about higher over all yields, but we have solved this problem by "marrying" coherent control with adiabatic passage[2] thereby enjoying the best of both worlds, high over all yields and high selectivities.

1 See *e.g.*, I. Levy, M. Shapiro, and P. Brumer, *J. Chem. Phys.*, 1990, **93**, 2493; C. K. Chan, P. Brumer and M. Shapiro, *J. Chem. Phys.*, 1991, **94**, 2688.
2 See, P. Kral, I. Tganopoulos, and M. Shapiro, *Rev. Mod. Phys.*, 2007, **79**, 53.

Professor de Vivie-Riedle opened the discussion of the paper by Dr Bryan: In the case of the N_2 experiment. What are the potential energy surfaces you are aiming at for the control of the electron dynamics? And have your already tried to use the carrier envelope phase of a few cycle pulse to control the asymmetry in the ion fragments.

Dr Bryan answered: The initial form of this experiment will be carried out in the lowest-lying orbitals as accessed by a strong NIR pulse, so should be dominated by the HOMO and HOMO-1, however I think there is a prospect of employing either a single XUV photon or a small number of UV photons to access a well-defined group of states. In principle, any permitted state in the neutral, cation, or dication could be accessed. The asymmetry measurement is yet to be attempted but will be carried out in the next month or so.

Professor Shapiro commented: The effects of the probe laser might be considerable in this case. We have recently presented a study[1] on the strong field dissociation of H_2^+ in a (successful) attempt to explain Silberberg *et al.*'s measured kinetic energy distributions[2] of the photofragments resulting from this process and have seen that very extensive continuum-mediated resonance Raman transitions between the vibrational and rotational states of H_2^+ take place. Thus, your "probe" laser is likely to take an active role in modifying the vibrational and rotational distributions of the molecule and your results are not merely due to the initial vib-rotational distribution established by the H_2 ionizing pulse.

1 Shapiro *et al.*, *Phys. Rev. A*, 2011, **83**, 033415.
2 V. S. Prabhudesai *et al.*, *Phys. Rev. A*, 2010, **81**, 023401.

Dr Bryan responded: In earlier theoretical studies carried out with Jim McCann (Queen's University, Belfast),[1] we investigated the influence of the probe on the wavepacket in a solution to the TDSE. I also investigated the influence of the probe in the currently applied QCM. I agree the probe will modify the rovibrational state to some extent as it is not a delta function, however we found that this modification essentially "smears" the wavepacket over a small number of states and that the effect is less significant than the influence of integrating over the laser focal volume.

1 D. S. Murphy, J. McKenna, C. R. Calvert, I. D. Williams and J. F. McCann, *New J. Phys.*, 2007, **9**, 260.

Professor Bearpark asked: What are the prospects for experimentally studying and controlling dynamic changes in the electron distribution of polyatomic molecular cations, as suggested for example by Cederbaum and Zobeley?[1]

1 L. S. Cederbaum and J. Zobeley, *Chem. Phys. Lett.*, 1999, **307**, 205–210.

Dr Bryan replied: The prospects for experimental study of electron distribution are currently advancing apace. Attosecond UV and XUV pulses will access electronic timescales while X-ray free electron lasers and single-harmonic sources will allow individual states to be accessed on femtosecond timescales. I see no reason that current coherent control pulse shaping and optimization methodologies should not be applied to electronic manipulation in polyatomics, and suggest that the rather challenging experimental implementation of such technologies will be far easier than corresponding theoretical predictions beyond the Born–Oppenheimer approximation. In terms of observing the influence of a shaped pulse to drive electron dynamics, current cold target recoil ion momentum spectroscopy, velocity map imaging, photoelectron spectroscopy and coulomb explosion imaging will all be applicable. The diffraction of XUV or electronic pulses will also be invaluable tools.

Dr Amitay opened the discussion of the paper by Dr Plenge: What was the pulse duration of the original XUV excitation pulse (HH13; 20.05 eV) before it had been phase shaped?

Dr Plenge answered: The pulse duration of the unshaped 13th harmonic has been determined by cross-correlation measurements in rare gases. Typically, a cross-correlation width of 100 fs has been measured indicating a pulse duration of 60 fs for the unshaped XUV pulse.

Professor Whitaker enquired: In your presentation you mentioned that you are implementing a velocity map imaging spectrometer to measure the angular distributions, but a crude estimate of the anisotropy might be obtained simply by recording the signals with the relative polarizations of the XUV and 800 nm either parallel or perpendicular. Have you tried such an experiment?

Dr Plenge responded: We did not try such an experiment.

Dr Stolow communicated: Time-resolved photoelectron angular distributions may also be used to probe coherent electronic dynamics in atomic and molecular systems.[1] Will measuring photoelectron angular distributions provide additional information in your experiments?

1 See, for example P. Hockett *et al.*, *Nature Physics*, 2011, **7**, 612.

Dr Plenge communicated in response: Time-resolved photoelectron angular distributions can provide information on the dynamics of orbital alignment of the 3d electrons in neon. Since the wavepacket is created from the singlet ground $^1S_0$, the ionic core and the excited electron have to be entangled. This is expected to be comparable to an experiment on krypton where time-dependent electron orbital alignment of coherently excited d electrons is observed during the recurrences of a Rydberg wavepacket.[1] However, it should be noted that the coherently excited 3d states in neon are built on a core with an angular momentum $J_c = 3/2$. Therefore, a more complex dynamics of the entangled states is expected compared to the experiments on krypton where the excited Rydberg states are built on a $J_1 = 1/2$ core.

1 Gilb et al., J. Phys. B, 2006, **39**, 4231.

Professor de Vivie-Riedle continued the discussion of the paper by Dr Stolow: The molecule used successfully for the dynamic Stark control was iodine bromide. This is a highly polarizable molecule. Larger molecules are coming more and more into the focus. Are they more or less polarizable? Is there a general trend? And in the case where the trend is towards less polarizability, will the higher laser power then needed complicate the control mechanism?

Dr Stolow replied: The excited state polarizability depends on both electronic level spacing and dipole matrix elements. As such, the strength of the second order NRDSE interactions may be very strong or relatively weak and I do not see a general trend. Furthermore, in general, both the first order and second order NRDSE may be operational. As fields become stronger still, higher order effects due to hyperpolarizabilities become operational, ultimately leading to electronic 'ladder climbing' and ionization. Nevertheless, the NRDSE remains the first and therefore simplest nonperturbative interaction and it participates in all strong field laser–matter interactions.

Dr Leibscher† commented: Inspired by the experiments on the non-resonant dynamic Stark effect of IBr by the group of Professor Stolow[1] and by investigations on the effect of static electric fields on the conical intersections of Merocyanine,[2] we, together with Steffen Belz (Freie Universität Berlin) and Shmuel Zilberg (Hebrew University of Jerusalem), are studying how electric fields influence the conical intersection and the subsequent radiationless decay of pyridinylidene-phenoxide (see our Fig. 1). We choose pyridinylide-phenoxide (PPO) because of its large dipole moment and polarizability anisotropy. PPO has a conical intersection between the ground electronic ground state S_0 and the first excited state S_1, which is connected with the torsion around the C–C bond that links together the two rings. The conical intersection has been localized for a torsion angle of using the Longuet–Higgins sign change theorem.[3] A one-dimensional cut of the adiabatic potential energy surfaces close to the conical intersection is shown in our Fig. 2(a) in blue lines. The blue line in Fig. 2(b) represents the corresponding non-adiabatic coupling term (NACT), $\left\langle \phi_0 \left| \frac{\partial}{\partial \varphi} \phi_1 \right. \right\rangle$, where $|\phi_0\rangle$ and $|\phi_1\rangle$ denote the two electronic states. All quantum chemical calculations were done with CAS (14,11) at MRCI level of theory using the aug-cc-pVDZ basis set. In order to investigate the effect of an electric field or a non-resonant laser pulse on the conical intersection, we also calculated the potential energies and NACTs in the presence of electric fields with field strength $F = 3 \times 10^8$ V m^{-1} (red lines in Fig. 2) and $F = 1 \times 10^9$ V m^{-1} (black lines in Fig. 2). Due to the Stark effect, the potentials are slightly shifted in the presence of an electric field. The consequence of a slight shift in energy is most pronounced in the vicinity of a conical intersection. Here, a small change of the gap between two potentials leads to a considerable change in the NACTs, as can be seen in Fig. 2(b). In order to demonstrate how the nuclear dynamics of photo-excited PPO can be controlled by varying the strength of the electric field, we show the population of the electronic ground state S_0 after vertical excitation of the initial torsional ground state—so far for a one-dimensional model with the torsion angle φ as nuclear coordinate. Without electric field, the conical intersection causes

---

† In collaboration with S. Belz and S. Zilberg.

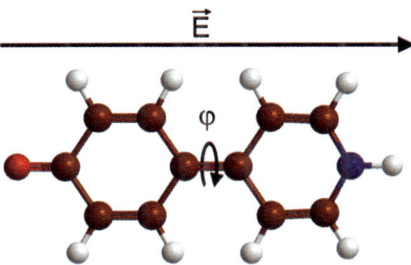

Fig. 1 Model of the pyridinylidene-phenoxide (PPO) molecule, indicating the torsion angle and the electric field that is applied to control the radiationless decay of photo-induced PPO.

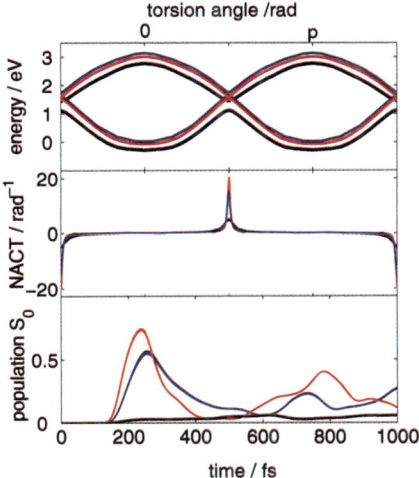

Fig. 2 Torsional potentials of PPO (a), non-adiabatic coupling terms (b) and the ground state population of photo-excited PPO (c). The strength of the electric field is $F = 0$ (blue lines), $F = 3 \times 10^8$ V m^{-1} (red lines) and $F = 1 \times 10^9$ V m^{-1} (black lines).

radiationless decay, which leads to an increase of the ground state population to approx. 50 % (blue lines in Fig. 2(c)). For a field strength of $F = 3 \times 10^8$ V m^{-1}, the ground state population increases to almost 80% (red line in Fig. 2(c)). A larger field strength (here with $F = 1 \times 10^9$ V m^{-1}) leads to a decrease of the NACTs at $\varphi = 90°$, and, as a consequence, the radiationless decay to the ground state is almost zero (black line in Fig. 2(c)). By changing the field strength it is thus possible to control the radiationless decay through a conical intersection. The simulations of the nuclear dynamics have been carried out for a one-dimensional model. Additional nuclear coordinates have to be taken into account for a more realistic model. However, the present investigations already suggest that electric fields can strongly influence the NACTs and the radiationless decay *via* conical intersections and can thus be applied for the control of non-adiabatic nuclear dynamics. We thank Prof. L. González, Dr. J. González-Vásquez, Prof. Y. Haas and Prof. J. Manz for stimulating discussions and advice and we acknowledge the financial support of the DFG in the framework of the project MA 515/22-3.

1 B. J. Sussmann, D. Townsend, M. Y. Ivanov and A. Stolow, *Science*, 2006, **314**, 278.
2 X. Xu, A. Kahan, S. Zilberg and Y. Haas, *J. Phys. Chem. A*, 2009, **113**, 9779.
3 Y. Haas, S. Cogan and S. Zilberg, *Int. J. Quant. Chem.*, 2005, **102**, 961.

Professor Ashfold remarked: I note that the schematic potentials for N-methylpyrrole (NMP) shown in Fig. 4 of paper 19 provide a nice illustration of the diversity of possible topologies at a conical intersection (CI). The CI at extended R_{N-CH_3} bond lengths is a local minimum, which might be expected to act as an efficient funnel for dissociating flux. Two of the displayed potentials also intersect at short R_{N-CH_3}. This would also be a CI in $3N-8$ dimensional configuration space, but its dynamical impact (in the displayed coordinate at least) would likely be hard to discern. I would also like to suggest a family of molecules whose dissociation dynamics might be more amenable to modification by the non-resonant dynamic Stark effect. My reservations about NMP centre on the likely multiplicity of (potentially difficult to distinguish) routes leading to 'slow' CH_3 fragments. Recent studies of S–H bond fission in a range of para-substituted thiophenols have revealed striking variations in the relative yields of ground (X^2B_1) and excited (A^2B_2) state thiophenoxyl radical products, which are sensitive to the choice of excitation wavelength and/or *para*-substituent.[1] Both sets of products arise as a result of dissociation on the $1^1\pi\sigma^*$ potential energy surface (PES), with the eventual branching established by coupling at the CI linking the $1^1\pi\sigma^*$ and S_0 PESs at extended R_{S-H}. Whether such variations extend to the loss of CH_3 fragments following photolysis of the corresponding thioanisoles remains to be established.

1 T. A. A. Oliver, PhD thesis, University of Bristol, 2011.

Professor Bearpark addressed Dr Stolow, Professor Ashfold, Professor de Vivie-Riedle and Dr Leonard: Conical intersections of potential energy surfaces can be both peaked and sloped.[1] Sloped crossings—with parallel gradients on upper and lower surfaces—have been linked with molecules that are inherently photostable,[2,3] as a result of having a single pathway back to reactants on the lower surface.

We've been reminded that conical intersections are extended structures or seams, and can exist for a range of molecular geometries. Computationally, it's known that crossing points connected along such a seam can change topology from peaked to sloped (for example, see ref. 4). But following on from Professor Ashfold's comment: has manipulation of decay from peaked to sloped regions along a single crossing seam been demonstrated experimentally yet?

1 G. J. Atchity, S. S. Xantheas and K. Ruedenberg, *J. Chem. Phys.*, 1991, **95**, 1862–1876.
2 B. E. Applegate, T. A. Barckholtz and T. A. Miller, *Chem. Soc. Rev.*, 2003, **32**, 38–49.
3 K. F. Hall, M. Boggio–Pasqua, M. J. Bearpark and M. A. Robb, *J. Phys. Chem. A*, 2006, **110**, 13591–13599.
4 D. Mendive–Tapia, B. Lasorne, G. A. Worth, M. J. Bearpark and M. A. Robb, *Phys. Chem. Chem. Phys.*, 2010, **12**, 15725–15733.

Dr Stolow replied: To the best of our knowledge, NRDSE control of dynamics at conical intersections has not been demonstrated experimentally.

Professor de Vivie-Riedle replied: In a combined experimental and theoretical approach we recently studied the photoinduced isomerization of hemithioindigo-hemistilbene in real time using UV/vis pump–probe experiments and revealed ultrafast processes for both directions[1] with significant different quantum yield and time constants. High-level quantum chemical calculations were performed to correlate the relevant time steps with the atomistic mechanisms.[1] As driving forces we identified the charge separation and charge balance recovery mediated by the electron donating or withdrawing character of the heteroatoms, also influencing the structure (sloped or peaked), energetic position, and accessibility of a number of CoIn. The experimental findings in quantum yields and time constants for forward and backward reaction is a direct consequence from the change in seam topology and might be regarded as a first experimental demonstration.

1 A. Nenov, T. Cordes, T. T. Herzog, W. Zinth and R. de Vivie-Riedle, *J. Phys. Chem. A*, 2010, **114**, 13016.

Professor Miller said: In reference to the suggested use of the non-resonant dynamic Stark effect in controlling passage through conical intersections of polyatomic molecules, I would like to ask the general question regarding how to experimentally identify a conical intersection in directing non-radiative relaxation processes. From a theoretical perspective, a conical intersection can be identified if a sufficiently accurate excited state surface can be calculated. Even in this case, I believe identifying a conical intersection is nontrivial. Highly simplified model structures generally need to be employed to approximate the system of interest for even moderate-sized molecules (>10 atoms). The real question here pertains to experimental evidence for conical intersections. I don't think it is possible to unequivocally state a process proceeds through a conical intersection just based on the observed excited state dynamics. Prior to the proliferation of explanations based on conical intersections, fast non-radiative dynamics were explained based on a high density of vibronic states and favorable Franck–Condon factors for fast relaxation processes between electronic surfaces. The old concepts based on the energy gap law for non-radiative relaxation were formulated for fairly rigid ring structures. However, for molecules with alkyl side groups, pendant groups, or flexible structures, there are high densities of low frequency modes to create a multitude of degenerate seams with good Franck–Condon factors to create fast non-radiative relaxation channels. See for example the dramatic effect of magnetically tuning the vibrational resonance between singlet and triplet manifolds of pentacene at low temperature on the fluorescence quantum yield as a simple case in point.[1] The overall transition depends on both the Franck–Condon vibrational overlap and resonant conditions. It is clear that the Franck–Condon overlap/factor is sufficient to yield picosecond dynamics for this single process that is dominated by CH modes never mind the high density of degeneracies involving lower frequency modes for most molecules of interest. It would seem to me that rapid non-radiative relaxation through the high density of degenerate vibronic states intersecting different electronic surfaces would be the general case for moderate to large molecular systems of greatest interest in photochemistry. How can we then experimentally distinguish unambiguously if the relaxation dynamics are proceeding through conical intersections as opposed to high density seams in the vibronic density of states connecting electronic surfaces? This is not a semantic issue as it pertains to the degree of coherent control possible for larger systems. It may well be that the degree of coherent control and pulse shapes will enable tracing out these details in the relaxation dynamics.

1 Astilean *et al.*, *Chem. Phy. Lett.*, 1994, **219**, 95–100.

Dr Stolow responded: In the limit of frozen nuclei, the Born–Oppenheimer approximation is exact. Its breakdown, therefore, is uniquely due to the motions of the atoms and the timescales of these motions relative to inverse electronic level spacings. The physics of nonadiabatic coupling is the same whether one considers either the 'conical intersection' or 'energy gap law' limiting case situations. The only qualitative difference lies in the coordinate dependence of the coupling matrix element. If the nonadiabatic coupling integral accumulates smoothly as a function of coordinates, one obtains the situations of the older 'energy gap law' as well as the 'high density of degenerate vibronic states' case you allude to, the only difference being one of strength of the coupling. However, if by contrast the nonadiabatic coupling integral accumulates rapidly near a particular locus, then this indicates an electronic degeneracy and a likely conical intersection. In this case, the 'radiationless relaxation' is often very rapid due to the real curve crossing (electronic degeneracy). The central role of conical intersections in the excited state dynamics of polyatomic molecules is well established in theoretical chemistry, as may be found for example in the book by Koeppel, Domcke and Yarkony.[1] It is the case that experiments studying dynamics at conical intersections have generally relied on quantum chemistry to characterize these. Of course, there is nothing wrong with that and current methods already permit direct *ab initio* calculations on reasonably large molecules. In order to experimentally discern conical intersections versus a less coordinate dependent coupling due to a 'high density of degenerate vibronic states', it would be necessary to study the excitation energy dependence of the rates. Again, the physics of the interaction (nuclear kinetic energy derivative coupling) is the same in either case so that there will be situations where both "pictures" may be relevant. Whether or not quantum control schemes can help to differentiate these limiting cases remains to be seen.

1 *Conical Intersections II: electronic structure, dynamics & spectroscopy*, ed. W. Domcke, D. Yarkony, H. Köppel, World Scientific, Singapore, 2011.

Professor Shapiro opened the discussion of the paper by Dr Amitay and Professor Kosloff: Contrary to the claim by the author, the topic of "hot" photoassociation is not new. Many researchers have contributed to it, and many papers, including works on the much more complicated atom-diatomic recombinations, such as O + CO to yield CO_2 in the excited B state (analyzed for example by Dixon[1]); O + NO → NO2 and O + SO → SO_2 have been published. There is an excellent review by Tucker Carrington Sr.[1] I fail to see what is so special about performing these experiments with femtosecond lasers. Is there any special advantage over longer pulses? My experience on ultracold atom photoassociation shows precisely the opposite, the yield goes up with the pulse duration because in this way more colliding atom pairs have a chance to be found at close enough inter-atomic distances during the laser pulse so as to be able to respond to the laser and be excited to a bound molecular state.

1 T. Carrington and J. C. Polanyi, "*Chemiluminescent Reactions*", in *Chemical Kinetics, Physical chemistry Series One*, ed. J. C. Polanyi, Butterworths, London, 1972 and references therein.

Dr Amitay replied: One of the main coherent control strategies for controlling photo-induced binary bimolecular reaction is the Tannor–Kosloff–Rice (TKR) pump–dump scheme using femtosecond laser pulses.[1,2] The first (pump) step of the TKR scheme is a photoassociation of the free colliding reactants into an electronically excited bound state that is induced by a femtosecond pulses, *i.e.*, femtosecond photoassociation (fs-PA). This step also sets the zero time for the whole scheme. At the "hot" temperature range, where the temperature of the reactants is in the range of hundreds of Kelvin, we refer to the process as hot fs-PA. Hence, hot fs-PA leading to chemical bond formation is a perquisite for coherent control of hot binary reactions along the TKR scheme and its variants. Despite its importance, previously there has been only a single observation of the hot fs-PA process in a pioneering experiment conducted by Dantus and his group more than 15 years ago.[3] In this experiment, they have demonstrated hot fs-PA of mercury atoms via a one-photon UV transition, and observed rotational coherence of the photoassociated molecules. Except for this single study, all the many previous photoassociation studies, including those mentioned in the question asked, were not conducted using femtosecond pulses. In order to utilize the hot fs-PA process as a basis for coherent control of binary reactions, it is, however, essential to induce coherent dynamics of the photoassociated molecules that is of vibrational nature. In the present work, as is also supported theoretically, we indeed demonstrate experimentally the formation of excited magnesium diatomic molecules with both rotational and vibrational coherence via free-to-bound two-photon hot fs-PA process. The significance of our work draws from all the above explanation. It also worth noting that femtosecond pulses are special in that they allow for a different efficient excitation process than longer pulses. A femtosecond pulse is shorter than the timescale of any motional dynamics of the diatom. Hence, to our benefit, the bandwidth of a femtosecond pulse can be translated into a Franck–Condon region around the Franck–Condon point where resonant excitation is possible, while for longer pulses the resonance condition is essentially fulfilled only at a single interatomic separation. The broad bandwidth of the femtosecond pulse thus allows for an efficient spatial filtering of those atom pairs which have the 'correct' interatomic separation for photoassociation, *i.e.*, for distinguishing those atom pairs from the remaining incoherent ensemble with a large distribution of interatomic separations. We emphasize that photoassociation at ultracold and hot temperatures proceeds in entirely different ways. In particular, the reactants are already pre-correlated at ultralow temperatures[4] while at high temperatures the ensemble is completely incoherent. In the ultracold regime, the initial ensemble is more pure due to the cooling process. In the hot regime, Franck–Condon filtering provides thus an alternative way to achieve any significant quantum purity. In summary, the initial state is different and also the excitation mechanism is different (short *versus* large interatomic separations, high *versus* low *J*), so it is not straightforward at all to apply insight from ultracold photoassociation to what happens at high temperatures.

1 D. Tannor, R. Kosloff, and S. A. Rice, *J. Chem. Phys.*, 1986, **85**, 505.
2 R. Kosloff, S. A. Rice, P. Gaspard, S. Tersigni, and D. Tannor, *Chem. Phys.*, 1989, **139**, 201.
3 U. Marvet and M. Dantus, *Chem. Phys. Lett.*, 1995, **245**, 393.
4 C. P. Koch and R. Kosloff, *Phys. Rev. Lett.*, 2009, **103**, 260401.

Professor de Vivie-Riedle asked: In the case of bimolecular reaction, the vibrational energy spacing in the dissociative regime is very small and thus the time evolution of the incoming wavepacket very slow. To adjust to these elongated time scales I had the impression that exprimentalist nowadays prefers longer pulses in the picosecond regime. Why do you use femtosecond pulses for the control?

Dr Amitay answered: Picosecond pulses fit the photoassociation (PA) dynamics best in the ultracold temperature range, while femtosecond pulses fit the PA dynamics best in the hot temperature range. Below is given a "back-of-an-envelope" classical calculation to qualitatively estimate the time scale characterizing the PA at the hot temperature case:
- For the collision energy of 0.5 KT with $T = 1000$ K as in our study (KT = 1.33×10^{-20} J), the relative velocity of the two colliding Mg atoms ($\mu_{reduced} = 1.99 \times 10^{-16}$ kg) along their internuclear axis is $v = 817$ m s^{-1}.
- As shown in Fig. 2 of our paper, the width of the present effective Franck–Condon window for PA (accounting also for the actual strong-field excitation) is about 1 Bohr. Hence, the corresponding timescale is $(0.529 \times 10^{-10}$ m) / (817 m s^{-1}) = 65 fs.
- Assuming that the velocity upon entering the Franck–Condon window is this calculated velocity v and it becomes zero at the inner edge of the Franck–Condon window (the classical turning point), then—the average velocity within the Franck–Condon window is approximately $v/2$, and the corresponding timescale is about 130 fs.
- The diatom stays twice this time within the Franck–Condon window (going in and out). So, overall, we get that the unbound magnesium diatom stays within the Franck–Condon window on the ground electronic state for a time of about 250–300 fs.
- A corresponding calculation that accounts for the exact shape of the Mg$_2$ electronic ground state for $J = 55$ (see Fig.1–2 of our paper), considering the Franck–Condon window to be located at the 6.8–7.8 Bohr, gives a result of about 270 fs.

Hence, indeed, it is the femtosecond time scale that best fits the photoassociation dynamics on the ground state at the hot temperature range. So, on one hand, the PA yield in our magnesium study is expected to be enhanced by femtosecond pulses having a duration of about 250–300 fs. However, on the other hand, for generating photoassociated molecules with high vibrational coherence the photoassociating pulses should be short with respect to the vibrational dynamics in the excited bound state. As seen in Fig. 4(b) of our paper, the corresponding vibrational time scale in our magnesium experiment is about 150 fs. Hence, overall, there is an interesting interplay here between the different control objectives of enhanced yield *versus* enhanced coherence.

Professor Kosloff commented:

I. PURIFICATION OF A HOT ENSEMBLE
The purity is defined as:

$$P = Tr\{\hat{\rho}^2\} \quad (1)$$

A pure state has $P = 1$ and a maximally mixed state $P = 1/N$ where N is the number of states.

The coherence is defined as:

$$C = Tr\{(\hat{\rho} - \hat{\rho}_D)^2\}, \quad (2)$$

where $\hat{\rho}_D$ is only the stationary diagonal part of the density operator in the energy representation $[\hat{\rho}_D, \hat{H}] = 0$. These measures allow us to compare different quantum ensembles. Alternative measures can be based on entropy.

A. Purity of the ground state $^1\Sigma_g^+$

We calculate the purity of a thermal initial ensemble of Mg atoms in temperature of 1000K and density of $\rho = 4.8 \times 10^{16}$ atoms cm^{-3} We define a small spherical volume $v = 4/3\pi R^3$.

1. We define P_2 as the probability of finding a pair of atoms in the volume v. $P_2 = 9.2 \times 10^{-8}$.

2. We calculate the initial purity of a thermal pair of angular momentum J within the volume v.

$$P_J^g = Tr\left\{\left(\frac{e^{-\beta \hat{H}_J^g}}{Z_J^g}\right)^2\right\}, \quad (3)$$

where $\hat{H}_J^g$ is the ground state Hamiltonian with angular momentum J and Z_J^g is the associated partition function.

3. The initial thermal purity then becomes:

$$P^g = P^2 \sum P_J^2 P_J^g \quad (4)$$

where P_J is the thermal weight of angular momentum J within the box.

We obtain $P^g = 2.6 \times 10^{-20}$. This means that the initial thermal density is spread on $N \sim 10^{20}$ elementary states!

B. Purity of the target state $^1\Pi_g$

$$P^e = \sum_J P_J^2 P_J^e, \quad (5)$$

where P_J is the probability of excitation from the ground to the excited state for angular momentum J:

We obtain: $P^e = 3.04 \times 10^{-4}$. The coherence measure of this ensemble is quite similar. This means that the pulsed photoassociation experiment purfies the ensemble by 16 orders of magnitude. A pure ensemble is important if we consider photoassociation as the first step in coherent control of a binary reaction.

1 L. Rybak, S. Amaran, L. Levin, M. Tomza, R. Moszynski, R. Kosloff, C. P. Koch, Z. Amitay, 2011, arXiv:1107.4755 [quant-ph].

Dr Amitay replied: We qualitatively interpret the physical mechanism, which underlies the generation of a photoassociated molecular ensemble with a higher purity as compared to the initial thermal atomic ensemble, as "Franck–Condon filtering".[1] In this mechanism, for a given initial J value there is only a very limited range of collision energies that allow the colliding Mg atoms to reach the Franck–Condon window for the photoassociation at short internuclear distances, *i.e.*, there is an efficient filtering of a narrow range of collision energies from which photoassociation takes place.

1 L. Rybak, S. Amaran, L. Levin, M. Tomza, R. Moszynski, R. Kosloff, C. P. Koch, Z. Amitay, 2011, arXiv:1107.4755 [quant-ph].

Dr Leonard addressed Dr Amitay, Professor Kosloff and Professor Shapiro: In photoassociation experiments in ultracold atomic vapors, molecules are produced in very high vibrational states, and very large interatomic distances due to Franck–Condon overlap with the original free pairs of atoms which do have large average interatomic distances. For these vibrationally excited molecules, the vibrational level spacing is very small and consequently vibrational dynamics is relatively slow as was pointed out by Professor de Vivie-Riedle. What is different in this respect in a hot atomic beam? Interatomic distances are presumably also very large, and I expect that bound states should also preferentially be created in very excited vibrational levels, for the same reason. Why isn't it the case? Since you seem to produce molecules in less excited vibrational states (as compare to cold vapor experiments), and since the proportion of free atom pairs having very small interactomic distances is presumably very small (even smaller than in cold vapor experiments?), what is the efficiency of the photoassociation process in a hot atomic beam?

Professor Kosloff replied: There is a profound difference between pulsed photoassociation of a hot or an ultracold ensemble of atoms. In a cold ensemble the initial density is determined by Wigner threshold laws.

The initial density lacks kinetic energy and therefore has a very long wavelength. As a result it is reflected from the downhill gradient of the interatomic potential. This means that a very small probability amplitude can be found in the region of the attractive well of the interatomic potential. Because of low kinetic energy S wave scattering (zero angular momentum) dominates. As a result photoassociation takes place at large interatomic distances. We have studied these processes and found that we can enhance the photoassociation by chirping the excitation pulse.[1,2] In hot photoassication the initial density has the largest amplitude at the classical turning point on the repulsive part of the interatomic potential energy surface. Photoassociation probability is determined by Frank–Condon overlap. We find that for Mg photoassociation to the $^1\pi_g$ surface very large angular momentum values dominate the process (l = 55).

1 C. P. Koch, R. Kosloff, E. Luc-Koenig, F. Masnou-Seeuws and A. Crubellier, *J. Phys. B: At. Mol. Opt. Phys.*, 2006, **39**, S1017–S1041.
2 M. J. Wright, J. A. Pechkis, J. L. Carini, S. Kallush, R. Kosloff, and P. L. Gould, *Phys. Rev. A*, 2007, **75**, 051401(R).

Dr Amitay replied: The photoassociation efficiency, both in the ultracold and in the hot regime, is determined entirely by the initial state of the scattering atoms since all atom pairs within the Franck–Condon window are easily excited. However, the initial state is fundamentally different for these two extrema of temperature. In the ultracold regime, only very few partial waves contribute to the density of atom pairs in the range of interatomic separations where photoassociation happens, and high partial waves are frozen out. Moreover, at short and intermediate interatomic separations, the potential energy dominates over kinetic energy, and all scattering wave functions show the same pattern of nodes. This is a signature of quantum correlations which persist in the ultracold regime, *cf.* ref. 1. The resulting pair density shows a distinct nodal structure at short and intermediate separations and a flat part at large separations, cf. Fig. 2 in ref. 2. Photoassociation in the ultracold regime is most efficient typically at interatomic separations of 40–150 bohr,[3] which is a compromise between sufficient pair density in the electronic ground state and sufficiently large binding energies in the electronically excited state. The picture is completely changed at high temperature where low partial waves play essentially no role, and partial waves around $J = 50$–100 carry the largest weight in the Boltzmann ensemble. The largest pair density is found roughly at the position of the maximum of the rotational barrier, see inset of Fig. 1 of our paper, at interatomic separations of about 10 bohr. This is where photoassociation happens for high temperatures—in strong contrast to ultracold photoassociation. The binding energies in the electronically excited state that are accessed in hot photoassociation depend on the potential energy surface in the Franck–Condon region. In our case, the electronically excited state displays a very deep minimum while the electronic ground state is of van der Waals type. This is in contrast to the previous study by de Vivie-Riedle, Sundermann and Motzkus where the potential well of the electronically excited state was less deep and broader.[4]

1 C. P. Koch and R. Kosloff, *Phys. Rev. Lett.*, 2009, **103**, 260401.
2 C. P. Koch *et al.*, *J. Phys. B*, 2006, **39**, S1017.
3 C. P. Koch, R. Kosloff, F. Masnou-Seeuws, *Phys. Rev. A*, 2006, **73**, 043409.
4 R. de Vivie-Riedle, K. Sundermann and M. Motzkus, *Faraday Discuss.*, 1999, **113**, 303.

Professor Weinacht asked: Can I consider the purity improvement via the Franck–Condon filtering the same as using a narrow slit to improve the transverse coherence of a molecular beam as in the Ulbricht experiments?

Dr Amitay answered: First, it is worth mentioning that, generally speaking, an existing coherence can usually be observed only within an ensemble of relatively high purity. Still, there is a difference between purity and coherence and their corresponding measures.[1,2] In our femtosecond photoassociation work we generate both purity and rotational–vibrational coherence in the photo-associated molecular ensemble.[1] Regarding the question asked, from the point of view of purity increase by filtering out of the initial thermal ensemble a sub-ensemble of higher purity, the two cases (*i.e.*, our femtosecond photoassociation case and the case mentioned in the question of using a narrow slit in a thermal molecular beam to select a narrow range of transverse velocities) are indeed qualitatively similar. Another qualitatively similar example is the selective molecular excitation using a high-resolution cw laser that is tuned to a single state-to-state molecular excitation. Then, out of an initial thermal ensemble of ground-state molecules one can create a sub-ensemble of high purity composed of molecules that populate a single rovibrational state on an electronically-excited state. We have previously used this technique in experiments of state-selective coherent control of molecular superpositions[3,4] as well as quantum information experiments.[5,6]

1 L. Rybak, S. Amaran, L. Levin, M. Tomza, R. Moszynski, R. Kosloff, C. P. Koch, and Z. Amitay, 2011, arXiv:1107.4755 [quant-ph].
2 U. Banin, A. Bartana, S. Ruhman, and R. Kosloff, *J. Chem. Phys.*, 1994, **101**, 8461.
3 R. Uberna, Z. Amitay, R. A. Loomis, and S. R. Leone, *Faraday Discuss.*, 1999, **113**, 385.
4 Z. Amitay, J. B. Ballard, H. U. Stauffer, and S. R. Leone, *Chem. Phys.*, 2001, **267**, 141.
5 J. Vala, Z. Amitay, B. Zhang, S. R. Leone, and R. Kosloff, *Phys. Rev. A*, 2002, **66**, 062316.
6 Z. Amitay, R. Kosloff, and S. R. Leone, *Chem. Phys. Lett.*, 2002, **359**, 8.

Dr Stolow enquired: The process of collisional photoassociation, at least in the weak field regime, can be thought of as the time-reversal of the process underlying 'extreme wing' line broadening (see, for example, ref. 1). In the 'extreme wing' case, a quasi-static picture obtains wherein spectroscopic transitions occur between levels of a 'quasi-stationary' collision complex. However, the applicability of the quasi-static picture depends on inverse Rabi rates compared to the 'duration' of the collision. Does the quasi-static picture apply in your situation?

1 N. Allard and J. Kielkopf, *Rev. Mod. Phys.*, 1982, **54**, 1103.

Dr Amitay responded: The current experiment of photoassociation of magnesium is in a very different regime than the quasi-stationary pressure broadening theories. First the collision time scale is on the same order as the pulse duration. Secondly the experiment is in the strong field two-photon regime. Finally our goals are different we are interested in generating molecules and not in the effect on lineshapes. In the experimental setup there is no simple perturbative approach. The system is not a two level system in the relevant time scale. Both nuclear and electronic dynamics are important. The strong field means that during the collision there is photon locking, the nuclear dynamics is strongly perturbed by the field. In addition there is a strong effect of orbital angular momentum. From calculations we can estimate that around $J = 55$ there are additional effects of shape resonances. Our advantage over the traditional pressure broadening studies is that we have very high quality potential energy surfaces including transition dipoles. This allows us to use a non-perturbative first principle time dependent quantum mechanical simulation for the process.

Professor Miller opened the discussion of the paper by Professor Ohmori: The linewidths in these high T_c materials are quite narrow relative to other crystals. Can you comment on this? Also in terms of inducing superconductivity, the driven mode must be connected to the electronic coupling between layers that increases the coherence and decreases the effective phonon scattering, as part of the superconducting state. In this case, wouldn't it only be possible to induce the superconducting state for timescales as long as the phonon mode that correlates the electrons? Even for such narrow linewidths, this is still picoseconds.

Professor Ohmori replied: If the superconductivity can actually be induced by the phonon motion, we believe that it should be induced only on the timescale of the phonon relaxation, which is picoseconds.

Dr Buckup asked: The relation between population transfer to the excited state and enhancement of vibrational coherence has already been shown for molecules in the liquid phase when excited with tailored pulses. By keeping the total excitation energy the same as the Transform-limited (TL) pulse, sequence of pulses with a separation period matching the vibrational period were able to bring more population from the ground- to the excited-state than when non-tailored pulses were applied. (See, for example, ref. 1 and ref. 2.) In the paper it is discussed that an enhancement of amplitude of oscillatory contributions (i.e. vibrational coherence?) can be observed. Is this a real enhancement (same total energy used as in the case of TL excitation) or it is more a kind of filtering effect, where the vibrational component out-of-phase, it suppressed from the signal? In other words, if a real enhancement of the vibrational coherence can be achieved, should someone expect some kind of enhancement of the superconductivity?

1 J. Hauer, T. Buckup and M. Motzkus *J. Chem. Phys.*, 2006, **125**, 061101.
2 T. Buckup, J. Hauer, C. Serrat and M. Motzkus *J. Phys. B*, 2008, **41**, 074024.

Professor Ohmori responded: The intensity of the phonon signal is doubled by the double-pulse excitation when the double-pulse delay is tuned to integral multiples of the phonon period. It is different from the enhancement of the population that we have previously observed for the vibrational wave packet in gas-phase molecules by a factor of four (see, for example, ref. 1 and ref. 2.).

1 K. Ohmori, H. Katsuki, H. Chiba, M. Honda, Y. Hagihara, K. Fujiwara, Y. Sato, and K. Ueda, *Phys. Rev. Lett.*, 2006, **96**, 093002.
2 H. Katsuki, H. Chiba, C. Meier, B. Girard, and K. Ohmori, *Phys. Rev. Lett.*, 2009, **102**, 103602.

Professor Miller enquired: Given the absorptivity at the excitation wavelength, even in the event that you could modulate the phonon mode sufficiently, wouldn't the photoinduced carrier injection kill the prospect of superconductivity as the excess energy involved at this excitation wavelength will strongly perturb the phonon distribution (in an uncorrelated manner, *i.e.* as heat)? Also could you comment on your approach, using effectively a stimulated Raman process, to that of Cavelleri *et al.*[1] who recently reported the use of IR excitation of low frequency optical phonon modes to optically induce superconductivity? The key modes are both Raman and IR active. With IR excitation, the absorbed energy goes entirely into the phonon mode of interest. I note that the superconducting state in the work of Cavalleri *et al.* was monitored by changes in reflectivity in the THz range where the Josephson Plasma Resonance gives a signature of superconductivity. I don't understand how the superconducting state can live longer than the phonon mode purportedly inducing the superconducting state.

1 D. Fausti, R. I. Tobey, N. Dean, S. Kaiser, A. Dienst, M. C. Hoffmann, S. Pyon, T. Takayama, H. Takagi and A. Cavalleri, *Science*, 2011, **331** (6014), 189.

Professor Ohmori replied: The 800 nm light is absorbed by the YBCO sample. There is a possibility that the photoinduced carriers kill the prospect of superconductivity AFTER the carrier relaxation. It may be better to employ the IR excitation to generate coherent phonons in order to avoid such carrier effects. However, in the present case, it may be possible to induce superconductivity on the picosecond timescale before the relaxation of coherent phonons and photo-induced hot carriers.

Mr Albert asked: In the superconductor you study, $YBa_2Cu_3O_7$, the Cu atoms have two different coordinations with respect to oxygen, Cu–O chains and Cu–O planes. One of these geometries, the chains, is not invariant upon rotation of 90 degrees. Would you/do you observe different reflectivity patterns if you were to rotate your sample?

Professor Ohmori answered: The observed Cu atoms are located in the Cu–O planes. We have not measured the polarization dependence, but it has previously been measured by Raman spectroscopy.[1]

1 See, for example, O. V. Misochko, E. Ya. Sherman, N. Umesaki, K. Sakai, and S. Nakashima, *Phys. Rev. B*, 1999, **59**, 11495.

Professor Tannor continued the discussion of the paper by Dr Amitay: Your paper described an effect of chirp on the photoassociation yield. Can you comment on this a little further, and perhaps comment somewhat more generally on the prospects for coherent control of photoassociation in these systems?

Dr Amitay responded: The coherent mechanism underlying the enhancement of the PA yield by linearly chirped pulses in our magnesium study is not yet completely clear to us, and we are currently studying it. It is a complicated two-photon strong-field free-to-bound excitation with an interplay between several important factors, including also the matching to the motional timescale of colliding Mg atoms on the ground state, the time scale of the excited state dynamics and intra-pulse pump–dump mechanism. In terms of the future prospects for PA control of atoms, our present work has shown that it is possible to generate high purity and rovibrational coherence within the sub-ensemble of photoassociated molecules. As such, the next natural direction to study, which seems feasible, is to control the generated coherence and its ensuing motion, *i.e.*, control the excited coherent superpositions and their corresponding dynamics. It seems that there is an interesting interplay here between this control objective and the objective of enhancing the total PA yield. Then, a continuation would be to harness the generated rovibrational coherence to control the products in a subsequent excitation or deexcitation leading, for example, to photodissociation or stabilization of the photoassociated molecules.

Professor Tannor asked: Can you give us a little more information on the accuracy of the potential energy surface calculations. Also, do these calculations include the effect of the dynamical Stark shift due to the field?

Mr Tomza replied: The van der Waals type ground $X^1\Sigma_g^+$ state potential was computed with the coupled cluster method restricted to single, double, and non-iterative triple excitations, CCSD(T). The theoretical dissociation energy of the ground $^1\Sigma_g^+$ state, $D_0 = 403.1$ cm^{-1},[1] is in an excellent agreement with the experimental value deduced from the RKR inversion of the experimental data, $D_0 = 404.1 \pm 0.5$ cm^{-1},[2] with a difference of only 0.25%. The number of bound vibrational states for $J = 0$ supported by the electronic ground state also agrees with the experimental number, $N_\nu = 19$. Finally, the root mean square deviation (RMSD) of the rovibrational levels computed with the potential energy curve from the CCSD(T) calculations from the experiment was 1.3 cm^{-1}, *i.e.* approximately 0.3% of the well depth.

The potential energy curves for the excited states were computed by using the equation of motion approach within the coupled-cluster singles and doubles framework, EOM-CCSD.[3,4] Thus far, only one excited state was studied by high-resolution spectroscopy, the $A^1\Sigma_u^+$ state.[5,6] Our spectroscopic parameters for the excited $A^1\Sigma_u^+$ state also agree well with the experimental values. The theoretical and experimental well positions agree within 0.07 bohr. The computed binding energy, $D_e = 9427$ cm^{-1}, is only 0.4% higher than the experimental value, $D_e = 9387$ cm^{-1}. The RMSD of the computed rovibrational levels is 30 cm^{-1}, again roughly 0.3% of the potential well depth. We expect a similar accuracy for other excited states.

The effective Hamiltonian for the nuclear motions in our model is correct to the second order of perturbation theory, *i.e.* it means that it contains both the two-photon transition dipole moments and the dynamic Stark shifts of the ground and excited states. The tensor elements of the two-photon electric transition dipole moment between the ground and excited states[7,8] and the tensor elements of the dynamic electric dipole polarizabilities[9,10] which are proportional to the Stark shifts were calculated as appropriate residues of the linear, quadratic or cubic response functions with the electric dipole operators within the CCSD framework.

1 W. Skomorowski *et al.*, in preparation.
2 W. J. Balfour and A.E. Douglas, *Can. J. Phys.*, 1970, **48**, 901.
3 H. Koch and P. Jørgensen, *J. Chem. Phys.*, 1990, **93**, 3333.
4 H. Sekino and R. J. Bartlett, *Int. J. Quantum Chem. Symp.*, 1984, **18**, 255.
5 H. Scheingraber and C.R. Vidal, *J. Chem. Phys.*, 1977, **66**, 3694.
6 C. R. Vidal and H. Scheingraber, *J. Mol. Spectrosc.*, 1977, **65**, 46.
7 C. Hättig, O. Christiansen and P. Jørgensen, *J. Chem. Phys.*, 1998, **108**, 8331.
8 C. Hättig, O. Christiansen and P. Jørgensen, *J. Chem. Phys.*, 1998, **108**, 8355.
9 O. Christiansen, A. Halkier, H. Koch, P. Jørgensen and T. Helgaker, *J. Chem. Phys.*, 1998, **108**, 2801.
10 C. Hättig, O. Christiansen, S. Coriani and P. Jørgensen, *J. Chem. Phys.*, 1998, **109**, 9237.

Professor Whitaker said: You showed a preliminary result in your presentation in which by means of a positive chirp you are able to enhance the fluorescent signal from the $(1)^1\Pi_u$ state by a factor 12 or 13 over the transform limited pulse (which is a lovely result), yet according to the excitation scheme presented in Fig. 1 of your paper the $(1)^1\Pi_u$ state is reached through a three-photon excitation. Since one and one and one is three, and three squared is nine, shouldn't the maximum expected enhancement for a totally in phase excitation be 9 and not 12?

Dr Amitay answered: The femtosecond photoassociation process (fs-PA) in our study takes place under strong-field conditions. In addition, the fs-PA process also generally involves nuclear wavepacket dynamics during the pulse excitation. Hence, the weak-field intensity dependence doesn't necessarily hold here and without calculations one cannot predict the maximal degree of PA enhancement that can be achieved by a proper positively chirped pulse relative to a transform-limited pulse.

Professor de Vivie-Riedle asked: What are the prospects for photoassociation, going beyond diatomics to larger systems. In case of triatomics the control of the angle between the approaching fragments becomes important and has to be controlled. Do you think that it is possible that the long range interactions experienced by the fragments will help to pre-orient the reactants? In early simulations on the collision dynamics between Na and H_2 we found these properties.

Professor Kosloff responded: We think that photoassociation is a route to control binary chemical reactions. For example the reaction $Mg + CO \rightarrow MgO + C$ or $Mg + CO \rightarrow MgC + O$. I agree that these chemical reactions are steriospecific. When strong electromagnetic fields are applied the photoassociation reaction will align the reactants according the the direction of the transition dipole. This photoselection will increase the purity of the products.
Long range polarization forces may also help in orienting the reactants.

Dr Offerhaus continued the discussion of the paper by Professor Ohmori: It looks like the 3.5 THz mode is certainly associated with superconductivity but is it clear whether there is a causal relation between that mode and superconductivity? Is there anything extra known about this phonon mode in the superconducting regime, does it have the same spatial coherence under superconducting circumstances, is it really the same vibration as in the non-superconducting state?

Professor Ohmori replied: The mechanism of superconductivity of the cuprate superconductors is still controversial, and there is no clear explanation of the relation between the phonon mode and the origin of superconductivity. However, phonons cause lattice deformation, which can change magnetism and some other properties to generate cooper pairs. The size of the spatial coherence is determined by the size of the excitation both above and below the critical temperature. We have no information on the spatial coherence of the phonons. The 3.5 THz mode is predominantly assigned to the Ba–O vibration with a small contribution by the Cu atom movement, whose degree of contribution changes as a function of temperature, so that the vibration in the superconducting state is not strictly the same as that in the non-superconducting state.

Professor Whitaker asked: In your summary remarks on this paper you mentioned that you have been able to observe an enhancement of a factor of four in a double pulse experiment on gas phase molecules but only a factor of two in a sample of bismuth. This is an intriguing result. Do you have any ideas as to why the former exhibits quantum coherence whilst the latter does not? What factors might be in play here that dictate the transition from the quantum to the classical world?

Professor Ohmori answered: At the moment it is not clear to us why this difference is seen between the vibrational wave packets of gas phase molecules and the coherent phonons of bulk solids. One important difference between these two cases is that we measure population in the wavepacket experiments, while it is the phonon displacement that is measured in the phonon experiments. I am not sure, however, this difference in the observables accounts for the difference in the enhancement factor.

A perspective on controlling quantum phenomena

Herschel Rabitz

Received 26th September 2011, Accepted 4th October 2011
DOI: 10.1039/c1fd00111f

Controlling dynamical processes at the atomic and molecular scales with laser radiation has been a long-standing dream. The Faraday Discussion presented a cross section of the current experimental and theoretical advances as well as the challenges for the field. This paper summarizes the current status of controlling quantum dynamics phenomena and provides a perspective on the future.

I. Background

Meeting many goals in chemical and material science requires manipulating atomic and molecular scale processes including reactions and state transformations. These manipulations are traditionally executed with chemical reagents along with the tuning of thermodynamic variables. Chemistry and material science are widely successful disciplines, but the operations involved remain arduous in various high value applications. Recognition of this situation provided the initial motivation for considering laser radiation as a tool for performing molecular scale transformations.

Lasers possess special characteristics for managing atomic and molecular scale dynamics. The goal may be formulated as seeking a laser whose application will steer a sample from an initial state to a final state, which has desirable properties. Original attempts along these lines focused on exploiting the monochromatic nature of cw lasers.[1] Molecular dynamical processes can be quite complex involving spectrally broad motion that often makes cw lasers inadequate for the tasks. In this regard, utilizing additional laser resources, starting with two frequencies, may excite interfering quantum mechanical amplitudes linking the initial and final states. Adjusting the phase and/or amplitude of the two frequency components can manipulate the quantum inferences thereby providing a means to control which final state is reached.[2] These concepts may be generalized by exploiting multiple amplitudes connecting the initial and final states. Capitalizing on this opportunity calls for spectrally rich laser resources ultimately leading to the use of shaped ultrafast laser pulses as controls.[1] The detailed structure of the shaped laser pulses depends on the specific application, and simple cases might even function with two or more sub-pulses where the time delays between them are used to control the evolving quantum dynamics. The complexity of commonly considered polyatomic molecules and material samples combined with the large number of variables needed to fully specify a broad bandwidth laser pulse has led to the use of adaptive feedback control (AFC) techniques for determining optimally shaped pulses directly in the laboratory.[3] Many successful control experiments have been performed in this fashion including applications with very demanding objectives.[1]

Notwithstanding the latter direct laboratory identification of effective controls, theory and simulations have a critical role to play in the development of the subject. Indeed, the AFC laboratory procedure is based on laboratory steps analogous to

Department of Chemistry, Princeton University, Princeton, NJ, 08544, USA

those involved in identifying optimal control fields through simulations.[3] This paper gives a synopsis of the current laboratory and theoretical aspects of the field, including how they may function synergistically as well as independently. These *modi operandi* were clearly evident in the Faraday Discussion.

II. The status of controlling quantum phenomena

The assessment below is separated into consideration of experimental and theoretical facets of controlling quantum dynamics, following the associated breakdown of many of the Faraday Discussion papers.

A. Experiment

Initial attempts at performing molecular manipulation with lasers were plagued by limited resources resulting in the need to often choose the molecules to meet the laser characteristics rather than the desired reverse circumstance. Ultrafast laser pulses provide the bandwidth to alleviate the situation and enable control in a wide class of molecular and material samples.[1] However, the commonly employed Ti:Sapphire laser operating at 800 nm with a bandwidth of typically ~25 nm is still limiting for many applications. In some cases, a high intensity shaped pulse can induce non-linear optical processes in the sample under control to effectively provide enhanced bandwidth.[4] In other situations, the sample cannot be driven in this fashion, although additional bandwidth may still be essential for successful control. With this need in mind, an alternative procedure is to first pass an intense unshaped laser pulse through a suitable medium (*e.g.*, an optical fiber) to provide the enhanced bandwidth. The resultant broad bandwidth pulse may then be shaped to form the control field for application to the sample. Notwithstanding these enhanced capabilities, laser resources still limit the laboratory advancement of quantum control. Attaining the necessary control field flexibility will require further improvements in laser technology, and a reasonable expectation is that better laser resources will become increasingly available.

Even when operating with the current laser resources, experiments often have available the option of considering alternative control scenarios in the laboratory. This freedom was not the case even a few years ago. Additionally, AFC provides the means to automatically optimize the laser pulse shape for a particular objective, with growing numbers of diverse illustrations in the literature.[1] The availability of broader bandwidth lasers has also led to considering guidance from quantum control simulations as a means to identify where to operate most effectively.

A successful control experiment implemented by AFC or other means, will result in the identification of a control field and the measured value for the achieved yield. Producing the desired product is essential before performing further dynamical analyses, and in some applications attaining the highest quality yield in itself may be sufficient. However, a natural desire is to understand the mechanisms involved in controlling quantum phenomena. The notion of mechanism in the context of quantum control is subject to definition just as for the analogous analysis of transformation mechanisms in the chemical and material sciences. A common approach to quantum control mechanism analysis is to perform simulations utilizing the observed experimental control field to determine the nature of the induced quantum dynamics.[5] Additionally, further experiments beyond those used to identify the optimal control field may also be carried out to reveal mechanistic insights directly in the laboratory. In this regard, the special capability of executing massive numbers of experiments with fields in the neighborhood of the optimal pulse may provide the means to extract mechanism information without the need for laborious modeling.[6] An issue related to mechanism concerns the quantum character of the controlled dynamics phenomena, as environmental disturbances are inherently present in the laboratory. Some control objectives may be satisfactorily met even when the ensuing dynamics is less than fully coherent. Simulations can be valuable to estimate the degree of coherence involved.

A prime goal for the field is the determination of broadly applicable systematic "rules" for controlling quantum phenomena, which goes beyond the identification of control mechanisms in specific applications. In this regard, performing analogous molecular manipulations in the chemical sciences frequently relies on empirical rules.[7] The latter rules have generally been discovered by drawing together the results from large numbers of experiments often from many laboratories. Currently, the number of successful optimal control experiments is small ($\sim$200), as the field is still young.[1] Nevertheless, upon combining basic principles of spectroscopy and dynamics some nascent rules are beginning to emerge in certain classes of control experiments.[8] As the laser resources expand in capability, the discovery of rules for controlling quantum phenomena will likely follow at an accelerated rate. The full understanding of quantum control mechanisms and rules open up the prospect of ultimately placing "photonic reagents" on an equal footing with traditional chemical reagents for manipulating atomic and molecular scale events.

B. Theory

Theoretical studies and simulations of quantum control are playing important roles in the advancement of the field.[1,5,8] Such analyses must contend with the fact that many experiments work with complex systems (*e.g.*, polyatomic molecules) often involving interactions with the surrounding environment. Under these circumstances, modeling can be difficult to perform due to a lack of precise knowledge about the Hamiltonian and the rapid scaling of computational effort with system complexity. Nonetheless, computations continue to be extremely valuable for assessing general quantum control behavior and estimating the feasibility of controlling new classes of quantum phenomena. Additionally, simulations can provide mechanistic insights into quantum control, and in many cases even a qualitative understanding can be sufficient.[5] At a fundamental level, theoretical analysis is also proving to be valuable for determining the topology of control landscapes, which relate the physical observable to the underlying control variables.[9] Another important goal of theory is the development of reduced models to capture the physical essence of control phenomena.

A prevalent request from the experimental community is for input from theory to guide the identification of appropriate control resources. Concomitantly, the theoretical community is equally interested in receiving experimental guidance on the nature of particular systems to model. This cross-fertilization is extremely positive for further development of the field. However, experiments need not wait for theory nor should modeling be confined by the current state of experimental capabilities. The latter point is especially important, as theory and simulation can provide a glimpse of what may lie ahead when laboratory resources are more advanced. A related matter is that simulations often indicate the attainment of high value control outcomes, while many experiments show more modest performance. The gap is likely a result of operating with limited laboratory resources as well as possibly in the presence of various physical constraints inherent in the experiments. For example, in the condensed phase there are limits to the intensity of the laser pulses that may be applied before unwanted ancillary excitations occur. In this regard, an important criterion for theory and simulation is to retain the physical essence of all the relevant phenomena involved, including that of the system and control field interacting with the environment.

III. Conclusion

The collective papers in the Faraday Discussion reflect the very significant advances in the field of quantum control in recent years. An important feature is the diversity of the applications, which continues to grow. In many respects, the field is still young with efforts now revealing the theoretical foundations of the field, and the necessary laboratory resources are just becoming readily available. With the collective laboratory, computational and human resources going into this effort it is anticipated that significant advances in the control of quantum phenomena will occur in the immediate years ahead.

References

1 C. Brif, R. Chakrabarti and H. Rabitz, *New J. Phys.*, 2010, **12**, 075008.
2 M. Shapiro and P. Brummer, *Rep. Prog. Phys.*, 2003, **66**, 869.
3 R. Judson and H. Rabitz, *Phys. Rev. Lett.*, 1992, **68**, 1500.
4 A. Assion, T. Baumert, M. Bergt, T. Brixner, B. Kiefer, V. Seyfried, M. Strehle and G. Gerber, *Science*, 1998, **282**, 919.
5 B. Schäfer-Bung, R. Mitri, V. Bonačić-Koutecký, A. Bartelt, C. Lupulescu, A. Lindinger, S. Vajda, S. Weber and L. Wöste, *J. Phys. Chem. A*, 2004, **108**, 4175.
6 R. Rey-de-Castro and H. Rabitz, *Phys. Rev. A*, 2010, **81**, 063422.
7 K. Moore, A. Pechen, X.-J. Feng, J. Dominy, V. Beltrani and H. Rabitz, *Phys. Chem. Chem. Phys.*, 2011, DOI: 10.1039/C1CP20353C.
8 T. Buckup, J. Hauer, J. Voll, R. Vivie-Riedle and M. Motzkus, *Faraday Discuss.*, 2011, **153**, DOI: 10.1039/C1FD00037C.
9 R. Chakrabarti and H. Rabitz, *Int. Rev. Phys. Chem.*, 2007, **26**, 671.

Poster titles

Time optimal control of two components Bose-Einstein condensate, **M. Lapert, G. Ferrini and D. Sugny**, *Université de Bourgogne, France*

Controlling quantum dynamics in polyatomic molecules: the OCT-MCTDH approach, **M. Schröder, M. Zhao and A. Brown**, *University of Alberta, Canada*

Coherent photoisomerisation in biomimetic molecular structures, **J. Léonard, J. Briand, V. Zanirato, I. Schapiro, M. Olivucci and S. Haacke**, *University of Strasbourg, France*

Effect of diatomic molecular properties on binary laser pulse optimizations of $ACNOT_1$ and NOT_2 quantum state operations, **R. R. Zaari and A. Brown**, *University of Alberta, Canada*

Photodissociation and control of ClNO, **K. Jones and B. J. Whitaker**, *University of Leeds, United Kingdom*

Two-photon parity and analytical approximations to the two-photon rabi hamiltonian, **V. V. Albert, G. D. Scholes and P. Brumer**, *University of Toronto, Canada*

Consistent treatment of coherent and incoherent energy transfer dynamics using a variational master equation, **D. P. S. McCutcheon and A. Nazir**, *Imperial College London, United Kingdom*

Controlling nonlinear luminescence in graphite and metal nanoparticle systems by pulse shaping, **G. Piredda, M. Handloser, R. de Vivie-Riedle and A. Hartschuh**, *Ludwig-Maximilians-Universität München, Germany*

Effects of molecular symmetry on quantum reaction dynamics: novel aspects of photoinduced non-adiabatic dynamics, **M. Leibscher and S. Al-Jabour**, *Freie Universität Berlin, Germany*

Electron steering and control of molecular dissociation by mid-infrared few-cycle pulses, **I. Znakovskaya, P von den Hoff, S. Zherebtsov, B. Bergues, X. Gu, Y. Deng, M. J. J. Vraking, F. Krausz, R. Kienberger, G. Marcus, R. de Vivie-Riedle and M. F. Kling**, *Max Planck Institute for Quantum Optics, Germany*

ClNO photodissociation dynamics *via* S_3 ($2^1A'$) ← S_0 ($1^1A'$) transition, **J. A. Milkiewicz, A. G. Sage, K. M. Jones and B. J. Whitaker**, *University of Leeds, United Kingdom*

Enhancing the probability of three-photon excitation in I_2, **N. T. Form, B. J. Whitaker and C. Meier**, *University of Leeds, United Kingdom*

Controlling the mechanism of fulvene S_1/S_0 decay, **D. Mendive-Tapia, B. Lasorne, G. A. Worth, M. J. Bearpark and M. A. Robb**, *Imperial College London, United Kingdom*

The Skinner Prize for the best poster was awarded to Ms Jadwiga Milkiewicz of the University of Leeds, United Kingdom, for her poster on ClNO photodissociation dynamics *via* S_3 ($2^1A'$) ← S_0 ($1^1A'$) transitions.

List of participants

Mr V Albert, *University of Toronto, Canada*
Dr S Amaran, *Hebrew University, Israel*
Dr Z Amitay, *Technion-Israel Institute of Technology, Israel*
Professor M Ashfold, *University of Bristol, United Kingdom*
Professor T Baumert, *Universität Kassel, Germany*
Professor M Bearpark, *Imperial College London, United Kingdom*
Dr A Brown, *University of Alberta Department of Chemistry, Canada*
Professor P Brumer, *University of Toronto, Canada*
Dr W Bryan, *Swansea University, United Kingdom*
Dr T Buckup, *Heidelberg University, Germany*
Dr M de Miranda, *University of Leeds, United Kingdom*
Professor R de Vivie-Riedle, *Ludwig-Maximilians-Universität München, Germany*
Professor G Engel, *University of Chicago, USA*
Professor H Fielding, *University College London, United Kingdom*
Dr S Hay, *University of Manchester, United Kingdom*
Professor M Janssen, *VU University Amsterdam, The Netherlands*
Miss K Jones, *University of Leeds, United Kingdom*
Professor V Kleiman, *University of Florida, U.S.A.*
Professor R Kosloff, *Hebrew University of Jerusalem, Israel*
Mr R Lane, *Photon Science Institute, University of Manchester, United Kingdom*
Mr M Lapert, *Université de Bourgogne, France*
Dr M Leibscher, *Freie Universität Berlin, Germany*
Dr J Leonard, *Institut de Physique et Chimie des Materiaux de Stasbourg, France*
Mr M Liebel, *University of Oxford, United Kingdom*
Dr G Marcus, *Max Planck Institute for Quantum Optics, Germany*
Dr P Marquetand, *Friedrich-Schiller-Universität Jena, Germany*
Professor S Matsika, *Temple University, USA*
Mr D Mendive-Tapia, *Imperial College London, United Kingdom*
Ms J Milkiewicz, *University of Leeds, United Kingdom*
Professor D Miller, *University of Toronto, Canada*
Professor M Motzkus, *University of Heidelberg, Germany*
Dr A Nazir, *Imperial College London, United Kingdom*
Dr M Nix, *University of Leeds, United Kingdom*
Dr H Offerhaus, *University of Twente, The Netherlands*
Professor K Ohmori, *Institute for Molecular Science, Okazaki, Japan*
Dr G Piredda, *Ludwig-Maximilians-Universität München, Germany*
Dr J Plenge, *Freie Universität Berlin, Institut fuer Chemie und Biochemie, Germany*
Professor H Rabitz, *Princeton Universityo, USA*
Dr G Richings, *University of Birmingham, United Kingdom*
Miss A Rondi, *University of Geneva, Switzerland*
Dr A Sage, *University of Leeds, United Kingdom*
Dr K Saita, *University of Leeds, United Kingdom*
Dr C Sanz-Sanz, *CSIC/University of Birmingham, Spain*
Professor R Sension, *University of Michigan, USA*
Dr D Shalashilin, *University of Leeds, United Kingdom*
Professor M Shapiro, *University of British Colombia, Canada*
Dr A Stolow, *Steacie Institute for Molecular Sciences, Canada*
Dr B Sussman, *National Research Council Canada, Canada*
Professor D Tannor, *Weizmann Institute of Sciencea, Israel*
Mr M Tomza, *University of Warsaw, Poland*

Dr D Townsend, *Heriot-Watt University, United Kingdom*
Dr J Tremblay, *Universitaet Potsdam, Germany*
Professor N van Hulst, *ICFO, Spain*
Mr P Venn, *University of Southampton, United Kingdom*
Mr P von den Hoff, *Ludwig-Maximilians-Universität München, Germany*
Professor T Weinacht, *Stony Brook University, U.S.A.*
Professor B Whitaker, *University of Leeds, United Kingdom*
Mr R Zaari, *University of Alberta, Canada*

Index of contributors*

Albert, V., 395
Amaran, S., **383**
Amitay, Z., 73, 189, **383**, 395
Ashfold, M., 189, 293, 395
Avisar, D., **131**
Baumert, T., **9**, 73, 189, 395
Bearpark, M., 395
Brinks, D., **51**
Brumer, P., **41**, 73, **149**, 189, 293
Bryan, W. A., **343**, 395
Buckup, T., 73, **213**, 395
Bustard, P. J., **321**
Calvert, C. R., **343**
Caram, J. R., **93**
de Miranda, M., 189
de Vivie-Riedle, R., 73, **159**, 189, **213**, 293, 395
Engel, G. S., 73, **93**, 189
Fielding, H., 189, 293, 395
González, L., **261**
González-Vázquez, J., **261**
Greenwood, J. B., **343**
Halpin, A., **27**
Hauer, J., **213**
Herek, J. L., **227**
Hildner, R., **51**
Irimia, D., **173**
Jafarpour, A., **227**
Janssen, M. H. M., **173**, 189
Jurna, M., **227**
Katsuki, H., **375**
King, R. B., **343**
Kleiman, V. D., **61**, 73, 293
Koch, C. P., **383**
Kosloff, R., 73, 189, 293, **383**, 395
Kotur, M., **247**
Kowalewski, M., **159**
Kuroda, D. G., **61**
Lane, R., 73
Lausten, R., **321**
Lehmann, C. S., **173**
Leibscher, M., 189, 293, 395
Leonard, J., 293, 395
Marquetand, P., 189, **261**, 293
Matsika, S., 189, **247**, 293
Mendive-Tapia, D., 189
Miller, R. J. D., **27**, 73, 189, 293, 395

Moszynski, R., **383**
Motzkus, M., 73, 189, **213**, 293
Nakagawa, Y., **375**
Nakamura, K. G., **375**
Newell, W. R., **343**
Offerhaus, H. L., 73, **227**, 293, 395
Ohmori, K., 73, 189, **375**, 395
Okano, Y., **375**
Peng, Z., **61**
Plenge, J., **361**, 395
Prokhorenko, V. I., **27**
Rabitz, H., 73, 189, 293, **415**
Ram, N. B., **173**
Raschpichler, C., **361**
Richings, G. W., **275**, 293
Richter, M., **261**
Rühl, E., **361**
Rybak, L., **383**
Sage, A., 395
Sanz-Sanz, C., **275**, 293
Sension, R. J., **117**, 189, 293
Shalashilin, D. V., 73, **105**, 189
Shapiro, M., **149**, 189, 293, 395
Singh, C. P., **61**
Singh, N., **41**
Sola, I., **261**
Stefani, F. D., **51**
Stolow, A., 189, 293, **321**, 395
Sussman, B. J., 293, **321**, 395
Takahashi, H., **375**
Tang, K.-C., **117**
Tannor, D. J., 73, **131**, 189, 293, 395
Tomza, M., **383**, 395
Townsend, D., **321**
Ulbricht, H., **237**
van Hulst, N. F., **51**, 73
van Rhijn, A. C. W., **227**
Venn, P., **237**, 293
Voll, J., **213**
von den Hoff, P., **159**, 189, 293
Walmsley, I. A., **321**
Wassermann, B., **361**
Weinacht, T. C., 73, 189, **247**, 293, 395
Whitaker, B., 73, 189, 293, 395
Williams, I. D., **343**
Wirsing, A., **361**

Wollenhaupt, M., **9**
Worth, G. A., **275**
Wu, G., **321**
Zhou, C., **247**

* The page numbers in **bold** type indicate papers submitted for discussions.